Implications of Nanoecotoxicology on Environmental Sustainability

Rafiq Lone
Central University of Kashmir, India

Javid Ahmad Malik
Guru Ghasidas Vishwavidyalaya, India

IGI Global
PUBLISHER of TIMELY KNOWLEDGE

A volume in the Advances in Environmental Engineering and Green Technologies (AEEGT) Book Series

Published in the United States of America by
IGI Global
Engineering Science Reference (an imprint of IGI Global)
701 E. Chocolate Avenue
Hershey PA, USA 17033
Tel: 717-533-8845
Fax: 717-533-8661
E-mail: cust@igi-global.com
Web site: http://www.igi-global.com

Library of Congress Cataloging-in-Publication Data

Names: Lone, Rafiq, 1982- editor. | Malik, Javd Ahmad, 1987- editor.
Title: Implications of nanoecotoxicology on environmental sustainability / Rafiq Lone, and Javid Ahmad Malik, editors.
Description: Hershey, PA : Engineering Science Reference, [2023] | Includes bibliographical references and index. | Summary: "This book provides relevant theoretical and practical frameworks and the latest empirical research findings in the nanotechnology and its implications discussing the benefits and consequences in further detail of nanoparticles in various fields, specifically on their toxic implications"-- Provided by publisher.
Identifiers: LCCN 2022024534 (print) | LCCN 2022024535 (ebook) | ISBN 9781668455333 (h/c) | ISBN 9781668455340 (s/c) | ISBN 9781668455357 (ebook)
Subjects: LCSH: Nanostructured materials--Environmental aspects. | Environmental protection--Materials. | Nanostructured materials--Toxicology. | Environmental toxicology.
Classification: LCC TD196.N36 I47 2023 (print) | LCC TD196.N36 (ebook) | DDC 620/.5--dc23/eng/20220725
LC record available at https://lccn.loc.gov/2022024534
LC ebook record available at https://lccn.loc.gov/2022024535

This book is published in the IGI Global book series Advances in Environmental Engineering and Green Technologies (AEEGT) (ISSN: 2326-9162; eISSN: 2326-9170)

British Cataloguing in Publication Data
A Cataloguing in Publication record for this book is available from the British Library.

All work contributed to this book is new, previously-unpublished material. The views expressed in this book are those of the authors, but not necessarily of the publisher.

For electronic access to this publication, please contact: eresources@igi-global.com.

Advances in Environmental Engineering and Green Technologies (AEEGT) Book Series

Sang-Bing Tsai
Zhongshan Institute, University of Electronic Science and Technology of China, China & Wuyi University, China
Ming-Lang Tseng
Lunghwa University of Science and Technology, Taiwan
Yuchi Wang
University of Electronic Science and Technology of China Zhongshan Institute, China

ISSN:2326-9162
EISSN:2326-9170

Mission

Growing awareness and an increased focus on environmental issues such as climate change, energy use, and loss of non-renewable resources have brought about a greater need for research that provides potential solutions to these problems. Research in environmental science and engineering continues to play a vital role in uncovering new opportunities for a "green" future.

The **Advances in Environmental Engineering and Green Technologies (AEEGT)** book series is a mouthpiece for research in all aspects of environmental science, earth science, and green initiatives. This series supports the ongoing research in this field through publishing books that discuss topics within environmental engineering or that deal with the interdisciplinary field of green technologies.

Coverage

- Waste Management
- Green Technology
- Green Transportation
- Cleantech
- Contaminated Site Remediation
- Radioactive Waste Treatment
- Air Quality
- Biofilters and Biofiltration
- Policies Involving Green Technologies and Environmental Engineering
- Sustainable Communities

IGI Global is currently accepting manuscripts for publication within this series. To submit a proposal for a volume in this series, please contact our Acquisition Editors at Acquisitions@igi-global.com or visit: http://www.igi-global.com/publish/.

The Advances in Environmental Engineering and Green Technologies (AEEGT) Book Series (ISSN 2326-9162) is published by IGI Global, 701 E. Chocolate Avenue, Hershey, PA 17033-1240, USA, www.igi-global.com. This series is composed of titles available for purchase individually; each title is edited to be contextually exclusive from any other title within the series. For pricing and ordering information please visit http://www.igi-global.com/book-series/advances-environmental-engineering-green-technologies/73679. Postmaster: Send all address changes to above address.

Titles in this Series

For a list of additional titles in this series, please visit: http://www.igi-global.com/book-series/advances-environmental-engineering-green-technologies/73679

Food Sustainability, Environmental Awareness, and Adaptation and Mitigation Strategies for Developing Countries
Ahmad Ni'matullah Al-Baarri (Diponegoro University, Indonesia) and Diana Nur Afifah (Diponegoro University, Indonesia)
Engineering Science Reference • © 2023 • 315pp • H/C (ISBN: 9781668456293) • US $110.00

Human Agro-Energy Optimization for Business and Industry
Pandian Vasant (MERLIN, Faculty of Electrical and Electronics Engineering, Ton Duc Thang University, Vietnam) Roman Rodríguez-Aguilar (Panamerican University, Mexico) Igor Litvinchev (Universidad Autónoma de Nuevo León, Mexico) and Jose Antonio Marmolejo-Saucedo (Panamerican University, Mexico)
Engineering Science Reference • © 2023 • 322pp • H/C (ISBN: 9781668441183) • US $240.00

Applying Drone Technologies and Robotics for Agricultural Sustainability
Pethuru Raj (Reliance Jio Platforms Ltd., Bangalore, India) Kavita Saini (Galgotias University, India) and Vinicius Pacheco (Office of the Prosecutor General, Brazil)
Engineering Science Reference • © 2023 • 318pp • H/C (ISBN: 9781668464137) • US $250.00

Intelligent Solutions for Optimizing Agriculture and Tackling Climate Change Current and Future Dimensions
Ahmed Karmaoui (Moulay Ismail University of Meknes, Morocco & Moroccan Center for Culture and Science, Morocco)
Engineering Science Reference • © 2023 • 269pp • H/C (ISBN: 9781668446492) • US $215.00

Recent Advances in Global Meliponiculture
Shamsul Bahri Abd Razak (Universiti Malaysia Terengganu, Malaysia) Tuan Zainazor Tuan Chilek (Universiti Malaysia Terengganu, Malaysia) and Jumadil Saputra (Universiti Malaysia Terengganu, Malaysia)
Engineering Science Reference • © 2023 • 257pp • H/C (ISBN: 9781668462652) • US $215.00

Biomass and Bioenergy Solutions for Climate Change Mitigation and Sustainability
Ashok Kumar Rathoure (M/s Akone Services, India) and Shankar Mukundrao Khade (Ajeenkya D.Y. Patil University, India)
Engineering Science Reference • © 2023 • 412pp • H/C (ISBN: 9781668452691) • US $250.00

Handbook of Research on Building Greener Economics and Adopting Digital Tools in the Era of Climate Change
Patricia Ordóñez de Pablos (The University of Oviedo, Spain)
Engineering Science Reference • © 2022 • 391pp • H/C (ISBN: 9781668446102) • US $350.00

IGI Global
PUBLISHER of TIMELY KNOWLEDGE

701 East Chocolate Avenue, Hershey, PA 17033, USA
Tel: 717-533-8845 x100 • Fax: 717-533-8661
E-Mail: cust@igi-global.com • www.igi-global.com

Table of Contents

Detailed Table of Contents

Environmental contamination is the most serious global threat to our ecosystem. Over the last few decades, there has been a rise in environmental sensitivity. Indiscriminate use of water resources beyond their regeneration limits has resulted in the current water crisis that the world is experiencing. Hazardous effluents from textile industries and resistant microbes are significant sources for water deterioration with detrimental impacts on both biotic and abiotic components. Effective solutions to water purification are the need of the hour. Hydrogels possess unique properties with potential as effective adsorbents for dyes and hazardous metal ions enrichment from wastewater. Nanocomposites are a new class of advanced materials for technology. Nanocomposite hydrogels with reinforcing nanofillers exhibit noticeably superior properties as compared to conventional hydrogels and have the potential to become efficient adsorbents for such pollutants. Present chapter provides a thorough overview of the role of nanocomposites in water pollution alleviation techniques.

The introduction of nanomaterials in biosensing technology has widened the applications of nano biosensors in several fields as in the precise detection of micropollutants. A huge burden has been driven out by the micropollutants in an environment that has resulted in an alarming situation for scientists to discover innovative methods for the precise or accurate detection of micropollutants that provides sensitivity and specificity at the same time. Apart from conventionally available methods, nano biosensors provide real-time analysis and high-frequency monitoring of pollutants without much sample preparation. Nano biosensors can detect and manipulate atoms or molecules by use of nanodevices resulting in the development of biosensor devices which in turn interact with small molecules named micropollutants. This chapter aims to glance over the advancement in the fields of nano biosensors-based functionalized nanoparticles, nanotubes, or nanowires and their interaction with biorecognition materials such as enzymes, antibodies, or aptamers for the detection of aptamers.

The agricultural industry is currently facing challenges of food production for the expanding human population. This will be worsened in the future by the negative impacts of climate change. The obstacles can be overcome by increasing the productivity of the available agricultural lands. The conventional fertilizers are, however, not very efficient in delivering nutrients to the plants. Formulation of nanofertilizers based on tiny particles (< 100 nm), prepared from common organic and inorganic sources, is a ground-breaking discovery that can ensure sustainable agriculture in an eco-friendly manner. Their unique properties make them effective at low concentrations. They can supply nutrients and help the plants withstand biotic and abiotic stress factors. This proves their efficiency over traditional fertilizers. However, they have certain drawbacks. Their high reactivity can be a source of toxicity to the plants as well as animals. Thus, this emerging field has high potentialities but requires extensive research.

Nanotechnology is a degree knowledge domain analysis field. In recent efforts, square measures were created to reinforce agricultural yield through a thorough analysis of technology. The revolution resulted in blind usage of pesticides and chemical fertilizers that caused loss of soil selection and developed resistance against pathogens and pests. Nanoparticles are experimenting as anti-infestation as opposed to morbific fungi. Anti-infestation venture nanoparticles of oxide, silver, and titanium oxide have been tested against Macrophomina phaseolina, a big soil-borne infectious agent of pulse and seed crops. A higher antifungal effect was discovered in silver nanoparticles at a decreased level than in flowers of metallic elements and titanic oxide nanoparticles.

In this chapter, the eco-friendly applications, disposal, and the risks of inorganic nanoparticles to human health and to the environment are discussed, including the advancements in the green applications of green nanomaterials for their sustainable and competitive energy systems. The first section summarizes an introduction with the importance of the nanomaterials in scientific and industrial applications. The second and third sections present green nanoparticles applications for energy generation (green

Foreword

Nanotechnology is the science of producing and utilizing nanometre sized particles. Nanotechnology can improve agricultural processes such as soil quality and the quality of agricultural products by using nanoparticle-based fertilizers. In addition, the use of fertilizers and pesticides using nanoparticle- based carriers and compounds is reduced without reducing productivity. Nanotechnology can provide specific applications for sustainable development through nanofertilizers, nanopesticides, and nanoherbicides by developing smart delivery systems. However, there are few consequences of using these nanoparticles in today's agriculture. This book focuses on the exploration of Nanoparticles in various fields. Moreover, their toxic implications are also being covered under the theme. These toxic manifestations if let undermined will be deteriorating the ecological balance of ample ecosystems. These need to be discussed in detail and further research need to be done to make this technology more useful and sustainable for future.

K. K. Koul
Jiwaji University, Gwalior, India

Preface

This book presents the latest research on nanotechnology, offering readers a detailed, yet comprehensive account of role of Nanoecotoxicology on environmental sustainability. It covers a diverse range of topics, including nanoparticles, types, Application and commercial products, their role in water pollution alleviation technologies, biomedical applications, sustainable agriculture etc. Further the book provides readers with an up-to-date review of this dynamic field and sets the direction for future research. Based on the authors' extensive experience and written in an engaging style, this highly readable book will appeal to scholars from various disciplines. Bringing together work from leading international researchers, it is also a valuable reference resource for academics, researchers, students and teachers wanting to gain insights into the implications of Nanoecotoxicology on Environmental Sustainability. The book has received 11 chapters from international authors that have first-hand expertise in the field of Nanotechnology.

The first chapter describes concerns and advantages regarding the impact of nanotechnology on the environment, health and society are a contested issue. It's too early to tell whether the growth of nanotechnology will be a boon or a bane.

The second chapter aim to review nanoparticles, their detailed classification, and commercialized products from nanoparticles.

The third chapter help us to understand the sources and occurrences of nanoparticles in aquatic ecosystems and need for future research in this direction.

The fourth chapter gives a thorough overview of role of nanocomposites in water pollution alleviation techniques.

The fifth chapter overviews the specificity of the system can be boosted by the use of nanoparticles/nanomaterial functionalized surfaces which makes Nano biosensors a more refined and reliable method. It also makes small devices for the detection of micropollutants having low sensitivity or specificity providing a new frontier between the biological, chemical, and physical properties of a substance.

The sixth chapter gives an overview of use of nanofertilizers in agriculture: advantages, disadvantages, and future implications.

The seventh chapter argues that the nanotechnological approaches in agriculture and its allied fields.

The eight chapter discuss the eco-friendly applications, disposal, and the risks of inorganic nanoparticles to human health and to the environment are discussed, including the advancements in the green applications of green nanomaterials for their sustainable and competitive energy systems.

The ninth chapter provides different mechanisms involved in the interactions were identified to propose solutions to reduce or eliminate the adverse effects of their cytotoxicity.

The tenth chapter review has discussed the recent advances and updates on various approach of green-synthesis of SNPs, their bio-medical applications, and their toxicological effects, this will help researchers and readers to investigate further economical and scale-up techniques of synthesis of SNPs and industrial applications.

The eleventh chapter includes the detailed account of the impact of heavy metal-based nanomaterials on environment and health: small things having big impacts.

Rafiq Lone
Central University of Kashmir, India

Javid Ahmad Malik
Guru Ghasidas Vishwavidyalaya, India

Acknowledgment

I would like to acknowledge my indebtedness and render my warmest thanks to all those contributors who made this book possible. I would also wish to express my gratitude to Professor Azra Kamili for extended discussions and valuable suggestions which have contributed greatly to the improvement of the books. We are highly thankful to Prof. K.K. Koul for his constant support and encouragement. My sincere thanks also go to Dr. Khursheed Ahmad Wani, for helping us to make things smooth in a right direction to accomplish the book well in time. We would like to thank our families for their love and support. With great Love and tears in our eyes, we dedicate this book to our beloved Sister Late Surayya.

Chapter 1
Nanotechnology:
Boon or Bane for the Environment?

Khursheed Ahmad Wani
https://orcid.org/0000-0001-5577-3455
Department of Environmental Science, Government Degree College Thindim Kreeri, India

Javid Manzoor
https://orcid.org/0000-0002-2821-206X
Department of Higher Education, Government of Jammu and Kashmir, India

Syed Javid Indrabi
Department of Water Management, SP College Srinagar, India

Tehmina Yousuf
Srinagar Womens College Zakura, India

ABSTRACT

In the field of nanoscale materials and electronics fabrication, the term "nanotechnology" refers to a broad range of techniques that may be used. Examples of nanotechnological processes include the production of nanowires, as well as techniques for fabricating semiconductors such as lithography, atomic layer deposition, molecular vapour deposition, focused ion beam machining, electron beam lithography, and molecular self-assembly. Concerns regarding the impact of nanotechnology on the environment, health, and society are a contested issue. It's too early to tell whether the growth of nanotechnology will be a boon or a bane.

DOI: 10.4018/978-1-6684-5533-3.ch001

INTRODUCTION

It is expected that Nano technological products, techniques, and applications will have a significant impact on environmental and climate protection through reducing greenhouse gas emissions and hazardous waste. There are some positive environmental and long-term implications of employing nanomaterials (Zhang and Lagally, 1997).

However, nanotechnology is now playing a very small role in environmental protection, whether in research or in real applications. For environmental engineering firms, nanotechnology has little practical use. One of the most commonly emphasized aspects of nanotechnology is its long-term potential. But in most cases, this is the outcome of irrational expectations. A product's whole life cycle, from raw material extraction to final disposal, must be investigated in order to assess its real environmental impact, both good and poor. Most of the time, environmental benefits are touted without taking into consideration the costs in terms of resources and energy used during production (Ngo et al., 2014).

In order for the environment to remain sustainable, we must address global climate change and pollutants, such as air, soil, and water. Even in large industrialized countries, climate change is still a contentious issue. Beyond technology's scope, direct proof from study, public awareness, and government should all work together to combat climate change (Srinivasan et al., 2016). Meanwhile, existing technology can provide direct control of pollution from a range of sources, including heavy metals, organic chemicals, biological or radiation dangers, oil spills, and micro plastics, among others. The use of nanomaterials has greatly advanced environmental technology, such as sensing and remediation. This demonstrated how environmental nanotechnology is developing and evolving (Tala-Ighil, 2015).

During the past decade, there has been an exponential rise in the number of articles devoted to nanomaterial's, reflecting the rapid growth of study into their production, uses, exposure, and risk. More than a dozen researches have investigated nanomaterial's' properties and potential hazards to human health and the environment, including their potential use as adsorbents, exchangers of ions and pathogens in water or air (Auplat et al., 2017). In order to highlight recent advances in Nano safety and ongoing research on the use of nanomaterial's for environmental remediation, such as the removal of ions and organic compounds from aqueous and air media, this chapter has been proposed.

NANOTECHNOLOGY FOR AIR POLLUTANTS

NOx (NO and NO_2) emissions from fossil fuel combustion have been reduced through the introduction of new technologies. Adsorbents often utilized to remove NOx at low temperatures include ion exchange zeolites, activated carbon, and FeOOH dispersed over active carbon fibre. NO can be effectively adsorbed on activated carbon due to the reactivity of surface functional groups, even if the quantity of adsorbed species is still tiny. According to Long and Yang, (2001) CNTs might be used as an adsorbent for the removal of NO.

Health problems such as heart disease and lung cancer have long been linked to exposure to particulate matter and heavy metal pollution over lengthy periods of time. Heavy metals may be found in a wide range of concentrations in urban areas, with particle sizes generally in the range of 100–300 nm (Filipponi and Sutherland, 2010). In addition, microorganisms cannot break down heavy metals (i.e. they are not biodegradable).

Detecting heavy metal ions before they reach dangerous levels is a pressing need since heavy-metal-contaminated soil is so difficult to clean up (Wang, 2000). Sensors that can identify toxins at the molecular level might enhance human health and the environment's capacity to be protected. For environmental decision-making and process management, more sensitive and less costly pollution detection technologies may lead to considerable gains. It is one of the focused technologies that may provide data on contaminants in a short length of time (Roduner 2006).

Some metal ions may be detected without the requirement for preconcentration using a nanocontact sensor. For on-site detection of heavy metal ions, such as radioactive contaminants, this sensor is a good fit. It is possible to manufacture nanocontact sensors in a tiny, automated form for usage on-site or in the field. To make things even more affordable, these sensors are made using standard microelectronics equipment and basic electrochemical techniques (cost-effective) (Arole and Munde, 2014).

Approximately 78 mg/g CNTs of NOx were absorbed. The peculiar architectures, electrical characteristics, and surface functional groups of CNTs may be responsible for the adsorption of NOx. NO_2 is oxidized to NO and adsorbed on the surface of nitrate species when NO and O_2 travel through CNTs. The oxidation of NO to NO_2 on activated carbon fiber by Mochida et al. (2009) supports this theory. CNTs may also adsorb SO_2 despite the low adsorption rate, but CO2 adsorption on CNTs is substantially lower than that of NO or NO_2 (Caterina, 2004). The collection and storage of carbon dioxide (CO_2) from fossil fuel-fired power plants has received considerable interest. Since the Kyoto Protocol came into force on February 16, 2005, the CO_2, NOx, and SO_2 adsorption/desorption profiles on carbon nanotubes at 25°C have attracted a lot of attention (Helmer and Anja, 2016).

Other CO_2 collection methods such as adsorption, cryogenics, and membranes have also been investigated. Among these technologies, the adsorption–regeneration process is the most advanced. Ammonia absorption is a term used to describe the amine-based absorption process. There are now a number of solutions being researched throughout the globe since the energy needed for absorption is still too high. If a new generation of material is developed capable of adsorbing CO_2 effectively, the Intergovernmental Panel on Climate Change (IPCC) predicts that adsorptive separation in flue gas applications will become more competitive (Bessekhouad et al., 2003). Carbon, zeolite, silica, SWNTs, and nanopore silica-based molecular baskets are among the adsorbents that may be used to remove toxins. Carbon dioxide (CO_2) may be captured using CNTs that have been chemically modified (Farahmandjou and Golabiyan, 2016).

Since 3-aminopropyltriethoxysilane, ethylene diamine (EDA), and polyethyleneimine (PEI) were added to the CNT, the value of qe rose (APTS). If there's no water present, the solution includes amine groups that react with CO_2 to form carbamate, raising qe. CNTs with EDA and PEI modifications grew more than CNTs with APTS modification. Increases in relative humidity lead to an increase in CO_2 adsorption efficiency on modified CNTs, although this impact diminishes as temperature increases (Ali and Mohammed, 2018).

As well as nitrous acid and polyaromatic compounds (Christian et al., 2008), volatile organic compounds, and soot, atmospheric processes create a wide range of chemicals, including nitrous acid, polyaromatic compounds, and soot (VOCs). Clean air regulations were tightened as a result of the increased risk to human health posed by these airborne particles. The bulk of modern air filtration systems employ photo catalysts, adsorbents like activated carbon, and ozonolysis. When it comes to eliminating organic contaminants at room temperature, conventional approaches don't work. Researchers in Japan have developed a new polymer that is exceptionally effective in removing volatile organic compounds (VOCs), nitrogen, and sulphur oxides from the air at room temperature (Guo et al., 2013).

Manganese oxide with gold nanoparticles embedded in it is the basic structure. This catalyst was studied by Sinha and Suzuki (2007) using acetaldehyde, toluene, and hexane, three of the most common organic contaminants in indoor air. Using this catalyst as a benchmark, researchers were able to show that it successfully eliminated and degraded all three airborne pollutants. A major factor in the success of porous manganese oxide is that it has a far greater surface area than any previously known chemical. The increased surface area aids in the adsorption of volatile chemicals. It also removes the pollutants that have been taken up. Surface deterioration is especially effective due to the presence of free radicals. When gold nanoparticles are present, the barrier to radical generation is reduced. After this operation, other Nano-metal components may be utilized.

NANOTECHNOLOGY FOR WATER TREATMENT

It's not as easy to get your hands on drinking water as you may think. According to the World Health Organization and other concerned authorities, water availability and affordability are becoming more difficult to come. There are an estimated 17–19 million individuals in the globe who do not have access to clean drinking water. Water, sanitation, and hygiene-related issues kill an estimated 3.4 million people each year, with 99 percent of the fatalities happening in underdeveloped countries (UNICEF, 2010). More than 5 million South Africans do not have access to clean drinking water, making their lives even more challenging. Availability, accessibility, quality, and sustainability are the four main issues underlying the current water problem. Science and technology may have a significant impact in each of these situations (Tian, 2017).

Both developed and developing nations are interested in nanotechnology applications in the water industry. Indeed, as indicated by research papers in this crucial area, nanotechnology innovation in the water industry has been a driving factor in many nations.

Adsorption

The ability of solids to attract molecules of gas or liquid to their surfaces is a trait shared by all solid substances. In the case of adsorbed molecules, an adsorbate is a collection of molecules that have been adsorbed onto the adsorbed material. Activated carbon in the form of granular or powdered form has a much slower rate of adsorption for organic chemicals than Nano adsorbents. These approaches have the potential to disinfect inorganic and organic contaminants, such as heavy metals and micro pollutants, more swiftly and efficiently. Not only does greater process efficiency allow for the use of smaller, more compact water and wastewater treatment equipment, but it also allows for the use of decentralized applications and point-of-use systems to be used. Currently, research is being conducted on the following types of Nano adsorbents (Ederth, 2003; Ali and Kawakib, 2013; McLachlan and Sauti, 2007)

Carbon Nanotubes

Carbohydrate nanotubes (CNTs) have a spherical nanostructure and are hence carbon allotropes (CA). Depending on how they are made, carbon nanotubes (CNTs) are either single-walled or multi-walled. There are a lot of adsorption sites on CNTs that may be assessed, as well as the ability to alter the chemical composition of the CNTs. It is necessary to stabilize CNTs in aqueous solution because to the hydrophobic surface, which reduces the active surface. Preconcentration and identification of pollutants, as well as adsorption of persistent contaminants, may be accomplished with these devices. CNTs are able to adsorb metal ions because of their electrostatic and chemical properties. Because CNTs generate oxidative stress in bacteria and destroy cell membranes, they have antimicrobial properties as well. A notable advantage over more typical disinfection treatments like chlorination and ozonation is that no hazardous byproducts are created during chemical oxidization. Regenerating them is as simple as altering the working parameters, such as the pH (Ali and Mohammed, 2017; Kulkarni et al., 2013).

Polymeric Nano Adsorbents

Organic and heavy metals may be removed from water using dendrimers (repetitively branched molecules). Organic compounds may be absorbed by the inside hydrophobic shells, whereas heavy metals can be absorbed by the specialized outside branches. Dendrimers in an ultrafiltration system were utilized by Diallo et al., (2005) to remove copper from water. Nearly 80% of the copper ions could be captured with this dendrimer-ultrafiltration combination. Renewing the adsorbent is as simple as adjusting the pH. Until recently, there were no commercial dendrimer providers owing to the multistage synthesis process; now, a few newly established Chinese companies now supply dendrimers.

Zeolites

Zeolites and silver atoms have been recognized since the 1980s. The porous nature of zeolite allows for the incorporation of nanoparticles such as silver ions. They are subsequently exchanged for other cations in solution, and the zeolite matrix is broken apart. Egger et al., (2014) studied zeolites and other nanosilver-containing materials. A tiny amount of silver ions are emitted when a metallic surface is in contact with liquids. Water disinfection was once again shown to be efficient.

The Water Research Commission Report No KV 297/12 shows how zeolites may be used as an adsorbing substrate for silver nanoparticles as a disinfecting source of silver ions. If nanoparticles aren't enough for you, zeolites may be employed as nanoparticles who used laser-induced fragmentation of zeolite Linde type A micro particles to create Nano zeolites. Zeolites and naturally occurring silver ions are used to make Agion's antibacterial component (Rice et al., 2013).

Nanometals and Nanometal Oxides

Metal oxides on the nanoscale have the potential to be excellent adsorbents for heavy metals and radionuclides, and they are a viable alternative to activated carbon. A large specific surface area is associated with a short intraparticle diffusion distance, and they have the ability to be compressed without losing a significant amount of surface area. Certain of these small metal oxides (such as nanomaghemite and nanomagnetite) are superparamagnetic, which means that they may be separated and recovered with the use of a low-gradient magnetic field. Both adsorption media filters and slurry reactors may benefit from the use of these materials (Bonevich, 2010; Kim and Deng, 2011). It is possible to absorb arsenic from waste and drinking water using nano iron hydroxide [-FeO(OH)], which is an abrasion-resistant adsorbent with a high specific surface area. ArsenXnp (SolmeteX Inc., Philadelphia, PA, USA) is a commercially available hybrid ion exchange medium that removes arsenic from water with high efficiency and little backwash, according to the manufacturer. Nanometals and nanometal oxides are mainly employed in industrial applications where they are compressed into porous pellets or used as powders (Kim et al., 2012; Jores et al., 2004; Bunaciu et al., 2015).

Magnetic Nanoparticles

Using magnetic nanoparticles (magnetite Fe3O4) to remove arsenic from groundwater has previously been shown in the field. "Pump-and-treat" groundwater treatment involves pumping groundwater to the surface, followed by further treatment, often with activated carbon, for ultimate purification. These additional operating hours and environmental cleanup costs may be avoided by using in situ technology. An easy method for extracting contaminated soil particles using a magnetic field is to introduce magnetic nanoparticles directly into the soil. An ideal usage for these nanoparticles would be to increase draw solutions' osmotic pressure, as well as ground water purification (Dorofeev et al., 2012).

Nano-Zero Valent Iron

The remediation of chlorinated hydrocarbon fluids and perchlorates in groundwater may be accomplished with the use of nano-zero valent iron. Groundwater may be treated in-situ by infusing a suspension of nano-zero valent iron into the water and allowing it to dissolve. Due to its high specific surface area, nano-zero valent iron is significantly more reactive than conventional granular iron; second, due to its high reactivity, nano-zero valent iron has an extremely short life span, necessitating additional research into surface modifications to stabilize these nanoparticles (Sharma et al., 2015).

Membranes and Membrane Processes

Increasingly, water and wastewater treatment processes are relying on membrane-based separation technologies. For substances, membranes provide a physical barrier, depending on the pore size and molecule size. Membrane technology has been generally accepted in the water and wastewater sector as a dependable and mostly automated process (Piebalgs and Potočnik., 2009).

Membranes Based on Nano Filtration

Particles and molecules having sizes of less than 0.5 nm to 1 nm are rejected by the membrane in Nano filtration technology. The unique charge-based repulsion mechanism provided by Nano filtration membranes allows for the separation of distinct ions. Hardness, color, smell, and heavy metals are some of the most prevalent contaminants that may be removed from groundwater using these devices. Since similar desalination methods are so expensive, desalination (the conversion of saltwater into drinkable water) has been a growing sector of usage (Yuan, 2013).

Membranes with Nano Fibers

Using an extremely fine material, such as polymers or ceramics, electro spinning produces fibres. Separation and filtering processes often make use of these nanofibers, which have a diameter of a few nanometers. A high surface to mass ratio is guaranteed as a consequence of the technological procedure's high specific surface porosity. These nanofibers are also customizable, such as their membrane thickness or connected open-pore structure. The electro spinning process is well-established in the air treatment industry, but it is completely unknown in the water and wastewater treatment industry. As a consequence, there is a great deal more research to be done (Ananda, 2017).

Membranes Based on Aquaporins

These pore-forming proteins, known as aquaporins, are present in all living things. Under certain circumstances, they may reject the majority of ionic molecules by creating highly selective water channels. High water permeability and selective rejection make them excellent for the construction of innovative high-flow biomimetic membranes. Stabilizing vesicles is accomplished by the incorporation of aquaporins. Due to their mechanically weak nature, vesicles such as these are either incorporated in a polymeric matrix or deposited on polymeric substrates such as Nano filtration membranes (Indarto, et al., 2010).

NANOTECHNOLOGY FOR SOIL REMEDIATION

Xenobiotics in soil include chemical fertilizers, pesticides, solid waste, heavy metals, solvents, and hydrocarbons. The kind of pollutant and the characteristics of the contaminated environment influence the biological treatment processes. There are several uses for nanomaterial's, such as soil remediation, where their tiny size and enormous surface area allow them to perform at a high level of efficiency. Nanotechnology in biological soil remediation might thus make a significant contribution to the improvement of the environment and the reduction of harmful pollutants found in the human environment. It is still a serious environmental problem, and cleanup has become a worldwide undertaking. Agricultural output and food safety are at risk due to the degradation of soil reservoirs, which must be addressed as soon as possible. It is exacerbated by the infiltration of soil by heavy metals (toxic residue), pesticides and POPs (persistent organic pollutants). Soil polluted with these substances raises the risk of toxicity in the food chain due to bio magnification. Both the urgent need to feed more people and the difficulty in preventing further land degradation has an influence on agricultural productivity. Immobilization or adsorption is a common therapeutic approach (Chermahini et al., 2018).

The efficacy, low cost, and environmentally benign method to removing metal pollutants from soils are some of the most important advantages of this sort of remediation technology. Soil pollutants may be immobilized using engineered nanomaterial's such as carbon nanotubes, metal oxides (ferric oxide and titanium oxide), and other Nano composites. For example, ferric oxide nanoparticles may collect and immobilize heavy metals like Cd and as from various medium samples (Wang, et al., 2014). This area's aquatic life suffers as a result of oily sewage. Iron nanoparticles are effective in removing waterborne Total Petroleum Hydrocarbons (TPHs) with improved results of 88.34% (Yan et al., 2013). As a consequence, nanotechnology-based treatments may provide high-performance treated water that contains fewer contaminants and harmful compounds and removes heavy metals. With the use of nanotechnology and biological therapies, environmental pollutants may be minimized. Organic dyes may be successfully decolored using industrial procedures aided by nanoparticles (Ajmal, et al., 2018). The accumulation of electrons on the particle surface and the subsequent diffusion of dye in a reduction reaction are the two steps that make gold and silver nanoparticles excellent dye degradation catalysts (Khataeea et al., 2017).

Magnetic nanoparticles may be used in pollution management in a variety of ways, including adsorption and catalysis (Baker et al., 1985). Polyacrylamide-modified magnetite nanoparticles may be used for soil erosion and arsenate leaching (White, et al., 2003). Soil pollution may be remedied in part by immobilizing contaminants on-site, which has gained a lot of attention as a practical and affordable solution. Various NM-based whole form alterations have been researched in order to find a viable material that combines low cost, high efficiency, increased stability, low environmental effect, and maximum reusability in a single component. Metal concentrations in the pore water of polluted sediments and soils were reduced by the immobilization of Nano hydroxyapatite particles (Cloete et al., 2010).

According to this investigation, sodium carboxymethyl cellulose-stabilized nZVI was able to remove 80% of soil-bound Cr (VI). Redox-sensitive elements may be treated as pollutants via abiotic reductive processes that affect the mobility and toxicity of these elements in soil alone or in conjunction with immobilization techniques. NZVI, for example, was applied to pyrene-affected soil to provide strong reducing conditions that allowed pyrene to be removed by reduction. Various oxidants are used in advanced oxidation processes (AOPs) to break down organic contaminants in contaminated soil using nanomaterial's (Cong et al., 2013). Organic contaminants may be effectively degraded by oxidative decomposition by using water radicals or sulphur dioxide radicals produced from the reaction of hydrogen peroxide or persulfate with soluble Fe (II). When it comes to fighting environmental pollution and treating specific regions polluted with toxic metal ions, green nanoparticles have enormous promise. They may be used to clean and rehabilitate trash areas. Heavy metal-contaminated water and soil may be disinfected with ferrous NP, which is also a good source of ecologically acceptable fertilizer. Immobilized Nano fertilizers reduce groundwater pollution while coordinating plant nutrient intake (Sharma, et al., 2015) by not transforming into permanent chemical or gaseous forms that are not accessible to developing plants. Magnetite (Fe_3O_4) and greigite (Fe_3S_4) with siliceous material produced by microorganisms have been used in optical coatings for solar energy applications (Karim et al., 2009).

Figure 1. Applications of nanotechnology

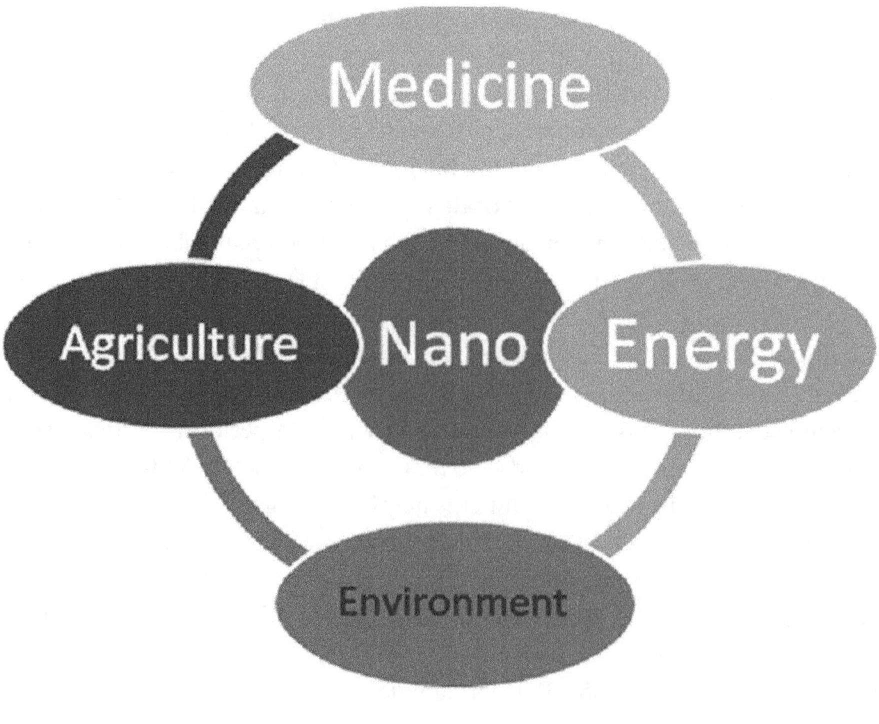

DELIMITATIONS OF NANOTECHNOLOGY

1. Concerns About Privacy and Security:

Nanotechnology opens the door to minuscule recording devices that are invisible. When in the wrong hands, atomic weapons would be more available and simple to build. It is much easier to carry out cellular destruction than it is to carry out mass destructions. Counter-attacking the idea of self-replicating weapons would be tough, especially if such safeguards have not been addressed.

2. It has the Potential to Render Current Energy Technologies Obsolete:

Fossil fuels are used in almost every industry of the planet. Because most families would switch to nano energy, the risk of current systems becoming obsolete is considerable. Such events would be terrible to sectors that are still reliant on fossil fuels, as mass manufacturing of commodities, food, oil, and precious metals might become worthless due to the ease with which molecular manipulation can reproduce them.

3. Consequences for the Environment:

Nanotechnology is made from toxic waste. It is possible that its development may have a negative impact on the environment since new toxins and pollutants will be emitted into the atmosphere. Because of the increase in respiratory disorders caused by air pollution, the death rate would rise.

Due to recent laboratory findings, scientists are concerned about the short and long-term consequences of nanotechnology on the environment; hence, another reason to be detrimental.

4. Economic Concerns:

Nanotechnology, like any other technology before it, would have a significant economic impact. Although the initial manufacturing method and goods may be costly, this is likely to change as the market adjusts. Companies would have to change to nanotech if there was a higher demand for Nano products. Some enterprises specializing in earlier technology and materials would be forced to close as a result of this massive market shift, as they would become obsolete.

5. Health Concerns:

Scientists are afraid that inhaling nanoparticles could result in inexplicable and incurable diseases. There is no guarantee that further progress will not result in new conditions that are more difficult to treat and control in the future.

6. Social Issues:

The possibility of a country stockpiling this technology would lead to social unrest and widening wealth disparities. It would result in a significant socioeconomic divide.

7. Ethical Issues:

Nanotechnology raises ethical concerns. The idea of designing specific features to change human genetic make-up is concerning.

8. Technological Concerns:

As with every computing system, high-tech breakthroughs and operations have limitations. Nanotech would also be vulnerable to hacking. This event could expose firm secrets and cause routine production operations to be disrupted.

9. Job Losses:

Nanotechnology would result in severe unemployment because most people working in manufacturing, food, and agriculture would lose their jobs. Human labor would be eliminated by the usage of nanobots. High research, production, maintenance, and advancement expenditures will be required for nanotechnology to succeed, posing a hurdle.

Figure 2. Delimitations of nanotechnology

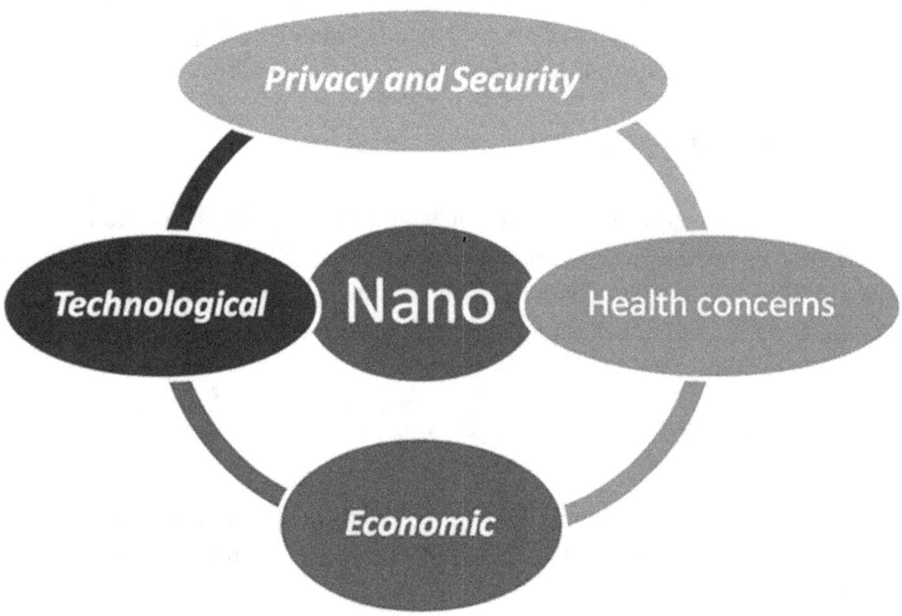

FUTURE PROSPECTS

This review, based on recent patents and scientific literature on the use of nanotechnology in cosmetics, reveals that this growing technology is on its way to becoming a critical instrument, not only for scientific research on novel cosmetic products, but also for industrial development. Patents and scientific investigations on nanotechnology in the cosmetics industry are concentrated in the formulation of ingredients for beauty products rather than in the development of new packaging materials or manufacturing equipment. It is possible that nanomaterials will be utilized in formulations, either as transporters of active substances, as more general formulation aids, or as active components in their own right in the future.

There are a wide range of nanomaterial's under investigation, including biological nanoparticles such as amino acid peptides and proteins, metals, oxides, carbon nanomaterial's, lipid-based NPs and a wide range of native, modified and synthesized polymers. A wide variety of cosmetic nanoparticles are expected to be accessible shortly if existing Nano toxicity and safety research incentives have had a chance to impact rules and guidelines for the production and use of nanomaterial's and if the appropriate economic backing is in place.

Public opinion and sentiments towards nanotechnology are crucial to the progress of this new science and business, in addition to toxicity and associated regulatory components. When it comes to items that come into close touch with our bodies but aren't important, our willingness to accept risks is likely to be lower. This means that trustworthy scientific findings confirming both the benefits and the safety of the innovations and inventions covered in this article, as well as those in the pipeline, are going to become more important when it comes to Nano cosmetics.

REFERENCES

Ajmal, Z., Muhmood, A., Usman, M., Kizito, S., Lu, J., Dong, R., & Wu, S. (2018). Phosphate removal from aqueous solution using iron oxides: Adsorption, desorption and regeneration characteristics. *Journal of Colloid and Interface Science*, *528*, 145–155. doi:10.1016/j.jcis.2018.05.084 PMID:29843062

Ali & Majeed (2013). *1st Annual International Interdisciplinary Conference*. AIIC.

Ali, A. S., & Mohammed, A. J. (2017). The optical properties of thin films of polymer PMMA/MR-Eosin prepared by casting methods. *Journal of International Academic Research for Multidisciplinary*, *5*(1).

Ali, A. S., Mohammed, A. J., & Saud, H. R. (2018). Hydrothermal synthesis of TiO2/Al2O3 nanocomposite and its application as improved sonocatalyst. *International Journal of Engineering & Technology*, *7*(4.37), 22-25.

Ananda, W. (2017, July). External quantum efficiency measurement of solar cell. In *2017 15th International Conference on Quality in Research (QiR): International Symposium on Electrical and Computer Engineering* (pp. 450-456). IEEE. 10.1109/QIR.2017.8168528

Arole, V. M., & Munde, S. V. (2014). Fabrication of nanomaterials by top-down and bottom-up approaches-an overview. *Journal of Materials Science*, *1*, 89–93.

AuplatC. A.BrochardG.Garidel-ThoronD. (2017). *Painting the Future: AMIPAINT Nanomaterials and the Safer-by-Design Approach for New Markets*. doi:10.2139/ssrn.3033840

Baker, M. D., Ozin, G. A., & Godber, J. (1985). Far-infrared studies of silver atoms, silver ions, and silver clusters in zeolites A and Y. *Journal of Physical Chemistry*, *89*(2), 305–311. doi:10.1021/j100248a026

Bayda, S., Adeel, M., Tuccinardi, T., Cordani, M., & Rizzolio, F. (2019). The history of nanoscience and nanotechnology: From chemical–physical applications to nanomedicine. *Molecules (Basel, Switzerland)*, *25*(1), 112. doi:10.3390/molecules25010112 PMID:31892180

Bessekhouad, Y., Robert, D., & Weber, J. V. (2003). Preparation of TiO2 nanoparticles by Sol-Gel route. *International Journal of Photoenergy*, *5*(3), 153–158. doi:10.1155/S1110662X03000278

Bonevich, J. E. (2010). Nanotechnology Characterization Laboratory. *National Cancer Institute-Frederick, MD*, *21702*(301), 846–6939.

Bunaciu, A. A., Udriştioiu, E., & Aboul-Enein, H. Y. (2015). X-ray diffraction: Instrumentation and applications. *Critical Reviews in Analytical Chemistry*, *45*(4), 289–299. doi:10.1080/10408347.2014.949616 PMID:25831472

Chermahini, S. H., Eslamian, S., Ostad-Ali-Askari, K., Singh, V. P., & Dalezios, N. R. (2018). New trends in self-cleaning materials for different purposes. *International Journal of Advances in Civil Engineering*, *1*(1).

Christian, P., Von der Kammer, F., Baalousha, M., & Hofmann, T. (2008). Nanoparticles: Structure, properties, preparation and behaviour in environmental media. *Ecotoxicology (London, England)*, *17*(5), 326–343. doi:10.100710646-008-0213-1 PMID:18459043

Cloete, T. E., de Kwaadsteniet, M., Botes, M., & Lopez-Romero, J. M. (2010). *Nanotechnology in Water Treatment Applications*. Caister Academic Press.

Cong, Q., Yuan, X., & Qu, J. (2013). A review on the removal of antibiotics by carbon nanotubes. *Water Science and Technology*, *68*(8), 1679–1687. doi:10.2166/wst.2013.420 PMID:24185047

Diallo, M. S., Christie, S., Swaminathan, P., Johnson, J. H., & Goddard, W. A. (2005). Dendrimer enhanced ultrafiltration. 1. Recovery of Cu (II) from aqueous solutions using PAMAM dendrimers with ethylene diamine core and terminal NH2 groups. *Environmental Science & Technology*, *39*(5), 1366–1377. doi:10.1021/es048961r PMID:15787379

Dorofeev, G. A., Streletskii, A. N., Povstugar, I. V., Protasov, A. V., & Elsukov, E. P. (2012). Determination of nanoparticle sizes by X-ray diffraction. *Colloid Journal*, *74*(6), 675–685. doi:10.1134/S1061933X12060051

Farahmandjou, M., & Golabiyan, N. (2016). Synthesis and characterization of Alumina (Al. *Int. J. Bio Inorg. Hybr. Nanomater*, *5*, 73–77.

Filipponi, L., Sutherland, D., & Center, I. N. (2010). *Fundamental Nano-effects*. Inano, Aarhus University.

Guo, D., Xie, G., & Luo, J. (2013). Mechanical properties of nanoparticles: Basics and applications. *Journal of Physics. D, Applied Physics*, *47*(1), 013001. doi:10.1088/0022-3727/47/1/013001

Guo, Z., & Tan, L. (2009). Fundamentals and applications of nanomaterials. *Artech House Canton Street, Norwood, MA, 02062*, 95.

Heera, P., & Shanmugam, S. (2015). Nanoparticle characterization and application: An overview. *International Journal of Current Microbiology and Applied Sciences*, *4*(8), 379–386.

Helmer, F., & Anja, O.S. (2016). *Synthesis; sol-gel.* Uni. Of Oslo, Dept. of Chemistry, Lectures at CUTN.

Indarto, A., Giordana, A., Ghigo, G., & Tonachini, G. (2010). Formation of PAHs and soot platelets: Multiconfiguration theoretical study of the key step in the ring closure–radical breeding polyyne-based mechanism. *Journal of Physical Organic Chemistry, 23*(5), 400–410.

Jores, K., Mehnert, W., Drechsler, M., Bunjes, H., Johann, C., & Mäder, K. (2004). Investigations on the structure of solid lipid nanoparticles (SLN) and oil-loaded solid lipid nanoparticles by photon correlation spectroscopy, field-flow fractionation and transmission electron microscopy. *Journal of Controlled Release, 95*(2), 217–227. doi:10.1016/j.jconrel.2003.11.012 PMID:14980770

Karim, M. R., Rhodes, E. R., Brinkman, N., Wymer, L., & Fout, G. S. (2009). New electropositive filter for concentrating enteroviruses and noroviruses from large volumes of water. *Applied and Environmental Microbiology, 75*(8), 2393–2399. doi:10.1128/AEM.00922-08 PMID:19218410

Khataee, A., Kayan, B., Gholami, P., Kalderis, D., & Akay, S. (2017). Sonocatalytic degradation of an anthraquinone dye using TiO2-biochar nanocomposite. *Ultrasonics Sonochemistry, 39*, 120–128. doi:10.1016/j.ultsonch.2017.04.018 PMID:28732928

Kim, E. S., & Deng, B. (2011). Fabrication of polyamide thin-film nano-composite (PA-TFN) membrane with hydrophilized ordered mesoporous carbon (H-OMC) for water purifications. *Journal of Membrane Science, 375*(1-2), 46–54. doi:10.1016/j.memsci.2011.01.041

Kim, E. S., Hwang, G., El-Din, M. G., & Liu, Y. (2012). Development of nanosilver and multi-walled carbon nanotubes thin-film nanocomposite membrane for enhanced water treatment. *Journal of Membrane Science, 394*, 37–48. doi:10.1016/j.memsci.2011.11.041

Kulkarni, A., Guney, D., & Vora, A. (2013). Optical absorption in nano-structures: Classical and quantum models. *International Scholarly Research Notices.*

Long, R. Q., & Yang, R. T. (2001). Carbon nanotubes as a superior sorbent for nitrogen oxides. *Industrial & Engineering Chemistry Research, 40*(20), 4288–4291. doi:10.1021/ie000976k

McLachlan, D. S., & Sauti, G. (2007). The AC and DC conductivity of nanocomposites. *Journal of Nanomaterials, 2007*, 2007. doi:10.1155/2007/30389

Minelli, C. (2004). *Bottom-up approaches for organizing nanoparticles with polymers* [Thesis]. EPFL.

Ngô, C., & Van de Voorde, M. (2014). *Nanotechnology in a Nutshell.* Atlantis Press.

Piebalgs, A., & Potočnik, J. (2009). *Photovoltaic Solar Energy.* Development and Current Research, European Communities. doi:10.2768/38305

Rice, S. B., Chan, C., Brown, S. C., Eschbach, P., Han, L., Ensor, D. S., & Grulke, E. A. (2013). Particle size distributions by transmission electron microscopy: An interlaboratory comparison case study. *Metrologia, 50*(6), 663–678. doi:10.1088/0026-1394/50/6/663 PMID:26361398

Roduner, E. (2006). Size matters: Why nanomaterials are different. *Chemical Society Reviews, 35*(7), 583–592. doi:10.1039/b502142c PMID:16791330

Sharma, K., Singh, G., Kumar, M., & Bhalla, V. (2015). Silver nanoparticles: Facile synthesis and their catalytic application for the degradation of dyes. *RSC Advances*, *5*(33), 25781–25788. doi:10.1039/C5RA02909K

Sharma, S., Jain, K. K., & Sharma, A. (2015). Solar cells: In research and applications—a review. *Materials Sciences and Applications*, *6*(12), 1145–1155. doi:10.4236/msa.2015.612113

Sinha, A. K., & Suzuki, K. (2007). Novel mesoporous chromium oxide for VOCs elimination. *Applied Catalysis B: Environmental*, *70*(1-4), 417–422. doi:10.1016/j.apcatb.2005.10.035

Srinivasan, S., Kannan, A. M., Kothurkar, N., Khalil, Y., & Kuravi, S. (2016). Nanomaterials for energy and environmental application. *Journal of Nanomaterials*, *2015*, 2.

Tala-Ighil, R. (2015). Nanomaterials in solar cells. Handbook of Nano electrochemistry, 1-18.

Tian, L. (2017). A short review on mechanical behavior of nanocrystalline materials. *Int J Metall Met Phys*, *2*(1), 1–13. doi:10.35840/2631-5076/9208

Wang, Y., Fang, Z., Liang, B., & Tsang, E. P. (2014). Remediation of hexavalent chromium contaminated soil by stabilized nanoscale zero-valent iron prepared from steel pickling waste liquor. *Chemical Engineering Journal*, *247*, 283–290. doi:10.1016/j.cej.2014.03.011

Wang, Z. L. (2000). Characterization of Nanophase Materials. Wiley-VCH.

White, C. M., Strazisar, B. R., Granite, E. J., Hoffman, J. S., & Pennline, H. W. (2003). Separation and capture of CO_2 from large stationary sources and sequestration in geological formations-coalbeds and deep saline aquifers. *Journal of the Air & Waste Management Association*, *53*(6), 645–715. doi:10.1080/10473289.2003.10466206 PMID:12828330

Yang, H. Y., Han, Z. J., Yu, S. F., Pey, K. L., Ostrikov, K., & Karnik, R. (2013). Carbon nanotube membranes with ultrahigh specific adsorption capacity for water desalination and purification. *Nature Communications*, *4*(1), 1–8. doi:10.1038/ncomms3220 PMID:23941894

Yuan, C. (2013). *Development of nanoparticle sensitized solar cells* [Doctoral dissertation]. KTH Royal Institute of Technology.

Zhang, Z., & Lagally, M. G. (1997). Atomistic processes in the early stages of thin-film growth. *Science*, *276*(5311), 377–383. doi:10.1126cience.276.5311.377 PMID:9103189

Chapter 2
Nanoparticles:
Types, Applications, and Commercial Products

Sameer Quazi
https://orcid.org/0000-0002-1258-4088
GenLab Biosolutions Private Limited, Bangalore, India & Department of Biomedical Sciences, Anglia Ruskin University, UK & School of Health Sciences, The University of Manchester, UK & SCAMT Institute, ITMO University, St. Petersburg, Russia

Sawani Rodrigo
Human Genetics Unit, University of Colombo, Sri Lanka

Zarish Fatima
Institute of Industrial Biotechnology, Pakistan

Aman Prakash
Jacob Institute of Biotechnology and Bioengineering, Sam Higginbottoms University of Agriculture, India

ABSTRACT

Nanomaterials are molecules with their size lying basically in nano range that are usually synthesized by proper arrangement of nanoparticles. These particles are basically solid phase materials with average size range 1 to 100 nm. They have gained prominent importance in technological development because of their biological and physiochemical properties. They are usually classified or characterized on the basis of size, shape, origin, and composition. The chapter reviews nanoparticles, their detailed classification, and commercialized products from nanoparticles. Because of enhanced production of nanomaterials and their increased applications in biological and industrial fields, the issues related to toxicity or contamination are inevitable. The differences between engineered nanoparticles and natural nanoparticles have been described along with their sources.

DOI: 10.4018/978-1-6684-5533-3.ch002

INTRODUCTION

Nanomaterials such as nanoparticles take into account the Nano technological based approach along with advancements in material synthesis and characterization developed in field of micro fabrication research. These nanomaterials have number of properties such as optical, electronic, physiochemical and mechanical (Roduner, E., 2006). On these basis nanomaterials are described as the materials having all dimensions in Nano scale being defined as the length ranging in between 1 to 100 nanometers which include not only the nanostructured materials but also their surface structure in between Nano scale (Peralta-Videa et al., 2011). The materials in nanoscale exhibit totally different properties than that of they depict at larger scale termed as macro scale. As for example, opaque materials get converted into transparent one in case of copper, inert materials become catalytically active for platinum, the most stable materials get combustion for aluminum and the solid materials get converted to liquid in terms of gold and last but not the least insulator become conductors. These materials are not only the miniaturizing of materials or particles but they require proper techniques for their conversion and operation. Keeping in mind, there are a number of techniques for this which can wither be bottom up or top down method (Chen et al., 2013).

They can easily be constructed from the top down technique thusresults in production of very small structures from the bulk of materials. They are prepared by bottom up mechanism from combination of atom or molecules. The one mechanism to do this is either by self-assembly where all other atoms or molecules tend to arrange themselves into a specific structure relying on the natural properties (Sharifi et al., 2012). The semi- conductor industry provides perfect example of chemical preparation of larger molecules. This assembly provides control for construction and is very laborious but found not to be suitable for industrial applications (Vollath D., 2008). Similarly based on available data nanoparticles can be engineered because of their unique properties and functions not previously found in biomedical science and industrial applications.The chapter aims to have idea about the nanomaterial's especially nanoparticles and their applications in a number of fields.

MAJOR SOURCES OF NANOPARTICLES

Engineered nanomaterials have been deliberately produced or engineered by humans in order to gain specific properties. Following this scenario legacy nanomaterials are described as those materials which have been commercialized before development of field of Nano biotechnology showing advancement over colloidal and other particle-based materials. They include particles such as carbon black or titanium oxide TiO2 (Landsiedel et al., 2012). Unintentionally nanoparticles have ben produces as a single by product of industrial or mechanical processes obtained through either combustion or vaporization process. These nanoparticles which are called as incidental nanoparticles have sources such as smoke or exhaust from vehicles, fumes of welding, combustion process being resulted from heating or cooking in homes (Rao and Cheetham, 2001).

Table 1. The types of nanoparticles along with its sources being listed

Nanoparticles type	Source
Natural nanoparticles	Dust storms
	Forest fires
	Volcanos
	Ocean and water evaporation
	Organisms
Anthropogenic nanoparticles	Diesel and engine
	Cigerrete smoke
	Demolition of buildings
	Cosmetics
	Indoor pollution
	Engineered nanoparticles
Metals and other dusts	Beryllium
	Lead and copper
	Asbestos
	Cadmium
	Aluminum
	Nickel
	Manganese
	Iron and silicon
	Organic dust
	Coal and polymer fumes

Occasionally, nanoparticle type such as fullerenes are produced by process of burning gases, biomass or in simple words candles. It can also be obtained as a byproductof other corrosion based products (Valsami-Jones and Lynch, 2015). These incidental nanoparticles are also termed as ultra-fine particles which are produced by chance without any specific intention but at the same time they take part in enhancing level of pollution. In the terms of biological system, only natural and functional nanoparticles are present.Chalks, waxcovering surface of leaf,spider silk, blue hue of tarantulas are all perfect examples (Rizwan et al., 2014).

Natural inorganic materials occurred via growth of crystal in certain chemical conditions on surface of earth. As for example clay show complexed nanostructures because of anisotropic behavior of crystal structure, volcanic activity can produce opals that are considered as naturally found volcanic crystals because of their Nano scaled structures. Similarly, fire also shows complex reaction thus giving rise to certain pigmented molecules, fumes and silica etc. Other natural sources of nanoparticles include forests on fire, volcanic eruptions and radon gas emission. These natural nanomaterials are also found during weathering process of certain metals or anions and other acid mines (Gogotsi and Presser, 2006).

DIFFERENT TYPES OF NANOPARTICLES

The nanomaterials are often categorized on the basis of their dimension in the Nano scale. The nanoparticle is defined as a nano based object having all the external dimensions in Nano scaleswhere the axes of particles don't differ from each other to a great extent. They can be embedded in the bulk form or solid in order to produce nanocomposites (Hasan S., 2015).There are a number of types of nanoparticle on the firm basis of size, shape, functionality and composition (Bhatia S.,2016). And these types of nanomaterials can be of following types.

Liposomes

The very first nanoparticle class discovered was liposomes and was being discovered in 1865 as a model for cell membrane. Since that they liposomes have been employed in the biomedical industry as the first mean for drug or gene delivery (Yih and Al-Fandi, 2006) These are the particles which are round or spherical in shape and contain single or double layer of lipid membranes in aqueous system (Wang and Wang, 2014). The most unique properties of liposomes that implement their use include their composition, capability to protect a number of biological molecules in addition to their compatibility in biological system and can be degraded very easily. These benefits have made them accurate for use in biological system acting as a host fortransfecting genetic material and the process is named as lipofection in biological research (Choi, et al., 2005). This method utilizes cations for the pairing with anionic genetic material during lipofection. Another application is their usage as a therapeutic carrier because their design allow entrapment of hydrophilic molecules in the core of hydrophobic drugs in the lipid bilayer itself (Ogolo et al., .2012). Liposomes are conjugated with other biocompatible polymers as for example polyethylene glycol in order to enhance their half lifr and also stability following in vivo systems. They can be functionalized with certain ligands in order to enhancediagnostic and therapeutic characters in desired cells. Till today 12 types of liposomes have been approved for use in biomedical applications for enhanced productivity (Youns et al., 2011).

Figure 1. Lipid based nanoparticles liposomes depicting their overall different classes and the size to droplet proportion of nanoparticle from 50 to 500nm

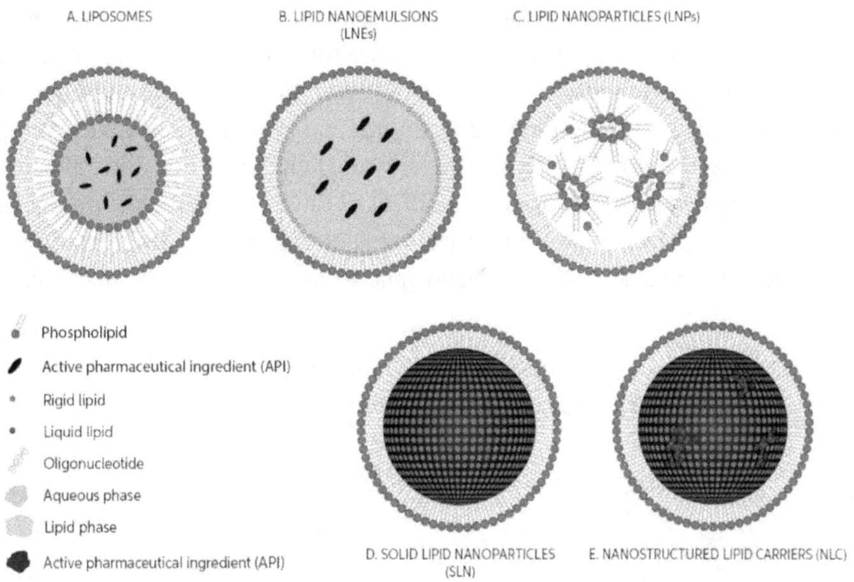

The lipid-based Nano emulsions consists of Nano sized droplets being stabilized by the surfactants for preventing overall aggregation in aquatic environment. Most common lines used in medical industry are plant based droplets with an average size less than 500 nm being stabilized by phospholipids and can be used in nutritional effects without any drug carrier functions.

Albumin Bound Nanoparticles

Albumin based nanoparticles utilize the endogenous mechanism of albumin for carrying out of the molecules in blood stream. These hydrophobic molecules get attached to hydrophobic molecules using non- covalent bonding in order to avoid the solvent-based toxicities in therapeutic pipeline. Because of these properties these particles have been employed successfully in the targeted drug delivery system (Elzoghby et al., 2012). As for example, Abraxane has been approved by Federal drug agency in 2005 in order to treat metastatic breast cancer having dimensions of about 130 nm in all directions (Tan and Ho, 2018). They attach to receptors present on surface of cells gp60-mediated and transported to endothelial cells with the purpose to concentrate. At the same time, it can also target the albumin targeted protein named as acidic protein rich in cysteine that are found to be over-expressed in tumor cells. An overall understanding of this mechanism can pave the ways for better targeting disease and revolution in therapeutic era by use of nab platform (Esim and Hascicek, 2021)

Polymeric Nanoparticles

The polymeric nanoparticles are actually the nanomaterials which lie in the size range in between 1 to 100 nm and are located in addition to the active compounds located at the surface core of polymer substances

(Vrignaud et al., 2011). There are two major types of the polymeric nanoparticles which can either be Nano sphere or nano capsules (Gumustas et al., 2017).They both differ in their method of preparation as well as the nano capsule being prepared via Nano precipitation while Nano spheres are prepared by either solvent extraction, emulsification, solvent diffusion and salting out. In most of techniques employed, organic compounds have been used as a first step for dissolving the polymer (Li et al., 2012). But at the same time these solvents result in toxicity and other environmental consequences. In addition to it all the solvent precursors must be removed from the final product (Yoshimatsu et al. 2012). Technique based on the polymerization of monomers are actually being used that allow the insertion with the greater efficiency in a single reaction step. But regardless of the method used the product are obtained as the colloidal solution or suspension. Another difference is that they both different systems as reservoir system in Nano capsule and matrix system in Nano spheres as shown in fig (Tuncel and Demir, 2010)

Figure 2. Depiction of the types of polymeric nanoparticles including Nano spheres and Nano capsule employing matrix and reservoir system especially in derma industry

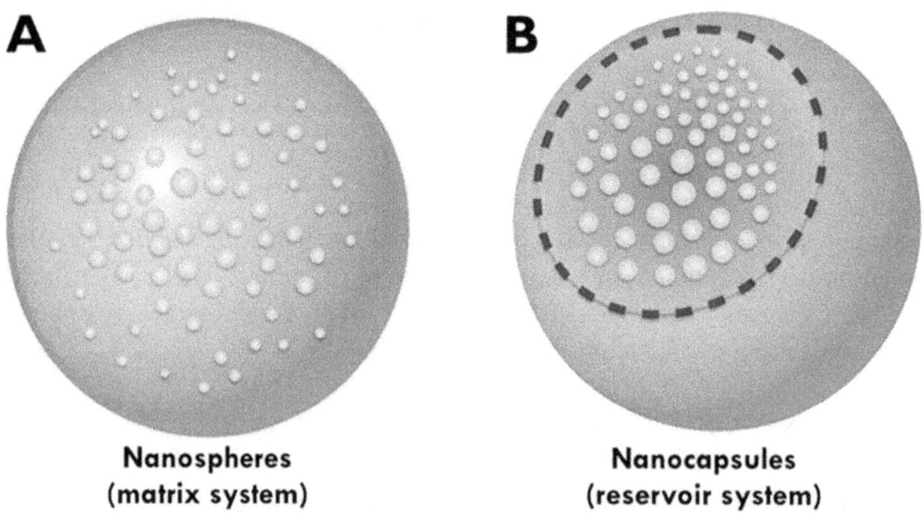

Dendrimers are another type of polymeric molecules that are regulatory branched macromolecules that have been synthesized from amino acids, sugar or other natural molecules. They are made up of core external branched layer and other exterior surfaces (Fischer, M. and F. Vögtle, 1999). The combination of these properties has resulted in production of dendrimers of well-definedsize, shape and length or density. On the basis of these characteristics these molecules can be used either as sensors, carriers or in gene delivery system. They can be loaded in the small cavities of core molecules by chemical; linking, hydrogen boding and other hydrophobic interactions. At the same time, their external surface can also be modified for providing functional groups, detecting or imaging agents and lastly in therapeutic regimen (Svenson and Tomalia, 2012).

Metal Based Nanoparticles

These include inorganic nanoparticles or nanomaterials such as quantum dots, nanowires or nano rods based on their electrical and optical properties and can be used as optoelectronic compound. Moreover, these properties depend on their size and hence can be tuned via other synthetic techniques. These possibilities have been used in production of organic material devices as for example solar cells and OLEDs (Sengul and Asmatulu, 2020). The operating principle are basically relayed on both optical and electronic processes such as electron and energy transfer. The devices efficiency depends on the photo induced process being responsible for functioning. Better understanding of photo induced composite system is significant property in optoelectronic devices (Luo et al., 2015).These nanoparticles or crystals are composed of metals, semiconductors and oxides that are of particular interest in mechanical, magnetic and other chemical properties. These particles have been used as quantum dots or as catalyst. These nanoparticles form a link in between the bulk materials and sub- atomic or molecular structures (Chen H., 2018). The bilk materials have specific physical properties irrespective of their size but these properties are different at nanoscale.

Metallic nanoparticles exhibit shows improved properties as compared to bulk materials. As for example in the case of copper nanoparticles, copper atoms bend in bulk followed by movement of copper atoms with the size range of 50 nm. Particles smaller than 50 nm are considered as super hard material showing less metallic properties and ductility. The variability in characteristics is not found to be done intentionally. Iron or ferrous based electric material smaller than 10 nm has ability to switch polarization temperature by using thermal energy making suitable for storage of memory (Zhang et al., 2019). Similarly, the suspensions or solution of nanoparticles are also possible as these particles interact with solvent material t be strong enough for overcoming density that makes the substances to sink in suspension (Quazi S., 2021). They also possess unexpected visual or optical properties as they are smaller to confine electrons thus giving out quantum effects. As for example gold nanoparticles appear red to black in color in suspension having high surface to volume ratio providing driving force for diffusion especially in the presence of high temperature. The sintering of nanoparticles is performed at lower temperature for shorter time lapse in case of large particles. On the basis of their dimension, they are categorized into three types:

1. Single- dimensional nanostructures
2. Bi-dimensional nanostructures
3. Bulk- nanostructured materials

Figure 3. Recent advantages in the types of metal nanomaterials which may be of any types (Ingle et al., 2014)

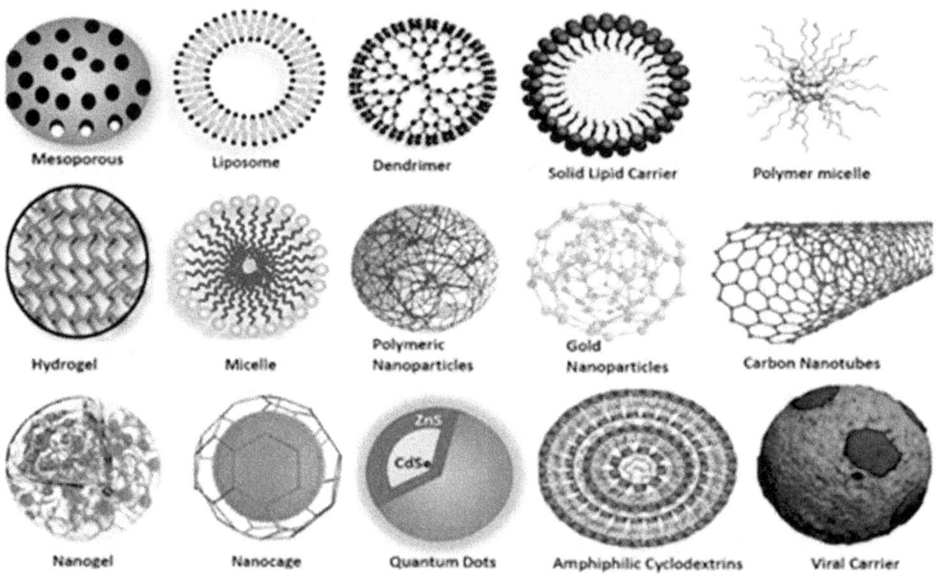

Quantum Dots

Quantum dots were first discovered in 1980 are semi- conductor materials with the size dimension less than or equal to 10nm. These nanomaterials exhibit optical and electrical properties based on their specific size (Lim et al., 2015). Most of the quantum dots are made up of zinc selenium and cadmium selenide cap. The absorption spectrum of these molecules are very broad in nature but their emission is just confined to a narrower bands (Quazi S., 2022). They show the capability of emitting bright colors having long half-lives, higher efficiencies and much stability again process of photo bleaching. They can be produced to generate a number of different biochemical efficiencies or specify and can be eventually detected or excited at the same time. Because of all these characters the quantum dots are superior to that organic polymers and organic fluorophore dyes employed for optical applications. In the biological system as an imaging tool such as biological labelling and tracking of biological molecules. The very small size ratio of quantum, dots make them suitable for use in biomedical and industrial applications (Reed MA., 1993)..

Iron Oxide Nanoparticles

These nanoparticles are particularly specialized for their passive targeting because they are super magnetic in nature. These nanoparticles usually have core made up of iron oxide coated with hydrophobic or hydrophilic material such as dextran or another in order to provide greater stability (Loss and DiVincenzo, 1998). The most widely used SPUIONS have core made up of magnetite Fe_2O_3 and show phenomenon of magnetism or super- magnetism based on their charge to size ratio that allows them to get magnetized upon exposure to magnetic field and depicting net magnetism when magnetic field is removed successfully (Reimann and Manninen, 2002). These particles have several advantages over other conventional particles adolinium-chelate contrast agents like improved sensitivity and specificity and lesser contamination problems. These oxide nanoparticles can be easily broken down into components such as iron oxide which are metabolized and can be stored easily in the form of ferritin and thus later can be incorporated into hemoglobin macromolecule (Woggon U., 1997). Following this almost SPIO agents femuroxide molecules having dimension to about 180 nm or carbon tubes with the dimensional values to 80nm have been approved by FDA to be use in magnetic resonance imaging and its applications such as molecular imaging for accurate detection of apoptosis and studying gene expression (Alhassid Y., 2000). These particles can also be functionalized with magnetic, optical and radionuclide ligands in order to be used in multimode imaging. At the same time they can also be used a potential candidate as non- invasive diagnostic tool and a vehicle for drug delivery vehicle (Bailey et al., 2004).

Gold Nanoparticles

Gold nanoparticles are found to have size range with a diameter between 1 to 100 nm and once they are dispersed in water they are termed as colloidal gold. They offer a number of size dependent and optical properties, biocompatibility and other facile surface modifications. They also have tendency to show enhanced level of absorption, scattering and ramen scattering because of interaction of nanoparticles with light photons. Summing all these all advantages have made these gold nanoparticles sufficient for use in biological sensing and detection, imaging diagnosis and other therapeutic applications (Sardar et al., 2009).. Sensing involve use of these nanoparticles made of gold in calorimetric analysis and as a substrate in SERS for enhancing scattering to a significant level thus allowing spectroscopic analysis and identification of proteins at surface of nanoparticles along with other industrial applications (Dreaden et al., 2012). In biological system they are not only used to detect the heart disease via heart biomarkers but also cancer biomarkers (Malik JA and Lone R, 2021). They can convert the absorbed light into heat and hence used in infrared phototherapy.

Silver Nanoparticles

They are either described as silver or silver oxide nanoparticles based on their large percentage of silver oxide and surface to bulk ratio having size range between 1- 100nm as described for gold NPs (Yu et al., 2013). They can be molded into any desired shape based on their applications and mostly these nanoparticles are spherical but diamonds, octagonal and sheet form of these nanoparticles is also present. They coordinate with a greater variety of ligands because of larger surface area and most common applications include biodegradation, biosafety, accession of potential efficacy and other animal studies (Rai et al., 2009). The most accurate method for synthesis of these nanoparticles is wet chemistry or nucleation of particles in the solution. This is done by reduction of silver nitrate or silver oxide into silver in the presence of reducing agents. When their concentration is increased a point reaches where dissolved silver aggregate to form bulk nanoparticles in order to create a stable surface (Quazi S, 2021). Another method is the reduction of monosaccharaides that include glucose, maltose but not sucrose following a simple method top reduce silver ion into silver nanoparticles involving a simple process (Abou El-Nour et al., 2010). There are a number of processes for chemical synthesis but all of them have drawbacks at same time. These drawbacks are overcome by the biological synthesis as they avoid use of harmful substrates such as sodium borohydride. They increase environmental foot prints by replacing strong reducing agents by using plants, bacteria, fungi (Khaydarov et al., 2009). These nanoparticles have been specifically gaining attention because of their antibacterial or medicinal properties along with catalytic properties for dyes or other compounds. Other metal nanomaterials and their oxides-based nanomaterials are given in table 02.

Table 2. The different metal-based nanoparticles along with their size range and color

Nanomaterial	Sample color	Size	Form of NPs (De, M., P.S. Ghosh, and V.M. Rotello, 2008) (approx.
PbO	Orange	Powdered	52
Ag_2O	Ash	Powdered	32
CeO_2	Yellow	Powdered	12
Alpha-Mn_2O_3	Black	Powdered	30.2
Ag-SnO_2	Yellowish black	Powdered	10
Ag-PbO	Brown orange	Powdered	51
Ag-Mn_2O_3	Brownish black	Powdered	29
SnO_2	Pale yellow	Powdered	12

TARGETED NANOPARTICLES

The concept of targeting using nanoparticles have been becoming a point of focus in the recent years. As in proteomics the targeted proteomics have gained its importance as a specific approach for addressing a biological question diverting focus towards protein molecule of interest (Pozo D., 2010). As in genetics, gene targeting has become a field of interest for determination of functional genes in mammalian cells. The development of monoclonal antibodies and small molecules in drug discovery as for example trastuzumab and rituximab has changed treatment regimen of cancer disease (Krutyakov et al., 2008). On the basis of this property, nanoparticles show any unique properties which that provide advantage over conventional techniques in biomedical research. In in- vivo conditions, nanoparticles can be targeted either by active or passive targeting.

1) Active targeting
2) Passive targeting

Active Targeting

It includes the use of specific targeting agents in order to enhance delivery of nanoparticles NPs at a specific site of infection. The targeting ligands typically include smaller molecules such as peptides, antibodies and their other fragments, nucleic acids or other aptamers. These ligands are always found to be in close coordination or conjugation with that of nanoparticles[65]. This type of conjugation can be performed by two methods which can either be covalent or non- covalent. In the terms of covalent conjugation the identical method is employed for the functionalization as well as characterization because the conjugation is wholly dependent on the functional groups present on the surface of nanoparticles

(Nguyen KT, 2011). Certain types of conjugation techniques are available or suitable for specific targeting of class of ligand. As for the conjugation of peptides, antibodies and other smaller fragments to surface of nanoparticles, malamide thiol is employed successfully. Following this reaction, maleic acid imides react with sulfhydral group at the pH ranging from 6.5 to 7. And this nanoparticle incorporated with sulfhydryl polymer can be further conjugated with other ligands being engineered with thiol or cysteine group or in presence of their side chain molecules (Pathak et al., 2015).

Another method that is manipulated in this conjugation process is carbodiimide-mediated amide present in between amino and carboxyl group of amino acids resulting in formation of bond. They can be further converted to active ester compounds in the presence of certain compounds such as ethyl-3--carbodiimide (EDC) and N-hydroxy-succinimide (NHS) or sulfo-NHS that are water soluble. These activated esters can react with further amine molecules in order to form a stronger amide bond. But this method has certain limitations and are only preferred for amines in order to form amine compounds and other smaller molecules which may contain either carbohydrates or other primary amines. The conjugation of the functional groups with high sensitivity and specify can be further done by click reactions under the presence of moderate reaction conditions. One such popular reaction is the 3 plus 2 additions of alkynes and azide molecules. It also facilitate the targeting of certain receptors with nanoparticle surface being present on surface of specific cells which leads to internalization or uptake via endocytosis using certain characters (Quazi S., 2022). In the term of therapeutics, they present specific advantages providing higher concentrations of drugs and decreased level of toxicity in comparison to other counterparts such as small molecules (Yun et al., 2013). This provides firm basis for identifying new targets for drug molecules with greater affinities or selectivity and decreased toxicities of cells. the nanoparticles bind to analytes, microbes and biomarkers giving amplified signals for detecting out using molecular imaging.

Passive Targeting

By passive targeting, nanoparticles can be used successfully for cancer therapy taking advantages from tumor biology. The biology of normal vascular tissue is simple and it has organized structures as well. On the other hand, the vasculature of tumors is irregularly branched and disorganized. The tumors also have higher density, vascularity with greater lymphatic drainage that is a very specific property of tumor cells. Altogether these properties make tumor cells make enhanced permeability and retention effect abbreviated as EPR which allow the nanoparticles to retain in the tumor cells (Bazak et al., 2014). Because of this retention in tumor tissues, the retention of NPs in other tissues also increases. They can be specifically targeted to a particular tissue or cell via perfect optimization of physicochemical properties (Devarajan, et al., 2010).

CARBON BASED NANOPARTICLES

Carbon Nanotubes

Carbon nanotubes are usually made up of carbon with their dimension being measured in the nanometers. They are further of two types such as single walled carbon nanotubes SWCNTs and multi-walled carbon nanotubes MWCNTs that are allotropic forms of carbon with the dimensions in between fullerene and graphene cages having diameter in nanometers(Ebbesen TW.,1994). SWCNT are idealized to be cut from bi dimensional hexagonal structure of carbon atoms being rolled up along vectors of lattice in order to result in formation of cylinder as a final structure. Periodic boundary conditions are usually improvised in the vector for yielding a helical lattice of consecutively bound carbon atoms in presence of cylindrical surface (Quazi S, 2022). On the other hand, multi-n walled carbon nanotubes usually made from accumulation of single- walled carbon nanotubes that are bound to each other by the weak forces of attraction such as vender walls forces forming a tree ring like structure (Dai H, 2002). If they are not identical or similar these nanotubes are parallel to carbon sheets or layers being arranged around a hollow tube like structure. They are also referred as double or triple walled carbon nanotubes with an undetermined carbon structure and diameter in between 100 nm. They show an appropriate conductivity of electricity in comparison to other nanomaterials which are basically non-conductor (Eletskii AV, 1997). In addition to it they also show a better thermal conductivity and tensile strength at same time because of nano based structure and the overall strength of bonds present between them. They can easily modified by attaching certain chemical groups and are a valuable tool in many areas of technology such as electrical, optics and other composite materials, nanotechnology and material based innovations (Endo et al., 1996)

These all the nanotubes are not helical because rolling of the hexagonal structures in different directions in order to give a long carbon tube form shows that thse are not only helical but they exhibit translational arrangement along axis of tube and have rotational type assembly about these axis at same time (Popov VN, 2004). Above all this, most of them are chiral which means that they cannot be super imposed and this cause labelling of single walled carbon nanotubes via integers. Similarly, a special type of SWNT are metallic in nature but all other of them are semiconductors having a perfect band gap. Such electrical properties is not dependent on weather a hexagonal structure is rolled from back to front or front to back providing a form of tube and mirror image. The remarkable properties of nanoparticles are amazing but a proper path to synthesize them was lacked till 1993 when some scientists discovered that co vaporizing carbon and transition metals as iron or cobalt can result in catalysis of SWCNTs. These discoveries thus showed success in increasing efficiency of a catalytic reaction leading to explosion for work regarding characterization and final applications of SWCNTs.

Graphene

This is the crystalline allotropic form of carbon which is transparent to visible light and is made up of sheet having thickness of about one atom. By weight it is about hundred times much stronger than that of steel and at the same time electrical or thermal conductivity showing a greater density of about 1000000 times higher than that of copper (Geim, A.K.2009). It was synthesized for the very first time in 2004 by Andre Grem and his coworkers who received a noble prize in field of physics because of their ground breaking experiment for the two dimensional grapheme material(Quazi S.,, 2021). Firstly they produced it from bulk graphene by lifting up graphene flakes with the help of an adhesive tape and then transferring it to a silicon based wafer.

Graphyne

Similarly, Graphyne is another allotropic form of carbon that is almost identical to that of graphene in structure and nature. In structure it can be seen as the benzene ring joined together by acetylene bonding. On the basis of acetylene group it can be considered as a mixed hybridization of spin where 1 is less than n which is further less than 2 in comparison to that of graphene which shows sp2 hybridization and diamond depicting sp3 type (Cranford and Buehler, 2011). Calculations based on first principle, ab- initio and mechanical or molecular stimulation has exhibited Graphyne and its boron nitride analogue to be stable in nature (Quazi, S, 2022). The presence of Graphyne was not known before 1960 and still it has not been synthesized properly. Although graphdiyne which is made up of Graphyne with diacetylene was prepared on substrate of copper. Recently, it has been emerged as a competitor for graphene because of directional dependent Diracones.

Thin- Film Nanoparticles

These films range in size from a fractions of nanometers to a few micrometers in thickness and its controlled synthesis is important as a fundamental step in a number of applications. The most familiar example of them is household mirror that possess a thin coating of metals on the back of glass sheet in order to form refractive index. Silvering was once used for producing mirrors while now a days a process named as sputtering has been used for coating metal on surface of mirror. Advances in 20 ^{ce}ntury have resulted in a major technological breakthrough in the areas such as magnetic recording, electronic or semi-conductor devices, optical coatings and coatings present on cutting tools for both purposes for energy storage as well as energy generation in thin film solar cells and batteries. In addition, these Nano films have an important role in the development of newer materials having unique properties. Examples of these films include multi- ferrous materials, super lattice which in turn allow phenomenon such as quantum phenomenon.

Nanocrystals

Nanocrystals are polycrystalline in nature having the size of few nanometers thus filling a wide gap in between amorphous and coarse-grained materials. The definitions of crystals may vary but the crystals have been defined as the materials with the size range below 100nm. The size of grains ranging from 100 to 500nm comes under ultra fine grain category and the size of grain can be estimated by X-ray diffraction. The peaks will be further broadened in the particles having very small size and this peak broadening is closely related to size of crystallite s estimated by Scherer equation which is usually applicable to 50nm. The crystal size can either be measured directly by X-ray diffraction or transmission electron microscopy(Xu et al., 2013). On the other hand, mechanical behaviors is dominated by certain flaws such as porosity, grain size and strengthening is also observed in high density strengthening materials. The larger volume grain fraction or grain boundaries associated to crystalline materials show interesting behaviors or properties in ceramics. The larger fraction allow more diffusional flow of atoms via creep similar to grain boundaries siding defamation phenomenon in Nano crystalline metals. This can provide super elasticity for processing of ceramics or its components because the materials may be get converted into conventional by additional thermal treatments after formation.

Fullerenes

Fullerenes are a class of allotropic carbons that are made up of graphene sheets that are rolled into tubes or spherical forms. These involve both carbon as well as silicon nanotubes that are of great interest because of their mechanical strength and electrochemical characters. The first fullerene molecule was discovered in 1985 by Richard Smallery. These molecules have ever been found in nature and have been detected in the outer space as well (Quazi S., 2022). From a last decade the chemical and physical properties of fullerene have revolutionized research and development and hope to continue in future as well (Wudl F., 2002). A study conducted in 2003 having extensive medical use in binding to a specific antibody, identifying structure of a specific bacteria and lastly certain type of cancerous cells. The 2005 research of biochemistry described use of fullerene as an activated anti-microbial agent. The intense properties of fullerene in nanotechnology is their super- conductivity and resistance. The best and easiest method to produce fullerene is to send the electric current in between graphite electrodes present in inert atmosphere. The result in creation of plasma that later into residue form from where fullerene can be isolated (Diederich and Gómez-López,1999).

APPLICATIONS OF NANOPARTICLES

These nanomaterials and nanoparticles have been used in a number of manufacturing process such as healthcare products including industry such as in paints, filters and lubricants. The overall applications of nanoparticles are summarized in Fig 04. Following health care system the Nano enzymes have been used in a number of applications such as that of bio sensing (De et al., 2008). They are mostly emerging type of enzyme-based systems that result in the production of high quality filters using nanostructures that are capable of removing a particle as small as that of virus by seldom technologies. The nanomaterial-based membranes in comparison to the conventional membrane have been recently used for treatment of wastewater. In the term of air pollution, nanotechnology has been recently used for treatment of middle east respiratory syndrome in Saudi Arabia IN 2012. These nanomaterials are also being used in insulation technologies as they were found in asbestos insulation. The applications of the nanoparticles are as follows:

Figure 4. Applications of nanoparticles in various fields such as environment, industrial, biomedical and food sciences

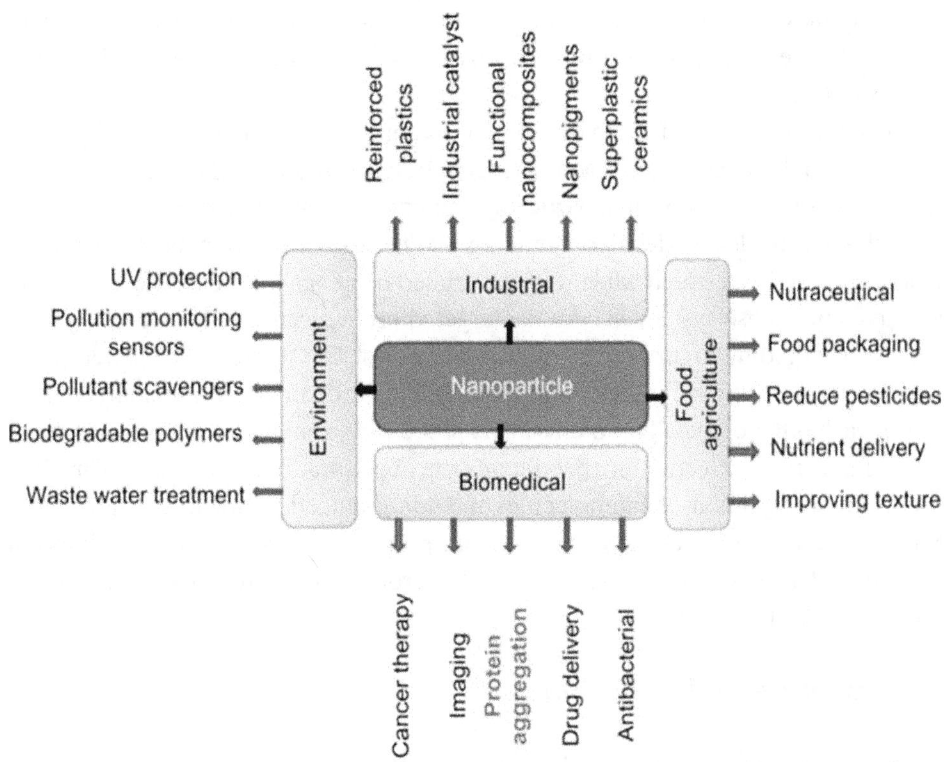

Nanoparticles as A Sensor

The development of better sensing techniques for biological analytes such as DNA, RNA or protein can provide more effective detection of biological analytes at molecular and sub- molecular levels. Most of these techniques have taken on advantage of the various characteristics of nanoparticles such as small size, high surface to volume ratio and other characteristics depending on composition which enable ligand to bind and amplify detection thus providing rapid detection of analytes. The functionalization of nanoparticles with specific ligand can aid in providing specifity and signaling of analytes giving out more efficient detection anyways. The research of developing nanoparticles is particularly focused on production of inorganic nanoparticles that can either be magnetic or metallic in nature (Baruwati et al., 2006). Because of their surface chemistry the nanoparticles can be extensively used as sensors. One of such method is biobarcode methodology that is provided by Mirkin group and this has ben extensively used for detection of DNA and protein. For detection of targeted DNA, sensitivity like polymerase chain reaction can be used (Li et al., 2009). This technique has extensively used in detection of cytokines at 31 attomolar concentration. And also as a soluble biomarker for diseases such as Alzheimer's. More recently a group of scientists used this process as the way for detection of prostate linked ligands or antigens that can be a biomarker of cancer in many tissues at elevated concentrations. Following bio-barcode assay gold nanoparticles functionalized with DNA are conjugated to PSA in order to generate PSA probe successfully. In this the DNA strands are bio barcodes. A magnetic micro particles having strong magnetism functionalized with antibodies to PSA are mixed with PSA targeted protein(Li et al., 2009). After separation by magnetic field and washing subsequently the probes are added to antibody bound PSA. Lastly these probes are released into solution and are further analyzed by scanometric assays that take advantage of gold nanoparticles catalyzed with silver nanoparticles. The assay results in sensitivity of 330fg/ ml.

This method employs the conjugation of DNA strand on nanoparticle surface followed by detection by technique such as polymerase chain reaction (PCR). Retello group has also presented another method for sensing of optical properties of nanoparticles. This include anionic molecules reversibly bound to cationic molecules via covalent molecular interactions. When the fluorescent molecule is bound, its fluorescence is quenched and is restored when it is dissociated from nanoparticle surface. The phenomenon is the basic of chemical nose system as here the targeted protein can be detected easily by the amount of fluorescence upon binding to surface of gold nanoparticles. The fluorescence pattern is recorded and statistically studied (Quazi et al., 2021).. The process detect, distinguish and analyze different target molecules from each other especially proteins. This process also enhance ramen scattering for accurate detection of biomarkers or detection of specific analytes by vibrational spectrum. Thus has been used for the ultra-sensitive detection of biomolecules including detection down to single molecule level. They have also been develop to employ the magnetic properties of NPs such as SPIONs and as most of biological molecules exhibit lower magnetism. This property is strongly used for most of biological samples (Jennings and Strouse, 2007).

Nanoparticles for Detection of Microbes

This has also been explored as a sensor for detection of pathogens using optical and magnetic properties of these nanoparticles. The most common method for detection of bacteria is by use of magnetic abilities of nanoparticles. Many groups have used functionalized nanoparticles for effectively labelling bacteria. As for example one group has used glycol magnetic nanoparticle based system that can detect or enable detection within the five minutes following by washing of sample with 88% removal rate by utilizing the interaction of bacteria to that of carbohydrates present on surface of mammalian cells (Theron et al., 2010). Another group has utilized vancomycin based SPOINs for bacteria such as gram positive or gram negative in a magnetic capture assay. At the same time they gave a demonstration that coverage on nanoparticle surface increases with the time required for labelling of bacteria to surface of nanoparticles reduction. Similarly optical bio sensing mechanism of bacteria has also been obtained by nanoparticles or quantum dots. This approach of bio barcode assay results in providing amplification and detecting of different targets in a single sample. nanoparticles have proved to be excellent imaging agent based on molecular properties that impart them sensitivity and monitoring of other molecular targets such as cellular responses associated with diseases such as cardiovascular ailments. For this purpose inorganic nanoparticles and quantum dots are preferred over other (Quazi S.,2022).

Nanoparticles in Food Sensing

The nanotextrured based food materials are being developed with the aim that they can provide with improved texture, taste and density or consistency. It also aids in the increment of half-life of food materials along with no wastage of food because of microbial infestation. The nanotechnology also takes its part in the delivering of food additives to the food without disturbing its overall morphology. Particle size is a critical thing that effect delivery of any bioactive compound to various sites in the body (Quazi S.,2022; Hajam et al.,2020; Malik et al., 2020. Most important food delivery system has following properties such as it is ideal for delivering targeted substance in a place, it also ensures targeted time period and rate and lastly it is easier to maintain the stability of compounds to a suitable level for longer period of time. It also aids in improving food quality by improving taste and flavor balance. Encapsulation of bioactive compounds such as carbohydrate, protein not only help them to resist adverse effects but also help them to assimilate foods easily which is hard to be obtain in non- capsulated form. For this purpose polymer nanoparticles are important for encapsulation of bioactive compounds to target structures and functions. They have also been used for detection of pathogens in food offering higher level of sensitivity and further novel attributes. It also works as indicator for environmental changes as for example humidity; microbial contamination etc. nanotubes have gained specific attention because of economical, simplicity and further detection.

Industrial Applications of Nanoparticles

Nanotechnology is improving the field of industry or consumer goods and a number of products have already been in use by general public that has not been even known by people that it contain nanoparticles products from easy clean to scratch resistance. Examples of this include car bumpers that are made lighter, clothing has become more stain repellent, cosmetics such as sun screen. Clothing has become more heat resistant ad packaging for drinks have resulted in much longer half-life. Using nanotechnology textiles will become smarter to be detectable by wearable electronics. The most important application is in households which make them easy to clean. And also Nano ceramics have resulted in improved resistance of common household equipment such as flat iron. The very first sunglasses using scratch resistant surfaces has been used as Nano composites. Nano fibers have been employed for precision of retinal, pupil repair and other eye surgeries. Lighter and stronger materials have been used by aircrafts that lead to increased performance and at the same time spacecraft will also be benefited where weight is a major factor. Nanotechnology thus help in decreasing size of equipment and also decreasing its fuel consumption to make it air borne. Hang gliders may intend to have their weights thus increasing toughness by use of nanomaterials. It also decreases mass of supercapcitor to provide increasing power to give electrical control for launching hand gliders off flattened to that of thermal chasing altitudes. The stronger and durable materials will be useful for creation of vehicles that are both faster and safe at same time. Combustion material or engineer may also get benefited that are more hard wearing and heat resistant at the same time (Wu et al., 2017).

Nanomaterial Use in Military

Nanomaterials can improve the military ability for detection of biological agents. Using this military is strongly able to create the biological agents and is the first most form of technology that military will use at its first place. These nanoparticles can be injected on military soldiers not to make only the uniform material of soldiers more durable but also aid in protecting them from different types of dangers such as temperature, chemicals and other effects at the same time. Another important aspect nanotechnology can improve soldiers or military uniform is to improve camouflage. Mobile pigment being injected into material can produce a much better form of camouflage material. Moreover the pigment will be able to change color of uniforms depending on the area of soldiers. A much amount of research is already being done on this form of camouflage. This technology can also improve thermal camouflage. Surfaces of different items can be designed in such a way. By reducing infrared signaling of both military vehicles being used by soldiers, it can result in providing better protection against guided weapons or other survivallance sensors at same time.

Nanomaterials in Construction Industry

Nanotechnology has the main ability to maker process of construction faster, cheaper and a much more different. Construction involves not only constructions of home but also sky scraper in meantime at much lower cost. The near future involve application of nanotechnology for sensing out cracks in architecture building and also can sense nanobots to repair them at same time. In construction it usually involve the use of nanoparticles as alumina and silica. Research has also been focused on development of Nano cements. If the cement with the nano sized particles can be processed or produced it will in turn open up a large number of opportunities in the terms of ceramic industry, strength, composites and electronics. A huge portion of research is focused on the ceramics and at same time a number of various techniques such as atomic force microscopy, scanning electron microscopy and focused ion beam etc. This in turn has proved to be a side benefit in general. After all concrete is a nanoparticle strongly effected by material properties and having understanding of it is actually yielding novel improvements in regards to strength or durability. Silica is present as a part of normal mix and that the particle packing at Nano scalecan be inherently improved by Nano silica which leads to densifying nanostructures that can provide improved mechanical properties in turn. It can control the degradation of fundamental C-S-H bonds that is reaction of concrete caused by calcium leaching in water as well as increasing durability (Kong et al, 2016). Another material having serious role in construction is steel. The use of nanotechnology helps in improving physical properties of steel. Cyclic loading can lead to structural failure and current designs of steel can lead to lowering of allowable stress, service life and further regular inspection of system. This on the whole produces a considerable effect on overall cost of structure limiting use of resources at same time. As clarified from research addition of copper oxide nanoparticle overcomes the unevenness of steel limiting number of stress risers and fatigue crackin alternatively. Following this regimen steel cables can be strengthen using carbon nanotubes (Kong et al, 2016). As the stronger cables are able to reduce cost and construction of bridges most specially in suspension bridges where the cables are run from end to end manner. Similarly wood is an important part of Nano fibers that are approximately as strong as that of steel. Using these Nano fibers paves ways for new construction paradigm as production and usage will be a part of renewable cycle. Some constructors have put forwarded their concept that if we want to build functionality over lingo- cellulosic surface can be an opportunity to get self sterilizing structures in construction era

COMMERCIAL PRODUCTS FROM NANOPARTICLES

Nanotechnology has been extensively used for enhancing the commercial applications of a number of consumer products. And these products are dependent on the modification or variation in physical characteristics as the sizes are minimized eventually. Following are the uses of nanoparticles in the industry and a very few products have already been commercialized.

Sunglasses by Nanoparticles

Sunglass use anti refractive or protective coating that has been the greater innovation of Nano biotechnology already in market. Following applications in optics, these products provide scratch resistant products or coatings based on nanoparticles. The coatings are found to be transparent in nature hence their optical performance is not effected at all which has make them ideal and simple to use. Their prices also offer durability and hence don't put economical burden on users (Simonet and Valcárcel, 2009).

Nanotechnology in the Field of Textiles

Increasing trend in textile sector is that the production of low cost clothes have been increased even in the under developing countries. Only the very high cost countries compete in the sector if they already develop higher sector clothes with the additional benefits to users the products include wind and water proof jackets where in nanotechnology has already major role to play on. In the terms of electronics, the smart clothes and wearable electronics have revolutionized whole world (McCall MJ.,2011).

Digital Sensors for Monitoring Body Functioning

Advanced technology such as that of nanotechnology can offer other features as for example development of sensors that detect functions of body and also the discharge of drugs into body according to requirement. Thus providing direct access to internet and that of self-repairing mechanism. At the same time simple realization techniques are also available that make wrinkle free clothes or also make them water repellent at same time. on these basis a sky jacket has been developed and manufactured by Franz Ziener GmbH & Co. company. These wind or water proof jackets are possible because of not only surface coatings but also due to involvement of Nano fibers at same time (Wang et al., 2013).

Nanotechnology Based Wrinkle Resistant Fabrics

Nano tex is a company that make stain repellent and wrinkle resistant fabric followed by attachment of certain molecular structures to cotton. These textiles are advised to be washed less frequently in presence of much lower temperature. Higher performance functional clothes are the requirement of a workplace. As for example, Gore-Tex has created highly resistant, protective, antistatic and outer wear fabric from nanotechnology. In this fabric minute carbon based nanoparticles have been incorporated and promises to provide full anti- current support to a person who wears.

Nanoparticles Based Sport Equipment

Ski wax a most reliable and supporting, hard gliding surface is present now a days. Their ultra thin coating is much more longer than that of conventional waxing system. Babolat a company who prepares tennis equipment has already provided rackets with nanotubes or nanoparticles inserted in it which provide better torsion ad flex resistance at same time. these rackets are usually more studier than that of present day rackets and provide more power. Similarly, tennis balls are also prepared with internal core made up of clay polymers to make them last longer. On the basis of this technology, the balls prepared by InMat have more half-life than that of conventional ones.

Nanoparticles in Sunscreen

One of the most important area of Nano biotechnology is their use in cosmetics especially sunscreens. Already available sunscreens are limited by their long time stability from UV rays. O the other hand, a sunscreen containing titanium dioxide has come forward at same time and has replaced already available technique (Nagarajan R., 2008). They provide us protection method better than any other material in bulk form and also lose unwantedly whitening t same time as a result of reduced particle size.L´Oréal has synthesized nanosome polymer based anti- wrinkle cream that usually transfers active agents such as vitamins proving nanoparticles to provide better care than other cosmetics.

Nanoparticles in Plastics and Colloids

Nanoparticles are the effective catalyst candidate because of increased surface area in presence of small size. They are used in manufacturing of plastics in order to enhance properties and versatility of the resulting material. Nanoparticles being also added in colloids that have further used ion painter ink and the printers providing long term effect.

Nanoparticles in Dietary Supplements

Dietary supplements is another aspect of nanotechnology where technology is hot enough to have its impact. Vitamin company Inc., of Fort Lauderdale, Florida, for one, boasts of the All Natural Patented Nanotechnology™ has been aimed to provide supplementary products that in turn maintain targeted drug delivery and active absorption of its components (Ugazio et al., 2002). The truth behind these claims is hard to find because of the fact that federal; drug agency has failed to approve, verify or test the labelling of supplements before they arrive in the market. The research has cleared that no adverse effects of supplements have been reported to FDA caused as a result of limited use of nanotechnology in nutrition or supplements. At the same time some of the antimicrobial or antibacterial nanoparticles against food borne pathogens have been already present offering their role in safe and quality food products by providing extended shelf life and protection (Malik et al., 2022) as listed in table 03.

Table 3. Antimicrobial nanoparticles against pathogens and their efficiency in either unit respectively

Nanoparticles	Antimicrobial agents	Efficacy
Gold	Not known	ZOI 10mm
Palladium	Not known	2.5 log CFU/ml
Cellulose	Tea, nisin	100% inhibition
Chitosan	Moringa oil	98% at 4°C
Copper	Not known	24.35MM
Magnesium	Thyme	99%
Platinum	Not known	2.0 logCFU/ml
Graphite	Konjac	1.3cm

CONCLUSION

Nanotechnology is improving our life by participating in strengthening efficiency and sensitivity of products of daily households. It also provides safer environment by providing access to safer water, air, renewable or biodegradable energy resources for a sustainable and better future. Not only it has revolutionized medical technology but also others such as construction, ceramics, military and others etc. because of this nanotechnology has gained pure attention of researchers where most of scientists and top institutions are putting up their efforts for development in every field even in food technology. It has been become that established field of science where extensive research has been carried out for implementing technology at such a nanoscale. Furthermore, it has been tested for the further new applications for increasing efficiency or performance of objects or processes in order to make this economical so that everyone can use it in a continuous manner. The technology is found to have brighter future because of its ecofriendly nature and affectivity at same time.

REFERENCES

Alhassid, Y. (2000). The statistical theory of quantum dots. *Reviews of Modern Physics*, *72*(4), 895–968. doi:10.1103/RevModPhys.72.895

Bailey, R. E., Smith, A. M., & Nie, S. (2004). Quantum dots in biology and medicine. *Physica E, Low-Dimensional Systems and Nanostructures*, *25*(1), 1–12. doi:10.1016/j.physe.2004.07.013

Baruwati, B., Kumar, D. K., & Manorama, S. V. (2006). Hydrothermal synthesis of highly crystalline ZnO nanoparticles: A competitive sensor for LPG and EtOH. *Sensors and Actuators. B, Chemical*, *119*(2), 676–682. doi:10.1016/j.snb.2006.01.028

Bazak, R., Houri, M., Achy, S. E., Hussein, W., & Refaat, T. (2014). Passive targeting of nanoparticles to cancer: A comprehensive review of the literature. *Molecular and Clinical Oncology*, *2*(6), 904–908. doi:10.3892/mco.2014.356 PMID:25279172

Bhatia, S. (2016). Nanoparticles types, classification, characterization, fabrication methods and drug delivery applications. In Natural polymer drug delivery systems. Springer.

Chen, G., Seo, J., Yang, C., & Prasad, P. N. (2013). Nanochemistry and nanomaterials for photovoltaics. *Chemical Society Reviews*, *42*(21), 8304–8338. doi:10.1039/c3cs60054h PMID:23868557

Chen, H. (2018). Metal based nanoparticles in agricultural system: Behavior, transport, and interaction with plants. *Chemical Speciation and Bioavailability*, *30*(1), 123–134. doi:10.1080/09542299.2018.1520050

Choi, W. S., Koo, H. Y., Park, J.-H., & Kim, D.-Y. (2005). Synthesis of two types of nanoparticles in polyelectrolyte capsule nanoreactors and their dual functionality. *Journal of the American Chemical Society*, *127*(46), 16136–16142. doi:10.1021/ja053981u PMID:16287301

Cranford, S. W., & Buehler, M. J. (2011). Mechanical properties of graphyne. *Carbon*, *49*(13), 4111–4121. doi:10.1016/j.carbon.2011.05.024

Dai, H. (2002). Carbon nanotubes: Opportunities and challenges. *Surface Science*, *500*(1-3), 218–241. doi:10.1016/S0039-6028(01)01558-8

De, M., Ghosh, P. S., & Rotello, V. M. (2008). Applications of nanoparticles in biology. *Advanced Materials*, *20*(22), 4225–4241. doi:10.1002/adma.200703183

Devarajan, P. V., Jindal, A. B., Patil, R. R., Mulla, F., Gaikwad, R. V., & Samad, A. (2010). Particle shape: A new design parameter for passive targeting in splenotropic drug delivery. *Journal of Pharmaceutical Sciences*, *99*(6), 2576–2581. doi:10.1002/jps.22052 PMID:20091830

Diederich, F., & Gómez-López, M. (1999). Supramolecular fullerene chemistry. *Chemical Society Reviews*, *28*(5), 263–277. doi:10.1039/a804248i

Dreaden, E. C., Alkilany, A. M., Huang, X., Murphy, C. J., & El-Sayed, M. A. (2012). The golden age: Gold nanoparticles for biomedicine. *Chemical Society Reviews*, *41*(7), 2740–2779. doi:10.1039/C1CS15237H PMID:22109657

Ebbesen, T. W. (1994). Carbon nanotubes. *Annual Review of Materials Science*, *24*(1), 235–264. doi:10.1146/annurev.ms.24.080194.001315

Eletskii, A. V. (1997). Carbon nanotubes. *Physics Uspekhi*, *40*(9), 899–924. doi:10.1070/PU1997v040n09ABEH000282

Elzoghby, A. O., Samy, W. M., & Elgindy, N. A. (2012). Albumin-based nanoparticles as potential controlled release drug delivery systems. *Journal of Controlled Release*, *157*(2), 168–182. doi:10.1016/j.jconrel.2011.07.031 PMID:21839127

Endo, M., Iijima, S., & Dresselhaus, M.S. (1996). *Carbon nanotubes.* Elsevier.

Esim, O., & Hascicek, C. (2021). Albumin-based nanoparticles as promising drug delivery systems for cancer treatment. *Current Pharmaceutical Analysis*, *17*(3), 346–359.

Fischer, M., & Vögtle, F. (1999). Dendrimers: From design to application—a progress report. *Angewandte Chemie International Edition*, *38*(7), 884–905. doi:10.1002/(SICI)1521-3773(19990401)38:7<884::AID-ANIE884>3.0.CO;2-K PMID:29711851

Geim, A.K. (2009). Graphene: Status and prospects. *Science, 324(*5934), 1530-1534.

Giljohann, D.A., & Seferos, D.S. (2020). Gold nanoparticles for biology and medicine. *Spherical Nucleic Acids,* 55-90.

Gogotsi, Y., & Presser, V. (2006). *Carbon nanomaterials*. CRC Press. doi:10.1201/9781420009378

Gu, F. X., & Karmik, R. (2007). Targeted nanoparticles for cancer therapy. *Nano Today*, *2*(3), 14–21. doi:10.1016/S1748-0132(07)70083-X

Gumustas, M., & Sengel-Turk, C. T. (2017). Effect of polymer-based nanoparticles on the assay of antimicrobial drug delivery systems. In Multifunctional systems for combined delivery, biosensing and diagnostics. Elsevier.

Hajam, Y. A., Rai, S., Kumar, R., Bashir, M., & Malik, J. A. (2020). Phenolic Compounds from Medicinal Herbs: Their Role in Animal Health and Diseases – A New Approach for Sustainable Welfare and Development. In R. Lone, R. Shuab, & A. Kamili (Eds.), *Plant Phenolics in Sustainable Agriculture*. Springer. doi:10.1007/978-981-15-4890-1_10

Hasan, S. (2015). A review on nanoparticles: Their synthesis and types. *Research Journal of Recent Sciences*, *2277*, 2502.

Ingle, A. P., Duran, N., & Rai, M. (2014). Bioactivity, mechanism of action, and cytotoxicity of copper-based nanoparticles: A review. *Applied Microbiology and Biotechnology*, *98*(3), 1001–1009. doi:10.100700253-013-5422-8 PMID:24305741

Jennings, T., & Strouse, G. (2007). Past, present, and future of gold nanoparticles. *Bio-Applications of Nanoparticles,* 34-47.

Khaydarov, R. R., & Khaydarov, R. A. (2009). Silver nanoparticles. In Nanomaterials: Risks and benefits. Springer.

Kholoud, M. M. (2010). Synthesis and applications of silver nanoparticles. *Arabian Journal of Chemistry*, *3*(3), 135–140. doi:10.1016/j.arabjc.2010.04.008

Kong, X., Squire, K., Li, E., LeDuff, P., Rorrer, G. L., Tang, S., Chen, B., McKay, C. P., Navarro-Gonzalez, R., & Wang, A. X. (2016). Chemical and biological sensing using diatom photonic crystal biosilica with in-situ growth plasmonic nanoparticles. *IEEE Transactions on Nanobioscience*, *15*(8), 828–834. doi:10.1109/TNB.2016.2636869 PMID:27959817

Koziara, J. M., Lockman, P. R., Allen, D. D., & Mumper, R. J. (2003). In situ blood–brain barrier transport of nanoparticles. *Pharmaceutical Research*, *20*(11), 1772–1778. doi:10.1023/B:PHAM.0000003374.58641.62 PMID:14661921

Krutyakov, Y. A., Kudrinskiy, A. A., Olenin, A. Y., & Lisichkin, G. V. (2008). Synthesis and properties of silver nanoparticles: Advances and prospects. *Russian Chemical Reviews*, *77*(3), 233–257. doi:10.1070/RC2008v077n03ABEH003751

Landsiedel, R., Fabian, E., Ma-Hock, L., Wohlleben, W., Wiench, K., Oesch, F., & van Ravenzwaay, B. (2012). Toxico-/biokinetics of nanomaterials. *Archives of Toxicology*, *86*(7), 1021–1060. doi:10.100700204-012-0858-7 PMID:22576463

Li, H., Cui, Z., & Han, C. (2009). Glutathione-stabilized silver nanoparticles as colorimetric sensor for Ni2+ ion. *Sensors and Actuators. B, Chemical*, *143*(1), 87–92. doi:10.1016/j.snb.2009.09.013

Li, K., Ding, D., Huo, D., Pu, K.-Y., Thao, N. N. P., Hu, Y., Li, Z., & Liu, B. (2012). Conjugated polymer based nanoparticles as dual-modal probes for targeted in vivo fluorescence and magnetic resonance imaging. *Advanced Functional Materials*, *22*(15), 3107–3115. doi:10.1002/adfm.201102234

Lim, S. Y., Shen, W., & Gao, Z. (2015). Carbon quantum dots and their applications. *Chemical Society Reviews*, *44*(1), 362–381. doi:10.1039/C4CS00269E PMID:25316556

Loss, D., & DiVincenzo, D. P. (1998). Quantum computation with quantum dots. *Physical Review A*, *57*(1), 120–126. doi:10.1103/PhysRevA.57.120

Luo, Y.-H., Chang, L. W., & Lin, P. (2015). Metal-based nanoparticles and the immune system: Activation, inflammation, and potential applications. *BioMed Research International*, *2015*, 143720. doi:10.1155/2015/143720 PMID:26125021

Malik, J. A., Bhadauria, M., Lone, R., & Hajam, Y. A. (2020). Exploitation of Plant Phenolics in Animal Farming. In R. Lone, R. Shuab, & A. Kamili (Eds.), *Plant Phenolics in Sustainable Agriculture*. Springer. doi:10.1007/978-981-15-4890-1_3

Malik, J. A., & Lone, R. (2021). Heat shock proteins with an emphasis on HSP 60. *Molecular Biology Reports*, *48*(10), 6959–6969. doi:10.100711033-021-06676-4 PMID:34498161

Malik, J. A., Mishra, G., Hajam, Y. A., Lone, R., & Quazi, S. (2022). Metabolome Analyses in Response to Diverse Abiotic Stress. In A. Roychoudhury, T. Aftab, & K. Acharya (Eds.), *Omics Approach to Manage Abiotic Stress in Cereals*. Springer. doi:10.1007/978-981-19-0140-9_6

McCall, M. J. (2011). Nanoparticles in the real world. *Nature Nanotechnology*, *6*(10), 613–614. doi:10.1038/nnano.2011.169 PMID:21979234

Nagarajan, R. (2008). *Nanoparticles: Building blocks for nanotechnology*. ACS Publications.

Nguyen, K.T. (2011). *Targeted nanoparticles for cancer therapy: Promises and challenge*. Academic Press.

Ogolo, N., Olafuyi, O., & Onyekonwu, M. (2012). Enhanced oil recovery using nanoparticles. In *SPE Saudi Arabia section technical symposium and exhibition*. OnePetro.

Pathak, R. K., Kolishetti, N., & Dhar, S. (2015). Targeted nanoparticles in mitochondrial medicine. *Wiley Interdisciplinary Reviews. Nanomedicine and Nanobiotechnology*, *7*(3), 315–329. doi:10.1002/wnan.1305 PMID:25348382

Peralta-Videa, J. R., Zhao, L., Lopez-Moreno, M. L., de la Rosa, G., Hong, J., & Gardea-Torresdey, J. L. (2011). Nanomaterials and the environment: A review for the biennium 2008–2010. *Journal of Hazardous Materials*, *186*(1), 1–15. doi:10.1016/j.jhazmat.2010.11.020 PMID:21134718

Popov, V. N. (2004). Carbon nanotubes: Properties and application. *Materials Science and Engineering R Reports*, *43*(3), 61–102. doi:10.1016/j.mser.2003.10.001

Pozo, D. (2010). *Silver nanoparticles*. BoD–Books on Demand. doi:10.5772/186

Quazi, S. (2021a). *Anti-cancer activity of human gastrointestinal bacteria*. Academic Press.

Quazi, S. (2021b). *TNFR2 antagonist and agonist: A potential therapeutics in Cancer Immunotherapy*. Academic Press.

Quazi, S. (2022a). Artificial intelligence and machine learning in precision and genomic medicine. *Medical Oncology*, *39*(8), 1–18. doi:10.100712032-022-01711-1 PMID:35704152

Quazi, S. (2022b). Elucidation of CRISPR-Cas9 application in novel cellular immunotherapy. *Molecular Biology Reports*, *49*(7), 1–9. doi:10.100711033-022-07147-0 PMID:35122203

Quazi, S. (2022c). An overview of CAR T cell mediated B cell Maturation Antigen therapy. *Clinical Lymphoma, Myeloma & Leukemia*, *22*(6), e392–e404. doi:10.1016/j.clml.2021.12.003 PMID:34992008

Quazi, S. (2022d). Telomerase gene therapy: A remission toward cancer. *Medical Oncology (Northwood, London, England), 39*(6), 1–20. doi:10.100712032-022-01702-2 PMID:35429243

Quazi, S. (2022e). *The potential implementation of biosensors for the diagnosis of biomarkers of various cancer*. Academic Press.

Quazi, S. (2022f). *Application of Biosensors in Cancers, An Overview*. Academic Press.

Quazi, S., & Malik, J. (2021). Discovery of potential drug-like compounds against Viral protein (VP40) of Marburg Virus using pharmacophoric based virtual screening from ZINC database. *bioRxiv*. Advance online publication. doi:10.1101/2021.05.13.444037

Rai, M., Yadav, A., & Gade, A. (2009). Silver nanoparticles as a new generation of antimicrobials. *Biotechnology Advances, 27*(1), 76–83. doi:10.1016/j.biotechadv.2008.09.002 PMID:18854209

Rao, C., & Cheetham, A. (2001). Science and technology of nanomaterials: Current status and future prospects. *Journal of Materials Chemistry, 11*(12), 2887–2894. doi:10.1039/b105058n

Reed, M. A. (1993). Quantum dots. *Scientific American, 268*(1), 118–123. doi:10.1038cientificamerican0193-118

Reimann, S. M., & Manninen, M. (2002). Electronic structure of quantum dots. *Reviews of Modern Physics, 74*(4), 1283–1342. doi:10.1103/RevModPhys.74.1283

Rizwan, M., & Singh, M. (2014). Ecofriendly application of nanomaterials: Nanobioremediation. *Journal of Nanoparticles*.

Roduner, E. (2006). Size matters: Why nanomaterials are different. *Chemical Society Reviews, 35*(7), 583–592. doi:10.1039/b502142c PMID:16791330

Sánchez-López, E., Gomes, D., Esteruelas, G., Bonilla, L., Lopez-Machado, A. L., Galindo, R., Cano, A., Espina, M., Ettcheto, M., Camins, A., Silva, A. M., Durazzo, A., Santini, A., Garcia, M. L., & Souto, E. B. (2020). Metal-based nanoparticles as antimicrobial agents: An overview. *Nanomaterials (Basel, Switzerland), 10*(2), 292. doi:10.3390/nano10020292 PMID:32050443

Sardar, R., Funston, A. M., Mulvaney, P., & Murray, R. W. (2009). Gold nanoparticles: Past, present, and future. *Langmuir, 25*(24), 13840–13851. doi:10.1021/la9019475 PMID:19572538

Sengul, A. B., & Asmatulu, E. (2020). Toxicity of metal and metal oxide nanoparticles: A review. *Environmental Chemistry Letters, 18*(5), 1659–1683. doi:10.100710311-020-01033-6

Sharifi, S., Behzadi, S., Laurent, S., Laird Forrest, M., Stroeve, P., & Mahmoudi, M. (2012). Toxicity of nanomaterials. *Chemical Society Reviews, 41*(6), 2323–2343. doi:10.1039/C1CS15188F PMID:22170510

Simonet, B. M., & Valcárcel, M. (2009). Monitoring nanoparticles in the environment. *Analytical and Bioanalytical Chemistry, 393*(1), 17–21. doi:10.100700216-008-2484-z PMID:18974979

Svenson, S., & Tomalia, D. A. (2012). Dendrimers in biomedical applications—Reflections on the field. *Advanced Drug Delivery Reviews, 64*, 102–115. doi:10.1016/j.addr.2012.09.030 PMID:16305813

Tan, Y. L., & Ho, H. K. (2018). Navigating albumin-based nanoparticles through various drug delivery routes. *Drug Discovery Today*, *23*(5), 1108–1114. doi:10.1016/j.drudis.2018.01.051 PMID:29408437

Theron, J., Eugene Cloete, T., & de Kwaadsteniet, M. (2010). Current molecular and emerging nanobiotechnology approaches for the detection of microbial pathogens. *Critical Reviews in Microbiology*, *36*(4), 318–339. doi:10.3109/1040841X.2010.489892 PMID:20879848

Tuncel, D., & Demir, H. V. (2010). Conjugated polymer nanoparticles. *Nanoscale*, *2*(4), 484–494. doi:10.1039/b9nr00374f PMID:20644748

Ugazio, E., Cavalli, R., & Gasco, M. R. (2002). Incorporation of cyclosporin A in solid lipid nanoparticles (SLN). *International Journal of Pharmaceutics*, *241*(2), 341–344. doi:10.1016/S0378-5173(02)00268-5 PMID:12100861

Valsami-Jones, E., & Lynch, I. (2015). How safe are nanomaterials? *Science*, *350*(6259), 388–389. doi:10.1126cience.aad0768 PMID:26494749

Vollath, D. (2008). Nanomaterials an introduction to synthesis, properties and application. *Environmental Engineering and Management Journal*, *7*(6), 865–870.

Vrignaud, S., Benoit, J.-P., & Saulnier, P. (2011). Strategies for the nanoencapsulation of hydrophilic molecules in polymer-based nanoparticles. *Biomaterials*, *32*(33), 8593–8604. doi:10.1016/j.biomaterials.2011.07.057 PMID:21831421

Wang, E. C., & Wang, A. Z. (2014). Nanoparticles and their applications in cell and molecular biology. *Integrative Biology*, *6*(1), 9–26. doi:10.1039/c3ib40165k PMID:24104563

Wang, Y., Xu, C., & Ow, H. (2013). Commercial nanoparticles for stem cell labeling and tracking. *Theranostics*, *3*(8), 544–560. doi:10.7150/thno.5634 PMID:23946821

Woggon, U. (1997). Optical properties of semiconductor quantum dots. Springer.

Wu, Z., Xu, E., Chughtai, M. F. J., Jin, Z., & Irudayaraj, J. (2017). Highly sensitive fluorescence sensing of zearalenone using a novel aptasensor based on upconverting nanoparticles. *Food Chemistry*, *230*, 673–680. doi:10.1016/j.foodchem.2017.03.100 PMID:28407965

Wudl, F. (2002). Fullerene materials. *Journal of Materials Chemistry*, *12*(7), 1959–1963. doi:10.1039/b201196d

Xu, G. R., Wang, J. N., & Li, C. J. (2013). Strategies for improving the performance of the polyamide thin film composite (PA-TFC) reverse osmosis (RO) membranes: Surface modifications and nanoparticles incorporations. *Desalination*, *328*, 83–100. doi:10.1016/j.desal.2013.08.022

Yih, T., & Al-Fandi, M. (2006). Engineered nanoparticles as precise drug delivery systems. *Journal of Cellular Biochemistry*, *97*(6), 1184–1190. doi:10.1002/jcb.20796 PMID:16440317

Yoshimatsu, K., Lesel, B. K., Yonamine, Y., Beierle, J. M., Hoshino, Y., & Shea, K. J. (2012). Temperature-Responsive "Catch and Release" of Proteins by using Multifunctional Polymer-Based Nanoparticles. *Angewandte Chemie*, *124*(10), 2455–2458. doi:10.1002/ange.201107797 PMID:22287126

Youns, M., Hoheisel, J. D., & Efferth, T. (2011). Therapeutic and diagnostic applications of nanoparticles. *Current Drug Targets, 12*(3), 357–365. doi:10.2174/138945011794815257 PMID:20955146

Yu, S. J., Yin, Y. G., & Liu, J. F. (2013). Silver nanoparticles in the environment. *Environmental Science. Processes & Impacts, 15*(1), 78–92. doi:10.1039/C2EM30595J PMID:24592429

Yun, Y., Cho, Y. W., & Park, K. (2013). Nanoparticles for oral delivery: Targeted nanoparticles with peptidic ligands for oral protein delivery. *Advanced Drug Delivery Reviews, 65*(6), 822–832. doi:10.1016/j.addr.2012.10.007 PMID:23123292

Zhang, C., Yan, L., Gu, Z., & Zhao, Y. (2019). Strategies based on metal-based nanoparticles for hypoxic-tumor radiotherapy. *Chemical Science (Cambridge), 10*(29), 6932–6943. doi:10.1039/C9SC02107H PMID:31588260

Chapter 3
Sources and Occurrence of Nano Particles in Aquatic Ecosystems

Khursheed Ahmad Wani
https://orcid.org/0000-0001-5577-3455
Department of Environmental Science, Government Degree College Thindim Kreeri, Baramulla, India

Tehmina Yousuf
Srinagar Women's College Zakura, India

Khalida Parveen
Government Degree College, Kalakote, India

ABSTRACT

Nanotechnology has garnered a lot of publicity because of its promising attributes and vast variety of applications in areas including structural components, ballistic protection, electronics, culinary science, and medication delivery. The sources of nanoparticles are multifaceted. The introduction of nano-particles into different ecosystems is a major concern and its traces have been found in marine ecosystems as well. This chapter explores the sources and occurrences of nanoparticles in aquatic ecosystems and the need for future research in this direction.

INTRODUCTION

The usage of plastics and polymer composites is on the rise in the construction and other sectors. Plastics manufacturing is expected to reach 300 million metric tonnes per year over the world. The European Union's plastic demand in 2010 was projected at 46.4 million tonnes, with the majority of that demand coming from the packaging of food and consumer goods, as well as from the construction sector (Plastics Europe 2013). Around 24.7 million metric tonnes of plastic garbage enters the waste stream each year, with a recycling rate of 57.9 percent. This includes littering, dumping trash in landfills, and using the

DOI: 10.4018/978-1-6684-5533-3.ch003

sewage system to dispose of the garbage. Between 4.8 and 12.7 million tonnes of plastic debris were released into the ocean in 2010, according to the study by Jambeck et al. Although these numbers are disturbing in terms of volume, it is not yet understood how much impact these substances cause to human health. In the food packaging business alone, plastic is consumed at a rate of 14.5 million tonnes annually. It's good to know that better packaging for food may help prevent bacterial illnesses, including **Salmonella** and other food borne diseases, while also reducing food waste and making it easier to distribute it. It is thought that human populations are exposed to pollutants associated with plastics mostly via food packaging migration (Grob et al. 2006), even though testing has only been conducted on a small proportion of the chemicals present (Claudio 2012). It is not apparent how food-contact compounds may be protected after they have been thrown away in the trash.

Chronic exposure to plastics and their additives can't be quantified because of the lack of data on rates of degradation and fragmentation, chemical leaching into environmental matrices, and entrance into the food chain. As a possible solution to this issue, scientists can discover the substances that make up the human body. Blood, breast milk, saliva, or urine may be used for human biomonitoring since they include concentrations of environmental pollutants and their metabolites. Environmental exposures may be accessed via biomonitoring, which is regarded the gold standard since it provides an integrated evaluation of an individual's exposure to pollutants from different sources (Sexton et al. 2004). This method has shown that chemicals used in plastics manufacturing are definitely present in the human population's body fat and urine. Public health concerns have been raised by the ubiquitous availability of several substances in the general population at amounts that have been shown to cause damage in animal models (Talsness et al. 2009; Melzer and Galloway 2010). Involving 345 investigations, the National Health and Nutrition Examination Survey (NHANES) is the largest and most extensive human biomonitoring project ever done in the United States (https://www.cdc.gov/nchs/nhanes.htm). Bisphenol A, phthalates, styrene, acrylamide, triclosan, and brominated flame retardants are among the chemicals studied by NHANES, and their prevalence in the general population is an important consideration for this research.

To increase their physical qualities, plastics may include complicated combinations of additives that leak from the polymer into the surrounding environment. There is the prospect of continual chemical diffusion from the core of the plastic particle to the surface of the particle, which will lead to leaching. Despite the fact that most chemicals aren't persistent and have limited half lives in the body, leaching from plastic particles might be a long-term source of toxins in tissues and bodily fluids (Engler 2012). Phthalates, bisphenol A, brominated flame retardants, triclosan, bisphenone, and organotins are some of the plastics additives that are harmful to human health and should be avoided at all costs.

Legislation has been enacted to control the possible migration of polymer components and additives into food and beverages because of their potential impact on human health. Typically, migration levels are estimated from measurements using different solvents to simulate the receiving environment or can be estimated using partitioning models that consider aspects such as the desportion rates from the polymers, dimensions of the polymer framework, and dimensions of the diffusing molecules (Helmroth et al. 2002).

OCCURRENCE OF MICRO- AND NANOPLASTICS IN THE ENVIRONMENT

Concerns have been raised that microplastic debris (microplastic) and acrylic fibres used in clothing closely resembled those found in coastal sediments that receive sewage discharges, suggesting that sewage effluents represent a significant source of microfibres from the washing of clothes, and that these are not wholly retained during wastewater treatment. Lake Garda, an alpine lake in Italy, was found to have microplastic pollution levels comparable to those observed in the Great Lakes of the United States (Imhof et al. 2013). However, they are within the range of recorded values for the abundance of plastic particles in maritime coastal sediments (0.21–77,000 particles/m^2), although being at the lower end of exposures. These levels of contamination are most likely derived from landfill, litter, and wastewater sources (Hidalgo-Ruz et al. 2012). Microplastics have been found in both agricultural soils and waterways, which is not encouraging. Due to the lack of effective techniques for the identification of nanoplastics in samples, our understanding of their presence in aquatic habitats and biota is severely restricted (Koelmans et al. 2015).

SOURCES OF NANOPLASTIC

The release of NPs from goods and applications in which nanoplastics are utilized or manufactured and that result in emissions to the environment over the product life cycle may be primary sources of NPs. Product categories may include aqueous paints, adhesives, coatings, redispersible lattices, biomedical devices, medication delivery, medical diagnostics, electronics, magnetics, and optoelectronics, among other potential product types. It was discovered not too long ago that the thermal cutting of polystyrene foam releases nanometer-sized polymer particles, namely in the range of 22–220 nm (Zhang et al. 2012). During their lifetimes, many polymers are subjected to heat treatments that are analogous to those described above. It has been shown that three-dimensional printing emits nanometer-sized polymer particles at significant rates, namely in the region of 11–116 nm (Stephens et al. 2013). Since polystyrene and polyethylene nanoparticles are simple to produce (Lu et al. 2009; Rao and Geckeler 2011), and since they have a variety of uses, including research, it is inevitable that some of them will be released into the environment.

Despite being biodegradable solid lipids, polymeric nanoparticles, nanospheres, and nanocapsules have found usage in a variety of medicinal applications (Guterres et al. 2007). These structures are used for drug delivery. Although this kind of nanoplastic falls inside the parameters of the study, our research suggests that it is very unlikely that it would be harmful to the environment due to the fact that it would not remain very long there. Nanoplastics are often brought up in conversation while discussing cosmetic items. However, current product inventories suggest that the smallest sizes of NPs, which are 4 m, are prevalent in exfoliating scrubs and skin cleansers (Fendall and Sewell 2009). This makes it unlikely that these items are the major source of NPs. The fragmentation of microplastic into smaller and smaller particles, which may potentially reach the nanoscale, is a second source of potential contamination (Andrady 2011). Mats with nanoscalefibres are produced by electrospinning of tailored polymers, which, when utilised in goods, may deteriorate further to the nanoscale (Lu et al. 2008). Polymers are made up of a combination of polymer chains that range in length from very short to very long. There is empty space in between the chains, but the chains are chemically coupled to one another via the formation of weak secondary bonds (such as hydrogen or Van der Waals bonding) or by the physical contact of the chains being entangled with one another. The weak interactions are prone to breaking down at low energy levels because of this vulnerability. This breakdown results in embrittlement, which, when combined with other external factors like as friction, may result in the production of tiny particles on the surface of the polymers that vary in size from nanometers to millimetres. Studies conducted by Shim et al. (2014) were the first of their kind to report the actual fragmentation of expanded polystyrene (EPS) beads into micro- and nano-sized EPS particles. The experiments lasted for one month and used rapid mechanical abrasion with glass beads and sand.

Utilizing scanning electron microscopy (SEM) and energy-dispersive X-ray spectroscopy (EDS), researchers were able to demonstrate that nanometer sized EPS were formed. These experimental circumstances may already approximate situations at beaches or river banks where continuous abrasion of macro- and microplastics by sand particles probably leads to the creation of NPs. This is the case even before taking UV exposure into consideration. At beaches, the combination of photo-oxidation caused by exposure to ultraviolet light, high temperature, and high humidity presumably increases fragmentation rates and decreases the average size of the plastic particles. On the other hand, the existence of this process as well as its relative significance still has to be proven in the field. Despite this, we believe that physical abrasion is a significant source of NPs, given the knowledge that is currently available.

Degradation of microplastics to the size of less than 100 nanometers may be considered a third source of NPs, however this hypothesis has not been verified. Since it is well known that all types of polymers experience slow weathering by photodegradation (Sivan 2011), nanocomposites use manufactured nanoparticles (nanofillers) to increase their resistance to oxidation (e.g. Grigoriadou et al. 2011; Bussière et al. 2013). This is done so that the nanocomposites will not degrade as quickly when exposed to sunlight. The photo-oxidation of low-density polyethylene (LDPE) that is assisted by UV-B irradiation has been shown to result in the formation of extractable oxygenated compounds as well as non-oxidized low-molecular weight hydrocarbons. These compounds were utilised by bacteria, which resulted in an 8.4 percent decrease in the mass of LDPE after 14 days (Roy et al. 2008). After then, the LDPE films became unmanageable due to their extreme fragility. In a recent environmental investigation, it was shown that in the space of 30 days, microorganisms that were isolated from marine waters and found to be present in high concentrations were able to degrade between 1% and 1.75% of the mass of PE in the laboratory (Harshvardhan and Jha 2013). Koelmans (2015) made the suggestion that a surface degradation based particle shrinkage model may be used to evaluate the time dependency of the volume loss in plastic. We computed the time scales necessary to reach the 100 nm nanoscale as a function of the starting plastic particle size by using this model in conjunction with the laboratory volume loss rate data that was provided by Harshvardhan and Jha (2013). It would appear that if the oxidation or degradation of the plastic surface were the rate-limiting process, the relatively ideal conditions in the laboratory would still predict that approximately 320 years are required to bring 1 mm (1000 m) microplastics to the 100 nm nanoscale. This is despite the fact that the microplastics have been exposed to rather optimal conditions. It is reasonable to believe that the process of deterioration moves at a considerably more glacial pace in the waters owing to the restricted supply of light, oxygen, and microbes. Therefore, photodegradation or microbial decomposition near the water surface may contribute to the production of smaller particles; nevertheless, getting to the nanoscale may take a significant amount of time.

Nano-fragmentation may thus entail two different mechanisms: (1) direct nano-fragmentation may occur at the surface of macro- and microplastics (the predominant process), and additional gradual size-reductions may occur as a result of degradation (minor process). Due to the disparity in time scales between the two processes, it is likely that the embrittlement of microplastics, followed by their physical abrasion, is the most significant mechanism that contributes to the production of NPs.

CONCLUSIONS AND FUTURE WORK

The purpose of this brief discussion was to identify some of the most common types of plastics that are used in daily life and to demonstrate some of the efforts that have been made to understand the source and burden of these particles in the aquatic ecosytems. These different routes of exposure involve plastic additives, micro- and nanoplastics from food items, and from discarded debris. It is abundantly obvious that our comprehension of the possible contamination of the human population by micro- or nanoplastics originating from the environment is in its nascent stages, leaving a great number of problems unanswered. It needs to be well understood that the environment supports considerable bioaccumulation and trophic transmission of micro and nanoplastics and if this is the case, what kinds of animals are in the greatest danger. This needs to be debated that how does the ageing of plastics impact the physicochemical characteristics of those polymers, and how does it in turn influence their toxicity. The argument lies in the fact that micro and nanoplastics are absorbed after they have been swallowed and to what extent does the binding of proteins to the surface of the particles result in the formation of a protein corona and how much change will be reflected in the organism under exposure. Which techniques should we use in order to localise, identify, and measure microplastics and nanoplastics in complex matrices, such as biological tissues? There is still a significant research gap that has to be filled in the future, and that gap is in the area of developing effective techniques for extracting micro- and nanoplastics from biological materials and for investigating them in situ.

Using produced NPs, certain first impact experiments have revealed that ingestion of NPs may have detrimental consequences on both marine and freshwater species. Despite this, the thresholds of toxicity seem to be greater than the quantities that may be anticipated in the environment based on a worst-case scenario assumption of a conservative breakdown of the microplastics that are present in the amounts that are now known. On the other hand, we contend that the possible adverse effects of NPs should not be evaluated in a vacuum. It is possible that NPs are an ecological stressor that adds to the numerous other anthropogenic stressors already present in the environment, including trace metals, organic pollutants, and non-polymer nanoparticles. As a consequence of this, the topic of the contribution that NPs make to the pool of other nano-sized materials that already exists. It has been shown that natural nanoparticles are present everywhere in the environment, even in harmful forms (Wiesner et al. 2011). It has been hypothesised that the contribution of manufactured nanoparticles to the concentrations of naturally occurring nanoparticles, such as soots, clays, or other colloids, that are already present may be very small, if not insignificant altogether (Koelmans et al. 2009). The origins, creation rates, and exposure levels of NPs, as well as the destiny of the particles in aquatic systems, may be the primary focuses of study that will be conducted in the future. We are in desperate need of methods that can identify NPs in both drinking water and natural waterways. After the ambient concentrations have been determined, predictive screening-level impact studies could be carried out in an effort to evaluate the severity of the risk. Harmonization and standardisation in the categorization of NPs, as well as in the approaches that are employed, would be very beneficial to this study.

REFERENCES

Andrady, A. L. (2015). Persistence of plastic litter in the oceans. In M. Bergmann, L. Gutow, & M. Klages (Eds.), *Marine anthropogenic litter* (pp. 57–72). Springer. doi:10.1007/978-3-319-16510-3_3

Araujo, P., Sayer, C., Poco, J., & Giudici, R. (2002). Techniques for reducing residual monomer content in polymers: A review. *Polymer Engineering and Science, 42*(7), 1442–1468. doi:10.1002/pen.11043

Awaad, A., Nakamura, M., & Ishimura, K. (2012). *Imaging of size-dependent uptake and identification of novel pathways in mouse Peyer's patches using fluorescent organosilica particles.* Academic Press.

Awaad, A., Nakamura, M., & Ishimura, K. (2012, July). Imaging of size-dependent uptake and identification of novel pathways in mouse Peyer's patches using fluorescent organosilica particles. *Nanomedicine; Nanotechnology, Biology, and Medicine, 8*(5), 627–636. doi:10.1016/j.nano.2011.08.009

Bonefeld-Jørgensen, E. C., Long, M., Hofmeister, M. V., & Vinggaard, A. M. (2007). Endocrine disrupting potential of bisphenol A, bisphenol A dimethacrylate, 4-n-nonylphenol, and 4-n-octylphenol in vitro: New data and a brief review. *Environmental Health Perspectives, 115*(Suppl 1), 69–76. doi:10.1289/ehp.9368 PMID:18174953

Browne, M. A., Crump, P., Niven, S., Teuten, E., Tonkin, A., Galloway, T. S., & Thompson, R. (2011). Accumulation of microplastic on shorelines woldwide: Sources and sinks. *Environmental Science & Technology, 45*(21), 9175–9179. doi:10.1021/es201811s PMID:21894925

Buchta, C., Bittner, C., Heinzl, H., Höcker, P., Mocher, M., & Mayerhofer, M. (2005). Transfusion-related exposure to the plasticizer di(2-ethylhexyl)phthalate in patients receiving plateletpheresis concentrates. *Transfusion, 45*(5), 798–802. doi:10.1111/j.1537-2995.2005.04380.x PMID:15847671

Burello, E., & Worth, A. P. (2011). A theoretical framework for predicting the oxidative stress potential of oxide nanoparticles. *Nanotoxicology, 5*(2), 228–235. doi:10.3109/17435390.2010.502980 PMID:21609138

Calafat, A. M., Ye, X., Wong, L.-Y., Reidy, J. A., & Needham, L. L. (2008). Exposure of the US population to bisphenol A and 4-tertiary-octylphenol: 2003–2004. *Environmental Health Perspectives, 116*(1), 39–45. doi:10.1289/ehp.10753 PMID:18197297

Cipelli, R., Harries, L., Okuda, K., Yoshihara, S., Melzer, D., & Galloway, T. S. (2013). Bisphenol A modulates the expression of Estrogen-Related Receptor-α in T-Cells. *Reproduction (Cambridge, England), 147*(4), 419–426. doi:10.1530/REP-13-0423 PMID:24231368

Claudio, L. (2012). Food packaging and public health. *Environmental Health Perspectives, 120*(6), A233–A237. doi:10.1289/ehp.120-a232 PMID:22659036

Cole, M., Lindeque, P., Goodhead, R., Moger, J., Halsband-Lenk, C., & Galloway, T. S. (2013). Microplastic ingestion by zooplankton. *Environmental Science & Technology, 47*(12), 6646–6655. doi:10.1021/es400663f PMID:23692270

Cole, M., Lindeque, P., Halsband-Lenk, C., & Galloway, T. S. (2011). Microplastic as a contaminant in the marine environment: A review. *Marine Pollution Bulletin, 62*(12), 2588–2597. doi:10.1016/j.marpolbul.2011.09.025 PMID:22001295

Crompton, T. (2007). *Additive migration from plastics into foods. A guide for analytical chemistry.* iSmithersRapra Publishing.

de Abreu, D. A. P., Cruz, J. M., Angulo, L., & Losada, P. P. (2010). Mass transport studies of different additives in polyamide and exfoliated nanocomposite polyamide films for food industry. *Packaging Technology & Science, 23*(2), 59–68. doi:10.1002/pts.879

De Azeredo, H. M. C. (2013). Antimicrobial nanostructures in food packaging. *Trends in Food Science & Technology, 30*(1), 56–69. doi:10.1016/j.tifs.2012.11.006

Dodds, E. C., & Lawson, W. (1936). Synthetic estrogenic agents without the phenanthrene nucleus. *Nature, 137*(3476), 996. doi:10.1038/137996a0

Dubaish, F., & Liebezeit, G. (2013). Suspended microplastics and black carbon particles in the Jade system, southern North Sea. *Water, Air, and Soil Pollution, 224*(2), 1–8. doi:10.100711270-012-1352-9

Eastman. (2010). *Eastman material data sheet.* http://www.eastman.com, http://www. eastman.com/lierature.center/T/TRS252.pdf

Ebel, J. P. (1990). A method for quantifying particle absorption from the small intenstine. *Pharmaceutical Research, 7*(8), 848–851. doi:10.1023/A:1015964916486 PMID:2235881

EFSA. (2006). Opinion of the scientific panel on food additives, flavourings, processing aids and materials in contact with food on a request from the commission related to 2,2-bis(4-hydroxyphenyl)propane (bisphenol A). *EFSA Journal, 428*, 13–75.

EFSA. (2011). Scientific opinion: Guidance on the risk assessment of the application of nanoscience and nanotechnologies in the food and feed chain. *EFSA Journal, 9*, 2140.

EFSA. (2015). Scientific Opinion on the risks to public health related to the presence of bisphenol A (BPA) in foodstuffs. *EFSA Journal, 13*(1), 3978. doi:10.2903/j.efsa.2015.3978

Ehrlich, S., Calafat, A. M., Hunblet, O., Smith, T., & Hauser, R. (2014). Handling of thermal receipts as a source of exposure to bisphenol A. *Journal of the American Medical Association, 311*(8), 859–860. doi:10.1001/jama.2013.283735 PMID:24570250

Engler, R. E. (2012). The complex interaction between marine debris and toxic chemicals in the ocean. *Environmental Science & Technology, 46*(22), 12302–12315. doi:10.1021/es3027105 PMID:23088563

European Commission. (2014). *EC guidance document. Guidance to hazard analysis and critical control point (HACCP) application and to regulation 852/2004 and 853/2004.* https://europa.eu/comm/food/food/biosafety/hygienelegislation/index_en.htm

Fischer, D., Li, Y., Ahlemeyer, B., Krieglstein, J., & Kissel, T. (2003). In vitro cytotoxicity testing of polycations: Influence of polymer structure on cell viability and hemolysis. *Biomaterials, 24*(7), 1121–1131. doi:10.1016/S0142-9612(02)00445-3 PMID:12527253

Florence, A. T., & Hussain, N. (2001). Transcytosis of nanoparticle and dendrimer delivery systems, evolving vistas. *Advanced Drug Delivery Reviews, 50*, S69–S89. doi:10.1016/S0169-409X(01)00184-3 PMID:11576696

Florence, A. T., Sakthivel, T., & Toth, I. (2000). Oral uptake and translocation of a polylysinedendrimer with a lipid surface. *Journal of Controlled Release*, *65*(1-2), 253–259. doi:10.1016/S0168-3659(99)00237-0 PMID:10699285

Fruijtier-Pölloth, C. (2012). The toxicological mode of action and the safety of synthetic amorphoussilica-a nanostructured material. *Toxicology*, *294*(2-3), 61–79. doi:10.1016/j.tox.2012.02.001 PMID:22349641

Galloway, T., Cipelli, R., Guralnik, J., Ferrucci, L., Bandinelli, S., Corsi, A. M., Money, C., McCormack, P., & Melzer, D. (2010, November). Daily bisphenolA excretion and associations with sex hormone concentrations: Results from the InCHIANTI adult population study. *Environmental Health Perspectives*, *118*(11), 1603–1608. doi:10.1289/ehp.1002367

Garrett, N. L., Lalatsa, A., Uchegbu, I., Schätzlein, A., & Moger, J. (2012). Exploring uptake mechanisms of oral nanomedicines using multimodal nonlinear optical microscopy. *Journal of Biophotonics*, *5*(5-6), 458–468. doi:10.1002/jbio.201200006 PMID:22389316

Gayrard, V., Lacroix, M. Z., Collet, S. H., Viguié, C., Bousquet-Melou, A., Toutain, P. L., & Picard-Hagen, N. (2013). High bioavailability of bisphenol A from sublingual exposure. *Environmental Health Perspectives*, *121*(8), 951–956. doi:10.1289/ehp.1206339 PMID:23761051

Geys, J., Nemmar, A., Verbeken, E., Smolders, E., Ratoi, M., Hoylaerts, M. F., Nemery, B., & Hoet, P. H. M. (2008). Acute toxicity and prothrombotic effects of quantum dots: Impact of surface charge. *Environmental Health Perspectives*, *116*(12), 1607–1613. doi:10.1289/ehp.11566 PMID:19079709

Grob, K., Biedermann, M., Scherbaum, E., Roth, M., & Rieger, K. (2006). Food contamination with organic materials in perspective: Packaging materials as the largest and least controlled source? A view focusing on the European situation. *Critical Reviews in Food Science and Nutrition*, *46*(7), 529–536. doi:10.1080/10408390500295490 PMID:16954061

Guart, A., Wagner, M., Mezquida, A., Lacorte, S., Oehlmann, J., & Borrell, A. (2013). Migration of plasticisers from TritanTM and polycarbonate bottles and toxicological evaluation. *Food Chemistry*, *141*(1), 373–380. doi:10.1016/j.foodchem.2013.02.129 PMID:23768370

Hanning, I. B., Nutt, J. D., & Ricke, S. C. (2009). Salmonellosis outbreaks in the United States due to fresh produce: Sources and potential intervention measures. *Foodborne Pathogens and Disease*, *6*(6), 635–648. doi:10.1089/fpd.2008.0232 PMID:19580447

Harush-Frenkel, O., Bivas-Benita, M., Nassar, T., Springer, C., Sherman, Y., Avital, A., Altschuler, Y., Borlak, J., & Benita, S. (2010). A safety and tolerability study of differently-charged nanoparticles for local pulmonary drug delivery. *Toxicology and Applied Pharmacology*, *246*(1-2), 83–90. doi:10.1016/j.taap.2010.04.011 PMID:20417650

Helmroth, E., Rijk, R., Dekker, M., & Jongen, W. (2002). Predictive modelling of migration from packaging materials into food products for regulatory purposes. *Trends in Food Science & Technology*, *13*(3), 102–109. doi:10.1016/S0924-2244(02)00031-6

Hengstler, J. G., Foth, H., Gebel, T., Kramer, P. J., Lilienblum, W., & Schweinfurth, H. (2011). Critical evaluation of key evidence on the human health hazards of exposure to bisphenol A. *Critical Reviews in Toxicology*, *41*, 263–291.

Hidalgo-Ruz, V., Gutow, L., Thompson, R. C., & Thiel, M. (2012). Microplastics in the marine environment: A review of the methods used for identification and quantification. *Environmental Science & Technology*, *46*(6), 3060–3075. doi:10.1021/es2031505 PMID:22321064

Imhof, H. K., Ivleva, N. P., Schmid, J., Niessner, R., & Laforsch, C. (2013). Contamination of beach sediments of a subalpine lake with microplastic particles. *Current Biology*, *23*(19), R867–R868. doi:10.1016/j.cub.2013.09.001 PMID:24112978

Israelachvili, J. (1992). *Intermolecular and surface forces* (2nd ed.). Academic Press.

Jambeck, J. R., Geyer, R., Wilcox, C., Siegler, T. R., Perryman, M., Andrady, A., Narayan, R., & Law, K. L. (2015). Plastic waste inputs from land into the ocean. *Science*, *347*(6223), 768–771. doi:10.1126cience.1260352 PMID:25678662

Jani, P. U., Florence, A. T., & McCarthy, D E. (1992). Further histological evidence of gastrointestinal absorption of polystyrene nanospheres in the rat. *International Journal of Pharmaceutics, 84*, 245–252.

Jani, P. U., Nomura, T., Yamashita, F., Takakura, A., Florence, A. T., & Hashida, M. (1996). Biliary excretion of polystyrene microspheres with covalently linked FITC fluorescence afteroral and parenteral administration to male wistar rats. *Journal of Drug Targeting*, *4*(2), 87–93. doi:10.3109/10611869609046266 PMID:8894968

Kim, B., Rutka, J., & Chan, W. (2010). Nanomedicine. *The New England Journal of Medicine*, *363*(25), 2434–2443. doi:10.1056/NEJMra0912273 PMID:21158659

Koch, H. M., & Calafat, A. M. (2009). Human body burdens of chemicals used in plastics manufacture. *Philosophical Transactions of the Royal Society of London. Series B, Biological Sciences*, *364*(1526), 2063–2078. doi:10.1098/rstb.2008.0208 PMID:19528056

Koelmans, A. A. (2015). Modeling the role of microplastics in bioaccumulation of organic chemicals to marine aquatic organisms. Critical review. In M. Bergmann, L. Gutow, & M. Klages (Eds.), *Marine anthropogenic litter* (pp. 313–328). Springer. doi:10.1007/978-3-319-16510-3_11

Koelmans, A. A., Besseling, E., & Foekema, E. M. (2014). Leaching of plastic additives tomarine organisms. *Environmental Pollution*, *187*, 49–54. doi:10.1016/j.envpol.2013.12.013 PMID:24440692

Koelmans, A. A., Besseling, E., & Shim, W. J. (2015). Nanoplastics in the aquatic environment. In M. Bergmann, L. Gutow, & M. Klages (Eds.), *Marine anthropogenic litter* (pp. 329–344). Springer. doi:10.1007/978-3-319-16510-3_12

Kreuter, J. (1991). Peroral administration of nanoparticles. *Advanced Drug Delivery Reviews*, *7*(1), 71–86. doi:10.1016/0169-409X(91)90048-H

La Mantia, F. (2002). Handbook of plastics recycling. iSmithersRapra Publishing.

Lagaron, J. M., & Lopez-Rubio, A. (2011). Nanotechnology for bioplastics: Opportunities, challenges and strategies. *Trends in Food Science & Technology*, *22*(11), 611–617. doi:10.1016/j.tifs.2011.01.007

Lang, I. A., Galloway, T. S., Scarlett, A., Henley, W. E., Depledge, M., Wallace, R. B., & .... (2008). Association of urinary bisphenolA concentration with medical disorders and laboratory abnormalities in adults. *Journal of the American Medical Association*, *300*(11), 1353–1355. doi:10.1001/jama.300.11.1303 PMID:18799442

Lee, H. J., Chattopadhyay, S., Gong, E. Y., Ahn, R. S., & Lee, K. (2003). Antiandrogeniceffects of bisphenol A and nonylphenol on the function of androgen receptor. *Toxicological Sciences*, *75*(1), 40–47. doi:10.1093/toxsci/kfg150 PMID:12805653

Lithner, D., Larsson, Å., & Dave, G. (2011). Environmental and health hazard ranking andassessment of plastic polymers based on chemical composition. *The Science of the Total Environment*, *409*(18), 3309–3324. doi:10.1016/j.scitotenv.2011.04.038 PMID:21663944

Lundqvist, M., Stigler, J., Elia, G., Lynch, I., Cedervall, T., & Dawson, K. A. (2008). Nanoparticle size and surface properties determine the protein corona with possible implications for biological impacts. *Proceedings of the National Academy of Sciences of the United States of America*, *105*(38), 14265–14270. doi:10.1073/pnas.0805135105 PMID:18809927

Lusher, A. (2015). Microplastics in the marine environment: distribution, interactions and effects. In M. Bergmann, L. Gutow, & M. Klages (Eds.), *Marine anthropogenic litter* (pp. 245–308). Springer. doi:10.1007/978-3-319-16510-3_10

Ma-Hock, L., Landsiedel, R., Wiench, K., Geiger, D., Strauss, V., Gröters, S., Ravenzwaay, B., Gerst, M., Wohlleben, W., & Scherer, G. (2012). Shortterm rat inhalation study with aerosols of acrylic ester-based polymer dispersions containinga fraction of nanoparticles. *International Journal of Toxicology*, *31*(1), 46–57. doi:10.1177/1091581811424778 PMID:22267870

Mauricio-Inglesias, M., Peyron, S., Guillard, V., & Gontard, N. (2010). Wheat gluten nanocomposite films as food contact materials: Migration tests and impact of a novel food stabilizing technology (high pressure). *Journal of Applied Polymer Science*, *116*, 2526–2535.

Melzer, D., & Galloway, T. S. (2010, Oct.). Burden of proof. New Scientist, 26–27.

Melzer, D., Osborne, N. J., Henley, W. E., Cipelli, R., Young, A., Money, C., McCormack, P., Luben, R., Khaw, K.-T., Wareham, N. J., & Galloway, T. S. (2012). Urinary bisphenol A: Concentration and risk of future coronary artery disease in apparently healthy men and women. *Circulation*, *125*(12), 1482–1490. doi:10.1161/CIRCULATIONAHA.111.069153 PMID:22354940

Melzer, D., Rice, N. E., Lewis, C., Henley, W. E., & Galloway, T. S. (2010). Association of urinary bisphenol. a concentration with heart disease: Evidence from NHANES 2003/06. *PLoS One*, *5*(1), e8673. doi:10.1371/journal.pone.0008673 PMID:20084273

Mielke, H., & Gundert-Remy, U. (2009). Bisphenol A levels in blood depend on age and exposure. *Toxicology Letters*, *190*(1), 32–40. doi:10.1016/j.toxlet.2009.06.861 PMID:19560527

Moriyama, K., Tagami, T., Akamizu, T., Usui, T., Saijo, M., Kanamoto, N., Hataya, Y., Shimatsu, A., Kuzuya, H., & Nakao, K. (2002). Thyroidhormone action is disrupted by bisphenol A as an antagonist. *The Journal of Clinical Endocrinology and Metabolism*, *87*(11), 5185–5190. doi:10.1210/jc.2002-020209 PMID:12414890

Muncke, J. (2011). Endocrine disrupting chemicals and other substances of concern in food contact materials: An updated review of exposure, effect and risk assessment. *The Journal of Steroid Biochemistry and Molecular Biology, 127*, 118–127.

Nel, A. E., Madler, L., Velegol, D., Xia, T., Hoek, E. M. V., Somasundaran, P., Klaessig, F., Castranova, V., & Thompson, M. (2009). Understanding bio physicochemical interactions at the nano- bio- interface. *Nature Materials, 8*(7), 543–557. doi:10.1038/nmat2442 PMID:19525947

Nel, A. E., Xia, T., Mädler, L., & Li, N. (2006). Toxic potential of materials at the nanolevel. *Science, 311*(5761), 622–627. doi:10.1126cience.1114397 PMID:16456071

Newbold, R. R., Padilla-Banks, E., Jefferson, W. N., & Heindel, J. J. (2008). Effects of endocrinedisruptors on obesity. *International Journal of Andrology, 31*(2), 201–208. doi:10.1111/j.1365-2605.2007.00858.x PMID:18315718

Nicholson, J. W. (1996). The chemistry of polymers. The Royal Society of Chemistry.

O'Hagan, D. T. (1996). The intestinal uptake of particles and the implications for drug and antigen delivery. *Journal of Anatomy, 189*, 477–482. PMID:8982819

OCED. (2004). *Emission scenario document on plastic additives. Series on emission scenariodocuments No. 3*. Environmental Directorate. OECD Environmental Health and SafetyPublications.

Okada, H., Tokunaga, T., Liu, X., Takayanagi, S., Matsushima, A., & Shimohigashi, Y. (2008). Direct evidence revealing structural elements essential for the high binding ability of bisphenol A to human estrogen-related receptor. *Environmental Health Perspectives, 116*(1), 32–38. doi:10.1289/ehp.10587 PMID:18197296

Pandurangi, R. S., Seehra, M. S., Razzaboni, B. L., & Bolsaitis, P. (1990). Surface and bulk infrared modes of crystalline and amorphorous silica particles: A study of the relation of surfacestructure to cytotoxicity of respirable silica. *Environmental Health Perspectives, 86*, 327–336. doi:10.1289/ehp.9086327 PMID:2169410

Pappo, J., Ermak, T. H., & Steger, H. J. (1989). Monoclonal antibody directed targeting of fluorescent microspheres to Peyer's patches M cells. *Immunology, 73*, 277–280. PMID:1879877

PlasticsEurope. (2013). *Plastics, the facts 2013*. www.plasticseurope.org

Poland, C. A., Duffin, R., Kinloch, I., Maynard, A., Wallace, W. A. H., Seaton, A., Stone, V., Brown, S., MacNee, W., & Donaldson, K. (2008). Carbon nanotubes introduced into the abdominal cavity of mice show asbestos-like pathogenicity in a pilot study. *Nature Nanotechnology, 23*(7), 423–428. doi:10.1038/nnano.2008.111 PMID:18654567

Ray, S. S., & Okamoto, M. (2003). Polymeric/layered silicate nanocomposites: A review frompreparation to processing. *Progress in Polymer Science, 28*(11), 1539–1641. doi:10.1016/j.progpolymsci.2003.08.002

Rochester, J. R. (2013). Bisphenol A and human health: A review of the literature. *ReproductiveToxicology, 42C*, 132–155. doi:10.1016/j.reprotox.2013.08.008 PMID:23994667

Ropero, A. B., Alonso Magdalena, P., García-García, E., Ripoll, C., Fuentes, E., & Nadal, A. (2008). Bisphenol A disruption of the endocrine pancreas and blood glucose homeostasis. *International Journal of Andrology, 31*(2), 194–200. doi:10.1111/j.1365-2605.2007.00832.x PMID:17971160

Schmidt, B., Petersen, J. H., Bender Koch, C., Plackett, D., Johansen, N. R., Katiyar, V., & Larsen, E. H. (2009). Combining asymmetrical flow field-flow fractionation with light-scattering andinductively coupled plasma mass spectrometric detection for characterization of nanoclayused in biopolymer nano-composites. *Food Additives and Contaminants A, 26*(12), 1619–1627. doi:10.1080/02652030903225740 PMID:19927249

Setälä, O., Fleming-Lehtinen, V., & Lehtiniemi, M. (2012). Ingestion and transfer of microplastics in the planktonic food web. *Environmental Pollution, 185*, 77–83. doi:10.1016/j.envpol.2013.10.013 PMID:24220023

Sexton, K., Needham, L., & Pickles, J. (2004). Human biomonitoring of environmental chemicals. *American Scientist, 92*(1), 38–45. doi:10.1511/2004.45.921

Silvestre, C., Duraccio, D., & Cimmino, S. (2011). Food packaging based on polymer nanomaterials. *Progress in Polymer Science, 36*(12), 1766–1782. doi:10.1016/j.progpolymsci.2011.02.003

Šimon, P., Chaudrey, Q., & Bakoš, D. (2008). Migration of engineered nanoparticles from polymer packaging to food—A physicochemical view. *Journal of Food and Nutrition Research, 47*, 105–113.

Talsness, C. E., Andrade, A. J. M., Kuriyama, S. N., Taylor, J. A., & vomSaal, F. S. (2009). Components of plastic: Experimental studies in animals and relevance for human health. *Philosophical Transactions of the Royal Society of London. Series B, Biological Sciences, 364*(1526), 2079–2096. doi:10.1098/rstb.2008.0281 PMID:19528057

Tenzer, S., Docter, D., Kuharev, J., Musyanovych, A., Fetz, V., Hecht, R., Schlenk, F., Fischer, D., Kiouptsi, K., Reinhardt, C., Landfester, K., Schild, H., Maskos, M., Knauer, S. K., & Stauber, R. H. (2013). Rapidformation of plasma protein corona critically affects nanoparticle pathophysiology. *Nature Nanotechnology, 8*(10), 772–781. doi:10.1038/nnano.2013.181 PMID:24056901

Thompson, R. C., Olsen, Y., Mitchell, R. P., Davis, A., Rowland, S. J., John, A. W. G., McGonigle, D., & Russell, A. E. (2004). Lost at sea, where is all the plastic? *Science, 304*(5672), 838. doi:10.1126cience.1094559 PMID:15131299

Vandenberg, L. N., Chauhoud, I., Heindel, J. J., Padmanabhan, V., Paumgartten, F. J., & Schoenfelder, G. (2010). Urinary, circulating and tissue biomonitoring studies indicate widespread exposure to bisphenol A. *Environmental Health Perspectives, 118*(8), 1055–1070. doi:10.1289/ehp.0901716 PMID:20338858

Varela, J. A., Bexiga, M. G., Åberg, C., Simpson, J. C., & Dawson, K. A. (2012). Quantifyingsize-dependent interactions between fluorescently labeled polystyrene nanoparticles andmammalian cells. *Journal of Nanobiotechnology, 10*(1), 39. doi:10.1186/1477-3155-10-39 PMID:23006133

Volkheimer, G. (1977). Passage of particles through the wall of the gastrointenstinal tract. *Environmental Health Perspectives, 9*, 215–225. PMID:4470938 vomSaal, F. S., Akingbemi, B. T., Belcher, S. M., Birnbaum, L. S., Crain, D. A., Eriksen, M., Farabollini, F., Guillette, L. J. Jr, Hauser, R., Heindel, J. J., Ho, S.-M., Hunt, P. A., Iguchi, T., Jobling, S., Kanno, J., Keri, R. A., Knudsen, K. E., Laufer, H., LeBlanc, G. A., ... Zoeller, R. T. (2007). Chapel Hill bisphenolA expert panel consensus statement: Integration of mechanisms, effects in animals and potential to impact human health at current levels of exposure. *Reproductive Toxicology (Elmsford, N.Y.)*, *24*(2), 131–138. doi:10.1016/j.reprotox.2007.07.005 PMID:17768031

Wang, J., Zhou, G., Chen, C., Yu, H., Wang, T., Ma, Y., Jia, G., Gao, Y., Li, B., & Sun, J. (2007). Acute toxicity and biodistribution of different sized titanium dioxide particles in mice after oral administration. *Toxicology Letters*, *168*(2), 176–185. doi:10.1016/j.toxlet.2006.12.001 PMID:17197136

Wright, S. L., Thompson, R. C., & Galloway, T. S. (2013). The physical impacts of microplasticson marine organisms: A review. *Environmental Pollution*, *178*, 483–492. doi:10.1016/j.envpol.2013.02.031 PMID:23545014

Zbyszewski, M., & Corcoran, P. L. (2011). Distribution and degradation of fresh water plasticparticles along the beaches of Lake Huron, Canada. *Water, Air, and Soil Pollution*, *220*(1-4), 1–8. doi:10.100711270-011-0760-6

Zubris, K. A. V., & Richards, B. K. (2005). Synthetic fibers as indicator of land application ofsludge. *Environmental Pollution*, *138*(2), 201–211. doi:10.1016/j.envpol.2005.04.013 PMID:15967553

Zweifel, H. (2001). *Plastics additives handbook* (5th ed.). Carl HanserVerlag.

Chapter 4
Role of Nanocomposite Materials for Water Pollution Alleviation Technologies

Nisha Sharma
https://orcid.org/0000-0001-6452-2965
Department of Physical Sciences, Sant Baba Bhag Singh University, Jalandhar, India

Mithu Maiti Jana
https://orcid.org/0000-0002-3178-4472
Department of Physical Sciences, Sant Baba Bhag Singh University, Jalandhar, India

Shweta Singh
Department of Life Sciences, Allied Health Sciences and Agriculture Sciences, Sant Baba Bhag Singh University, Jalandhar, India

ABSTRACT

Environmental contamination is the most serious global threat to our ecosystem. Over the last few decades, there has been a rise in environmental sensitivity. Indiscriminate use of water resources beyond their regeneration limits has resulted in the current water crisis that the world is experiencing. Hazardous effluents from textile industries and resistant microbes are significant sources for water deterioration with detrimental impacts on both biotic and abiotic components. Effective solutions to water purification are the need of the hour. Hydrogels possess unique properties with potential as effective adsorbents for dyes and hazardous metal ions enrichment from wastewater. Nanocomposites are a new class of advanced materials for technology. Nanocomposite hydrogels with reinforcing nanofillers exhibit noticeably superior properties as compared to conventional hydrogels and have the potential to become efficient adsorbents for such pollutants. Present chapter provides a thorough overview of the role of nanocomposites in water pollution alleviation techniques.

DOI: 10.4018/978-1-6684-5533-3.ch004

INTRODUCTION

The biggest global threat to our environment is pollution. The need for environmental protection has increased during the last few decades. Technological advances in the twenty-first century and the industrial revolution have led to a sharp rise in water contamination. The globe is currently experiencing a water crisis as a result of the indiscriminate use of water resources beyond their capacity for regeneration and the massive amounts of waste water that are produced every day. Water pollution has an impact on both the biotic and abiotic elements of the environment. The power of chemical companies to harm flora and fauna sends a worrisome message to society. Aquatic biosystems have substantially degraded in recent years due to the rise of pathogenic microorganisms, synthetic dyes, and the buildup of dangerous metal ions (Abu Elella et al., 2021b). The majority of these effluents contain hazardous compounds that are non-biodegradable and harm exposed organisms and marine life. Untreated water contaminants are frequently released into water reservoirs and are one of the most prominent sources are hazardous effluents from the textile sector. Contamination of water is a serious environmental issue that limits access to safe drinking water.

Heavy Metals and Their Pollution

Heavy metals are poisonous to humans and non-biodegradable contaminants. Transition metals, certain metalloids, lanthanides, and actinides all contain elements with an ill-defined group. Metals such as Pb, Cd, Cu, Al, Co, Hg, and others have been considered contaminated when their concentration in a medium exceeds a specific threshold and has a negative influence on the environment. Different metal processing sectors, alloy-based businesses, and mining operations are the main sources of contamination of this metal in the environment. Heavy metals in drinking water have permitted limits, such as 0.0I for As, 0.400 for Mn, 0.015 for U, 0.003 for Cd, and 0.006 for Hg (WHO,1990). Irrigation is one of the primary sources of heavy metal contamination in water bodies. These bodies of water employ various aspects of the environment and, as a result, have an impact on the health of living things via the food chain (Meade,1995)

Technology Used for Removal of Heavy Metals

Toxic metal ions and organic contaminants are the main causes of environmental pollution. Various research groups are looking for powerful tactics and practical ways to eliminate these pollutants. Biological degradation, ion exchange, coagulation-flocculation, chemical precipitation, adsorption, enhanced oxidation processes, photo catalysis, membrane filtration, solvent extraction, and adsorption methods have all been utilized to remove these contaminants from water bodies and make the water fit for consumption (Zhong et al.,1996). Industrial wastewaters can be cleaned up by using re-precipitation techniques to get rid of heavy metal ions like hydroxides, carbonates, sulphides, and sulphates (Feng et al.,2000). Water contamination is typically treated and reduced using membrane filtration and ion exchange resins (Oliva et al.,2011., Korus & Loska.,2009). The above solutions, however, have gotten less attention because of their high operational expenses. Another strategy for separating heavy metals is solvent extraction. A wide range of extractants and chemical ligands can be used with this simple operation method. The process of adsorption is another way to eliminate metals and their ions from water (Al-Asheh et al.,2008). The adsorption method is the most dependable and effective technology because of its simplicity, affordability, and lack of harmful sludge production. Utilizing a range of adsorbents, substantial research has been done on the use of adsorption as a water treatment technology to remove colours and other impurities. Adsorption is viewed as a potential strategy for eliminating dye wastes due to its good removal capabilities at low concentrations (Pai et al., 2021).

Polymeric Materials for Separation of Heavy Metal Ions

Solvent extraction and ion exchange are the good technology for the removal of impurities from waste water but hydrogels include various functional groups are effective candidates for removing heavy metals from various contaminated water. Hydrogels have a number of advantages over other techniques, including high adsorption capacity, end-product degradability, and low consumption of organic solvents (Dinu & Dragan., 2010). Table 1describes the applications of different hydrogels in waste water treatment (Dafia et al., 2019; Gupta & Ako,2005; Rahimdokht et al., 2019).

Table 1. Wastewater Treatment using Different Polysaccharide Based Hydrogels

S.No	Natural polysaccharide based hydrogels
1.	In industrial wastewater treatment Polyacrylamide applied as coagulants
2.	Guar gum is a natural polymeric matrix which is applied as to lower the amount of turbidity of water.
3.	Gum Arabic has been used for various applications for water treatment and also applied to reduce the alum concentration needed for water treatment.
4.	Gum ghatti, a popular hydrogel matrix has been used for waste water treatment and as adsorbents for removing different contaminants from wastewater, like synthetic dyes and toxic heavy metal cations.
5.	Xanthan gum is applied as a novel flocculants.
6.	Alginate based matrix is employed for removal of native microorganism from wastewater which also has a characteristic of forming a physical barrier against microorganisms,
7.	Heavy metals and dyes is mainly removed from water by chitosan

Hydrogels as Toxic Metal Ion/Dye Enrichment Devices

Hydrogels are advanced materials possess unique characteristics such as biocompatibility, various synthesis processes, variety of components, and desired physical qualities, make them efficient materials of choice for variety of industries. Hydrogels have qualities such as swelling,semi-wet, three-dimensional environments, elasticity, soft consistency, biological tissue-like elasticity, mechanical, and rheological properties. Hydrogels can alter their characteristics in response to a variety of environmental factors, such as temperature, ionic strength, solvent polarity, electric/magnetic field, light, or small (bio)molecules. These attributes make hydrogels effective adsorbents for removing organic dyes from waste fluids. Since hydrogels have functional groups that are chemically active, they can take in metal ions and colours from wastewater and release them when the aqueous solution conditions change. The ability to generate a flexible network of polymer chains that enables solutes to swiftly infiltrate the network with water as well as make stable three dimensional networks with functional groups is a property of hydrophilic materials, such as hydrogels. Natural polymer composites have been studied as effective adsorbents to adsorb colours and heavy metal ions as a substitute for wastewater treatment because they contain amin, carboxyl, and hydroxyl groups, which can operate as active sites. Natural polymers must be altered to have more surface area and include functional groups in order to improve their ability to sorb substances. In this regard, a number of modifications have been described to improve the ability of metal ions and dyes to adsorb substances (Kayan et al.2021; Hussain et al. 2020). Apart from diverse characteristics and usability of hydrogels, there have some disadvantages, including as weak mechanical and tensile capabilities, an uncontrolled swelling profile, and low stability, which limit their use in advanced applications where such behaviour is desirable and reusability is required. Mechanical and chemical properties of hydrogels are increasingly being tuned for specific applications. The incorporation of various strength-inducing components in hydrophilic networks has recently been used to develop a novel type of hybrid matrix with enhanced characteristics.

NANOCOMPOSITE HYDROGELS: TYPE, PROPERTIES

A nanocomposite hydrogel is made by physically/chemically joining polymeric chains with different nanoscale structures, resulting in a network with unique properties. New properties and behaviours develop after filler dispersion at the nanoscale in the composite. A variety of fundamental hydrogel actuation features, including quick response, high efficiency, and directional motion, are enhanced by the incorporation of zero-dimensional (0D), one-dimensional (1D), or two-dimensional (2D) shaped additives. The performance of hydrogel actuators could be considerably improved by using functional nanoparticles as additives (Rana &Sharma 2020).

Nanocomposite hydrogels are water-swelling polymers made up of a polymer matrix with organic and inorganic nanoparticles. Natural polymers, synthetic polymers, and mixtures of natural and synthetic polymers are utilised to create nanocomposite hydrogel matrix with a variety of properties. In nanocomposite hydrogels, one of the components must be in the nanoscale range (1-100 nm). Biopolymeric nanocomposite hydrogels are a type of homogeneous natural polymeric network created by reinforcing the polymer matrix with nanoparticles, resulting in a hydrogel with more tensile strength and toughness than conventional hydrogels. Synthetic polymers can be replaced with polysaccharide-based nanocomposite hydrogels, which are more ecologically friendly and cost less. The following qualities of natural polymers have an edge over synthetic ones: biodegradability, biocompatibility, permeability, hydrophilicity, non-toxicity, non-immunogenicity, and usability. They are distinguished from artificial polymer matrix-based nanocomposite hydrogels by their biodegradability, biocompatibility, permeability, hydrophilicity, lack of toxicity, lack of immunogenicity, and simplicity of surface modification. Functionalized polysaccharides-based nanocomposites have recently come to be recognised as an innovative method for the remediation of dyes and other contaminants in aquatic environments in the current scenario, with the advent of nanotechnology in every industry, with the global goal of making potable water accessible globally (Sadiq et al. 2021).

Various approaches are utilised from time to time to improve the physicochemical and structural properties of hydrophilic networks in order to create novel materials. In the literature, many types of reinforcing materials have been used as nonofillers in synthetic or biopolymeric matrixes. Metals, such as metallic nanoparticles, metal oxides, inorganic inorganic/ceramic nanoparticles, organic particles, polymerics nanostructures, and carbon-based nanomaterials CNTs and inorganic/clay based nanomaterial are all known to improve polymer performance. Because of their small size, reinforcing materials have special features that change the behaviour of polymeric matrices. Each type of reinforcing material has characteristics that make it ideal for a certain application. This is mostly due to better interactions between nanoparticles and polymeric chains, which leads to an increase in desired characteristics for target applications (Rana &Sharma 2020). The size/dimensions of reinforcing material employed in hydrophilic networks have widened the applicability bounds. These reinforcing materials modulate the inherent behaviour of gel matrix as well as incorporate unique stimuli responsiveness such as stimulus sensitivity to radiation, temperature, and electric or magnetic fields, as well as multi-responsiveness, improved mechanical strength/tensile strength etc in hybrid matrix, all thanks to reinforcing chemicals (Figure 1). Due to their special characteristics, metal-organic frameworks—which are comprised of metal clusters and organic connectors—are one of the most promising antecedents for the creation of improved adsorbents (Mahmoodi & Javanbakht 2021).

Figure 1. Different types of nanocomposite hydrogels

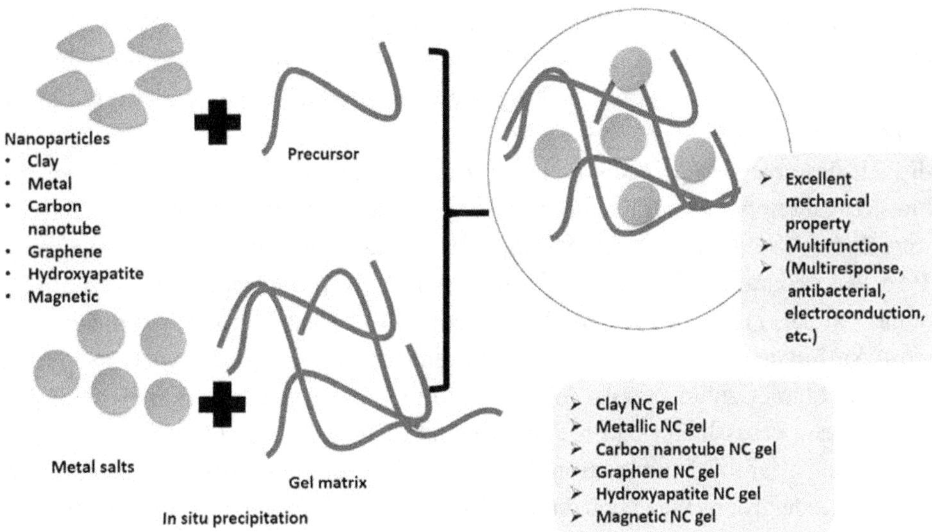

Role of Nanocomposite Hydrogels for Mitigation of Water Pollution

Polysaccharide-based nanocomposites are eco-friendly and have a wide range of uses in waste water treatment. Such systems serve as dual-purpose devices for water pollution mitigation solutions, since they are derived from natural sources on one hand and employed to remove environmental toxins on the other. The removal of heavy metal ions from contaminated areas utilising hydrogels manufactured from both biomaterials and synthetic grade materials is currently the focus of several research organisations. New adsorbent materials are being developed and tested in order to recover these metals from industrial effluents.

Role of Nanocomposite Hydrogels in Metal Ion Removal & Enrichment

Many scientists have demonstrated that biopolymeric nanocomposite gels can remove heavy metals and organic pigments from water bodies, and that hydrogels have the highest adsorption capability. Denizli et al., 2000 have created dithiocarbamate-modified polystyrene microspheres. The greatest Hg(II) adsorption (Yan et al., 2011) generated surface customized chitosan hydrogel beads by $ClCH_2COOH$ may be revealed in these microspheres. In aqueous solutions, cupper ions is more selectively adsorbed than lead and magnesium ions by the customised beads. At pH 5.0, these beads demonstrate maximum cupper ions uptake. Say et al., 2022 used radical suspension polymerization of methacrylamidohistidine and 2-HEMA in an aqueous dispersion media to make poly(hydroxyethyl methacrylate-co-methacrylamidohistidine) beads. The largest amount of Cu adsorption attained by the beads was 122.7 mg.g-1 (II). Kasgo et al.,2008 created heavy metal ion removal nanocomposite hydrogels based on AAm-AMPSNa/clay and in acidic pH at the range of 3-4.5, Cd^{2+}, Pb^{2+} and Cu^{2+} ion absorption was highest amount. Sol-gel procedure was used successfully to produce magnetic ion-imprinted polymer. The adsorption capacity and selectivity coefficient of these surface ions imprinting polymers were both increased. Composite hydrogel mixing with clays have large adsorption ability because of easily accessibility of dynamic functional groups. Mineral clay are used to prepared composite hydrogel, as they are large ionic exchange capabilities, surface areas, and chemical/mechanical stability (Sharaf et al.,2009, Raval et al.,2016a & Bhattacharyya et al.,2008). Interpenetrating polymer network (IPN) hydrogels are synthesized by cross linking of two or more polymer chains. This types of cross linking occurs in two backbone by covalent bonds, hydrogen bonds, or van der Walls forces.

IPN hydrogels have recently gained popularity in wastewater treatment due to their reversible phase change at low critical temperatures, making it easy to collect and separate hydrogels from wastewater. Zhou et al. created a CS/NIPAAm hydrogel doped with nano-TiO_2 for ionic dye adsorption (Zhou et al. 2017). Pd^{2+} was separated using a new form of poly(N-isopropylacryl amide-co-allyl mercaptan) gel (Tokuyama et al. 2017). Morán-Quiroz et al. recently used a redox initiator to make superabsorbent poly(AAc-co-AAm) hydrogels, which they used to remove Cu^{2+} ions from aqueous solutions. At pH 5.0, free-radical copolymerization of poly(AAm-co-sodium methacrylate) hydrogels using poly(ethylene glycol) diacrylate as a cross-linker can remove $Cu^{2+}$and Cd^{2+}ions. Poly(AAm-AAc) hydrogels could bind metal ions like $Cu^{2+}$and Cd^{2+}more strongly than alkali or alkaline earth ions. . Ionogenic hydrogels containing ionising side groups may contribute to large variations in the degree of swelling after a modest change in the medium's pH values (Kumar et al 2007; Peppas 1991). The hydrogel's outstanding physicochemical characteristics have created a deep interest among researchers to fabricate hydrogel based advanced materials for variety of applications. Lakouraj and colleagues developed a novel sodium alginate-supported tetrasodium thiacalix[4]arene tetrasulfonate (TSTC[4]AS-s-SA) nanogel. A sodium alginate-based super paramagnetic nanocomposite was synthesized by coprecipitating sodium alginate-supported tetrasodium thiacalix[4]arene tetrasulfonate with in situ generated Fe_3O_4 nanoparticles. Different spectroscopic techniques were used to examine the magnetic characteristics of the nanoadsorbents. Superparamagnetism is a feature of this hybrid nanocomposite that aids in the effective adsorption of hazardous metal ions. The following was the order of adsorption percentages: At pH = 7, Pb(II) > Cd(II) > Cu(II) > Cr(III) > Co(II) > Ni(II). Different pollutant compounds, such as ionic pigments, herbicides, and fungicides, are encapsulated using these hydrogel systems (Lakouraj et al., 2014). Hong Zhang et al. have created composite hydrogel beads made of magnetic bentonite/carboxymethyl chitosan/sodium alginate (Mag-Ben/CCS/Alg) and investigated their ability to absorb copper ions. The use of composite systems like these can help with copper ion enrichment. Mag-Ben/CCS/Alg reaches equilibrium after 90 minutes of adsorption at pH 5 (Zhang et al. 2019). Gangapuram et al. created a colorimetric probe made of functionalized carboxymethyl gum karaya capped gold nanoparticles selective recognition of Cu(II) ions. This nanocomposite hybrid possess good pH, salt, and shelf life stability, as well as a highly selective colorimetric response to Cu^{2+} ions (Gangapuram et al. 2017). Fosso-Kankeu et al used the free radical graft co-polymerization approach to create a gum karaya (GK)-grafted poly(acrylamide-co-acrylic acid) hydrogel nanocomposite with iron oxide magnetic nanoparticles that can be used as flocculants to remove hazardous metal ions. When compared to inorganic coagulants such as polyferric chloride, $Al_2(SO_4)_3$, $FeCl_3$, and $Mg(OH)_2$, these Fe_3O_4 incapsulated hydrogel nanocomposite performs better acquiring 100% enrichment with in 5 minutes as well as show stimuli responsive behaviour. Metal was removed in the following order: Pb^{2+}>Cr^{6+}>Ni^{2+} (Fosso-Kankeu et al. 2016)

ROLE OF NANOCOMPOSITE HYDROGELS IN DYE REMOVAL & ENRICHMENT

Textile and associated industries' dye industry is a key economic sector. Every day, billions of pounds of untreated dye-based waste are dumped into water bodies, putting biotic and abiotic components of the environment under severe stress. Water pollution has a significant impact on the environment's biotic and abiotic components. The majority of these effluents contain toxic, non-biodegradable chemicals that affect exposed organisms and marine life. Heavy metals and organic dyes infiltrate living creatures at low amounts, causing sickness and other issues. Taking them out of industrial waste is a significant problem in the current circumstances. The dye's difficult removal is due to the dye's complicated structure. For heavy metal ions and organic dyes, polysaccharide-based nanocomposite hydrogels work well as adsorbents. Several research on the role of nanocomposites and nanoparticles in dye enrichment have been conducted during the last two decades. The adsorption capacities of nanoparticle and nanocomposite adsorbents have been demonstrated (Raval et al.,2016b). Among the several adsorbents employed, hydroxyapatite is a biocompatible adsorbent that removes anionic and cationic dyes with reasonable efficiency. Functionalized hydroxyapatite has also been used to boost the removal effectiveness of anionic and cationic dyes by doping with other materials to construct various hydroxyapatite nanocomposites in the literature (Pai et al., 2021). Clay based nanocomposites occupy an integral position for dye enrichment (Islam et al., 2015). Using K+ ions to gel after freezing-thawing to integrate sodium montmorillonite nanoclay, Hosseinzadeh and coworkers studied cationic crystal violet (CV) dye enrichment using hydrogels made of kappa-carrageenan and poly(vinyl alcohol). Nanoclay drastically alter the swelling behaviour of the functionalized nanocomposite matrix declining it from 1200% to 320% due to extensive crosslinking. The effects of contact time, nanoclay concentration, pH, temperature, and ion strength on the uptake of crystal violet dye have also been investigated. The nanocomposites had a higher adsorption than the clay-free hydrogel in the same batch setup. Presence of $CaCl_2$ and $AlCl_3$ salt decreased the adsorption capacity. Adsorption kinetics followed pseudo-second-order model for dye uptake. Langmuir model predicted maximal adsorption capacity upto 151 mg/g using the (Hosseinzadeh et al., 2015). When sepiolite clay is present, acrylamide can be copolymerized by being grafted onto the biopolymer kappa-carrageenan to create nanocomposite hydrogels in another work for the enrichment of CV. Without any chemical treatment, sepiolite nano clay was added to the hydrogel matrix. Sepiolite exists as distinct needle shapes, as depicted from the TEM images. Swelling parameters, adsorption kinetics and isotherms for dye uptake have also been evaluated as a function of reaction parameters. CV adsorption through newly synthesized nanocomposites followed pseudo-second-order adsorption kinetic. The adsorption data best suit the Langmuir isotherm, according to the results. The pH of the reaction medium had no effect on the behaviour of the adsorption in the clay-free hydrogel, but nanocomposites revealed pH-dependent adsorption (Mahdavinia & Asgari, 2013). Malachite green (MG) uptake was studied using carboxymethyl cellulose/montmorillonite nanocomposites. The concentration of montmorillonite nanoparticles injected into the hydrogel matrix that had the best MG dye adsorption efficiency was found to be 10% by weight. The maximum monolayer adsorption capacity was calculated using the Langmuir isotherm model and the adsorption process kinetic was better described by a pseudo-second-order kinetic model. Functionalized nanocomposite display higher adsorption as compared to simple hydrogel. The effectiveness of the adsorption process decreased as ion concentration increases (Peighambardoust et al., 2020).

Inorganic oxides such as silica, titania and graphene oxide have also been utilized extensively for functionalization of polysaccharides and further analysed for dye enrichment studies. Ghorai et al. developed nanosilica-incorporated polyacrylamide-grafted xanthum gum-based nanocomposites that were used to enrich and remove methylene blue (MB) and methyl violet (MV). Synthesis of nanosilica into the polymeric matrix have been carried out through sol-gel reaction, in which biopolymer matrix act as new template for nanosilica incorporation. Similarly, dye adsorption followed pseudo-second-order kinetics and a Langmuir adsorption isotherms. These nanocomposites have a high adsorption capacity (99.4% for MB and 99.1% for MV), as well as a strong regeneration capacity. H-bonding, dipole-dipole as well as electrostatic interations among dye molecules and anionic matrix were responsible for high uptake (Ghorai et al.,2014). Thakur et al. used sol-gel synthesis of dehydroxylation condensation of tetraethoxysilane (TEOS) with xanthan gum under basic conditions to make nanocomposite adsorbent and further evaluated for enrichment of methylene blue (MB) and bismarck brown Y (BB). XG/SiO_2 nanocomposite possess macro- and meso-porous structures with excellent surface area, chemical/mechanical stability, and binding capability. Adsorption capacity of the nanocomposite for MB and BB was very high upto 432.90mgg-1 and 448.43mgg-1, respectively. With in 35 minutes, 99.4% of MB and 87.2% of BB from 100mgL-1 dye solutions have been adsorbed (Thakur et al., 2017). Another study found that modified xanthan gum/silica hybrid nanocomposites worked well a brilliant adsorbent material for enrichment of Congo red (CR) dye. The efficacy of CR dye adsorption is controlled by various reaction parameters such as temperature, pH of reaction medium, equilibrium time, stirring speed, amount of initial dye and adsorbent. Adsorption capacity upto (Q0=209.205 mg g(-1)) have been achieved. Adsorption was controlled by intraparticle diffusion and pseudo-second order kinetic models and was in correlation with the Langmuir isotherm. The dye-loaded nanocomposite has a good regeneration ability as verified via desorption research (Ghorai et al., 2013). Mittal et al used graft copolymerization to create a functionalized gum ghatti/TiO_2 nanoparticles incorporated polymeric networks and further evaluated for removal of MB. These matrix depicted strong pH responsive behaviour for dye uptake as well as dependence on ionic strength of the medium, and the adsorbent loading. The dye-adsorption process was spontaneous, endothermic, and adopt physisorption. This matrix was proved effective for enrichment of cationic dyes and could be used for 5 consecutive cycles without losing adsorption capacity (Mittal & Ray, 2016). Molavi and coworkers have used the solvothermal approach to build an ultra-fast adsorbent from thermally oxidised nanodiamond (OND) functionalized with UiO-66 and its composite nanoparticles and further evaluated for removal of methyl red (MR) and malachite green (MG). These nanoparticle showed a harmonious impact in enriching 59% of MR and 43% of MG concurrently from a mixture of both dyes with in 2 minutes as compared individual UiO-66 and OND nanoparticles. UiO-66-OND nanoparticles represents that the adsorption process is endothermic and spontaneous, with an entropy shift rather than an enthalpy effect directing it, according to thermodynamic studies. Due to their quick adsorption kinetics, high water stability, and outstanding reusability, these nanocomposites have proven to be effective dye uptake matrix (Molavi et al., 2020). Hosseinzadeh and Ramin created a malachite green dye-removing hydrogel adsorbent made of starch/polyacrylamide matrix functionalized with graphene oxide, and hydroxyapatite. The MG sorption confirmed to be feasible, impulsive, and heat-absorbing. These functionalized matrix achieved maximum adsorption capacity of MG dye upto 297mgg^{-1} as well as good regeneration capacity was achieved even after five successive cycles of dye adsorption-desorption (Hosseinzadeh & Ramin, 2018). Nanocomposite hydrogel based on xanthan gum and acrylic acid has been functionalized with reduced graphene oxide. Such hydrogel composite have been synthesized using green microwave aided synthesis and then infused with reduced graphene oxide which was further evaluated for enrichment of methyl violet (MV) and MB. Reaction parameters affecting dye uptake includes contact time, feed dye concentration, pH of medium and adsorbent dose. Maximum adsorption capacities (Qmax) of 1052.63 mg/g for MV and 793.65 mg/g for MB respectively have been achieved from these nanocomposites at 25°C (Makhado et al., 2018a). Carbon nanotubes are another reinforcing materials used for nanocomposite formulations. In another study, carbon nanotube infused in xanthan gum based matrix have been synthesized using microwave aided copolymerization. These carbon nanotubes based hydrogel nanocomposite have been evaluated for dye uptake studies of MB. Negatively charged groups (COO-) present on polymeric matrix interact strongly with the positive charges of dye molecules. Optimum dye uptake was achieved within 30 minutes above pH 6 representing good adsorption potential, with adsorption capacities of 521.0 mg/g at 30°C. These nanocomposites possess excellent regeneration capacity and could be used several times (Makhado et al., 2018b). Similarly, due to their selectivity and efficient regeneration capabilities, stimuli responsive nanocomposites have a distinct advantage when utilised in dye/metal ion enrichment. Magnetic iron oxide nanoparticles (MIONs) and cress seed mucilage (CSM) were combined in situ to generate a magnetic nanocomposite adsorbent, which was subsequently used to eradicate MB from aqueous solutions. This adsorbent possesses high dye adsorption capacity as well as a strong magnetic character. The dye adsorption process was spontaneous and exothermic as evaluated from thermodynamic data. Adsorption mainly follow Langmuir kinetic model (R2 > 0.99). CSM-MIONs adsorbent showed strong adsorption performance up to five desorption-adsorption consecutive cycles (Allafchian et al., 2019). In literature, magnetic field responsive hydrogel nanocomposite have been fabricated using carboxymethyl cellulose graft copolymerized with vinylic monomers (CMC-cl-pAA) and functionalized using / Fe_3O_4 nanoparticles. Homogenous dispersion of the Fe_3O_4-$C_{30}B$ nanocomposite improve mechanical/tensile strength, regeneration stability and thermal stability of the polymeric matrix. Both functionalized gel matrix hydrogel and hydrogel nanocomposite exhibits maximal adsorption capacities of 1109.55 and 1081.60 mg/g, for MB enrichment respectively. The adsorption thermodynamic analyses depicts endothermic as well as spontaneous adsorption process (Malatji et al., 2020). Moghaddam et al. created a magnetic adsorbent made of chitosan and nickel ferrite for uptake of Congo red and MO from an aqueous medium. Response surface approach was used to analyse parameters influencing dye uptake. Langmuir model was used to investigate the adsorption isotherm, yielding adsorption capacities of 551.2 mg g-1 for methyl orange and 274.7 mg/g for Congo red, respectively. These nanocomposites proved as fast adsorbents since 50 percent of dyes could be removed with in 5 minutes. Adsorption kinetics shows film diffusion and intra-particle diffusion mechanism (Moghaddam et al., 2018). Kazemi et al. used sodium alginate and chitosan to make a dual polysaccharide-based magnetic bionanocomposite that was impregnated with natural zeolite and further evaluated for MB enrichment from aqueous medium. Magnetic responsiveness is present in these bionanocomposites. Dye adsorption is governed by the pseudo-second-order kinetic model and the Freundlich isotherm model and dye adsorption follows exothermic spontaneous process (Kazemi & Javanbakht, 2020). Bayat et al. used a co-precipitation approach to create a sodium alginate-based magnetic nanocomposite of zeolite/nickel ferrite, which they then used for MB enrichment. The magnetic core is shielded from oxidation by a layer of sodium alginate. Adsorption thermodynamics, adsorption kinetics, and isotherms have been evaluated using Minitab software. Reaction parameters have been optimized through experimental design approach, response surface methodology, and Box-Behnken design. At an ideal pH of 5, MB could be removed from an aqueous solution with a maximum capacity of 54.05 mg/g. Results of adsorption were found consistent with Langmuir isotherm model and pseudo-second-order kinetic model (Bayat et al., 2018). Recently, azo dye enrichment studies have been carried out using bionanocomposite of basil seed mucilage functionalized with magnetic zeolitic imidazolate framework-8. Because of the emulsifying and stabilising capabilities of basil seed mucilage, there was a negatively charged magnetite surface, with metal-organic frameworks, yielded superior ZIF-8 crystals. Maximum adsorption capabilities achieved for MB and Eriochrome Black T were 9.09 and 13.21 mg/g respectively. The nanocomposite demonstrates excellent reusability and magnetic separation property (Mahmoodi & Javanbakht, 2021). In another study, magnetic oxidised multiwalled carbon nanotube (OMWCNT)-Fe_3O_4 and OMWCNT—carrageenan- Fe_3O_4 nanocomposites have been evaluated for enrichment of MB. These nanocomposites possess superparamagnetic properties. Magnetic OMWCNT-carrageenan- Fe_3O_4 nanocomposite has a superior adsorption ability for removing MB as compared to magnetic OMWCNT-Fe_3O_4 nanocomposite (Duman et al., 2016). Masilompane et al. created new nanoadsorbent nanocomposites for enriching Brilliant Black dye (BB) from Chitosan-lignin and titania. These nanoadsorbents can withstand temperatures of up to 300°C. Larger particles with well-defined edges were found in lignin, while non-uniform, short fibrous microstructures were found in chitosan. In the matrix, there was a polycrystalline anatase phase with crystalline nanoparticles. The Freundlich isotherm was followed by the batch adsorption data (r2 = 0.91), and Langmuir isotherm 15.8 mg/g for monolayer adsorption (Masilompane et al., 2018). Nanolinin/chitosan-based nanocomposites were produced and used by Sohni et al to remove the dye MB. These chitosan/nano-lignin composite material removed MB with a high efficiency (83 percent). Enrichment of MB was found better with nonolignin based composite as compared to simple chitosan/lignin composites. This may be substantially responsible for the collaborative effects of nano size and integrated functional groups introduced. Composite reusability has been demonstrated for up to four cycles. These composites have also been shown to provide up to 89 percent dye enrichment (Sohni et al., 2019). Some nanocomposites also introduce antimicrobial character in the matrix thus making them suitable candidate for purification as well as fabrication of safer packaging materials. Abu Elella et al. have prepared antimicrobial adsorbents from grafted xanthan gum/SiO_2 nanocomposites using ultrasonication and further utilized for MG dye enrichment and inhibiting the growth of bacteria like E.coli and S.aureus. Various spectroscopic techniques have been utilized to characterize the nanoadsorbent. Maximum adsorption capacity upto 99.5 percent (Q_{max} = 588.2 mg/g) was achieved by these nanocomposites. This matrix also followed intraparticle diffusion and pseudo-first-order adsorption mechanism. These nanocomposites demonstrated superior inhibition against the growth of S. aureus and E. coli, showing an inhibition diameters of about 23.5 mm and 25.33 mm for E. coli and S. aureus, respectively(Abu Elella et al., 2021b). Abu Elella et al. have also fabricated dual nano clay based nanocomposite adsorbents and evaluated their biofilm inhibition in water bodies. Antibacterial behaviour of these nanocomposites had increased dramatically as MMT nanoclay concentrations were raised (Abu Elella et al., 2021a). A magnetic field sensitive bioadsorbent Fe_3O_4 embedded CS/PVA polymer matrix was used for enrichment of congo red dye from waste water (Zhu et al., 2012). A starch/polyaniline based bionanoadsorbent was tested for colour removal from aqueous solution. The composite's intermolecular hydrogen bonds were broken, allowing dye molecules to interact freely (Janaki et al., 2012). Zeng and colleagues encapsulated Fe NPs in tea extract and disseminated them within sodium alginate beads and utilized for MG degradation studies. Hybrid nanocomposite hydrogels increased dye adsorption as compared to unbound nanoparticles. There was 92% dye enrichment using Fe NPs encapsulated sodium alginate beads as compared to free Fe NPs which could remove upto 52% only. It has been demonstrated that these nano-composite hydrogel beads can act as a helpful matrix for MG degradation (Zeng et al., 2019). Charradi et al. created a polysaccharide-based photocatalyst made of a spinel zinc ferrite ($ZnFe_2O_4$)/cellulose composite and tested it for photocatalytic activity in an aqueous solution against MB degradation. The $ZnFe_2O_4$/cellulose composite successfully eliminated (100%) MB within 180 minutes at a pH 6.5. These nanocomposites possess good reusability up to 4 cycles. After the fourth cycle, the loaded composite was additionally regenerated with an alcohol/water solution, resulting in reusability with varied efficiency. These matrix show The photocatalytic behaviour as well (Charradi et al., 2022). Ghorai et al. developed nano silica included polyacrylamide grafted xanthum gum based nanocomposites that were used to enrich and remove MB and MV. Using a sol-gel process, nanosilica has been incorporated into a gel matrix which acts as a new template for nanosilica formation. High regeneration capacity and excellent adsorption capacity of 99.4% and 99.1% for MB and MV, respectively, were attained (Ghorai et al., 2014).

Role of Nanocomposite Hydrogels in Inhibition of Biofilm Formation

Antibiotic-resistant bacteria in ground water have been widely reported which are responsible for ground water contamination, and they are a major public health concern (Mahmoodi et al.,2021). Cross contamination and Multi drug resistance [MDR] is a major challenge that researchers of 21st century currently facing. In the search of solution for the MDR and biofilm formation, researchers are looking towards the development of novel antimicrobial nanomaterials that can prevent the medical device/instrument-based infections. Inhibition of biofilm formed by the virulent bacterial pathogens is a novel approach to combat with the bacterial infections. Various nano-emulsions based hydrogels can be utilized to inhibits the bacterial virulence and biofilm formed by the bacteria (Barik et al., 2019)].

Recent advancement in nanotechnology and nanoscience paves way to develop novel antibacterial agents which can prevent the bacterial colonies formation and discourages the biofilm formation. Metals such as copper, zinc and silver already having the inherent antibacterial properties. Metal based novel nanomaterials having significant impact on the elimination of medical devices/instruments-based infections (Agnihotri et al., 2017). Recently, nanocomposite functionalized biopolymers have been employed as novel materials for bacterial contact killing in water. Nanoparticles with multifunctional ligands have also been used to prevent bacterial cell adhesion, disrupt membranes, and reduce particular immune reactions. Both planktonic and biofilm formation conditions showed that Silver Nanoparticle included nanocomposites had significant antibacterial activity against *E. coli* (Gram-negative) and *S. epidermidis* (Gram-positive). Recent developments in the application of novel materials for bacterial contact killing in water environments include nanocomposite functionalized biopolymers. It is known that nanoparticles coated with multifunctional ligands can bind to bacterial cells, damage membranes, and lessen some immune reactions (Peighambardoust et al.,2020). As a more environmentally friendly alternative for water purification, functionalized cellulose fibres and polyelectrolyte surface charged cellulose were used (Hosseinzadeh et al., 2018).

BACTERIAL BIOFILM

Bacterial biofilms are amorphous in nature and they have mono or poly microbial structure associated with living or non-living surfaces (Jamal et al., 2018). Biofilm formation is the strategy of bacterial cells to shield themselves from antibacterial agents and host immune system. Bacterial biofilm is made up from proteins, nucleic acid, water and polysaccharides (Jamal et al., 2018; Singh et al., 2017). Hydrogels provides effective way to treat the MDR and biofilm formation due to hydrophilicity, Complex 3-Dimensional structure, Cell adhesion and unique biocompatibility (Miculescu et al., 2017). Biofilm formation by virulent microbes poses a greater risk for the hospital acquired infections. Biofilm formation will take place in multiple steps, first step involves initial attachment of the microbial population to the surface where they can accumulate in multiple layers and leads to formation of biofilm. Subsequently maturation of the biofilm takes place further microbial population detach to spread the infection in other part of the material. The biofilm formation makes implants were more susceptible towards many infections. Direct contamination of the biomaterial surface will make the process of implantation complicated (Campoccia et al., 2013). Microbial infection is a major problem that is world facing today due to emergence of multi drug resistance biofilms and no effective strategy against it. Stable nano encapsulations (NEs) of *G. fragrantissima* essential oil and its bioactive compounds showed better biofilm elimination capabilities than the pure Essential oils against *E. coli* (Prateeksha et al. 2019).

The application of nanomaterials as a carrier agent gains momentum in food industries to improve the self-life and improving the effectiveness of preservatives at low dose (Prakash et al., 2018). The use of essential oils in food and pharmaceutical industries is limited due to its inherent properties e.g. hydrophobicity, instability and exposure to environmental condition i.e. oxygen, temperature, and light may cause degradation. These factors resulted in restricted use of EOs in food and pharma sector. To tackle with this problem nano encapsulation of EOs is the best method to employ. Liposomes, Microcapsules, Nanospheres, Nano emulsions and molecular inclusion complexes seems to be the solution of the problem (Shaaban et al. 2022).

Strategies for Application of Nanomaterials Against Biofilm Formation

In literature there are numerous nanoscale drug delivery devices based on organic, inorganic, metallic based, biopolymer microgels, cyclodextrins, smart responsive materials polymer-based nanoparticles for effective drug delivery (Jafari et al., 2018). The stability, bioavailability, cellular absorption, and pharmacokinetic profile of medications can be improved, and their toxicity can be decreased, by encapsulating them in nanocarriers (Eleraky et al., 2020). Nano technology-based lipid drug delivery system and metal oxides nano particles shows antimicrobial activity and can be utilized to control the resistant bacteria (Santos et al., 2019).

The functions of the efflux pump and transporters present in bacterial membranes are related to multidrug resistance. These efflux pumps and transporters have a big impact on drug intake and drug activity suppression. Efflux transporters activity effectively inhibited by the application of nano particles as encapsulating agents, this strategy is responsible for the enhanced accumulation of the drugs and it provides effective way to combat bacterial resistance. Nano particles along with natural or synthetic antimicrobials provides effective strategy against the MDR bacteria (Dashtbani-Roozbehani et al., 2021; Jampilek et al. 2022). Biofilm formation by the MDR bacteria poses greater risk for the healthcare sector. Biofilm contains the microcolonies of the bacteria. Microbes in biofilm forms divers microbial communities.

They may have the same gene composition but different expression profile. These biofilm microbes are usually having the more virulence than their natural variant. Various components of the biofilm make bacteria more resistant to the immune system of the host and conventional drugs (Schulze et al., 2021).

Several nano formulations which include micelles, solid liquid nanoparticles and quantsomes can be used as an effective strategy against the biofilm formation. The capacity of MDR bacteria to infect wounds and cause persistent wound infections, which in turn cause serious illness or even death (Fasiku et al., 2021). Nanotechnology plays an important role in combating with the bio film based antibiotic resistance (Malaekeh-Nikouei et al., 2020). Various nano technologies and biomaterials are utilised to administer antimicrobial and anti-biofilm chemicals to target resistant bacteria in an efficient manner (Bianchera et al., 2020). Different nanoparticles, such as organic nanomaterials, are utilised successfully to prevent biofilms from sticking to surfaces. It lessens the formation of biofilm and increases the permeability of the antibacterial drugs (Li et al., 2021). Recent research emphasised the efficiency of the drug-encapsulated nanofibers in accelerating wound healing (Tiwari et al., 2021). Colloidal solution of widely accessible metal-based nanoparticles including silver nanoparticles (10 nm and 40 nm), gold nanoparticles (20 nm), platinum nano particles (4 nm) and Zinc oxide nanoparticles, Titanium oxide nanoparticles, Al_2O_3 nanoparticles, Y_2O_3 nanoparticles and ZrO2 nanoparticles of 100 nm showed efficient antibacterial activity against MRSA and methicillin-resistant S. aureus strains. These nanoparticles are applied as coatings on biodegradable polymers produced through 3D printing, such as catheters and bandages for persistent wounds (Rezić et al., 2021).

Biologically safe hydrogels possess a porous structure, which can effectively used against the biofilm formation through sustained release of the antimicrobial drug and effectively used for the topical application (Fasiku et al., 2021). Polymer based drug delivery system which includes polymeric liposomes, polymeric nanogels, polymeric dendrimers and polymeric micelles, improves the effective usage of the drugs in treating bacterial infections in comparison of standard antibiotics (Wang et al., 2021).

Folate derived lipid nanoparticles showed an improved bactericidal effect against the Multi drug resistant *S. Aureus* and showed the excellent inhibition of the biofilm formed. It is showing a reduced accumulation of the drugs in the kidney. These nano particles effectively work against the MDR and also reducing the side effects of the drug on kidneys (Van mala et al., 2021).

Tobramycin and glycoside hydrolase-loaded liquid crystalline nanoparticles are effective at preventing the growth of *P. auginosa* biofilms. It offers a powerful method that can be used to combat hospital acquired illnesses (Jampilek et al., 2022). Recent developments in responsive polymeric nanoparticles that are efficient for treating infections with drug resistance are able to prevent the formation of biofilms and exhibit increased efficacy in the removal of mature biofilms as well (Su et al., 2021)].

In recent years, nanocomposites and nanoparticle embedded polymeric systems have proved a good alternative in water pollution alleviation technologies. These functionalized systems could also be utilized as sensors to detect numerous pollutants in the environment and as detecting devices to assess ecotoxicity which is a major concern in current scenario. At the same time nanoparticle sometime got leached form the functionalized matrix and result itself in toxicity. This could be modulated by manipulating the reaction conditions of nanofabrication, dimensions/ type of the nanoparticle incorporated according to the application. Present study provides an guiding path for the young researchers to modulate nanotechnology and nanoparticles for the betterment of society in best possible way.

CONCLUSION

Ecotoxicity has grown to be a serious problem for the ecosystem's sustainability in the current situation. Environmental sensitivity has increased over the past few decades as a result of excessive pollution and careless use of chemicals, medicines, and other substances. The current global water problem and the production of enormous amounts of waste water are the results of the indiscriminate exploitation of water resources past their regeneration limits. One of the major sources of untreated water pollutants released into bodies of water is hazardous effluents from the textile sector. The environment is under a lot of stress because of toxic chemicals and other pollution. The content of poisonous dyes in water must therefore be reduced to acceptable levels by the treatment of industrial effluents. Nanocomposites possess diverse array of properties as compared to their base matrix due to incorporation of nanoparticles integrated with in. Such functionalized materials have been exploited for the removal and enrichment of toxic metal ions, organic pollutants as well as have ability to minimize the pathogenic contamination of water bodies. Such nanocomposites act as efficient adsorbents for dye and metal ion enrichments. Nanocomposites made of metal oxides mostly exhibit effective biofilm inhibition behaviour in aquatic environments. Inhibition of biofilm formation on various surfaces by nanocomposites provides a best possible solution for most of basic commodities in food security as well as in health sector. Polysaccharide based functionalized nanocomposites have proved as efficient materials for water pollution alleviation technologies.

REFERENCES

Abu Elella, M. H., Goda, E. S., Abdallah, H. M., Shalan, A. E., Gamal, H., & Yoon, K. R. (2021a). Innovative bactericidal adsorbents containing modified xanthan gum/montmorillonite nanocomposites for wastewater treatment. *International Journal of Biological Macromolecules, 167*, 1113–1125. doi:10.1016/j.ijbiomac.2020.11.065 PMID:33197479

Abu Elella, M. H., Goda, E. S., Gamal, H., El-Bahy, S. M., Nour, M. A., & Yoon, K. R. (2021b). Green antimicrobial adsorbent containing grafted xanthan gum/SiO_2 nanocomposites for malachite green dye. *International Journal of Biological Macromolecules, 191*, 385–395. doi:10.1016/j.ijbiomac.2021.09.040 PMID:34537301

Agnihotri, S., & Dhiman, N. K. (2017). Development of nano-antimicrobial biomaterials for biomedical applications. *Advances in Biomaterials for Biomedical Applications, 66*, 479–545. doi:10.1007/978-981-10-3328-5_12

Al-Asheh, S., Banat, F. R., Al-Omari, Z., & Duvnjak, Z. (2000). Predictions of binary sorption isotherms for the sorption of heavy metals by pine bark using single isotherm data. *Chemosphere, 41*(5), 659–665. doi:10.1016/S0045-6535(99)00497-X PMID:10834365

Allafchian, A., Mousavi, Z. S., & Hosseini, S. S. (2019). Application of cress seed musilage magnetic nanocomposites for removal of methylene blue dye from water. *International Journal of Biological Macromolecules, 136*, 199–208. doi:10.1016/j.ijbiomac.2019.06.083 PMID:31201917

Barik, S. K., & Singh, B. N. (2019). Nanoemulsion-loaded hydrogel coatings for inhibition of bacterial virulence and biofilm formation on solid surfaces. *Scientific Reports, 9*(1), 1–13. PMID:30626917

Bayat, M., Javanbakht, V., & Esmaili, J. (2018). Synthesis of zeolite/nickel ferrite/sodium alginate bionanocomposite via a co-precipitation technique for efficient removal of water-soluble methylene blue dye. *International Journal of Biological Macromolecules*, *116*, 607–619. doi:10.1016/j.ijbiomac.2018.05.012 PMID:29738864

Bhattacharyya, K. G., & Gupta, S. S. (2008). Adsorption of a few heavy metals on natural and modified kaolinite and montmorillonite: A review. *Advances in Colloid and Interface Science*, *140*(2), 114–131. doi:10.1016/j.cis.2007.12.008 PMID:18319190

Bianchera, A., Buttini, F., & Bettini, R. (2020). Micro/nanosystems and biomaterials for controlled delivery of antimicrobial and anti-biofilm agents. *Expert Opinion on Therapeutic Patents*, *30*(12), 983–1000. doi:10.1080/13543776.2020.1839415 PMID:33078643

Campoccia, D., Montanaro, L., & Arciola, C. R. (2013). A review of the biomaterials technologies for infection-resistant surfaces. *Biomaterials*, *34*(34), 8533–8554. doi:10.1016/j.biomaterials.2013.07.089 PMID:23953781

Charradi, K., Ahmed, Z., & BenMoussa, M.A. (2022) A facile approach for the synthesis of spinel zinc ferrite/cellulose as an effective photocatalyst for the degradation of methylene blue in aqueous solution. *Cellulose, 29*(9), 1-12. doi:10.1007/s10570-021-04334-3

Daifa, M., Shmoeli, E., & Domb, A. J. (2019). Enhanced flocculation activity of polyacrylamide-based flocculant for purification of industrial wastewater. *Polymers for Advanced Technologies*, *30*(10), 2636–2646. doi:10.1002/pat.4730

Dashtbani-Roozbehani, A., & Brown, M. H. (2021). Efflux pump mediated antimicrobial resistance by staphylococci in health-related environments: Challenges and the quest for inhibition. *Antibiotics (Basel, Switzerland)*, *10*(12), 1502. doi:10.3390/antibiotics10121502 PMID:34943714

Denizli, A., Kesenci, K., Arica, Y., & Piskin, E. (2000). Dithiocarbamate-incorporated monosize polystyrene microspheres for selective removal of mercury ions. *Reactive & Functional Polymers*, *44*(3), 235–243. doi:10.1016/S1381-5148(99)00099-1

Dinu, M. V., & Dragan, E. S. (2010). Evaluation of Cu2+, Co2+, and Ni2+ ions removal from aqueous solution using a novel chitosan/clinoptilolite composites: Kinetics and isotherms. *Chemical Engineering Journal*, *160*(1), 157–163. doi:10.1016/j.cej.2010.03.029

Duman, O., Tunç, S., Polat, T. G., & Bozoğlan, B. K. (2016). Synthesis of magnetic oxidized multiwalled carbon nanotube-κ-carrageenan-Fe_3O_4 nanocomposite adsorbent and its application in cationic Methylene Blue dye adsorption. *Carbohydrate Polymers*, *147*, 79–88. doi:10.1016/j.carbpol.2016.03.099 PMID:27178911

Eleraky, N. E., Allam, A., Hassan, S. B., & Omar, M. M. (2020). Nanomedicine fight against antibacterial resistance: An overview of the recent pharmaceutical innovations. *Pharmaceutics*, *12*(2), 142. doi:10.3390/pharmaceutics12020142 PMID:32046289

Fasiku, V. O., Omolo, C. A., Devnarain, N., Ibrahim, U. H., Rambharose, S., Faya, M., Mocktar, C., Singh, S. D., & Govender, T. (2021). Chitosan-based hydrogel for the dual delivery of antimicrobial agents against bacterial methicillin-resistant Staphylococcus aureus biofilm-infected wounds. *ACS Omega*, *6*(34), 21994–22010. doi:10.1021/acsomega.1c02547 PMID:34497894

Feng, D., Aldrich, C., & Tan, H. (2000). Treatment of acid mine water by use of heavy metal precipitation and ion exchange. *Minerals Engineering*, *13*(6), 623–642. doi:10.1016/S0892-6875(00)00045-5

Fosso-Kankeu, E., Mittal, H., Waanders, F., Ntwampe, I. O., & Ray, S. S. (2016). Preparation and characterization of gum karaya hydrogel nanocomposite flocculant for metal ions removal from mine effluents. *International Journal of Environmental Science and Technology*, *13*(2), 711–724. doi:10.100713762-015-0915-x

Gangapuram, B. R., Bandi, R., Dadigala, R., Kotu, D. M., & Guttena, V. (2017). Facile green synthesis of gold nanoparticles with carboxymethyl gum karaya, selective and sensitive colorimetric detection of copper (II) ions. *Journal of Cluster Science*, *28*(5), 2873–2890. doi:10.100710876-017-1264-3

Ghorai, S., Sarkar, A., Raoufi, M., Panda, A. B., Schönherr, H., & Pal, S. (2014). Enhanced removal of methylene blue and methyl violet dyes from aqueous solution using a nanocomposite of hydrolyzed polyacrylamide grafted xanthan gum and incorporated nanosilica. *ACS Applied Materials & Interfaces*, *6*(7), 4766–4777. doi:10.1021/am4055657 PMID:24579659

Ghorai, S., Sarkar, A. K., Panda, A. B., & Pal, S. (2013). Effective removal of Congo red dye from aqueous solution using modified xanthan gum/silica hybrid nanocomposite as adsorbent. *Bioresource Technology*, *144*, 485–491. doi:10.1016/j.biortech.2013.06.108 PMID:23896441

Gupta, B. S., & Ako, J. E. (2005). Application of guar gum as a flocculant aid in food processing and potable water treatment. *European Food Research and Technology*, *221*(6), 746–751. doi:10.100700217-005-0056-4

Hosseinzadeh, H., & Ramin, S. (2018). Fabrication of starch-graft-poly(acrylamide)/graphene oxide/hydroxyapatite nanocomposite hydrogel adsorbent for removal of malachite green dye from aqueous solution. *International Journal of Biological Macromolecules*, *106*, 101–115. doi:10.1016/j.ijbiomac.2017.07.182 PMID:28778526

Hosseinzadeh, H., Zoroufi, S., & Mahdavinia, G. R. (2015). Study on adsorption of cationic dye on novel kappa-carrageenan/poly(vinyl alcohol)/montmorillonite nanocomposite hydrogels. *Polymer Bulletin*, *72*(6), 1339–1363. doi:10.100700289-015-1340-5

Hussain, M. S., Musharraf, S. G., Bhanger, M. I., & Malik, M. I. (2020). Salicylaldehyde derivative of nano-chitosan as an efficient adsorbent for lead (II), copper (II), and cadmium (II) ions. *International Journal of Biological Macromolecules*, *147*, 643–652. doi:10.1016/j.ijbiomac.2020.01.091 PMID:31931059

Islam, M. S., Rahaman, M. S., & Yeum, J. H. (2015). Electrospun novel super-absorbent based on polysaccharide–polyvinyl alcohol–montmorillonite clay nanocomposites. *Carbohydrate Polymers*, *115*, 69–77. doi:10.1016/j.carbpol.2014.08.086 PMID:25439870

Jafari, S. M., & McClements, D. J. (2018). *Nanoemulsions: formulation, applications, and characterization*. Academic Press.

Jamal, M., Ahmad, W., Andleeb, S., Jalil, F., Imran, M., Nawaz, M. A., & Kamil, M. A. (2018). Bacterial biofilm and associated infections. *Journal of the Chinese Medical Association*, *81*(1), 7–11. doi:10.1016/j.jcma.2017.07.012 PMID:29042186

Jampilek, J., & Kralova, K. (2022). Advances in Nanostructures for Antimicrobial Therapy. *Materials (Basel)*, *15*(7), 2388. doi:10.3390/ma15072388 PMID:35407720

Janaki, V., Vijayaraghavan, K., Oh, B., Lee, K., Muthuchelian, K., Ramasamy, A. K., & Kamal-Kannan, S. (2012). Starch/polyaniline nanocomposite for enhanced removal of reactive dyes from synthetic effluent. *Carbohydrate Polymers*, *90*(4), 1437–1444. doi:10.1016/j.carbpol.2012.07.012 PMID:22944400

Kasgo, H., Durmus¸, A., & Kasgoz, A. (2008). Enhanced swelling and adsorption properties of AAm-AMPSNa/clay hydrogel nanocomposites for heavy metal ion removal. *Polymers for Advanced Technologies*, *19*(3), 213–220. doi:10.1002/pat.999

Kayan, G. Ö., & Kayan, A. (2021). Composite of natural polymers and their adsorbent properties on the dyes and heavy metal ions. *Journal of Polymers and the Environment*, *29*(11), 3477–3496. doi:10.100710924-021-02154-x

Kazemi, J., & Javanbakht, V. (2020). Alginate beads impregnated with magnetic Chitosan@Zeolite nanocomposite for cationic methylene blue dye removal from aqueous solution. *International Journal of Biological Macromolecules*, *154*, 1426–1437. doi:10.1016/j.ijbiomac.2019.11.024 PMID:31733254

Korus, I., & Loska, K. (2009). Removal of Cr(III) and Cr(VI) ions from aqueous solutions by means of polyelectrolyte-enhanced ultrafiltration. *Desalination*, *247*(1-3), 390–395. doi:10.1016/j.desal.2008.12.036

Kumar, A., Srivastava, A., Galaev, I. Y., & Mattiasson, B. (2007). Smart polymers: Physical forms and bioengineering applications. *Progress in Polymer Science*, *2007*(32), 1205–1237. doi:10.1016/j.progpolymsci.2007.05.003

Lakouraj, M. M., Mojerlou, F., & Zare, H. (2014). Nanogel and superparamagnetic nanocomposite based on sodium alginate for sorption of heavy metal ions. *Carbohydrate Polymers*, *106*, 34–41. doi:10.1016/j.carbpol.2014.01.092 PMID:24721048

Li, X., Chen, D., & Xie, S. (2021). Current progress and prospects of organic nanoparticles against bacterial biofilm. *Advances in Colloid and Interface Science*, *294*, 102475. doi:10.1016/j.cis.2021.102475 PMID:34280601

Mahdavinia, G. R., & Asgari, A. (2013). Synthesis of kappa-carrageenan-g-poly(acrylamide)/sepiolite nanocomposite hydrogels and adsorption of cationic dye. *Polymer Bulletin*, *70*(8), 2451–2470. doi:10.100700289-013-0966-4

Mahmoodi, M., & Javanbakht, V. (2021). Fabrication of Zn-based magnetic zeolitic imidazolate framework bionanocomposite using basil seed mucilage for removal of azo cationic and anionic dyes from aqueous solution. *International Journal of Biological Macromolecules*, *167*, 1076–1090. doi:10.1016/j.ijbiomac.2020.11.062 PMID:33186651

Makhado, E., Pandey, S., Nomngongo, P. N., & Ramontja, J. (2018b). Preparation and characterization of xanthan gum-cl-poly(acrylic acid)/o-MWCNTs hydrogel nanocomposite as highly effective re-usable adsorbent for removal of methylene blue from aqueous solutions. *Journal of Colloid and Interface Science*, *513*, 700–714. doi:10.1016/j.jcis.2017.11.060 PMID:29216578

Makhado, E., Pandey, S., & Ramontja, J. (2018a). Microwave assisted synthesis of xanthan gum-cl-poly (acrylic acid) based-reduced graphene oxide hydrogel composite for adsorption of methylene blue and methyl violet from aqueous solution. *International Journal of Biological Macromolecules*, *119*, 255–269. doi:10.1016/j.ijbiomac.2018.07.104 PMID:30031820

Malaekeh-Nikouei, B., Bazzaz, B. S. F., Mirhadi, E., Tajani, A. S., & Khameneh, B. (2020). The role of nanotechnology in combating biofilm-based antibiotic resistance. *Journal of Drug Delivery Science and Technology*, *60*, 101880. doi:10.1016/j.jddst.2020.101880

Malatji, N., Makhado, E., Ramohlola, K. E., Modibane, K. D., Maponya, T. C., Monama, G. R., & Hato, M. J. (2020). Synthesis and characterization of magnetic clay-based carboxymethyl cellulose-acrylic acid hydrogel nanocomposite for methylene blue dye removal from aqueous solution. *Environmental Science and Pollution Research International*, *27*(35), 44089–44105. doi:10.100711356-020-10166-8 PMID:32761344

Masilompane, T.M., Chaukura, N., Mishra, S.B., & Mishra, A.K. (2018). Chitosan-lignin-titania nanocomposites for the removal of brilliant black dye from aqueous solution. *International Journal of Biological Macromolecules, 120*(Pt B), 1659-1666. . doi:10.1016/j.ijbiomac.2018.09.129

Meade, R. H. (1995). Contaminants in the Mississippi River. *Geological Survey Circular*, 1133.

Miculescu, F., Maidaniuc, A., Voicu, S. I., Thakur, V. K., Stan, G. E., & Ciocan, L. T. (2017). Progress in hydroxyapatite–starch based sustainable biomaterials for biomedical bone substitution applications. *ACS Sustainable Chemistry & Engineering*, *5*(10), 8491–8512. doi:10.1021/acssuschemeng.7b02314

Mittal, H., & Ray, S. S. (2016). A study on the adsorption of methylene blue onto gum ghatti/TiO2 nanoparticles-based hydrogel nanocomposite. *International Journal of Biological Macromolecules*, *88*, 66–80. doi:10.1016/j.ijbiomac.2016.03.032 PMID:26997239

Moghaddam, A.Z., Ghiamati, E., Pourashuri, A., & Allahresani, A. (2018). Modified nickel ferrite nanocomposite/functionalized chitosan as a novel adsorbent for the removal of acidic dyes. *International Journal of Biological Macromolecules*, *120*(Pt B), 1714-1725. . doi:10.1016/j.ijbiomac.2018.09.198

Molavi, H., Neshastehgar, M., Shojaei, A., & Ghashghaeinejad, H. (2020). Ultrafast and simultaneous removal of anionic and cationic dyes by nanodiamond/UiO-66 hybrid nanocomposite. *Chemosphere*, *247*, 125882. doi:10.1016/j.chemosphere.2020.125882 PMID:32069713

Oliva, J., & Pablo, J. (2011). Removal of cadmium, copper, nickel, cobalt and mercury from water by Apatite II (TM): Column experiments. *Journal of Hazardous Materials*, *194*, 312–323. doi:10.1016/j.jhazmat.2011.07.104 PMID:21871722

Pai, S., Kini, M. S., & Selvaraj, R. (2021). A review on adsorptive removal of dyes from wastewater by hydroxyapatite nanocomposites. *Environmental Science and Pollution Research International*, *28*(10), 11835–11849. doi:10.100711356-019-07319-9 PMID:31867692

Peighambardoust, S. J., Aghamohammadi-Bavil, O., Foroutan, R., & Arsalani, N. (2020). Removal of malachite green using carboxymethyl cellulose-g-polyacrylamide/montmorillonite nanocomposite hydrogel. *International Journal of Biological Macromolecules, 159*, 1122–1131. doi:10.1016/j.ijbiomac.2020.05.093 PMID:32422256

Peppas, N. A. (1991). Physiologically responsive hydrogels. *Journal of Bioactive and Compatible Polymers, 6*(3), 241–246. doi:10.1177/088391159100600303

Prakash, B., Kujur, A., Yadav, A., Kumar, A., Singh, P. P., & Dubey, N. K. (2018). Nanoencapsulation: An efficient technology to boost the antimicrobial potential of plant essential oils in food system. *Food Control, 89*, 1–11. doi:10.1016/j.foodcont.2018.01.018

Prateeksha, B., Barik, S. K., & Singh, B. N. (2019). Nanoemulsion-loaded hydrogel coatings for inhibition of bacterial virulence and biofilm formation on solid surfaces. *Scientific Reports, 9*(1), 6520. doi:10.103841598-019-43016-w PMID:31019240

Rahimdokht, M., Pajootan, E., & Ranjbar Mohammadi, M. (2019). Titania/gum tragacanth nanohydrogel for methylene blue dye removal from textile wastewater using response surface methodology. *Polymer International, 68*(1), 134–140. doi:10.1002/pi.5706

Rana, V. S., & Sharma, N. (2020). A Review on polysaccharide based nanocomposite hydrogel systems fabrication using diverse reinforcing materials. *Journal of Polymer & Composites, 8*(1), 6–17.

Raval, N. P., Shah, P. U., & Shah, N. K. (2016a). Adsorptive removal of Ni^{2+} ions from aqueous environment: A review. *Journal of Environmental Management, 179*, 1–20. doi:10.1016/j.jenvman.2016.04.045 PMID:27149285

Raval, N.P., Shah, P.U., & Shah, N. K. (2016b) Adsorptive amputation of hazardous azo dye Congo red from wastewater: a critical review. *Environ Sci Pollut Res Int, A, 23*(15), 14810-14853. . doi:10.1007/s11356-016-6970-0

Rezić, I., Majdak, M., Ljoljić Bilić, V., Pokrovac, I., Martinaga, L., Somogyi Škoc, M., & Kosalec, I. (2021). Development of Antibacterial Protective Coatings Active against MSSA and MRSA on Biodegradable Polymers. *Polymers, 13*(4), 659. doi:10.3390/polym13040659 PMID:33672118

Sadiq, A. C., Olasupo, A., Ngah, W. S. W., Rahim, N. Y., & Mohd Suah, F. B. (2021). A decade development in the application of chitosan-based materials for dye adsorption: A short review. *International Journal of Biological Macromolecules, 191*, 1151–1163. doi:10.1016/j.ijbiomac.2021.09.179 PMID:34600954

Santos, A., Veiga, F., & Figueiras, A. (2019). Dendrimers as pharmaceutical excipients: Synthesis, properties, toxicity and biomedical applications. *Materials (Basel), 13*(1), 65. doi:10.3390/ma13010065 PMID:31877717

Say, R., Garipcan, B., Emir, S., Patir, S., & Denizli, A. (2002). Preparation of poly(hydroxyethyl methacrylate-co-methacrylamidohistidine) beads and its design as a affinity adsorbent for Cu(II) removal from aqueous solutions. *Colloids and Surfaces. A, Physicochemical and Engineering Aspects, 196*(2-3), 199–207. doi:10.1016/S0927-7757(01)00891-3

Schulze, A., Mitterer, F., Pombo, J. P., & Schild, S. (2021). Biofilms by bacterial human pathogens: Clinical relevance-development, composition and regulation-therapeutical strategies. *Microbial Cell, 8*(2), 28–56. doi:10.15698/mic2021.02.741 PMID:33553418

Shaaban, H. A., & Farouk, A. (2022). Encapsulation of Essential Oils and Their Use in Food Applications. In M. a. de Oliveira & E. e. de Aguiar Andrade (Eds.), *Essential Oils -Advances in Extractions and Biological Applications*. Intech Open. doi:10.5772/intechopen.103147

Sharaf, M. A., Arida, H. A. S., Sayed, A., Younis, A. A., & Farag, A. B. (2009). Separation and preconcentration of some heavy-metal ions using new chelating polymeric hydrogels. *Journal of Applied Polymer Science, 113*(2), 1335–1344. doi:10.1002/app.30137

Singh, B. N., Prateeksha, N., Upreti, D. K., Singh, B. R., Defoirdt, T., Gupta, V. K., & Vahabi, K. (2017). Bactericidal, quorum quenching and anti-biofilm nanofactories: A new niche for nanotechnologists. *Critical Reviews in Biotechnology, 37*(4), 525–540. doi:10.1080/07388551.2016.1199010 PMID:27684212

Sohni, S., Hashim, R., Nidaullah, H., Lamaming, J., & Sulaiman, O. (2019). Chitosan/nano-lignin based composite as a new sorbent for enhanced removal of dye pollution from aqueous solutions. *International Journal of Biological Macromolecules, 132*, 1304–1317. doi:10.1016/j.ijbiomac.2019.03.151 PMID:30922916

Su, L. Z., Liu, Y., Li, Y. F., An, Y. L., & Shi, L. Q. (2021). Responsive Polymeric Nanoparticles for Biofilm-infection Control. *Chinese Journal of Polymer Science, 39*(11), 1376–1391. doi:10.100710118-021-2610-3

Thakur, S., Pandey, S., & Arotiba, O. A. (2017). Sol-gel derived xanthan gum/silica nanocomposite-a highly efficient cationic dyes adsorbent in aqueous system. *International Journal of Biological Macromolecules, 103*, 596–604. doi:10.1016/j.ijbiomac.2017.05.087 PMID:28528942

Tiwari, R., Tiwari, G., & Lahiri, A. (2021). Localized delivery of drugs through medical textiles for treatment of burns: A perspective approach. *Advanced Pharmaceutical Bulletin, 11*(2), 248. PMID:33880346

Tokuyama, H., Onodera, M., & Ban, T. (2017). Selective temperature-swing adsorption of Pd(II) ions onto apoly(N-isopropylacrylamide-co-allyl mercaptan) gel. *Separation and Purification Technology, 182*, 166–170. doi:10.1016/j.seppur.2017.03.049

Vanamala, K., Bhise, K., Sanchez, H., Kebriaei, R., Luong, D., Sau, S., Abdelhady, H., Rybak, M. J., Andes, D., & Iyer, A. K. (1791). Folate. Functionalized Lipid Nanoparticles for Targeted Therapy of Methicillin-Resistant Staphylococcus aureus. *Pharmaceutics, 13*(11), 1791. doi:10.3390/pharmaceutics13111791

Wang, T., Rong, F., Tang, Y., Li, M., Feng, T., Zhou, Q., & Huang, W. (2021). Targeted polymer-based antibiotic delivery system: A promising option for treating bacterial infections via macromolecular approaches. *Progress in Polymer Science, 116*, 101389. doi:10.1016/j.progpolymsci.2021.101389

WHO. (1990). Guidelines for drinking-water quality, 3rd edition: Volume 1 - Recommendations incorporating the first and second addenda. WHO.

Yan, H., Dai, J., Yang, Z., Yang, H., & Cheng, R. (2011). Enhanced and selective adsorption of copper(II) ions on surface carboxymethylated chitosan hydrogel beads. *Chemical Engineering Journal*, *174*(2-3), 586–594. doi:10.1016/j.cej.2011.09.064

Zeng, S., Jin, X., Rajarathnam, D., & Chen, Z. (2019). Enhanced degradation of malachite by iron nanoparticles encapsulated in sodium alginate beads. *Journal of Industrial and Engineering Chemistry*, *77*, 238–242. Advance online publication. doi:10.1016/j.jiec.2019.04.043

Zhang, H., Omer, A. H., Hu, Z., Yang, L., Ji, C., & Ouyang, X. (2019). Fabrication of magnetic bentonite/carboxymethyl chitosan/sodium alginate hydrogel beads for Cu (II) adsorption. *International Journal of Biological Macromolecules*, *135*, 490–500. doi:10.1016/j.ijbiomac.2019.05.185 PMID:31145956

Zhong, X., Wang, Y. X., & Wang, S. C. (1996). Pressure dependence of the volume phase-transition of temperature-sensitive gels. *Chemical Engineering Science*, *51*(12), 3235–3239. doi:10.1016/0009-2509(95)00344-4

Zhou, J., Hao, B., Wang, L., Ma, J., & Cheng, W. (2017). Preparation and characterization of nano-TiO_2/chitosan/poly(N-isopropylacrylamide) composite hydrogel and its application for removal of ionic dyes. *Separation and Purification Technology*, *176*, 193–199. doi:10.1016/j.seppur.2016.11.069

Zhu, H. Y., Fu, Y. Q., Jiang, R., Yao, J., Xiao, L., & Zeng, G. M. (2012). Novel magnetic chitosan/poly (vinyl alcohol) hydrogel beads: Preparation, characterization and application for adsorption of dye from aqueous solution. *Bioresource Technology*, *105*, 24–30. doi:10.1016/j.biortech.2011.11.057 PMID:22189074

Chapter 5
Nano(bio)sensors in Detection of Micropollutants

Sameer Quazi
https://orcid.org/0000-0002-1258-4088
GenLab Biosolutions Private Limited, Bangalore, India & Department of Biomedical Sciences, Anglia Ruskin University, UK & School of Health Sciences, The University of Manchester, UK & SCAMT Institute, ITMO University, St. Petersburg, Russia

Javed Ahmad Malik
Guru Ghasidas University, India

Aman Prakash
Jacob Institute of Biotechnology and Bioengineering, Sam Higginbottoms University of Agriculture, India

Pragalbh Tiwari
Jacob Institute of Biotechnology and Bioengineering, Sam Higginbottoms University of Agriculture, India

ABSTRACT

The introduction of nanomaterials in biosensing technology has widened the applications of nano biosensors in several fields as in the precise detection of micropollutants. A huge burden has been driven out by the micropollutants in an environment that has resulted in an alarming situation for scientists to discover innovative methods for the precise or accurate detection of micropollutants that provides sensitivity and specificity at the same time. Apart from conventionally available methods, nano biosensors provide real-time analysis and high-frequency monitoring of pollutants without much sample preparation. Nano biosensors can detect and manipulate atoms or molecules by use of nanodevices resulting in the development of biosensor devices which in turn interact with small molecules named micropollutants. This chapter aims to glance over the advancement in the fields of nano biosensors-based functionalized nanoparticles, nanotubes, or nanowires and their interaction with biorecognition materials such as enzymes, antibodies, or aptamers for the detection of aptamers.

DOI: 10.4018/978-1-6684-5533-3.ch005

INTRODUCTION

Nanotechnology is now an integral part of almost every field including health, medicine, electronics food, communication, etc. basically nanotechnology has been defined as the production of functional materials or devices in the range of 1-100nm (Parolo and Merkoçi, 2013). The involvement of nanotechnology is increasing day by day especially the use of Nanofibers, Nano balls, and nanoparticles in the development of Nanosensors. This includes the development of both biosensors and chemical sensors (Christopher et al., 2020) A huge collection of nanomaterials having different morphology and properties are available nowadays. The most important feature of nanomaterials is their size which has revolutionized fields of science and technology at the same time (Pumera et al., 2007). Particularly they can tailor size hence providing the excellent ability for designing novel sensing systems and also the performance of certain chemical and biological assays at the same time. The particles which are smaller than their characteristic length often display specific chemistry and physics that lead to new properties being dependent on the size of those particles (Xu,et al., 2020). Surface to volume ratio increases considerably with a decrease in size at the same time and the surface resonance phenomenon predominates in the meanwhile. The reduction of sensing parts is important for miniaturizing devices at the nanoscale (Solaimuthu et al., 2020). Nanotechnology deals with such a new phenomenon dealing with the creation of biosensors eventually. Sensitivity can also be increased for obtaining better conduction properties providing properties such as small amounts of samples that can be analyzed directly, detections without the use of labels, and elimination of labels thus, in turn, reducing the use of chemicals and waste products (Nair & Alam, 2006). The transduction mechanism is significant in the Nanosensors as they have the ability to conversion of bio responses of the interaction of analytes to a reproducible manner (Chiu & Huang, 2009). Specifically, they convert biochemical reaction energy to electrical form. The nanoparticles are found to be excellent material in this sense as they possess a high surface-to-volume ratio that allows proper usage of the surface more functionally. Moreover, electrochemical and optical properties are important in terms of biosensors (Quazi et al., 2021). Nanomaterials have also updated the field of molecular biology providing opportunities for the manipulation of atoms and molecules and the monitoring of cellular structures with high precision and sensitivity.

For understanding the technology behind Nano biosensors one must be able to understand the idea of biosensors. Nano biosensors are the modification of biosensors that employ the use of nanomaterials in such an analytical device (Hammond et al., 2016). Nanomaterials employ the use of a variety of nanomaterials that help in the detection of toxic contaminants in food as well as soil. This chapter aims to provide an overview of the Nano biosensors and their applications in environmental analysis involving mechanisms used by them and the detection of micropollutants at a very early stage providing solutions to environmental pollution issues.

COMPONENTS OF NANO BIOSENSORS

Each of the biosensors is further classified into Nanosensors based on their properties such as biological, chemical, and surgical characteristics. So Nanosensors are Nanoscale devices with an overall dimension of 10 nm and few angstrom mass i.e. 10^{-18} g approximately (Elahi et al., 2019). They are designed in such a way that they can provide the properties like that of proteins, deoxyribonucleic acid DNA, and membranes to detect changes in samples such as soil or food by different sorts of mechanisms (Rai et al.,

2012). A biosensor is usually defined as a device that is designed for estimating biological interactions in material and thus converting it into readable form via transduction and electrochemical phenomenon interpretation (Mishra et al., 2022). Three major components of biosensors are given in figure 01. Based on their mode of operation, these are divided into three main parts such as bio receptors, transducer, and detector. The major function is to sense biologically specific materials for example immunological materials. It is usually done by incorporating biological sensitive material taking part in baroreceptors. An antibody is screened by using an antibody and vice versa (Chen et al., 2014). Hence receptors can be the component of biosensor that acts as a template for the material to be detected. The second component is the transducer that converts the interaction of biological components into electrical form (Touhami A, 2014).. As the word indicate showing trans means change and ducer indicates energy as it results in the transformation of one form of energy into another. The first is biochemical that is resulted from the coupling ofbioanalyteandreceptors while the second is electrical in nature being mediated with help of a transducer (Solaimuthu et al., 2020). The last component is a detectors system that receives electrical signals from the transducer component and converts them into a specific response that can be read easily. In addition to all of this, the most important thing in terms of Nano biosensors is the immobilization scheme that immobilizes baroreceptor so that its reaction with bioanalyte becomes more efficient and sensitive (Otles and Yalcin, 2012). This process makes the overall process very cheap and is affected by the parameters e.g. pH, any contaminant, and variation in physiochemical characteristics (Soy et al., 2022).

Figure 1. Overall three major components of biosensors include an analyte which gets attached to the receptor or sensory layer followed by a transducer that amplifies the signal and converts the signals into the electrically readable form (Perumal and Hashim, 2014)

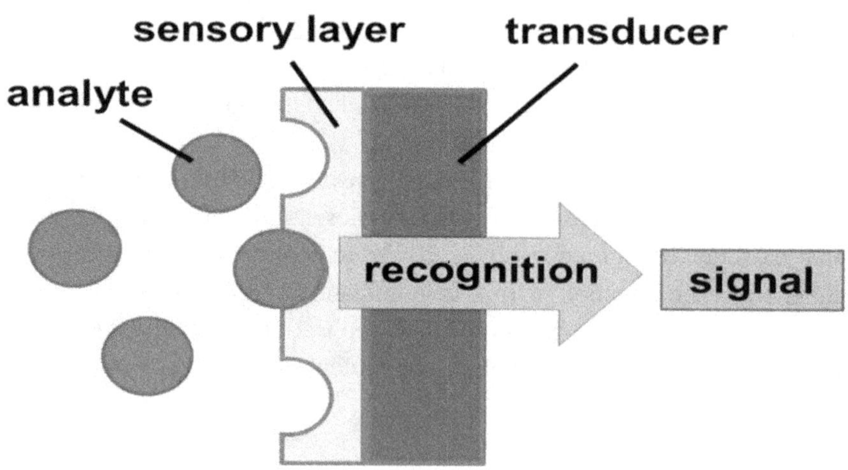

RECOGNITION ELEMENTS/ BIO ANALYTES FOR NANOBIOSENORS

Selectivity and sensitivity are two major paradigms in designing nanobiosenors and for this purpose a number of elements have been proposed that may include antibodies, aptamers and other functional proteins. The most widely used agents are as follows:

Antibodies

Antibody is the protein produced by human immune system in response to the foreign agents. They are very specific following binding to a single antigen and thus aid in the selection of specific microorganism along with its labelling generating a specific immune response. Three types of antibodies are used as a recognition element or analyte detector namely polyclonal, monoclonal and engineered antibody fragments. At the same time there are a number of drawbacks of using antibodies which include expensive or developmental cost, pH sensitivity, and easy transfer from batch to batch and lastly shorter shelf lives. Despite of all these disadvantages antibodies are still proved to be most specific and selective recognition agent. The sensors which use one or more antibodies incorporated in them are termed as immunosensors or immunoassays. The most common descriptor of electrochemical sensor is that it is label free as the variation in the transducer surface can be measured directly based on direct interaction of antigen and antibody.

Aptamers

These are actually flexible short oligonucleotide strands which can either be single stranded DNA or RNA that binds to a specific element. They can be produced naturally or can be produced synthetically for the recognition of toxic and non- immunogenic substances. The production cost is about 30 to 0 times less than that of antibodies. These substances possess lower batch to batch variability, longer shelf life and are thermally stable. De novo synthesis is used for the production of nucleic acid aptamers due to either use of systemic evolution by enrichment SELEX process. It usually starts by the exposing of longer library of oligonucleotides for recognizing targeted sequences. Using other biochemical techniques such as polymerase chain reaction and amplification, sequence with highest binding ability can be found followed by its de novo incorporation in biosensors.

SIGNAL TRANSDUCTION

Three types of signal transduction mechanisms are being followed in the nanobiosenors which include optical, magnetic and electrochemical. Other types of technique are also employed by public such as optical which produces signals in visible spectrum. On the other hand, the example of calorimetric method is general pregnancy test. On the other hand these methods such as electrochemical have higher specifity and can be miniaturized easily. As compared to electrochemical and calorimetric, magnetic has shown lower background noise that makes it ideal for very small concentrations of samples. For this purpose many sensors use magnetic property to pre- concentrate analyte prior to electrochemical and optical transduction mechanism.

Magnetic

This one is preferred method for the use in biological pathway because of lower background signals and magnetic nanoparticles can be collected under the presence of magnetic field irrespective of all other properties of solution. The use of magnetic nanoparticles for purification, separation and concentration of analyte of interest so that it can be easily detected in a detection zone is termed as magnetic transduction. However another method electrochemical methods have also been suggested a hence the magnetic method can also be miniaturized. Magnetic switches that involve super magnetic iron oxide nanoparticles that is found to be purest form of nanoparticles detected so far. The mechanism underlying is the accumulation or clustering together of individual nanomegnatic probes into larger probes followed by interaction with probes. Binding of analyte results in the formation of clusters and change in phase of water proton present around. Magnetic relaxation has been successfully used for detection of nucleic acids such as DNA, RNA and other protein molecules.

Electrochemical

This phenomenon is as a result of change in current or potential providing string interaction between analyte and electrode. A multi steps methods have been used for observing these changes as for example chronoamperometry, spectroscopy and transistor based methods. These Nano sensors also face modifications such as that of electrode with nano carbon being functionalized with recognition elements such as antibodies etc. the direct contact between electrode and recognition elements provide larger amplification of signals as compared to electrochemical sensor. In addition to other electrode properties, size and morphology has deeper effect on function of sensors. Improved detection can be obtained because of much lower hindrance and least steric hindrance phenomenon.

Optical

Optical mechanism is based on the interaction between electromagnetic radiations and that of analyte. Thus this provides emission or absorption of under observance by UV radiations and infrared light. Two methods employed in this process are surface plasmon resonance and fluorescence. The fluorescence one is based on emission of fluorophore when it returns back to the normal condition after excitation to a specific stage. The process involve quantum dots, silicon or other polymer nanoparticles as they are more robust and give accurate results than that of traditional one. The change in design is followed by change in pattern of signals of florescence and upon interaction with analyte its either turns on or turn off. Quenching sometimes also result in change in confirmation in the sensors or interaction between analyte and particles. Surface plasmon resonance is optical properties transduction mechanism relying on spectroscopy. It is based on antiparticle distance. It can also be used in conjugation with other types or secondary spectroscopy techniques surface enhanced fluorescence for enhancing ramen properties.

CONCEPT MERGING OF NANOTECHNOLOGY AND BIOSENSORS

The triggering question is why nanomaterials are used in making biosensors as they are a unique gift of nanotechnology having a dimension between 1 to 100 nanometers. And this size makes them special as most of their atoms are located on the surface possessing all important physiochemical properties better than those available in bulk form. Along with it, the integrated devices of nanomaterial along with electrical systems are also involved in the electrical transduction phenomenon (Lowry et al., 2019). There is some nanomaterialbeing explored for their mechanical as well as electrical properties so that they can be easily employed in biosensing or transduction mechanism. Some of them are nanotubes, Nanorods, nanowires, and nanofilms made up of non-crystalline matter. Among all of them, the use of nanoparticles has been described as the most potent and useful to date. These nanosensors have widened roads in biosensor technology that is basically due to Nano technological implications of matter. A various number of bio-sensingdevices employing nanoparticles have been explored in many research studies. Such as they can be employed for enzymatic detection of glucose and use of quantum dots as a fluorescent agent in order to detect binding and lastly use of bio conjugated nanomaterials for applications such as immunolabelling and immunosensing (Elmer and White, 2008). These devices are also employed for enhancing the applications of microscopic detection at the same time. Moreover, these metal-based nanoparticles prove to be very excellent for electric as well as optical properties and are thus involved in nucleic acids detection based on optoelectronic characters (Chauhan et al., 2019). The nanomaterials being improvised in the sensing mechanism in biosensor technology are given in table 1.

Table 1. The nanomaterial is used most commonly in the Nano sensing technology along with specific functions in sensing technology

Nano biosensor	Function
Nanoparticles	Excellent loading of bioanalyte, help in immobilization and also have good catalysis properties
Quantum dots	Better Fluorescence properties, quantum confinement of charges, affordable size, and catalytic characteristics
Nanowires	Increased versatility,
Nanorods	Very efficient plasma material coupling size and sensing properties for regulation of energy can generate certain field responses.
Nanotubes	Specific functional ability, higher enzyme loading capacity, higher aspect ratio, and better electric transducer

Above all iron and iron oxide nanoparticles are being used in magnetism events and other interactions such as magnetic resonance imaging MRI. These particles can further be coupled with fluorescent molecules or with magnetic particles to give magnetism effects (Quazi et al., 2022). Similarly, zinc and zinc oxide nanoparticles can be used in a similar manner for sensing biochemical reactions in a very precise manner. More specifically it can be used for cholesterol and other metabolites. Continuing on the same path carbon biosensors can be used for rapid interaction between analytes and baroreceptors. The carbon Nanorods have been extensively used for detecting glucose and insulin (Pirzada and Altintas, 2019).

OPTIMIZATION OF NANOMATERIAL FOR BIOSENSING TECHNOLOGY

There are a number of factors that are to be considered while using a specific type of nanomaterial in particular biosensor technology. These factors are major components of both physical and chemical properties giving specific responses (Syu et al., 2018). The very first thing to consider is the design of nanomaterial i.e. nanofabrication which targets two main operational components i.e. manufacturing and design of Nano adhesive surfaces via integrated circuits and micromachining (Tian et al., 2021). The technique thus used for biosensingtechnology uses a combination of four processes namely photolithography, thin film etching/growth, surface etching strategies, and chemical bonding parameters. Lithographic techniques have put forward the light ion the development of Nano electrodes and have enhanced the accuracy of Nano sensing technology (Morales-Narváez and Merkoçi, 2019). A glucose biosensor that uses glucose oxidase uses the same innovation. The use of platinum nanoparticles over carbon nanotubes has enhanced the immobilization of the enzyme system required for analytes detection (Chung et al., 2021). This system is so efficient that it can detect glucose concentration from sources other than that blood. In the same way, a number of coupled biosensors have been developed as a result of deep research which includes the coating of thin film over sensing surfaces thus enabling faster detection of analytes system (Xu and Geng, 2021).

Another factor while designing nano biosensors is the optimization of optical properties. The surface resonance phenomenon is of prime importance in this scenario providing sharp optical properties using incident light (Lichtenberg et al., 2019). Nanoparticles are specialized for this property due to their photonic character and capability to use as a fluorophore (Quazi et al., 2021). This involves total internal reflection that takes place for the angle of incidence reaching a surface at a critical value or angle. In terms of nanoparticles, the phenomenon is termed localized surface plasmon resonance and is dependent on the refractiveindex of the medium governing light through the medium (Aman et al., 2020). For this purpose, the Nano biosensor is aimed to detect the minutest change in the interacting phenomenon in Nano biosensors in comparison to biosensors alone. Hence biosensors are needed to be optimized for better performance being implemented for biosensing purposes. Nano-structured semiconductors can be employed efficiently for the detection of the desired goal or neurological responses by coupling sensing material with a biomolecule. This can also be coupled with the arrangement of peptides with a wide range of nonmaterial and this self-assembly saves a lot of time in current technologies or methods. This can also rapidly detect the stimulus for example DNA segment or nucleotide sequences paired up with RNA or protein. A key concept in the shaping of nanomaterials in biosensors involves tuning their surface by micromachining procedures. Other factors such as aspect ratio functionalized with other materials and compatibility issues to materials are critical for the application of nanomaterials in biosensing technology.

Types of Biosensors

The classification of biosensors is a very diverse discipline and is dependent on the type of nanomaterial used in bio-sensing technology. Theclassification in addition is not as simple as in the case of biosensors (Su et al., 2019). As in the biosensors we classify based on two criteria i.e. material to be utilized and signal transduction mechanism employed. For example, if we are screening any antibody or enzymes through this technology we verify the term enzyme or antigen biosensor for them based on the nature of the analytes (Magar et al., 2021). Based on sensing mechanisms, the types are electrochemical, calorimetric, optical, and others. All of these depend on the transduction mechanism being followed and include a number of sensors involved in it(George et al., 2018). If we consider the area of nan biosensors, the criteria for classification are the nature of nanomaterials being involved in the improvement of sensing mechanisms (Quazi et al., 2021). For example, nanoparticle-based Nano sensors include all the metallic Nanosensors that produce enhanced biochemical signals. Similarly, nano biosensors are also called nanotube-based sensors if they involve carbon nanotube for enhancing reaction specificity using nanowires as a charge carrier termed nanowire biosensors (Quazi et al., 2021). In the same way, there are quantum, dots-based Nano sensors using quantum dots as a contrasting agent for improved optical responses (Dai and Liu., 2019). Based on practical usage the Nano biosensors are classified as

1. Magnetic biosensors
2. Acoustic Wave Biosensors
3. Electrochemical biosensors
4. Nanotube based biosensors
5. Nanowire based biosensors
6. Bio nanomaterial-based biosensors
7. Quantum dots-based biosensors

Magnetic Biosensors

Magnetic biosensors include specially designed magnetic nanoparticles and these are the most potent nanomaterial being used alone or in the combined form. These nanobiosensors are very special for use in biomedical applications which provide a huge variety for analytical applications. The reason behind this is that the screening compounds in iron along with other transition metals have different characteristics (Magar et al., 2021). With the incorporation of nanomaterial devices biosensing devices have become more specific and sensitive, in addition to it alloys of metals with iron and other metals are highly versatile regarding their magnetic behaviors (George et al., 2018). A very common material employing magnetic bioassay has been discovered to be used for the isolation of magnetically labeled targets using a magnetometer. For this purpose special devices are used named superconducting quantum interference devices (SQUID) for early detection of biological targets using the paramagnetic nature of nanoparticles (Dai and Liu, 2019). These are typically used for screening antigens from a specific mixture by the use of antibody-bound nanoparticles. These use super magnetic behavior of magnetic material observed in Nanoscale particles.

Acoustic Wave Biosensors

They have been used for enhancing biosensor responses thus increasing the overall precision of biodetection. There can be a number of stimuli in these effects. The mass-based variant of these sensors involves conjugation of modified sol particles that bind themselves on the surface of the electrode complexed with analytes surface so that the antibody molecules are immobilized over the electrode surface. The increased mass of bound sol particles results in a change in the vibrational frequency of the quartz-based sensing platform providing the basis for detection. The given or preferred diameter of antibody or analytes is between 5 to 100nm and preferred particles are gold, platinum, cadmium, titanium, etc.

Electrochemical Biosensors

This is used to analyze biochemical reactions, especially in terms of electrical properties, and is based on metallic nanoparticles. The reaction between biomolecules can be efficiently carried out by metallic nanoparticles that help in the immobilization of one of the reactants. This makes the reaction very specific and eliminates the chances of any available undesired products. In this reference colloidal gold nanoparticles have been used to enhance the immobilization of DNA on biosensors that increased the efficiency of biosensors via lowering the detection limit. The biosensors have been developed using enzymes conjugated with gold nanoparticles for the detection of glucose, xanthine, and hydrogen peroxide. Xu et al. have described an enzyme-based system consisting of horse reddish peroxidase immobilized on gold electrodes of carbon nanoparticles thus producing a faster response, improved sensitivity, and earlier detection as compared to those without nanoparticles. Similarly, Nano-sized semiconductor crystals are used to improve the efficiency of chemical reactions that are photochemical in nature which can be tagged to biological entities like enzymes and other precursors for designing novel photoelectrochemical systems. Similarly, Currie et al have also used immobilized CDS Nanocrystalline structures utilizing the approach of self-assembly for the preparation of an enzymatic detection system based on immobilized formaldehyde dehydrogenase onto the gold electrodes with the purpose to carry out catalytic oxidation of formaldehyde.

Nanotube Based Biosensors

Nanotubes are the most common nanomaterials employed in the world of material science and they have attracted attention because of their extraordinary properties most vital of them are electrical conductivity along with other geometric features and other physiochemical characteristics. Moreover, they have higher aspect ratios making them much more functionalized materials. Both singles well as multiwall nanotubes are of great interest regarding this aspect and have been used in designing biosensors for efficient parameters. The most important bio-sensing device which has been coming forward is the glucose biosensor which uses nanotubes as an immobilizing agent for enzyme glucose oxidase i.e. enzyme used for estimation of glucose from various body fluids. Usually, the sensors use an assembly of nanotubes for estimation of the glucose even from body fluids in a very small amount i.e. saliva or tears (Sobhan et al.,2018). In one such arrangement, single-walled nanotubes have been efficiently used for efficient enzymatic detection and the results have depicted an increase in enzymatic activity at the same time. The higher performance of Nano biosensors is not due to higher enzyme loading capacity but also due to better conductivity at the same time. because of their structural flexibilities, these nanotubes have been used in electrical sensing technology because of their smooth electron transfer property. Improvement in catalytic biosensors has been explored in a number of studies but one study has exhibited better oxidoreductase properties in both glucose oxidase and flavin adenine precursor bound to their substrate more efficiently. The coupling of carbon nanotubes with sensing elements has provided betterconductance controlling their flow characteristics.

Nanowire-Based Biosensors

Nanowires are the cylindrical arrangement of nanotubes having a length of a few millimeters to centimeters lying within nano orange. They are very good one-dimensional materials with excellent electrical properties. The movement of charge in the nanowires is more efficient than that of other materials in bulk. They are very small in number but literature demonstrates very good examples where nanowires have shown improved performance and at the same time detection of biological materials. As a study has reported the use of silicon nanowires doped with boron provides detection of both biological and chemical properties Similarly, semiconductor nanotubes have been used extensively for coupling biosensors to identify their biosensors. Similarly in another study, streptavidin molecules from a mixture have been detected using silicon nanowires coated with biotin. The small size and other capabilities of nanowires have made them better candidates as the sensing materials usually work in very precise dimensions they have been prescribed to use in in vivo applications and also operate in the environment as well. Similarly, a study has used Nanofibers with a diameter of nanometers and have been coated with antibodies to detect toxins. Cullum et al. have synthesized zinc oxide nanoparticles being coated on the gold electrode and have been used for the detection of hydrazine based on certain responses. These sensors have reported greater sensitivity, lower response rate, and detection limit as compared to conventionally available sensors present at the same time. these nanowires are very versatile in the terms of their performance as compared to those of nanotubes because of two things i.e. they allow certain modifications in their structural properties and secondly they give much more functionalized assemblies because of the presence of compatible material on their surface. Although they have several advantages at the same time, their use in sensor technology has met several challenges. The very first is that it is very difficult to incorporate these nanowires into sensing systems to improve their conduction properties. In one of the studies, the Liber group synthesized nanowires by the combination of already available methods. They have developed a one-dimensional structure with over 200 nanowires and have used them for the detection of low-level cancer antigens. Huang et al. have mainly focused on the phenomenon of plasmon surface resonance which can be incorporated into sensing devices to improve sensing abilities. Stern et al. have developed nanowires with a complementary material oxides semiconductor approach and have been very useful in controlling and optimizing synthesis procedures in terms of nanowires for analyzing serum fluids for detecting pathogens and proteins in the crude form.

Bio Nanomaterial-Based Biosensors

When a biomolecule is combined with a suitable transducer element, it produces signals after interaction with certain target molecules. The presence of elements makes the sensor system more specific and effective thus providing an upper edge over conventionally available biosensors (Maduraiveeran, G., 2020). For a number of years, a variety of bio elements have been used in biosensor technology but the most important of them are enzymes, dendrimers, etc. In enzyme-based biosensors, the most important biological element is enzymes which selectively react with its substrate molecule. It is the most commonly used biological element that enables the detection of any analytes in various ways. Enzymatic reactions are followed by the production of certain substances or transducer molecules which are helpful in the detection as well as a correlation of these species to that of substrate. Dendrimers are on the other hand organic molecules which have three dimensions as well as diverse functionality (Quazi, S., 2022). Crooks was the first one who reported the stability and integrity of these metallic nanoparticles. The process of development of microelectrodes for oxygen and hydrogen peroxide measurement is based on the stabilization of silicon substrate using microfabrication technology. The photoresist is the phenomenon for fixation of the biomolecules or enzymes on the surface of these Nano biosensors providing greater power of detection having intrinsic properties. These intrinsic properties can be easily explained at the organization level when the molecular arrangement is obtained at a solid conductor substrate. At the same time, the detection limit has also been increased by the use of compatible materials such as natural polymers (Singh et al., 2021).

Pesticide Nano biosensors have also been reported on the liposomes for investigation of organophosphate pesticides. Following this system, prions are usually embedded in lipid membrane allowing for free substrate or lipid transportation into liposomes. Using this prion concentration down to 10^{-10}moll per liter can be detected efficiently by inhibition of fluorescence biosensors (Patolsky et al., 2006).

Figure 2. Depiction of the Nanobiosensors classification along with their applications in optical, mechanical and electrochemical phenomenon

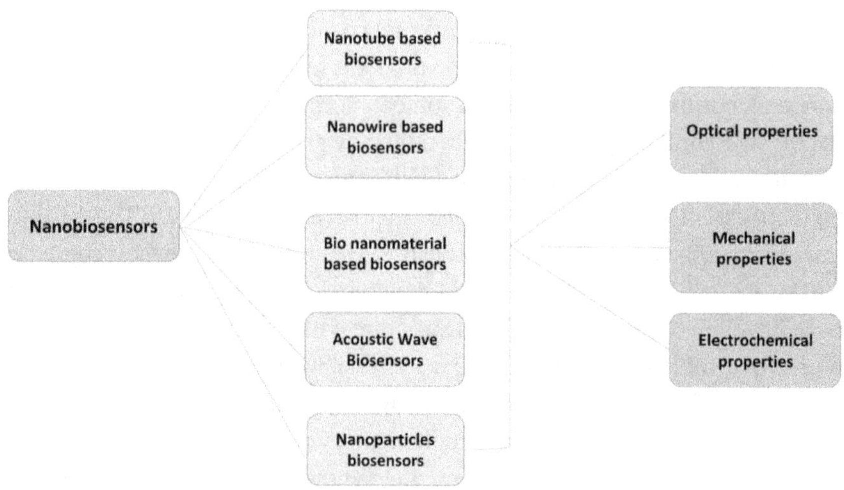

Quantum Dots-Based Biosensors

These are usually described as semiconductor type Nano crystals formed from a combination of metals which can be either cadmium or zinc combined with selenium, sulfur or tellurium, etc. these are further coated by the combination or alloy of two or more metals that in turn creates a shell, core providing protection as well as highly tunable properties.There are certain characteristics or properties of quantum dots that make them suitable for use in nano biosensor technology, for example, they have narrower energy emission bands but they are absorbance bands at the same time that in turn make them excellent transducer molecule (Zhu et al., 2015). Changes in the size shape and composition of the quantum dots can lead to a drastic change in their wavelength pattern and hence these quantum dots have become an ideal choice for the detection of several analytes. They are different with respect to their size, shape, and composition and thus they can be excited by a single energy source because of the property that they exhibit absorption spectrum. Moreover, as we know that about 70% of the earth's surface is covered by water that is consumed by humans, animals, and plants which are striving for their survival. However, most of the water is insufficiently clean and the major source of them is toxic heavy metals which have to be removed as early as possible (Gao et al., 2015). Apart from conventional techniques available, graphene-based quantum dots have been designed as highly powerful nanomaterials which in turn enhance the optical detection of metal ions in water at every level.

MICRO POLLUTANTS

These are contaminants of organic materials whose persistence cannot be eliminated by the traditional wastewater treatment procedures in other words they are not biodegradable hurting organisms and the environment as well (Akgönüllü et al., 2021). Hey, are present in the products of daily usage such as drugs, insecticides, and pesticides at home and other industries at the same time. when they are released into wastewater they can cause long-term hazards in short because they may bioaccumulate and can cause further products in water. They enter the natural environment via municipal waste, agricultural or industrial activities, transport or machinery, and lastly direct fallout (Huang et al., 2021; Malik et al., 2022).

The determination of the existence and type of micropollutants is of prime importance as they cause serious consequences to human health as well as other organisms as reported in the literature. Hence there is a need for the development of Nanosensors either chemical or biological for determining micro pollutants because they provide the best alternative to conventionally available methods for detecting and treating micropollutants at every level for example spectroscopy or chromatography (Adetunji et al., 2021). In this regard, considerable efforts have been made for Nano biosensors which are economical, reliable, and robust at the same time. Nano biosensors have provided considerable advantages such as increased sensitivity or specificity thus providing greater efficiency for the detection of micropollutants (Schwarzenbach et al., 2006). Potential applications of Nanosensors are found in wastewater treatment and thus classified into metals, metalloids, radioactive and organic micropollutants, and pharmaceutical products (Luo et al., 2014).

TYPES OF MICROPOLLUTANTS

Organic Micropollutants

Most organic micropollutants include hydrocarbon, solvents, detergents, cosmetics, etc. One common example is pesticides being based on the mechanism of action of the enzyme acetyl cholinesterase (AChE). The organophosphate and carbonate moiety gets bind to serine moiety at the active site of an enzyme thus in turn preventing deacetylation of acetylcholine. The problem with this procedure is that the other compounds such as enzymes or detergents also inhibit the enzyme. Also, the enzyme is unstable outside their natural environment and each pesticide has different affinities for the enzymes. In these types of biosensors gold nanoparticles and multiwalled carbon nanotubes were used for promoting electron transfer and catalyzing the oxidation of thiocholine for detecting monocrotophos (Virkutyte and Varma, 2010). The flow-based system used glassy carbon nanotubes modified with gold electrodes and MWCNTs. These nanotubes contain chitosan for enhancing the stability and immobilization of enzymes. Similarly, polyaniline (PANI) was also used as a biosensor for the atmosphere. For this purpose Nano biosensor was constructed for its dual nature i.e. for immobilization as well as electrocatalytic activity for biosensing purposes. It shows that the enzyme is further catalyzed by AChE forming thiocholine and acetic acid which is also electro-reactive and also gets oxidized in the reaction.

Another universally used pesticide 2,4-Dinitrophenol (DNP) has also been investigated using nano biosensor technology. It is based on the hybridization of latex microspheres with that gold nanoparticle giving chromogenic effects. A toxin analog named DNP bovine serum albumin was attached to the gold nanoparticle which help in the binding of DNP to that of antibody DNP on the microsphere surface giving out pinkish red color. Another model toxin was also identified in the same way referred to as DNP- glycine where the DNP binds with the anti-DNP antibody on the surface of a gold nanoparticle in presence of an analog (Yang et al., 2004). The displacement of gold nanoparticles from the microsphere surface has resulted in a visible change in color from pinkish red to white instantly.Dichlorodiphenyltrichloroethane (DDT) is one of the most widely used pesticides containing chlorine in its organic form being aimed to be detected by the use of Nano biosensors. The nano biosensor was developed by competitive immunoassay or ELISA based on gold nanoparticles in dipstick format. The dipstick containing nanoparticles with a specific size range was designed and coupled with an anti-DDT antibody to detect at the Nanogram level. Immuno complex was resulted by the free DDT with an anti-DDT antibody which later on forms a complex with DDT-BSA conjugate (Ten Hulscher and Cornelissen, 2005). The results were identified by the change in color to red and the intensity of the color is inversely proportional to DDT concentration with maximum intensity at zero concentration (Stuer-Lauridsen, F., 2005).

Another phenomenon that can be used for the detection of pesticides is the plasmon surface resonance which uses molecular properties such as plastic antibodies. Molecular imprinting was done by attachment of imprinted model pesticide on nanoparticle on the surface of a gold chip. The results declared high sensitivity and selectivity for atrazine a model pesticide (Alsbaiee et al., 2016). For having an idea of the selectivity of pesticides, the competitive adsorption of other pesticides was also done but the results showed that atrazine had higher selectivity than that of other pesticides at the same time. Polyaromatic hydrocarbons can also be detected using nanosensors (Khanzada et al., 2000). These nanosensors have been developed using single-walled carbon nanotubes (Quazi, S., 2021). There are a number of examples where other nanomaterials such as graphene, and quantum dots have been utilized for the fluorescence detection of hexanes containing benzene rings. This optical Nano sensor has shown higher sensitivity, lower detection limit, high response time, and wide linear range based on quantum dots made up of cadmium, selenium, zinc, etc. the detection limit as low as 5nm was achieved using this strategy.

Neonicotinoids/ Insecticides

Another class of neuroactive insecticides was being introduced in 1980 1nd is the major class of insecticide ion use now a days. But with increase in use the hazardous impacts of them on human health has also been increased. The nanosensors designed for insecticides are aimed to detect acetamiprid along with the aptamer being regarded as most important recognition element in the literature by previous studies.Weerathunge et al. has employed aptamer functionality in order to create nanosenor as peroxidase like activity of gold nanoparticles. The reporter which is colorless has the ability to convert purplish compound into blue with the process of oxidation for creating of sensors and observance by UV light. The introduction of target molecule results in lowering detection limit to about 10 minutes.

Figure 3. The process for schematic quantitative detection of acetamiprid involving combination of aptamer with nanomaterial's and then both optical and electrochemical phenomenon is utilized for detection

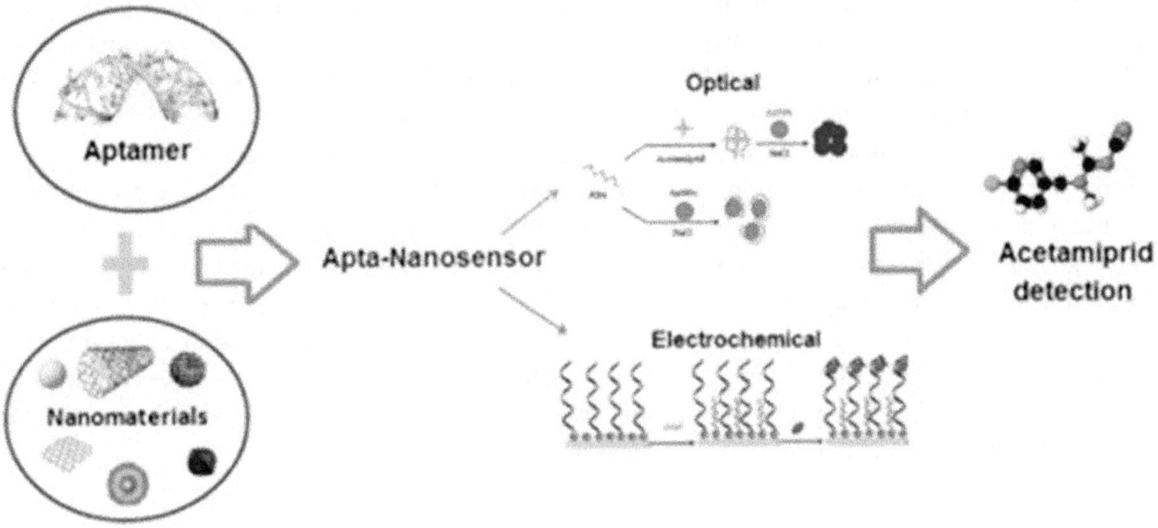

Herbicides

Triazine is another nitrogenous based detector used for investigating atrazine as it is the most versatile used herbicide in United States. A huge collection of the labeled and non- labelled based biosensors have been available for detecting atrazine.Liu et al has designed competitive immunoassay on basis of electrochemical phenomenon. The electrode is gold electrode coated with gold nanoparticles and at the same time functionalized with monoclonal antibodies anti- atrazine. Voltammetry measurement were then used for direct measurements in electrode surfaces as a result of interaction between antigen and antibody. The sensors provided the detection limit up to 74Pm.

Another label free method was also described in literature where a gold or bacterial nanocellulose particle was prepared by in situ oxidation of gold salt in the presence of bacteria cellulose. Atrazine was attached on basis of PH by lowering of PH below Pka value less than 1.7 and was further confirmed by SER technique in gold/ bacterial nanoparticles. The group was able to obtain the detection limit to 11nm with EPA concentration to about 4ug/L but the magnitude is greater than that of labelled based detection.

Toxins/ Pathogens

Since 1940, Cholera was the pathogen which had been spread successfully due to intake of contaminated water as a water borne disease causing organism in research. On the other hand, World health organization has recognized about 8 viruses, 7 protozoans and 2 plathihelminths. The detection of pathogen usually focuses on identification of whole analyte or in other words detection of epitopes on cell membrane surface followed by the detection of the genetic material and lastly product or toxin detection successfully. WHO has also published the list on the basis of antimicrobial gene resistance basis pathogens. Cholera toxin has been removed successfully from the United States but it has been costing a serious health problem throughout the world because of 4 million cases every year. And about 2 lac deaths. The disease is caused by a bacterium which enters through contaminated food thus residing mucosa of intestine and start to secrete toxin CT there. The nanosenors are aimed to detect the uptake of bacterium or in other words CT toxin and more specifically subclass of toxin CT-B that is thought to be toxicogenic compared to other subtypes. The labeled based detection of toxin can be performed via antibodies or with the help of galactose having stronger affinity for CT. This method provides detection limit to about 280Pm.Similarly, P. aeruginosa based aptamer was discovered in 2011 and that has been used in a number of sensors or nanosenors successfully. It highlights involvement of the two sensors that involve optical transduction mechanism. Hu et al., successfully fabricated localized surface plasmon resonance phenomenon LSPR chips (Hu et al., 2017). On the other hand, Yoo et all used a three step technique for the production of Nano biosensors. In the first step the gold was deposited on glass slide containing silica nanoparticles which is further added by addition of another gold layer. While Hu et al., used simple lithographic technique for formation. These two groups also used different materials for functionalization of sensors where one group attaching aptamer to that of senor directly with help of thiol bond. Polyethylene glycol PEG has been used as a spacer element to steric hindrance with the objective to reduce steric hindrance and also lower level of detection.

Other Pharmaceuticals

Pharmaceuticals are any type of drugs that have a higher rate of persistence, are toxic, and have adverse effects on the animal as well as the environment for example chemotherapy drug molecules, analgesics or anti-depressants, etc. The antibiotics and phenolics used for bacterial or viral infections for example ampicillin are a good example that has also been used in agriculture and animal growth. (Mlaik et al., 2020; Hjam et al., 2020). A nanoparticle using gold and other calorimetric detection methods has been developed successfully. The aptameric Nano sensor developed in this way had higher sensitivity and selectivity than that of ampicillin. For example, AMP17 ssDNA aptamer was used for absorption on the surface of gold nanoparticles (Xiong et al., 2018). Because of the interaction of aptamer and ampicillin, the ampicillin causes the release of gold nanoparticles, and the released particle was aggregated with salt to provide a color change from red to purple. The technology behind this is that as more free gold nanoparticles are present in the structure, more ampicillin is present thus providing the best detection strategy for ampicillin. Volumetric-based nanosensors were synthesized by amoxicillin nanosensors using gold nanoparticles with ethylenediamine (en) as a cross-linker (Lee et al., 2020). The increased electron flow property along with the catalytic property of gold nanoparticles is responsible for the electrocatalytic performance of the screen-printed electrode abbreviated as SPE. The results detected in this way declared amoxicillin as a micropollutant in real samples as well.

Metals and Metalloids

The trend has recently shifted towards the use of nanomaterial in biosensor technology for the detection of metals and their alloys or metalloids most importantly heavy metals (McGrath and Zhao, 2003). The nanomaterial used for this purpose is multiwall carbon nanotubes, gold or silver nanoparticles, and silica Nanospheres (Quazi S., 2022). Multiwall carbon nanotubes being modified with 2-amino thiophenol have been used to detect lead in wastewater samples by using sensors such as potentiometric sensors (Mogren and Trumble, 2010). These sensors exhibited stronger analytical performance i.e. stronger covalent bonds between the amino group of amino thiophenol and the carbonyl group of MWCNTs for the creation of the stronger solid frame for binding of lead. Researchers have studied the use of multi-walled carbon nanotubes being functionalized with triphenylphosphine providing detection on the monolayer surface for binding to heavy metal ions with the detection limit up to sub-molecular levels too. Similarly, the literature has also reported the presence of bismuth or carboxylic acid functionalized multi-walled carbon nanotubes beta cyclodextrin Nafion was discovered for detection of metals like lead, and mercury. Due to the synergistic effects of all these things, these nanomaterials exhibited more advantages as compared to traditional ones providing larger surface area, stability, and more electrical conductivity. These nanosensors have resulted in the detection of lead and cadmium successfully with specific recovery results holding promise for the detection of micropollutants at the nanoscale level.

Gold nanoparticles have also been involved in the detection of mercury as a pollutant in wastewater treatment on the nanoscale technique based on the colorimetric method with the detection limit up to lower than that of nm which was much lower than that determined from data given by World health organization regarding the level of mercury in drinking water[108]. The result was attributed to a change of color from ruby red to purple resulting from the aggregation of gold nanoparticles when no mercury was present at all. If gold nanoparticles were present the color of the solution can be reverted successfully. The electrochemical biosensor of gold is also present for the determination of lead and copper at ppb levels. Another electrochemical sensor derived from a monolayer of chitosan-based graphene oxide carbon electrode for detection of lead using square wave volumetric with the detection limit further less than that of 1 ppb. For the functionalized group acting as a recognition element, urine has also been used as a sensing fluid (Vodyanitskii, Y.N., 2013).. The creatinine and uric acid present in urine not only can selectively bind gold nanoparticles but at the same time absorb mercury ions at the same time. thereby low-cost sensor and further lower detection up to 50nm can be obtained in this case. Zinc and copper can be detected very easily with the specific color change of 20 nm chitosan-coated gold nanoparticles. As chitosan has proved to be a very specific chelating agent for metallic particles causing colloidal instability and aggregation of gold nanoparticles. The phenomenon also causes a rapid change in color due to the concentration of metal. Lead up to 200 u M to 500 u M can be detected following this aggregation and disassociation procedure using DNAzyme gold nanoparticles cleaving in presence of lead and thus resulting in the change of color from blue to red.

Other than gold nanoparticles, silver nanoparticles have also been explored for their optical sensing of lead, chromium, mercury, and other metal analysis in water.4-mercaptobenzoic acid (4-MBA) functionalized silver nanoparticles act as a nanosensor for the detection of copper. The presence of copper ion results in aggregation of 4-MBA and has resulted in a change in color from yellow to purple with the detection of another nanosensor based on citrate ion coated silver nanoparticles was also reported with a detection limit to nanometers. The Nano sensor is a simple and easy method with a much more detection limit of up to 10 nm (Gonzales et al., 2014).

DISRUPTORS FOR ENDOCRINE

A wide variety of substances can be used as endocrine disruptors. The substances which can disrupt endocrine tissues include polychlorinated biphenyls (PCBs), DDT including a number of pesticides. They are very toxic to the environment as well as increase the risk of cancer because of their persistence in the environment for a large time and also they can't be degradable. Sensitive and rapid dioxins detection can be performed by the use of quartz crystal microbalance (Quazi S., 2021). The synthesized polypeptides were modified by cysteine residues or other amino acids such as glycine at both N and C terminals. This results in lowering of detection limit up to 10 nanometers by lowering electrostatic interaction between amino acids (Bredhult et al., 2007). .Other chlorine disruptors are polychlorinated biphenyls (PCBs), which have found their applications, especially in the production of condensers, transformers, and capacitors. These synthetic substances have deleterious effects on the endogenous hormone which in turn promotes cellular malignancies. Conventional methods have been reported for the detection of PCBs including gas chromatography, mass spectrometry, and HPLC (Lu et al., 2020). It can be the best alternative to simple methods available for detection of PCBs by use of immune assay with emission of fluorescence, electrochemical methods and d piezoelectric techniques, and surface resonance technique.

SPR Nano biosensors on gold film modified with cytochrome C have been used to detect PCBs. The changed configuration of PCBs via immobilized cytochrome C was followed by detection with SPR spectroscopy. The nanosensors showed greater sensitivity with a much lower detection limit to about less than 1 ppb. Endo et al. (2005) have also used antibodies being grafted on the substrates such as paper-based SPR which was prepared by dipping paper into gold nanoparticle solution. And the Nano-rods were absorbed efficiently on the surface of filter paper because of the electrostatic force of attar tion in between cellulose which is negatively charged banded with positively charged CTAB-based gold nanoparticles or Nano rods which provides the advantage of easy operation for collection of samples from the solid surface. The chemical was absorbed on the surface by swabbing paper with the surface having chemical on it and their ramen spectrum was also obtained easily (Rice, V.A., 1942)

Inorganic Arsenite

Inorganic arsenites as well as HM ions including cadmium, mercury, and lead have determinable effects on the environment because of their toxicity, ability to persist in a number of different mediums, and deposition to the biological system along with tropics (Ali et al., 2021). When human beings intake these arsenic compounds through the water and contaminated food it results in a number of diseases including various types of cancers and immune diseases (Riedel et al., 2006). Electrochemical-based biosensors are very versatile in the detection of these arsenic and their types along with other anthropogenic pollutants such as that HM ion (Zamfir et al., 2020). They have been preferred over other available techniques based on their specificity, sensitivity, discrimination ability, and cost effectivity. Moreover, their rapid operation capabilities ad responses, no need for pre-treatment, and miniaturized devices make them useful for the use n detection of these toxic metal (Quazi S., 2022).. There are various drawbacks of using bare electrodes during the analysis procedure such as passivation of the electrode and lower electron transfer. To overcome these drawbacks they are coated with certain materials such as nanomaterials such as nanoparticle polymers, metal oxides, nanotubes, etc. among the carbon-based materials are carbon nanotubes, graphene carbon nanotubes and they are preferred because of their environmentally friendly nature, higher surface area to volume ratio, conductivity and faster transfer of electron and hence they have become a point of attraction for energy harvesting and environmental analysis[130].

Figure 4. Biosensing technologies used in nanobiosenors for detection of contaminants or bioanalytes including colorimetric sensing, fluorescence sensing followed by emission of fluorescence, SER sensing based on ramen scattering and lastly LSPR sensing based on wavelength of light

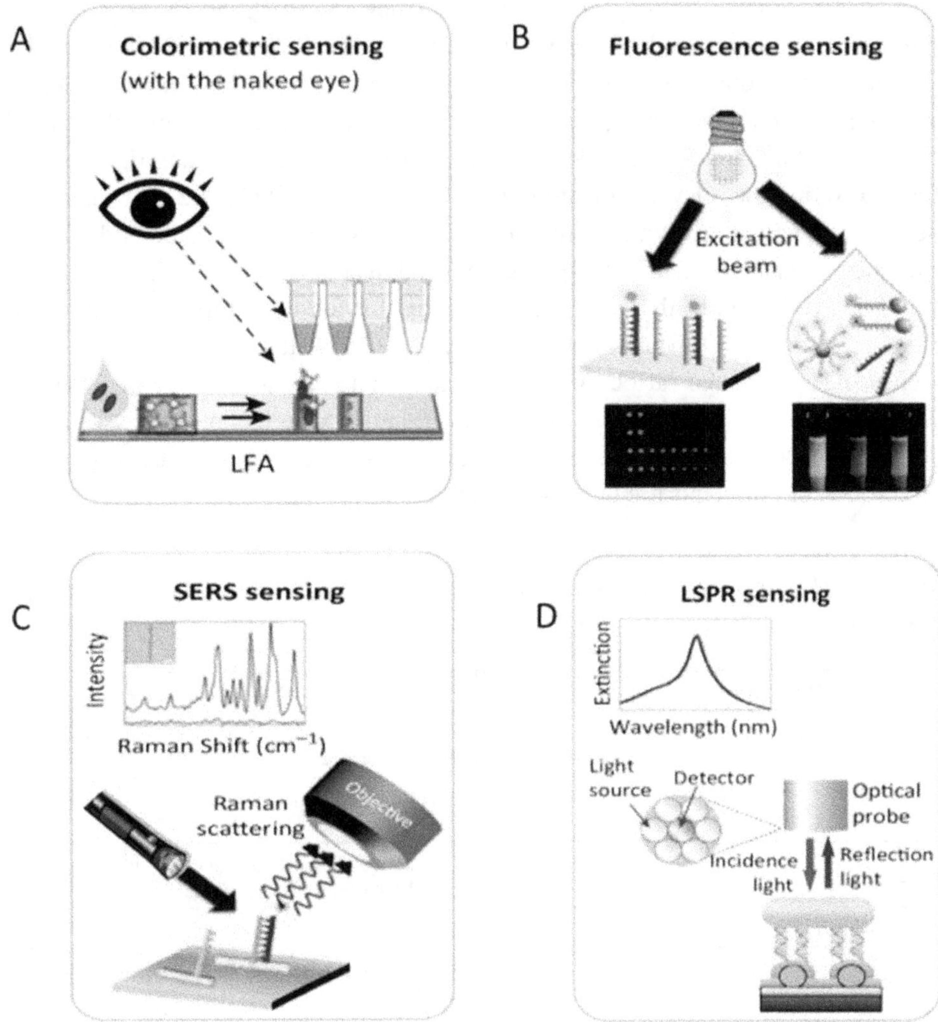

NANOBIOSENSING TECHNOLOGY FOR WASTEWATER EPIDEMIOLOGY

Nanobiotechnology merges the applications of both biotechnology and nanotechnology fields and the research has shifted from the development of molecular diagnostics to that of nanotechnology-based biosensors. Biosensors are mainly designed after deciding on the analytes or biomarker forwhich the sensor is to be designed. The development of biosensors at the basic level involves the use of a specific material or a combination of materials to make nanocomposites bio composites followed by the recognition of elements for binding of target molecules and lastly a signal transduction mechanism[132]. For nanobiotechnology, direct sensing methods such as polymerase chain reaction PCR, sequencing, and CRISPR are of prime importance and are later miniaturized or converted into the paper chip for the detection of analytes (Quazi S. 2022)

A study has demonstrated the use of methicillin-resistant Staphylococcus aureus (MRSA) detection by use of a magneticmicrofluid system with LAMP technology for detection of MRSA DNA. These target molecules interact with recognition molecules for example proteins, aptamers or antibodies, etc-generating detectable signals with help of transducers (Quazi, S., 2022; Malik JA nad Lone R, 2021). Also, the implantation of their nano biosensor requires a whole combination of biological, chemical, and physiological techniques. Immediate and easy detection of the pollutants present in drinking water is of immense importance as they may result in serious hazards following use by animals or humans(Quazi S., 2022). The use of the appropriate tool is of prime importance for the quantificationof biomarkers in the wastewater to find the relevant concentrations(Albuquerque, D.C., 2021). Thus nanotechnology provides the best alternative involving interaction between the target molecule and analytes and thus modifying unique properties such as optical-electrical and magnetic or so on. This can also be used for the detection of analytes alone with the advantages suchas low cost and rapid detectionup to a nano or sub-nano range providing real-time analysis of analytes in wastewater.

The presence and detection of various analytes can help in making decisions and designing a specific policy for the public health sector not only for the propagation of infectious diseases in humans such as SARC –CoV 19 in a community. The targeted biomarkers which can be either organic or inorganic chemicals, microbes, or pollutants are given in the table.

Table 2. The types of biosensors along with their targets presented by WBE including biological, chemical or microbial contaminants

WBE targets	Representative contaminants/ Micropollutants
Heavy metal	Cadmium, chromium, copper, nickel, zinc, lead, mercury, aluminum arsenic
Non-metal ions	Sulphate, chlorine, per- chlorate, nitrate or nitrite and arsenic and phospahatase ion
Pesticide formulation	Diazinon, glyphosate deet, malathion, dursbonacephate, metaldehyte, boric acid and DDT.
Pharmaceutical compounds	
Disruptors of endocrine	Bisphenol A, BIPHENYL, Diethylstilbestrol, doxins, triclosan, polychlorinated biphenyl
Surfactant molecules	Soap, benzene sulfite, alkyl ether phosphate and benzal kaonium
Polycyclic aromatic hydro-carbons (PAHs)	Pyrene, chrysene, benz anthracene, Flurene, Naphthalene, Acenapthene, Perylene
Synthetic based dyes	Eosin, oil red, picric acid, acid fuchsin, light green, aromatics and Azo dye, Fenton
Microbes/ pathogens	Cholera, Fungi, bacteria, protists, viruses and parasitic worm, listeria, Pestis, Salmonella, Brucella
Infectious Virus	Coronavirus, hepatitis A,B, Dengue, Influenza, Chicken pox, Ebola
Nucleic acid material	Deoxyribonucleic acid DNA, Ribonucleic acid RNA
Anti- microbial or antibiotic resistant genes	Methicillin resistant Staphylo coccus aureus (MRSA), Vancomycin resistant genes
Disinfectant substances	Alcohol, formaldehyde, lodophors, hydrogen per oxide, paraacetic acid and gluteraldehyde
Micro- plastics	Various others

Nucleic acid obtained from wastewater can provide biomarkers or analytes that are found to be appealing for WBE that can be later used for checking out the prevalence, spread, and scale of infection in savage water (Gacem et al., 2021). Additionally, ARGs and mobile gene elements (MGEs) also aid in checking out antimicrobial genes. As PCR has been successfully used for surveillance of SARS-CoV-2 in wastewater treatment thus showing a lower rate of variability in 24 hours sample using real-time polymerase chain reaction q RT- PCR. Similarly, for the detection of AMR, certain genes in wastewater nucleic acid have been detected such as tet, sulI, sulII, and blaTEM. LAMP was successfully employed for the detection of human mitochondrial DNA from an untreated wastewater sample in an environment where mt DNA was the main nucleic acid showing the presence of carcinogenic elements in wastewater detecting 40 copies per reaction volume. The results depicted that LAMP-based detection can directly detect a virus in wastewater avoiding viral load at the same time and other extraction steps at the same time. Additionally, next-generation sequencing NGS provides an opportunity for the detection of the whole sequence of WBE targets for the investigation of viral variants in wastewater accurately.Nano pore sequencing technology is an emerging technology enabling real-time analysis of long-read DNA fragments excluding those of 20-kilo bases in length. They use a Nanopore channel on the surface of the membrane immersed in an electrolyte solution thus providing a measurement of electrical current magnitude (Lim et al., 2020)

Recently SER has also gained interest in biosensing applications due to ultra-sensitive detection limit and simple implementation. Continuous development in the development of Nano composites have to lead to the development of SER substrates of a wide range thus they have been accepted widely for biosensing and environmental applications. Various spectroscopic techniques combined with electrochemistry have been used for the detection of DNA. For example, Au nanodot-modified indium tin substrates are used for the detection of hepatitis C and the influenza A virus. But at the same time, SEC remained an evolving field for improved understanding of SEC field and sensing applications as well. A new SEC technique named EC-SERS has been coming forward which not only provides spectral information but electrochemical changes at the same time providing detection, more sensitivity being aimed for detection of analytes via WBE. Moreover, there is a need to standardize the analytical procedures for the detection of analytes to endure yield and other specificities of a given method. Thus giving an easy way to implement these biosensors and their applications in the real world effectively.

CONCLUSION

Recently biosensor technology has been revolutionized by the integration of nanotechnology to develop nano biosensors in a very small size. Nanosensor production for the removal of environmental contaminants has been increasing as stated by this chapter also. The analytes and recognition materials are combined in a successful manner in presence of nanosensors or Nano surface in a creative manner. Nano sensors have employed various nanomaterial's which make the reaction of biological and chemical components react with the target molecules which are then sent to transducer component to convert these signals into detectable forms providing detection of micro pollutants being classified into several classes such as metals, non-metals, metalloids, organic, inorganic and pharmaceutical products. They help in the detection of this micro pollutant which in turn can lead to preventive actions if needed if required. The development of Nanosensors requires the development of technologies for contribution to daily life. The development of nanosenors is with the objective to follow up variation in size, aggregation phenomenon and stability of nanoparticles. These technologies take the form of Nano engineering or biologically active surfaces and necessary tools for their characterization. Lastly, a well-programmed system is needed to involve bioscience scientists, chemists, and engineers for revealing more refined and reliable biosensors in near future. The sensitivity and robustness of the nanosenors is necessary if we are having aim to analyze environment successfully.

REFERENCES

Adetunji, C. O. (2021). *Bionanomaterials for green bionanotechnology. In Bionanomaterials.* IOP Publishing.

Adhikary, R. K. (2021). Risk of Human Illness from Recreational Exposure to Microbial Pathogens in Freshwater Bodies: A Systematic Review. *Exposure and Health*, 1–19.

Akgönüllü, S., Bakhshpour, M., & Denizli, A. (2021). Molecularly imprinted bionanomaterials and their biomedical applications. Bionanomaterials Fundamentals and Biomedical Applications, 9-27.

Albuquerque, D.C. (2021). Quan. *TTEES 5 Sponsors,* 34.

Ali, S. A., Mittal, D., & Kaur, G. (2021). In-situ monitoring of xenobiotics using genetically engineered whole-cell-based microbial biosensors: Recent advances and outlook. *World Journal of Microbiology & Biotechnology*, *37*(5), 1–24. doi:10.100711274-021-03024-3 PMID:33843020

Alsbaiee, A. (2016). Rapid removal of organic micropollutants from water by a porous β-cyclodextrin polymer. *Nature, 529*(7585), 190-194.

Aman, R. A., Mahas, A., & Mahfouz, M. (2020). Nucleic acid detection using CRISPR/Cas biosensing technologies. *ACS Synthetic Biology*, *9*(6), 1226–1233. doi:10.1021/acssynbio.9b00507 PMID:32159950

Bredhult, C., Bäcklin, B. M., & Olovsson, M. (2007). Effects of some endocrine disruptors on the proliferation and viability of human endometrial endothelial cells in vitro. *Reproductive Toxicology (Elmsford, N.Y.)*, *23*(4), 550–559. doi:10.1016/j.reprotox.2007.03.006 PMID:17493787

Chauhan, A., Sillu, D., & Agnihotri, S. (2019). Removal of pharmaceutical contaminants in wastewater using nanomaterials: A comprehensive review. *Current Drug Metabolism*, *20*(6), 483–505. doi:10.2174/1389200220666181127104812 PMID:30479212

Chiu, T. C., & Huang, C. C. (2009). Aptamer-functionalized nano-biosensors. *Sensors (Basel)*, *9*(12), 10356–10388. doi:10.339091210356 PMID:22303178

Christopher, F. C., Kumar, P. S., Christopher, F. J., Joshiba, G. J., & Madhesh, P. (2020). Recent advancements in rapid analysis of pesticides using nano biosensors: A present and future perspective. *Journal of Cleaner Production*, *269*, 122356. doi:10.1016/j.jclepro.2020.122356

Chung, S., Revia, R. A., & Zhang, M. (2021). Graphene quantum dots and their applications in bioimaging, biosensing, and therapy. *Advanced Materials*, *33*(22), 1904362. doi:10.1002/adma.201904362 PMID:31833101

Dai, Y., & Liu, C. C. (2019). Recent advances on electrochemical biosensing strategies toward universal point-of-care systems. *Angewandte Chemie*, *131*(36), 12483–12496. doi:10.1002/ange.201901879 PMID:30990933

Elahi, N., Kamali, M., Baghersad, M. H., & Amini, B. (2019). A fluorescence Nano-biosensors immobilization on Iron (MNPs) and gold (AuNPs) nanoparticles for detection of Shigella spp. *Materials Science and Engineering C*, *105*, 110113. doi:10.1016/j.msec.2019.110113 PMID:31546438

Elmer, W., & White, J. C. (2008). The future of nanotechnology in plant pathology. *Annual Review of Phytopathology*, *56*(1), 111–133. doi:10.1146/annurev-phyto-080417-050108 PMID:30149792

Gacem, M. A., Telli, A., & Khelil, A. O. E. H. (2021). Nanomaterials for detection, degradation, and adsorption of pesticides from water and wastewater. Aquananotechnology, 325-346.

Gao, Z., Wang, L., Su, R., Huang, R., Qi, W., & He, Z. (2015). A carbon dot-based "off–on" fluorescent probe for highly selective and sensitive detection of phytic acid. *Biosensors & Bioelectronics*, *70*, 232–238. doi:10.1016/j.bios.2015.03.043 PMID:25829220

George, J. M., Antony, A., & Mathew, B. (2018). Metal oxide nanoparticles in electrochemical sensing and biosensing: A review. *Mikrochimica Acta*, *185*(7), 1–26. doi:10.100700604-018-2894-3 PMID:29974265

Gonzales, P., Felix, O., Alexander, C., Lutz, E., Ela, W., & Eduardo Sáez, A. (2014). Laboratory dust generation and size-dependent characterization of metal and metalloid-contaminated mine tailings deposits. *Journal of Hazardous Materials*, *280*, 619–626. doi:10.1016/j.jhazmat.2014.09.002 PMID:25222928

Gupta, S., Murthy, C., & Prabha, C. R. (2018). Recent advances in carbon nanotube based electrochemical biosensors. *International Journal of Biological Macromolecules*, *108*, 687–703. doi:10.1016/j.ijbiomac.2017.12.038 PMID:29223757

Hajam, Y. A., Rai, S., Kumar, R., Bashir, M., & Malik, J. A. (2020). Phenolic Compounds from Medicinal Herbs: Their Role in Animal Health and Diseases – A New Approach for Sustainable Welfare and Development. In R. Lone, R. Shuab, & A. Kamili (Eds.), *Plant Phenolics in Sustainable Agriculture*. Springer. doi:10.1007/978-981-15-4890-1_10

Hammond, J. L., Formisano, N., Estrela, P., Carrara, S., & Tkac, J. (2016). Electrochemical biosensors and nanobiosensors. *Essays in Biochemistry*, *60*(1), 69–80. doi:10.1042/EBC20150008 PMID:27365037

Hua, Y., Heal, K. V., & Friesl-Hanl, W. (2017). The use of red mud as an immobiliser for metal/metalloid-contaminated soil: A review. *Journal of Hazardous Materials*, *325*, 17–30. doi:10.1016/j.jhazmat.2016.11.073 PMID:27914288

Huang, X., Zhu, Y., & Kianfar, E. (2021). Nano biosensors: Properties, applications and electrochemical techniques. *Journal of Materials Research and Technology*, *12*, 1649–1672. doi:10.1016/j.jmrt.2021.03.048

Khanzada, N. K., Farid, M. U., Kharraz, J. A., Choi, J., Tang, C. Y., Nghiem, L. D., Jang, A., & An, A. K. (2020). Removal of organic micropollutants using advanced membrane-based water and wastewater treatment: A review. *Journal of Membrane Science*, *598*, 117672. doi:10.1016/j.memsci.2019.117672

Lee, W., Goh, P. S., Lau, W. J., & Ismail, A. F. (2020). Removal of pharmaceutical contaminants from aqueous medium: A state-of-the-art review based on paracetamol. *Arabian Journal for Science and Engineering*, *45*(9), 7109–7135. doi:10.100713369-020-04446-1

Lichtenberg, J. Y., Ling, Y., & Kim, S. (2019). Non-specific adsorption reduction methods in biosensing. *Sensors (Basel)*, *19*(11), 2488. doi:10.339019112488 PMID:31159167

Lim, S. A., Lim, T. H., & Ahmad, A. N. (2020). The Applications of Biosensing and Artificial Intelligence Technologies for Rapid Detection and Diagnosis of COVID-19 in Remote Setting. In Diagnostic Strategies for COVID-19 and other Coronaviruses. Springer.

Lu, X., Sun, J., & Sun, X. (2020). Recent advances in biosensors for the detection of estrogens in the environment and food. *Trends in Analytical Chemistry*, *2020*(127), 115882. doi:10.1016/j.trac.2020.115882

Luo, Y., Guo, W., Ngo, H. H., Nghiem, L. D., Hai, F. I., Zhang, J., Liang, S., & Wang, X. C. (2014). A review on the occurrence of micropollutants in the aquatic environment and their fate and removal during wastewater treatment. *The Science of the Total Environment*, *473*, 619–641. doi:10.1016/j.scitotenv.2013.12.065 PMID:24394371

Maduraiveeran, G. (2020). Bionanomaterial-based electrochemical biosensing platforms for biomedical applications. *Analytical Methods*, *12*(13), 1688–1701. doi:10.1039/D0AY00171F

Magar, H. S., Hassan, R. Y., & Mulchandani, A. (2021). Electrochemical impedance spectroscopy (EIS): Principles, construction, and biosensing applications. *Sensors (Basel)*, *21*(19), 6578. doi:10.339021196578 PMID:34640898

Malik, J. A., Bhadauria, M., Lone, R., & Hajam, Y. A. (2020). Exploitation of Plant Phenolics in Animal Farming. In R. Lone, R. Shuab, & A. Kamili (Eds.), *Plant Phenolics in Sustainable Agriculture*. Springer. doi:10.1007/978-981-15-4890-1_3

Malik, J. A., & Lone, R. (2021). Heat shock proteins with an emphasis on HSP 60. *Molecular Biology Reports*, *48*(10), 6959–6969. doi:10.100711033-021-06676-4 PMID:34498161

Malik, J. A., Mishra, G., Hajam, Y. A., Lone, R., & Quazi, S. (2022). Metabolome Analyses in Response to Diverse Abiotic Stress. In A. Roychoudhury, T. Aftab, & K. Acharya (Eds.), *Omics Approach to Manage Abiotic Stress in Cereals*. Springer. doi:10.1007/978-981-19-0140-9_6

McGrath, S. P., & Zhao, F. J. (2003). Phytoextraction of metals and metalloids from contaminated soils. *Current Opinion in Biotechnology*, *14*(3), 277–282. doi:10.1016/S0958-1669(03)00060-0 PMID:12849780

Mogren, C. L., & Trumble, J. T. (2010). The impacts of metals and metalloids on insect behavior. *Entomologia Experimentalis et Applicata*, *135*(1), 1–17. doi:10.1111/j.1570-7458.2010.00967.x

Morales-Narváez, E., & Merkoçi, A. (2019). Graphene oxide as an optical biosensing platform: A progress report. *Advanced Materials*, *31*(6), 1805043. PMID:30549101

Nair, P., & Alam, M. (2006). Performance limits of nanobiosensors. *Applied Physics Letters*, *88*(23), 233120. doi:10.1063/1.2211310

Parolo, C., & Merkoçi, A. (2013). Paper- based nanobiosensors for diagnostics. *Chemical Society Reviews, 42*(2), 450–457. doi:10.1039/C2CS35255A PMID:23032871

Patolsky, F., Zheng, G., & Lieber, C. M. (2006). *Nanowire-based biosensors.* ACS Publications. doi:10.1021/ac069419j

Pirzada, M., & Altintas, Z. (2019). Nanomaterials for healthcare biosensing applications. *Sensors (Basel), 19*(23), 5311. doi:10.339019235311 PMID:31810313

Pumera, M., Sanchez, S., Ichinose, I., & Tang, J. (2007). Electrochemical nanobiosensors. *Sensors and Actuators. B, Chemical, 123*(2), 1195–1205. doi:10.1016/j.snb.2006.11.016

Quazi, S. (2021a). TNFR2 antagonist and agonist: A potential therapeutics in Cancer. *Immunotherapy.*

Quazi, S. (2021b). *Anti-cancer activity of human gastrointestinal bacteria.* Academic Press.

Quazi, S. (2021c). In-silico pharmacophore and Molecular Docking based drug discovery against Marburg Virus's Viral Protein 35; A potent of MAVD. *bioRxiv.* Advance online publication. doi:10.1101/2021.07.01.450693

Quazi, S. (2021d). In-silico Structural and Molecular Docking-Based Drug Discovery Against Viral Protein (VP35) of Marburg Virus: A potent Agent of MAVD. *bioRxiv.* Advance online publication. doi:10.1101/2021.02.09.430405

Quazi, S. (2021e). Discovery of potential drug-like compounds against Viral protein (VP40) of Marburg Virus using pharmacophoric based virtual screening from ZINC database. *bioRxiv.* Advance online publication. doi:10.1101/2021.05.13.444037

Quazi, S. (2022a). An overview of CAR T cell mediated B cell Maturation Antigen therapy. *Clinical Lymphoma, Myeloma & Leukemia, 22*(6), e392–e404. doi:10.1016/j.clml.2021.12.003 PMID:34992008

Quazi, S. (2022b). *Application of Biosensors in Cancers, An Overview.* Academic Press.

Quazi, S. (2022c). Artificial intelligence and machine learning in precision and genomic medicine. *Medical Oncology (Northwood, London, England), 39*(8), 1–18. doi:10.100712032-022-01711-1 PMID:35704152

Quazi, S. (2022d). Elucidation of CRISPR-Cas9 application in novel cellular immunotherapy. *Molecular Biology Reports, 49*(7), 1–9. doi:10.100711033-022-07147-0 PMID:35122203

Quazi, S. (2022e). Telomerase gene therapy: A remission toward cancer. *Medical Oncology (Northwood, London, England), 39*(6), 1–20. doi:10.100712032-022-01702-2 PMID:35429243

Quazi, S. (2022f). *The potential implementation of biosensors for the diagnosis of biomarkers of various cancer.* Academic Press.

Quazi, S. (2022g). Artificial intelligence and machine learning in medicinal chemistry and validation of emerging drug targets. *Advancements in Controlled Drug Delivery Systems,* 27-43.

Rai, V.S., Acharya, C.C., & Dey, N. (2012). *Implications of nanobiosensors in agriculture.* Academic Press.

Rice, V. A. (1942). Breeding and Improvement of Farm Animals (3rd ed.). Academic Press.

Riedel, K., Kunze, G., & Baronian, K. H. (2006). *Biosensors and bioassays based on microorganisms*. Research Signpost.

Schwarzenbach, R. P., Escher, B. I., Fenner, K., Hofstetter, T. B., Johnson, C. A., von Gunten, U., & Wehrli, B. (2006). The challenge of micropollutants in aquatic systems. *Science*, *313*(5790), 1072–1077. doi:10.1126cience.1127291 PMID:16931750

Singh, K.R., Nayak, V., & Singh, R.P. (2021). Introduction to bionanomaterials: An overview. In *Bionanomaterials: Fundamentals and biomedical applications*. IOP Science.

Sobhan, A., Oh, J.-H., Park, M.-K., Kim, S. W., Park, C., & Lee, J. (2018). Single walled carbon nanotube based biosensor for detection of peanut allergy-inducing protein ara h1. *Korean Journal of Chemical Engineering*, *35*(1), 172–178. doi:10.100711814-017-0259-y

Solaimuthu, A., Nishitha, A., Murali, P., & Sai, K. P. (2020). Nano-biosensors and their relevance in tissue engineering. *Current Opinion in Biomedical Engineering*, *13*, 84–93. doi:10.1016/j.cobme.2019.12.005

Stuer-Lauridsen, F. (2005). Review of passive accumulation devices for monitoring organic micropollutants in the aquatic environment. *Environmental Pollution*, *136*(3), 503–524. doi:10.1016/j.envpol.2004.12.004 PMID:15862404

Su, S., Sun, Q., Gu, X., Xu, Y., Shen, J., Zhu, D., Chao, J., Fan, C., & Wang, L. (2019). Two-dimensional nanomaterials for biosensing applications. *Trends in Analytical Chemistry*, *119*, 115610. doi:10.1016/j.trac.2019.07.021

Syu, Y. C., Hsu, W. E., & Lin, C. T. (2018). Field-effect transistor biosensing: Devices and clinical applications. *ECS Journal of Solid State Science and Technology: JSS*, *7*(7), Q3196–Q3207. doi:10.1149/2.0291807jss

Ten Hulscher, T. E., & Cornelissen, G. (2005). Effect of temperature on sorption equilibrium and sorption kinetics of organic micropollutants-a review. *Chemosphere*, *32*(4), 609–626. doi:10.1016/0045-6535(95)00345-2

Tian, X., Murfin, L. C., Wu, L., Lewis, S. E., & James, T. D. (2021). Fluorescent small organic probes for biosensing. *Chemical Science (Cambridge)*, *12*(10), 3406–3426. doi:10.1039/D0SC06928K PMID:34163615

Virkutyte, J., & Varma, R. (2010). *Treatment of micropollutants in water and wastewater*. IWA Publishing. doi:10.2166/9781780401447

Vodyanitskii, Y. N. (2013). Contamination of soils with heavy metals and metalloids and its ecological hazard (analytic review). *Eurasian Soil Science*, *46*(7), 793–801. doi:10.1134/S1064229313050153

Xiong, J. Q., Kurade, M. B., & Jeon, B. H. (2018). Can microalgae remove pharmaceutical contaminants from water? *Trends in Biotechnology*, *36*(1), 30–44. doi:10.1016/j.tibtech.2017.09.003 PMID:28993012

Xu, L., Shoaie, S., Jahanpeyma, F., Zhao, J., Azimzadeh, M., & Al-Jamal, K. T. (2020). Optical, electrochemical and electrical (nano) biosensors for detection of exosomes: A comprehensive overview. *Biosensors & Bioelectronics*, *161*, 112222. doi:10.1016/j.bios.2020.112222 PMID:32365010

Xu, T., & Geng, Z. (2021). Strategies to improve performances of LSPR biosensing: Structure, materials, and interface modification. *Biosensors & Bioelectronics*, *2021*(174), 112850. doi:10.1016/j.bios.2020.112850 PMID:33309521

Yang, W., Zhou, H., & Cicek, N. (2004). Treatment of organic micropollutants in water and wastewater by UV-based processes: A literature review. *Critical Reviews in Environmental Science and Technology*, *44*(13), 1443–1476. doi:10.1080/10643389.2013.790745

Zamfir, L. G., Puiu, M., & Bala, C. (2020). Advances in electrochemical impedance spectroscopy detection of endocrine disruptors. *Sensors (Basel)*, *20*(22), 6443. doi:10.339020226443 PMID:33187314

Zhu, G., Yang, K., & Zhang, C. Y. (2015). A single quantum dot-based biosensor for telomerase assay. *Chemical Communications*, *51*(31), 6808–6811. doi:10.1039/C5CC01600B PMID:25788014

Chapter 6
Use of Nanofertilizers in Agriculture:
Advantages, Disadvantages, and Future Implications

Debraj Biswal
https://orcid.org/0000-0002-7317-6774
Government General Degree College at Mangalkote, India

ABSTRACT

The agricultural industry is currently facing challenges of food production for the expanding human population. This will be worsened in the future by the negative impacts of climate change. The obstacles can be overcome by increasing the productivity of the available agricultural lands. The conventional fertilizers are, however, not very efficient in delivering nutrients to the plants. Formulation of nanofertilizers based on tiny particles (< 100 nm), prepared from common organic and inorganic sources, is a ground-breaking discovery that can ensure sustainable agriculture in an eco-friendly manner. Their unique properties make them effective at low concentrations. They can supply nutrients and help the plants withstand biotic and abiotic stress factors. This proves their efficiency over traditional fertilizers. However, they have certain drawbacks. Their high reactivity can be a source of toxicity to the plants as well as animals. Thus, this emerging field has high potentialities but requires extensive research.

INTRODUCTION

According to reports, the escalating population will push up the demand for food supply by as much as 70% even before 2050 is reached (Mandal, 2021). Currently, the agricultural sector is under the pressure of two additional factors – availability of lands for cultivation of crops and global water crisis. The situation will be aggravated in the future by the changing climate and its negative impacts on agricultural production (Hussein and Abou-Baker, 2018). Under these circumstances, the productivity will largely be dependent on the efficiency of the fertilizers being used (Zulfiqar et al., 2019). Accordingly, the

DOI: 10.4018/978-1-6684-5533-3.ch006

fertilizer consumption has augmented in the recent years to meet the growing pressures of increasing food production. Fatima et al. (2021) reported that the fertilizer use has shot up by almost 13 folds from 1950 to 2020. However, continuous use of traditional fertilizers has resulted in several environmental issues like air pollution, water pollution, eutrophication of aquatic bodies and degradation of the soil quality (Verma et al., 2022). Furthermore, the economic burden of synthetic fertilizers decreases the profit margins of the cultivators and inflates the cost of agricultural products (Zulfiqar et al., 2019).

Despite the increased utilization of conventional fertilizers, the agricultural output has not been boosted enough. This is because of the tendency of chemical fertilizers to get immobilized in the soil depriving the plants of the required amounts of nutrients. Additionally, they may leach to the groundwater causing contamination or may be washed off by the rain to nearby aquatic bodies causing eutrophication (Bindraban et al., 2020). The problem tends to worsen under the present conditions of indiscriminate use of synthetic fertilizers (Bindraban et al., 2020). According to several workers, the nutrient utilization efficiency or NUE of the conventional fertilizers is very poor (El-Saadony et al., 2021; Verma et al., 2022). This calls for improvisation of the agricultural sector for ensuring the future food security while keeping in mind the production costs and environmental safety.

The recent years have witnessed tremendous growth of research on nanofertilizers. They are based on the use of nanoparticles that differ from their source bulk materials. Their miniature size, high surface-to-volume ratio and excessive power of ionization makes them tremendously reactive (Shang et al., 2019). Additionally, their resistance to high temperatures, pH fluctuations and several chemicals make them stable. Furthermore, they can be readily absorbed making them highly effective. Their efficiency can be multiplied several times by combining them with agrochemicals. This can be achieved by encapsulating them or adsorption or utilizing ionic bonds for attaching them to the nanomatrix of the active elements of nanoformulations (Pandey, 2018).

Chemical or biofertilizers can be coated with nanoparticles to increase their efficiency. The nanomaterials can be processed in such a manner that they liberate the nutrients slowly to be taken up by the plants according to their requirements thereby preventing nutrients loss (Janmohammadi et al., 2016). Sustained release of nutrients also prevents their undesirable interactions with the microbes in the rhizosphere which may have negative outcomes on plant productivity (Janmohammadi et al., 2016). The importance of nanoparticles is not only restricted to the supply of micronutrients to plants but also in their roles of stimulating the biofactors, reducing the toxic effects of heavy metals and counteracting the abiotic stress conditions (El-Saadony et al., 2021). In some cases, nanofertilizers boost plants resistance to certain diseases by forming binary films on microbial cell walls hampering their infectivity (Rastogi et al., 2019). Nanoparticles can not only accelerate the growth of different plant parts like the roots and shoots but also help in the development of seeds (Verma et al., 2022). Sometimes, nanofertilizers can help to improve the water retention capacity of the soil demonstrating their roles as buffers during drought conditions (Karunakaran et al., 2016). However, the impacts of nanoparticles vary with the nanomaterial being used as well as their concentrations during application (Fatima et al., 2021). Furthermore, the outcomes also depend on the plant species being treated and their application techniques (Fatima et al., 2021).

Nanotechnology is often regarded as the future of agriculture because it has the potentialities of developing precision farming and sensors for disease detection in plants (Prasad et al., 2017). Precision farming is based on employing nanosensors to know the conditions of the weather and nutrient status of the soil apart from identifying the plant health issues that can aid farmers to take necessary steps for improving crop production (Fraceto et al., 2016). According to Guru et al. (2015), nanofertilizers can be a viable alternative to conventional fertilizers because of their sustainability, low budget, high efficiency and eco-friendly properties. However, nanoparticles can get deposited on the surface of cells or within the organelles and trigger signals of oxidative stress in plants (El-Saadony et al., 2021). Additionally, since nanoparticles have the capability to affect the plant physiology at very low concentrations, researchers are concerned about their probable effects on human and environmental health (Fatima et al., 2021). Though they can enter the food chain there are insufficient data on their toxic effects. As such, their effects on humans and other organisms should be properly evaluated before using them in large-scales.

The current chapter discusses the advantages and drawbacks of nanofertilizers. It also highlights their manufacturing processes and the various ways by which they can be formulated to benefit the agricultural sector. Their modes of entry within the plant body and the molecular interactions have also been reviewed. The drawbacks and safety concerns that have prevented them from gaining mass acceptance have been elaborated as well. Finally, the chapter highlights the areas that need further exploration to draw the attention of the research community for ensuring the future food security.

ADVANTAGES OF NANOFERTILIZERS OVER CONVENTIONAL FERTILIZERS

One of the greatest disadvantage of conventional fertilizers is the tendency of the inorganic substances to get fixed or lost restricting the availability of nutrients to plants. According to available data, as high as 80% to 90% of nitrogen (N), phosphorus (P), and potassium (K) present in chemical fertilizers may not be available to the plants due to one reason or the other (Ombódi and Saigusa, 2000). This accounts for the poor nutrient delivery capacity of the conventional fertilizers. Moreover, the fertilizers may leach down and pollute the groundwater or get washed off and cause eutrophication of the water bodies nearby (Liu and Lal, 2015). The crop production can be negatively affected by increased soil salinity due to excessive use of fertilizers (Feregrino-Pérez et al., 2018). The agricultural yields can also be decreased due to nutrient imbalances in the soil arising from erratic application patterns of fertilizers (Feregrino-Pérez et al., 2018). This calls for precise nutrient delivery systems using low amounts of fertilizers to alleviate the drawbacks of conventional fertilizers.

The unique properties of nanoparticles such as large surface-to-volume ratio, sustained-release kinetics and high sorption capacity provide immense possibilities for the evolution of nanofertilizers (Feregrino-Pérez et al., 2018). They have earned the reputation of smart fertilizers due to their sluggish nutrient release properties achieved by coating or encapsulating the nutrients with nanomaterials (Zuverza-Mena et al., 2017). The high reactivity of nanomaterials allows them to elicit responses even at low concentrations (Prasad et al., 2017). Thus, nanofertilizers can overcome the problem of nutrient loss by slowing down the release of essential supplements. Additionally, the desired effects can be obtained by using small quantities of nanofertilizers. Furthermore, the solubility and dispersion of the nutrients in the soil can be enhanced by using nanocomposites (i.e. a combination of nanoparticles) which facilitate the uptake of nutrients by plants (El-Saadony et al., 2021).

Slow release rates of nanofertilizers can be attained by coating them with natural (e.g. chitosan and alginate) or synthetic polymers (e.g. polyacrylates, polylactide and polycaprolactones) (El-Saadony et al., 2021). For instance, bioactive compounds when coated with nanomaterials can withstand unfavourable circumstances (like elevated temperatures, oxidation and U.V. radiation) better, thereby, increasing their longevity. Nanoencapsulation employs different techniques like diffusion, dissolution or biodegradation for delivering the bioactive compounds to plants. Thus, it enables the nutrients and other supplements to be delivered in a sustained manner (Saifullah et al., 2019). Besides, active components can also be delivered in the form of nanopowders and nanoemulsions (Naraghi et al., 2018).

Using natural minerals like zeolite which possesses a large surface area with numerous sites that can strongly bind several nutrients in the form of charged ions can also slow down the release of nutrients (Selva and Balakrishnan, 2017). For instance, nanoformulations of phosphorus (P) fertilizer using zeolite have been reported to result in a sustained release of phosphorus (P) for about 50 days compared to the longevity of conventional phosphorus (P) fertilizer for approximately 12 days (Selva and Balakrishnan, 2017). Similarly, slow releasing nitrogen (N) fertilizer can be manufactured as nanohybrids composed of urea and hydroxyapatite. They last longer and thereby remain active for a greater period of time compared to pure urea (Kottegoda et al., 2017).

Micronutrients like copper (Cu), iron (Fe), zinc (Zn) and silica (SiO_2) can be manufactured at nanoscales and delivered to plants. They can be applied individually or in combination with a microbe or other metals (as nanoparticles). For instance, plant growth-promoting rhizobacteria (PGPR) when combined with FeNPs (iron nanoparticles) increase the grain yield and physiological functioning of maize (Heidari et al., 2018). The same combination when applied to young plants has also been reported to augment their heavy metal phytoremediation properties, especially in the zone of the roots and leaves (Mokarram et al., 2019). Application of Zn-Fe_2O_3NPs (Zn-Fe oxide nanoparticles) to wheat plants suffering from salt stress due to high soil salinity reportedly helps them to withstand the stress factors and improve their growth and grain yields (Babaei et al., 2017). Combination of ZnNPs (zinc nanoparticles) and *Bradyrhizobium japonicum* inoculum has been reported to increase the oil content of soybean (Sharifi, 2016). Salama (2012) reported the positive physiological impacts of silver nanoparticles on protein and carbohydrate content of wheat together with chlorophyll content and surface area of their leaves. Similar positive effects of silicon and titanium dioxide nanoparticles on maize and spinach plants have also been reported (Hong et al., 2005; Suriyaprabha et al., 2012). Thus, nanofertilizers when applied appropriately help to enhance agricultural productivity both quantitatively and qualitatively.

Under the current context of climate change, plants have to deal with several abiotic stress factors which have been hypothesized to rise in the years to come and can, therefore, be considered as an emerging threat to crop production (Wu and Ma, 2015). Since nanofertilizers have been found to be effective in helping plants withstand biotic and abiotic stress factors, they have a promising future in the agricultural industry (Ashraf et al., 2018). For example, nano calcium has been reported to alleviate salinity stress in *Solanum lycopersicum* (Tantawy et al., 2014). SiO_2NPs (silica nanoparticles) and iron nanoparticles (FeNPs) help plants withstand drought stress (Tripathi et al., 2015; Mozafari et al., 2018). Nanocomposites consisting of TiO_2 and gibberellin (GA_3) can also enable plants withstand conditions of water scarcity (Hatami, 2017). SiNPs (silicon nanoparticles) have been observed to make plants resistant to brief events of heat or cold stress (Seghatoleslami et al., 2015). Furthermore, the application of SiO_2NPs stimulates the photosynthetic pigments, carbonic anhydrase enzyme and transpiration or water-use efficiency of plants. Thus, they help plants to overcome conditions of poor water availability (Rajput et al., 2021).

There are several other advantages of using nanofertilizers. For example, their use drastically reduces the transport and application costs involved (Zulfiqar et al., 2019). Additionally, since they are required in small amounts they neither raise the soil salinity nor they cause nutrient overload in the soil (Seleiman et al., 2021). Moreover, nanofertilizers can be manufactured as per the nutrient requirement of the plant species and the conditions of soil health (Kah et al., 2018). This can prevent crops and fruits from nutrient deficiency diseases (Srivastava and Malhotra, 2017). It has been observed that the nutrient delivery system of nanofertilizers can be enhanced by incorporating biosensors that tune the release of nutrients according to the health status of the soil and the plants as well as the environmental conditions during the growing season of the concerned crop (Seleiman et al., 2021). Small size, high reactivity and large surface-to-volume ratio are the three important attributes of nanoparticles that increase the bioavailability of nutrients (Liu and Lal, 2015). Thus, they have the potentialities to attenuate the negative impacts of climate change on agriculture.

While unrestrained application of chemical fertilizers has impaired the soil microbial population, chemical pesticides have triggered the development of resistance in several plant pathogens. This necessitates innovative agricultural strategies to boost plant health and increase crop production. Few recent studies highlight that nanoparticles can be applied in the form of nanochips or nanosensors to enhance the need-based nutrient delivery to plants (Duhan et al., 2017). These nanochips can also aid in early detection of plant pathogens to enable and liberate the required nanoparticle(s) to combat plant diseases. Copper (Cu), silver (Ag) and gold (Au) nanoparticles find application as biosensors for detecting the health status of plants (Fatima et al., 2021). Additionally, these nanobiosensors can detect the physicochemical conditions of the soil like pH and moisture content and regulate the release of substances required to buffer the adversities (Rajonee et al., 2017). Application of conventional fertilizers cannot be manipulated similarly.

SYNTHESIS OF NANOFERTILIZERS

Nanofertilizers are derived from nanomaterials with particle size intermediate between that of the bulk materials from which they are derived and the corresponding atoms or molecules. This endows them with some special physical, chemical and biological properties (Mastronardi et al., 2015; Reda et al., 2021). The size usually ranges between 1 nm to 100 nm (Reda et al., 2021). The nanomaterials can be obtained by bottom-up or top-down methods or can be manufactured by biological processes (Ahmed et al., 2016). The bottom-up techniques of manufacturing nanoparticles are based on producing combinations of atoms and molecules to obtain the desired nanoparticle. The reactions involved in the process are chemically regulated and reduce the inclusion of contaminants. For example, synthesis of silver nanoparticles by this method involves the reduction of Ag^+ (silver ions) in aqueous or non-aqueous media by organic or inorganic reducing agents like sodium citrate, poly(ethylene glycol) block copolymers, Tollen's reagent, etc. (Iravani et al., 2014). For size stabilization of the nanoparticles, capping agents are used. By this technique huge amounts of nanoparticles can be produced within a short time period. However, the chemicals involved are poisonous often exerting adverse environmental effects (Ahmed et al., 2016).

The top-down method, on the other hand, depends on reducing the bulk materials to nanoscale particles by applying shearing forces or pulverizing them. Since the reactions governing the top-down method are regulated by physical force, the size of the nanoparticles cannot be managed properly. Besides, there are huge possibilities of contamination in the process (Pradhan and Mailapalli, 2017). Evaporation-condensation technique is most commonly applied for synthesis of nanoparticles by the top-down approach. The reactions are made to occur within a tube furnace at atmospheric pressure. The bulk material is placed in a boat at the centre of the furnace and allowed to vaporize into a carrier gas. Gold (Au) and silver (Ag) nanoparticles can be obtained by this technique. However, this is not a preferred method because of huge energy consumption by the tube furnace and high temperatures generated during the reaction. Additionally, it is very slow resulting in the structural imperfections of the nanomaterial surface which interferes with the physical properties and efficiency of the final product (Ahmed et al., 2016).

Besides, the nanofertilizers can also be obtained by milling of a substrate or an adsorbent (e.g. zeolite) for a long time to obtain nanosized particles. These tiny particles are endowed with large surface areas that can be utilized for adsorption and desorption of nutrients. The nanoparticles need to be stabilized by surfactants or polymers to inhibit their aggregation. The final product, however, depends on the affinity of the adsorbent. For instance, zeolites have greater affinity for cations (e.g. K^+, $NH4^+$) compared to their negligible affinity for anions (e.g. NO_3^-, SO_4^{2-}) making them efficient delivery systems for cationic nutrients. Modifying the surface of zeolites with surfactants can, however, facilitate the adsorption of anions on to them (Manikandan and Subramanian, 2014; Subramanian et al., 2015). Similarly, they can also be synthesized from nanomaterials like clay particles with sites for adsorption of nutrient ions which thereby act as a storehouse of nutrients. Actually, the nanoparticles get adsorbed to the clay lattice in their zero-valent states which prevents the fixation of nutrient ions. The nanoparticles also inhibit the precipitation of nutrient ions that are freely mobile. By these means, the nanomaterials can result in the formation of a dynamic pool of nutrients to be utilized by the plants (Subramanian et al., 2015).

Biosynthesis of nanoparticles utilizes the reducing or antioxidant properties of biological agents like microbes or plant extracts for reducing metal salt solutions. Since use of microbes for such reactions is expensive, requires aseptic conditions and often poses biohazards, plant extracts are preferred. They are better than the physical and chemical methods because they do not depend on the use of harmful chemicals or produce toxic wastes. Moreover, they do not consume large amounts of energy or generate heat. Furthermore, natural derivatives can be used as capping agents for stabilizing the nanoparticles which greatly improves their efficiency (Kalishwaralal et al., 2010; Dhuper et al., 2012).

The nanoparticles can be derived from both organic and inorganic sources. The organic nanoformulations commonly include carbon nanotubules, polymers and lipids. The metal oxide nanoparticles usually form the inorganic nanomaterials. Some popular ones are zinc oxide nanoparticles (ZnONPs), silica nanoparticles (SiO_2NPs), titanium dioxide nanoparticles (TiO_2NPs) and magnesium oxide nanoparticles (MgONPs). The metal nanoparticles like ZnNPs (zinc nanoparticles), AgNPs (silver nanoparticles) and AuNPs (gold nanoparticles) are also inorganic nanomaterials (Zulfiqar et al., 2019).

More than one nutrient can be combined together to form nanocomposites. The surface physical and chemical properties of these nanocomposites are known to regulate the soil biotic and abiotic reactions to some extent which influence the bioavailability of nutrients to the plants (Navrotsky, 2004). They can improve the soil quality by regulating the moisture holding capacity, adjusting the C/N ratio, stimulating the soil microbes and boosting the soil organic composition. They can also regulate the plant physiology directly (Liu and Zhang, 2005). Nanocomposites consisting of both organic and inorganic substances can be synthesized by utilizing hydrogen bonds. For instance, organic substances introduced in-between the layers of kaoline clay nanoparticles can be used to deliver organic substances to plants (Liu et al., 2005).

Based on the characteristics of the nutrients that the nanoparticles are made to deliver, they can be categorized into two groups – micronutrient nanofertilizers and macronutrient nanofertilizers. As the name suggests, micronutrient nanofertilizers supply the elements required by plants in trace amounts (also termed as micronutrients) for maintaining their normal metabolic functioning (Sharonova et al., 2015). For instance, boron (B) is a micronutrient which helps in the biosynthesis and lignification of plant cell walls besides stimulating plant growth and aiding some vital physiological reactions within the plant body (Navarro-León et al., 2016). Zinc (Zn) is also an important micronutrient which regulates the life-sustaining enzymatic reactions in the plant tissue by acting as a cofactor of several enzymes and proteins (Noreen et al., 2018). It also develops the resistance of plants to pathogenic infections (Broadley et al., 2007). Application of zinc (Zn) and boron (B) nanoparticles to a number of crop and fruit plants have increased their yields (Zulfiqar et al., 2019).

Iron (Fe) is another micronutrient which regulates plant growth and development. Its deficiency lowers crop yields (Palmqvist et al., 2017). Manganese (Mn) is also an important micronutrient essential for regulating metabolic functions of plants like biosynthesis of chlorophyll, proteins, fatty acids and ATP. It is also required for the production of secondary metabolites like flavonoids and lignin. Additionally, it acts as a cofactor for various enzymes and helps in photosynthesis (Palmqvist et al., 2017). Beneficial effects of application of the corresponding nanoparticles have been demonstrated experimentally by Zulfiqar et al. (2019).

Substances required by plants in large amounts for sustaining their life are referred to as macronutrients. Phosphorus (P), nitrogen (N), potassium (K), magnesium (Mg) and calcium (Ca) are some common macronutrients. One or more of these nutrients can be formulated together within capsules or coatings and applied to plants. These nanocomposites have high efficiency and can therefore exert their effects in low amounts compared to the huge amounts of traditional fertilizers (Chhipa, 2017). For example, urea-modified zeolites and nanoscale water-phosphorite suspension (60 nm to 120 nm) are promising nanofertilizers used to supply N and P to the plants respectively (Patra et al., 2016).

Biofertilizers are formulations of beneficial microbes that have been found to be very effective because they provide nutrition to plants, protect them from infections and buffer them against stress factors like drought and high salt or heavy metal concentrations (Glick, 2012). Their ease of applicability, lack of toxicity and absence of adverse effects on the environment enabled them to be recommended as better alternatives to conventional fertilizers. Despite these advantages they could not gain popularity because of the complications involved in their manufacturing processes and problems in their storage (Chittora et al., 2020). However, many of these drawbacks can be overcome by using them in conjunction with nanomaterials.

Nanobiofertilizers are fabricated by combining microbes (e.g. phosphate solubilizing bacteria, nitrogen fixing bacteria, plant growth promoting bacteria, etc.) with nanoparticles to boost plant growth and productivity (Simarmata et al., 2018). However, the formulation of nanobiofertilizers is complicated as one has to evaluate the outcomes of interactions between the microbes being used and the accompanying nanoparticle(s) to understand their effects on plants. For instance, plant growth promoting rhizobacteria can be combined with AuNPs (gold nanoparticles) because the interactions between them result in positive outcomes (Shukla et al., 2015). However, AgNPs (silver nanoparticles) cannot be used similarly because they hamper the normal functioning of the microbes by altering the structure and function of their cell membranes (Duhan et al., 2017). Nanobiofertilizers increase the shelf life of biofertilizers by helping them withstand U.V. radiation, heat and desiccation (Simarmata et al., 2018). Besides, nanoformulations help to deliver the beneficial microbes in a better way. The nanoscale size (usually smaller than even a cell) of these formulations is a major drawback of manufacturing biofertilizers. In this regard, macroscopic filters composed of radially arranged carbon nanotube walls have been proposed to be useful to trap and deliver microbes from fermentation chambers to the plants (Zulfiqar et al., 2019). However, manufacture of these nanoconstructs need further research and development.

Apart from nutrients, nanotechnology can be used to deliver pesticides or other plant protection agents. They are usually applied in the form of nanocapsules which can enter the plants through stomata (during foliar mode of application). The capsule walls disintegrate and their contents are released on being acted upon by stress enzymes produced by plants in response to biotic and/or abiotic stress factors. In this way, the nanomaterials can improve the plant defence responses to various stress factors including the infectious agents like bacteria and fungi (Uzu et al., 2010). The nanocapsules when applied to soil are acted upon by the chemicals in the root exudates degrading the capsule walls followed by liberation of nutrients and plant protection agents. It has been reported that root exudates are released by plants during their growth which indirectly helps to boost plant productivity by breaking down the nanocapsules (Corredor et al., 2009). Moreover, they can be made as liquid formulations or be applied in the form of powder having diameter less than 100 nm (Josef and Katarína, 2015). Interestingly, the engineered nanoparticles have been found to increase the dimensions and number of pores on plant cell walls facilitating their uptake (Subramanian et al., 2015). Thus, nanoparticle-based fertilizers may be used to deliver nutrients, health stimulants and other anti-stress factors to plants and are manufactured accordingly.

APPLICATIONS AND MODE OF ACTIONS OF NANOFERTILIZERS

Delivery of Nanoparticles to Plants

Mastronardi et al. (2015) categorized three modes of synthesis of nanofertilizers. The fertilizer directly (e.g. urea, ammonium salts) or the nutrient or any other plant health supplement (e.g. Zn, Se, Fe, Mg) can be manufactured as nanoscale derivatives and applied to the plants. They may also be combined with some nanoscale additives [e.g. carbon nanotubes (CNT), silica (SiO_2), titanium dioxide (TiO_2)] for improving the delivery of the nanofertilizer(s) or stimulating the plant defence mechanisms against infectious agents and different stress factors. Moreover, the nanofertilizers (including their combinations with nanoscale additives) can be surrounded or encapsulated by nanoscale coatings or nanothin membranes or any other nanoporous substances (e.g. zeolite) that can slow down the release of the nutrients or supplements helping to reduce nutrient wastage and increasing their longevity. Some extensively studied nanoscale additives include carbon nanotubes (CNT), silica (SiO_2) and titanium dioxide (TiO_2). CNT generate certain channels and capillaries that facilitate the uptake and retention of water in the plant tissues. Additionally, they stimulate the activity of genes related to transport of water, cell division and cell wall extension (Mastronardi et al., 2015). In these ways, they help the plants withstand conditions of water scarcity.

Nanoclay derivatives obtained from smectites, kaolinites and halloysites have been used in nanofertilizer formulations as agents for adsorption of nutrients. These nanoclay derivatives or other nanofilms and nanolayers (with nanosized pores) housing the nanofertilizers act as protective layers and prevent degradation caused by heat, light and microbial actions. Moreover, the layers also inhibit the leaching away of nanofertilizers and help in sustained release of nutrients thereby checking nutrient losses. This improves the efficiency of nanofertilizers (Park et al., 2004). For example, nanocomposites made of nanokaolin and polystyrene mixtures with pores measuring approximately 10 to 20 nm when blended with urea are able to act as outstanding sustained release fertilizer compared to the traditional urea fertilizer (bin Hussein et al., 2002). Several other substances like humic acid, cellulose, polystyrene, polyvinyl alcohol particles and lignosulfonate particles can be employed in preparing polymer films of nanosize thickness (Mastronardi et al., 2015).

Another way of preparing the fertilizers is by coating them on nanoparticles (that act as carriers) or by placing them within nanotubes. For example, halloysite nanotubes having diameters less than 100 nm and approximately 500 nm to 1.2 μm long can be filled with nanomaterials and used as fertilizing agents. The halloysite nanotubes help to maintain sustained release of supplements preventing fertilizer losses (Price and Wagner, 2008). The nanotube carriers can also be applied as cochleate nanotubes such that they are multilayered spiral structures made of lipid bilayers containing the fertilizers. The cochleate structures not only act as protective layers but also aid in improved solubility of the fertilizers. They can gain easy entry through the stomata of leaves and supply their contents following the fusion of their outer layer with the plasma membrane (Gould-Fogerite et al., 2003). Kottegoda et al. (2011) suggested the use of hydroxyapatite nanoparticles as carriers to deliver fertilizers. For example, hydroxyapatite nanoparticles modified with urea have been found to be more effective than the conventional urea fertilizer (Kottegoda et al., 2011).

Layered double hydroxides (LDHs) have also been explored as important agents of fertilizer conveyance. They are multilayered nanomaterials with nanoscale thick crystalline inorganic layers that are positively charged having negatively charged ions localized within the interlayers. The ease with which the anions are liberated by slight modifications of the external conditions (like pH alterations) enables these LDHs to be utilized as excellent sustained release fertilizer delivery systems (bin Hussein et al., 2002). The nanomaterials can also be delivered in the form of nanoemulsions consisting of droplets or colloids of nanoscale dimensions. For instance, the sludge obtained during paper manufacture can be blended with salts of magnesium, ammonium and phosphate and finally mixed with cellulose to formulate nanosized micelles which can serve as efficient fertilizers (Inada et al., 2007).

The basic principles involved in the delivery of nanofertilizers while retaining their effectiveness have been summarized by Fatima et al. (2021). Accordingly, the following points are to be considered while using them:

1. they should be stable,
2. they should be soluble in water,
3. they should have target-specific activities,
4. they should be efficient at low concentrations,
5. they should have sustained nutrient release properties to prevent nutrient losses,
6. their toxic effects on the environment should be minimum if not entirely lacking, and
7. they should possess simple techniques of delivery and disposal.

Modes of Application and Routes of Entry Within the Plant Cells

The nanofertilizers enter plants primarily through two routes – the roots and the surface of leaves as shown in Figure 1. The routes, in turn, depend on the mode of application. They can be applied by two ways:

1. **Soil**: The high reactivity of the nanoparticles enable them to interact with various inorganic and organic compounds in the soil helping them to ameliorate the conditions of soil salinity making it conducive for plant growth (Yassen et al., 2017). For instance, silica nanofertilizers (SiO_2NPs) can reduce the adverse consequences of soil salinity on plants (Yassen et al., 2017). They can improve the availability of nitrogen (N), phosphorus (P), potassium (K) and other minerals to the plants thereby enhancing their health. They can also boost soil health by controlling soil moisture content, soil pH and its cation exchange capacity (Rajonee et al., 2017). Nanofertilizers such as hydroxyapatite nanorods, when applied hydroponically, can stimulate seed germination and plant growth (Bala et al., 2014). Due to the same reasons, they can enter within the plant tissues and may interact in ways that can modify their physicochemical properties, behaviour, toxicity and their

final impacts (Bais et al., 2006). For instance, nickel hydroxide nanoparticles ($NiOH_2NPs$) when applied to the roots of mesquite plants demonstrate $NiOH_2NPs$ in the root tissue while the leaves and shoots exhibit nickel nanoparticles (NiNPs) indicating biotransformation of nanoparticles within the plant tissue (Parsons et al., 2010). According to Ma et al. (2017), biotransformations of nanoparticles are influenced by certain factors (e.g. root exudates) in the rhizosphere.

When applied to soil, the nanofertilizers enter the plants through their roots (Figure 1). This is facilitated by the presence of very small nanometre-sized pores in the plant roots meant for the uptake of ions and molecules (Mastronardi et al., 2015). Their uptake can be facilitated by combining them with root exudates or molecular transporters or by stimulating the ion channels and/or the endocytic pathway (Rico et al., 2011). The nanoparticles entering plants through the roots travel to different parts through the xylem vessels. Then, they pass through the pores in the cell wall (measuring approximately 3nm to 8 nm in diameter), move across the plasma membrane and finally enter the cells (Carpita and Gibeaut, 1993). To facilitate the entry of nanoparticles, they should have size lesser than 8 nm in diameter. For instance, AuNPs of 18 nm diameter fail to be taken up by the roots of tomato plants while those measuring 3.5 nm diameter are absorbed readily (Zhang et al., 2015). The elaborate pathway of nanomaterials uptake and their physiological impacts on plants, however, varies with the plant species (Odzak et al., 2015). Sometimes, plants have been reported to initiate defence pathways in response to certain nanoparticles, especially the metallic oxide ones (White and Gardea-Torresdey, 2018). In these cases, the plants are exposed to the metal oxide nanoparticles as well as the ions derived from them (Lowry et al., 2012). Soil mode of application of the nanofertilizers are, however, avoided when the weather conditions are unfavourable and the soil status is extremely poor. Under such conditions, the foliar mode of application is preferred (Mittal et al., 2020).

2. **Foliage:** According to some workers like Schwab et al. (2016) and Semida et al. (2021), the foliar application of nanoparticles have been found to be more beneficial than the soil mode of application because it leads to better utilization of nutrients. Moreover, it also reduces nutrient loss (El-Saadony et al., 2021). Often, nanocarriers are used to enhance the delivery of nutrients to plants (El-Saadony et al., 2021). For example, various nanoparticles when applied to the leaves in combination with nitrogen (N), phosphorus (P) and potassium (K) promotes plant growth, increases the number of branches, augments seed production and improves pod yields (Marzouk et al., 2019). When applied to leaf surfaces, the nanoparticle-based fertilizers enter the plants through the cuticular surface, the stomata or the hydathodes and are distributed across their interiors through the phloem vessels (Mittal et al., 2020) (Figure 1).

The absorption and distribution of nanofertilizers within the plants depend on various factors like their size, combination (with other nanoparticles in nanocomposites), the physiological effects they generate and the dimension of the cell wall pores (Verma et al., 2022). Usually, the nanoparticles should be lesser than 5 nm to 20 nm i.e. smaller than the size of the pores present on the cell wall to enable them to enter the cell and exert their effects (Navarro et al., 2008). To cross the cuticular layer of plants, a particle should be smaller, measuring about 2 nm to 2.4 nm in diameter (Pérez-de-Luque, 2017). However, the size exclusion limit of nanoparticles to enter through the stomata vary slightly among different plant species (Wang et al., 2013).

The trafficking of nanofertilizers across the cell wall and the plasma membrane is controlled by the receptors on the cell surface, transporters and receptor proteins (Verma et al., 2022). Nair et al. (2010) documented that nanoparticles penetrate the plant cells either by endocytosis or through aquaporins by binding to carrier proteins or by moving across the ion channels. Kurepa et al. (2010) stated that the nanoparticles can form complexes with the membrane transport proteins and be carried to different parts of the plants. In general, the negative charge of the plant cell wall surface promotes the entry of positively charged (cationic) nanoparticles over the negatively charged (anionic) ones (Meychik et al., 2005). However, the anionic nanoparticles have greater transport efficiency and increased rates of internalization and translocation within the plant tissues (Zhu et al., 2012). For instance, CeONPs (cerium monoxide nanoparticles) exhibit high degrees of adsorption on to the surfaces of the roots by virtue of interactions between the positively charged NPs and negatively charged root surface. CeO_2NPs (cerium dioxide nanoparticles), on the other hand, demonstrate lesser root acquisition and greater shoot internalization because of their negative charge that helps them to overcome the electrostatic resistance within the plant body (Liu et al., 2019). After the nanoparticles gain entry within the plant cells their distribution across the plant tissue occurs through the plasmodesmata (Mittal et al., 2020).

Figure 1. Routes of distribution of nanofertilizers following their application by (a) Foliar Mode or (b) Hydroponic/Soil Mode. When applied through foliar mode, the nanomaterials enter the plants and travel via the 'foliar pathway.' When applied through hydroponic/soil mode, the nanomaterials enter the plants and travel via the 'root pathway.' (Based on Seleiman et al., 2021)

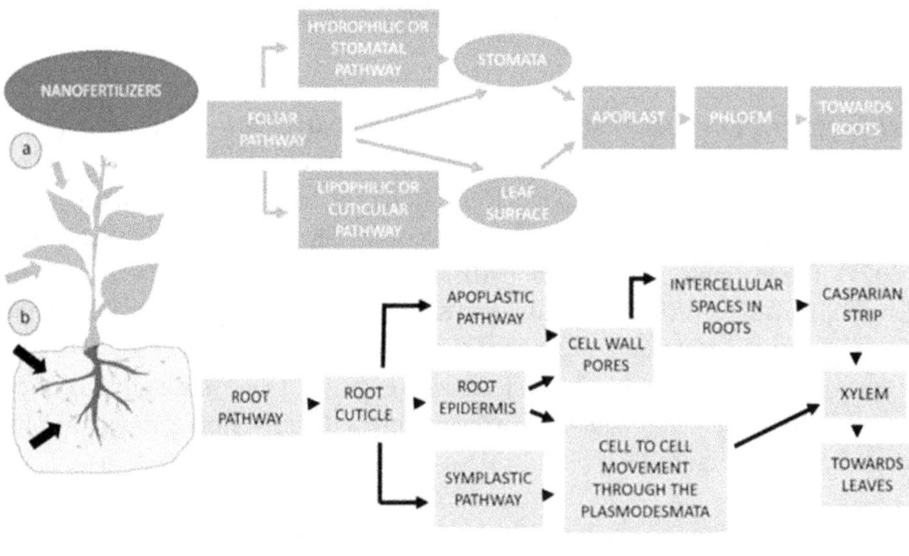

Mechanism of Action of Nanoparticles

The basic mechanism by which nanoparticles exert their effects after entering the plant cells is by mimicking the metal ions in their reactions with the carboxyl or thiol groups of proteins. The nanoparticles primarily interact with proteins related to the regulation of cell signalling pathways, cellular stress, cell metabolism and those involved in protein synthesis or degradation (Hossain et al., 2015). They can also interact with cell organelles like the chloroplasts (Rajput et al., 2021). They may activate the growth-promoting pathways in plants by stimulating the levels of miRNA (Boykov et al., 2019). They have the ability to make the plants more stress resistant by stimulating their stress-related enzymes. Their capacity to do this arises from their large surface areas (Verma et al., 2022). The nanoparticles can also invade through the tough coating of the seeds making way for water which ultimately helps in germination (Acharya et al., 2020). Thus, the nanoparticles can regulate plant physiology by controlling important biochemical pathways. The properties and applications of some common nanoparticles have been depicted in Table 1.

ZnNPs, for instance, stimulate plant growth, production of greater bulk of total dry matter and increases the amount of total soluble proteins in leaves (Tarafdar et al., 2014). They also increase the chlorophyll content, enhance the antioxidant pathways, augment auxin production, activate the enzymes of protein and carbohydrate metabolism pathways, induce the formation of pollen, maintain the integrity of biological membranes and influence the metabolic pathways related to plant growth (Rezaei and Abbasi, 2014). In savory plants, they can stimulate the production of essential oils (Vafa et al., 2015). Similarly, they can also enhance the oil contents of sunflower seeds and pod yields of groundnuts (Prasad et al., 2012; Singh, 2015).

TiO_2NPs can stimulate nitrogen (N) assimilation in plants, removal of ROS and augment the photoreduction properties of photosystem II which ultimately increases the dry biomass of plants (Raliya et al., 2015; Janmohammadi et al. 2016). TiO_2NPs can enhance the chlorophyll structure increasing its capacity to trap sunlight and induce the activity of ribulose-1, 5-bisphosphate carboxylase-oxygenase thereby augmenting photosynthesis (El-Saadony et al., 2021). Additionally, they can also increase the contents of carotenoids and anthocyanin pigments in plants like maize and barley which also have roles in boosting photosynthesis (Morteza et al., 2013; Janmohammadi et al., 2016). The increased photosynthetic activities can improve the agricultural outputs (Janmohammadi et al., 2016).

Metallic oxide nanoparticles like CuONPs (cupric oxide nanoparticles) and MgONPs (magnesium oxide nanoparticles) can result in increased yields of cotton (Sohair et al., 2018; El-Saadony et al., 2021). Boron nanoparticles (BNPs) can increase fruit production per tree in pomegranates (Davarpanah et al., 2016). Iron nanoparticles (FeNPs) can increase the pod yields of black pea and soybean (Sheykhbaglou et al., 2010; Delfani et al., 2014). Manganese nanoparticles (MnNPs) can enhance the yield of peanuts and *Vigna radiata* (Ghafariyan et al., 2013; Mekkdad, 2017). Tahmasbi et al. (2011) reported that AgNPs can not only increase the productivity of potato tubers but also aid in preserving the seed tubers for longer time periods because of their antimicrobial effects (Table 1). Molybdenum nanoparticles (MoNPs) can enhance the pod and seed yields of peanuts (Mehrangiz et al., 2014).

Nanofertilizers help to increase the quantity of crops as well as their quality when compared to conventional fertilizers. For instance, application of FeNPs (iron nanoparticles) and ZnNPs (zinc nanoparticles) to forage corn results in elevated levels of soluble carbohydrates and crude proteins in the grains (Sharifi et al., 2016). ZnNPs also help to increase the oil contents of sunflower seeds while FeNPs increase the protein contents of black-eyed pea seeds (Safyan et al., 2012; Delfani et al., 2014). Experiments demonstrate that Zn stimulates protein synthesis within the plant tissue by enhancing the growth of roots which in turn facilitates nitrogen (N) absorption. Moreover, Zn has a pivotal function in starch formation (El-Saadony et al., 2021).

Abiotic stress factors like soil salinity and drought can have adverse effects on crop production. These conditions often lead to osmotic imbalance hindering the nutrient uptake from soil and inhibiting enzymatic functions in plants (Hasanuzzaman et al., 2013). Elevated soil salinity, particularly, can affect both the water and nutrient absorption by plants exerting negative impacts on the quantity and quality of agricultural yields (El-Saadony et al., 2021). Nanofertilizers can help to overcome these constraints and boost quality crop production. They do so by imitating the common antioxidant catalases and peroxidases which help to remove the reactive oxygen species (Upadhyaya et al., 2015). Or, they (e.g. ZnNPs) may increase the absorption of nutrients by facilitating them to penetrate the roots (Hussein and Abou-Baker, 2018). It has been found that zinc and iron oxide nanoparticles when applied appropriately can increase plants' tolerance to salinity by augmenting the enzymatic reactions involved in the process (Wang et al., 2018). Silica nanoparticles (SiO_2NPs) have also been reported to enhance plants tolerance to salinity stress by reducing the acquisition of sodium (Na) and stimulating the uptake of phosphorus (P) and nitrogen (N) from the soil (Siddiqui et al., 2014). They can enhance antioxidant enzymes and thereby promote seedling growth and help them withstand biotic and abiotic stress factors (Li et al., 2020; Younis et al., 2020). Thus, SiO_2NPs have the ability to enhance plant growth and development under unpredictable weather conditions such as irregular rainfall, drying up of soil and temperature extremes (Younis et al., 2020).

The presence of excessive amounts of heavy metals within plant tissues has often been associated with reduced availability of essential nutrients and/or enzyme activity which eventually lowers plant productivity. For instance, cadmium (Cd) acquisition in plant tissues has been found to decrease the contents of nutrients like Zn, Mg, and Fe together with glutathione and chlorophyll a (Wang et al., 2012). Foliar application of SiO_2NPs has been recommended as a good remedy for Cd toxicity. They mainly act by preventing the translocation of Cd from the plant roots to the shoots thereby reducing their build-up within plant tissues (El-Saadony et al., 2021). Additionally, Cd can exert toxic effects by generating oxidative stress and elevating the rates of lipid peroxidation which negatively impacts plant growth. Some nanomaterials like Fe_3O_4NPs (magnetite or iron oxide nanoparticles) can stimulate the antioxidant activities thereby helping to combat the conditions of oxidative stress in plants (Konate et al., 2017). Furthermore, Fe_3O_4NPs can also help to overcome the root growth inhibiting effects of certain heavy metals like cadmium (Cd), lead (Pb), copper (Cu) and zinc (Zn) (Konate et al., 2017). Thus, application of appropriate nanofertilizers can not only help plants to overcome heavy metal toxicity but also enhance their phytoremediation capabilities.

Table 1. Properties and applications of some commonly used nanoparticles for synthesis of nanofertilizers (based on earlier reports)

Nanoparticle	Properties	Application(s)	Reference(s)
Silver nanoparticles (AgNps)	• Antibacterial and antifungal properties. • Promotes growth of plant roots, shoots and surface area of leaves. Can enhance the protein content of beans and chlorophyll content of leaves. • Can modify the diversity of microbes in the rhizosphere.	• Can be used as pesticides for plant disease management. Have been successfully applied against *Trichoderma* sp., *Fusarium oxysporum* f. sp. *cicero, Bacillus cereus* and *Staphylococcus aureus*. • Plant growth stimulant.	Salama, 2012; Fatima et al., 2015; 2016; 2018; Qing et al., 2018; Kaur et al., 2018
Zinc Oxide nanoparticles (ZnONPs)	• Stimulates the overall plant physiology by acting as cofactors and regulatory components of several enzymes. • Helps to overcome zinc deficiency in plants. • Stimulates seed germination. • Enhances the pod yields, chlorophyll content of leaves, growth of plant roots and shoots and can trigger early flowering.	• As plant growth stimulant since it mimics the micronutrient, zinc (Zn). • Used to protect plants against infections. • Used to improve seed vigour.	Prasad et al., 2012; Hafeez et al., 2013; Jamdagni et al., 2018; Fatima et al., 2021
Titanium dioxide nanoparticles (TiO_2NPs)	• Can help to trigger germination of seeds. • Stimulates plant growth by increasing the lengths of roots and shoots, the surface area of leaves and the dry weight of plants. • Enhances plants tolerance to saline soils.	• Used to improve seed and seedling vigour. • Can be applied as plant growth stimulant.	Mahmoodzadeh et al., 2015; Fatima et al., 2021
Copper nanoparticles (CuNPs)	• Biologically synthesized CuNPs exhibit antibacterial and antifungal properties. Has been shown to be effective against *Alternaria alternata* and *Phytophthora cinnamomi*. • Stimulates seed germination and plant growth at low concentrations. • Can enhance fruit properties like its firmness and abscisic acid and antioxidant contents when applied on foliage.	• As fungicides and bactericides they can defend plants from infections. • As a promoter of plant growth by imitating the activities of the micronutrient, copper (Cu). • To enhance the bioactive contents of fruits.	Banik and Pérez-de-Luque, 2017; López-Vargas et al., 2018
Iron oxide nanoparticles (Fe_2O_3NPs) or Iron nanoparticles (FeNPs)	• Stimulates plant growth. • Enhances the antioxidant enzymes and phytohormone contents. • Iron nanoparticles (FeNPs) can remediate polluted groundwater and soil water.	• As a promoter of plant growth to improve crop quality and production. • To remediate contaminated soil and groundwater.	Rui et al., 2016; Mukherjee et al., 2016; Jeyasubramanian et al., 2016

Continued on following page

Table 1. Continued

Nanoparticle	Properties	Application(s)	Reference(s)
Nickel nanoparticles (NiNPs) or nickel ferrite nanoparticles ($NiFe_2O_4$NPs)	They exhibit antimicrobial activities against plant pathogenic fungi and bacteria. Have been found to be effective against *Klebsiella pneumonia*, *Salmonella typhi*, *Staphylococcus epidermidis*, *Bacillus subtilis*, *Aspergillus fumigatus*, *A. niger*, *Colletotrichum gloeosporioides* *Candida tropicalis* and *C. albicans*.	Applied to protect plants from diseases. Primarily used for plant disease management.	Sharma et al., 2017; Fatima et al., 2021
Organic nanoparticles	They are synthetically produced from atoms or molecules and made available in various forms such as capsules, vesicles, liposomes, micelles and the likes. They may be formulated in ways that they contain several organic acids, microhumates, amino acids, vitamins, probiotics, biological substances, soil microfloras, organic micronutrients, etc. They can also possess chelated nitrogen (N), phosphorus (P) and potassium (K). Thus, they can stimulate the physiological activities of plants and promote their overall growth.	Can be used as an all-rounder nanonutrient fertilizer for several varieties of crops.	Ekinci et al., 2014; Manikandan and Subramanian, 2016; Fatima et al., 2021

DRAWBACKS OF NANOFERTILIZERS

According to Kah (2015), applying nanotechnology to agriculture can result in unwanted irreversible effects because many aspects of nanofertilizers are still unknown. The effects of nanomaterials on plants depend upon their concentration which necessitates standardization of nanomaterial concentrations before their application (Ashkavand et al., 2018). For instance, AgNPs, at low concentrations, can promote germination and development of seeds besides boosting plant growth. However, at high concentrations the effects of AgNPs are reversed (Zulfiqar et al., 2019). Similarly, the inhibitory effects of high concentrations of CuNPs on seed germination and plant growth have been reported by Kasana et al. (2017). In another study, Feizi et al. (2012) reported that high concentrations of TiO_2NPs can inhibit wheat growth.

Though the high reactivity of nanomaterials is considered as an advantage it can actually be a cause of concern because of the various cytotoxic and genotoxic effects they may trigger in biological organisms following their exposure (Nair, 2018). For example, Yang and Watts (2005) reported that Al_2O_3NPs (aluminum oxide nanoparticles) can exert their toxic effects on plants by hampering their root growth. Moreover, the possibility of nanomaterials being introduced in the food chains raises concerns about their bioavailability in large-scale applications (Feregrino-Pérez et al., 2018). This is further heightened by their tendencies to get stored within the plant tissues leading to food safety issues. For instance, López-

Moreno et al. (2018) reported that metal oxide nanoparticles can accumulate in the outer layer of carrot while the metal ions liberated by them tend to gather in the fleshy consumable part. This becomes a potential route of exposure of nanoparticles and/or their derivatives to humans. As such, Zulfiqar et al. (2019) recommended extensive studies of the mode of entry of NPs within plant tissues, their routes of distribution, their transformation and accumulation.

One of the primary concerns of researchers is the accumulation potential of the nanoparticles within the edible parts of the plants that can not only threaten human health but also prove toxic to the plants themselves (Lowry et al., 2012). The tendencies of these nanoparticles to get piled up in plant tissues depends on factors like their size, type, the plant species and the concerned tissue (Lowry et al., 2012). For example, accumulation of CeO_2NPs (cerium dioxide nanoparticles) in soybean can not only inhibit nitrogen fixation and lower their yields but also become a source of human health issues in the future (Zulfiqar et al., 2019). In another experiment, multi-walled carbon nanotubes accumulation was seen to exert toxic effects in red spinach by generating oxidative stress-induced cell death and inhibition of plant growth (Reddy Pullagurala et al., 2018). The plant parts that are more prone to accumulation of specific nanoparticles need to be deciphered to ascertain the safety of the food products. Moreover, the bioavailability of these accrued nanoparticles to the consumers should also be evaluated to prevent them from entering the food chain (Zulfiqar et al., 2019). Furthermore, the nanoparticles can accumulate on the surface of leaves causing foliar heating leading to modification of the gas exchange process due to blockage of the stomata. This exerts negative impacts on plant physiology (Fernandez and Eichert, 2009).

Deciphering the toxic effects of nanoparticles helps to determine their safety on biological organisms. It has been observed that their residues may activate genotoxic pathways leading to DNA damage or inflammatory responses with deleterious effects on the biological system (Bahadar et al., 2016). Moreover, when nanoparticles are applied to the plants at high concentrations they may exert negative effects. For instance, ZnONPs at high concentrations can inhibit the absorption of macronutrients and micronutrients by plant roots (Nair and Chung, 2017). They have been reported to exert toxic effects on the microflora of the marine environment (Lyon et al., 2005). As such, their toxicity should be thoroughly assessed before they can be applied on a large-scale.

Several researchers have reported about the phytotoxic effects of nanoparticles. For example, nanoparticles can reduce oxygen molecules and generate reactive oxygen species (ROS) within the cells leading to severe oxidative stress which impacts the cell signalling pathway, cell motility and can cause damage to the nucleic acids. Additionally, the ROS can also induce cytotoxicity, apoptosis or can provoke the formation of cancers (Verma et al., 2022). The ROS generate their cytotoxic effects by causing DNA breakage or by cross-linking the nucleic acids. They can also induce sugar or base adducts of nucleic acids or even cause DNA mutations (Klaine et al., 2008). For example, fullerenes can bind to DNA leading to deformation of its strands (Klaine et al., 2008). DNA damage ultimately affects the stability and proper functioning of the cells (Verma et al., 2022).

The nanoparticles can interact with and deform or destabilize the structure of proteins. They can do so by producing ROS which can disrupt the iron-sulphur cofactors of several enzymes or the ROS can bind to the sulphur containing amino acids through disulphide bonds severely impairing the structure and function of proteins (Klaine et al., 2008). These may ultimately disrupt cell growth and metabolism (Wang et al., 2016). Low concentrations of nanoions have been reported to trigger the uncoupling of ATP synthesis and respiration by interacting with the enzymes of the respiratory chain. They can also bind to the transport protein leading to leakage of protons thereby impairing the proton motive force (Holt and Bard, 2005). These can have dire impacts on plant physiology.

Nanoions or nanoparticles can impair DNA replication or gene expression with severe impacts on cell functioning (El-Saadony et al., 2021). For instance, AgNPs can enhance the expression of nitrification genes without affecting the nitrogen fixation or denitrification genes in bacteria such as *Nitrosomonas europaea*, *Azotobacter vinelandii* and *Pseudomonas stutzeri* which may ultimately affect the nitrogen (N) availability to plants (Yang et al., 2013; El-Saadony et al., 2021). Besides, the AgNPs have been reported to affect the bacterial transcriptome or even the genes that control cellular metal homeostasis which disturbs the normal cell functioning (Minghetti and Schirmer, 2019). Metal nanoparticles can also affect several other cellular proteins that regulate tolerance to oxidative stress, degradation of proteins, detoxification of metals, remodelling of the cell cytoskeleton and even cell division (Mirzajani et al., 2014).

Nanoparticles exert their toxic effects on the bacterial cells because of their tendencies of intracellular accumulation due to impairment of the cellular efflux mechanism (El-Saadony et al., 2021). According to some, the metal ions released by the metal oxide nanoparticles (when they dissolve in water) are responsible for their toxic effects while some other researchers say that the metal oxide particles themselves can have toxic properties (Xiao et al., 2015). For instance, silver (Ag) ions released from AgNPs can bind to the sulphydryl groups of proteins disabling several important enzymes (Klaine et al., 2008). Again, the AgNPs themselves can bind with the cell membrane proteins and induce the cell signalling pathways with inhibitory effects on cell division (Roh et al., 2012). Furthermore, AgNPs can bring about changes in the protein structure of the vacuoles and endoplasmic reticulum impairing their functions (Vannini et al., 2013). ROS have been found to impair the membrane permeability of microbial cells. They can also augment the expression of oxidative proteins ultimately causing apoptosis. Moreover, the normal functioning of the periplasmic enzymes are inhibited with direct adverse effects on the physiology and morphology of microbes (Wang et al., 2017). Thus, the rhizospheric microbial population may be negatively affected which may interfere with the nutrient acquisition by plants and protection against pathogenic microbes.

Apart from these, there are various problems associated with the fabrication of nanofertilizers. Their manufacture, especially the nanoscale coatings and encapsulations, is very expensive (Mastronardi et al., 2015). Moreover, the synthesis and application of nanofertilizers have not been standardized yet. The nanomaterials are very sensitive because their properties are largely dependent on their size and surface chemistry which may vary with their modes of production. As such, the same nanomaterial obtained from different sources by different methods may differ in properties and thereby their outcomes. Research in this field should, therefore, focus on uniformity in the synthesis procedures of nanomaterials to prevent these anomalies (Mastronardi et al., 2015). Moreover, confusion exists regarding the mode of application of nanofertilizers because their effectiveness have often been reported to vary by several workers (Mastronardi et al., 2015). Advanced research in this field shows that inclusion of molecular agents like aptamers and antibodies in nanofertilizer formulations may help to improve the nutrient delivery system (Jongeijk and Verheesen, 2011; Mastronardi et al., 2014). This 'smart delivery system' is, however, largely unknown.

FUTURE PERSPECTIVES

Despite the potentialities of nanofertilizers and their advantages over the conventional ones there are many aspects which need detailed study before they can be used for large-scale agricultural purposes. One of the primary aspect involves their pathways of transport within the plant tissues together with their abilities of biotransformation and their bioavailability to the next trophic level that need elaboration through extensive research. As stated earlier, the high reactivity of the nanoparticles may induce toxic effects in animals consuming the plants containing nanomaterial deposits. Techniques to overcome this problem can go a long way in helping them gain mass acceptance. Besides, the smart nutrient delivery and plant disease detection systems need further developments since nanofertilizers offer mitigation strategies against the negative impacts of climate change on agricultural yields.

CONCLUSION

Considering the problems of population growth, poor availability of cultivable agricultural lands and the negative impacts of global climate change, nanotechnology-based fertilizers are the future of the agricultural industry. They are functional in small amounts and are thereby cost-effective and less toxic to the environment. The greatest advantage is that they can be manufactured as various formulations depending on the requirements of the crop plants, their health status, the environmental conditions and the status of soil health. As such, they can deliver nutrients to the plants, protect them from diseased conditions and help them withstand stress factors. The net outcome is increased agricultural production, both quantitatively and qualitatively. However, they are not free from drawbacks. Their probable toxic properties have raised doubts in the research community regarding the safety of the final food products. Moreover, lack of standardization of the manufacturing processes of these nanofertilizers and their application often leads to different outcomes with the same product. As such, this field needs extensive research before the nanofertilizers can become popular.

AUTHOR CONTRIBUTION

Debraj Biswal conceived, designed, and wrote the book chapter along with the survey of literature.

FUND DISCLAIMER

For this work no grant was received from any funding agency in the public, commercial, or not-for-profit sectors.

COMPETING INTERESTS

No competing interests exist.

REFERENCES

Acharya, P., Jayaprakasha, G. K., Crosby, K. M., Jifon, J. L., & Patil, B. S. (2020). Nanoparticle-Mediated Seed Priming Improves Germination, Growth, Yield, and Quality of Watermelons (*Citrullus lanatus*) at multi-locations in Texas. *Scientific Reports, 10*(1), 5037. doi:10.103841598-020-61696-7 PMID:32193449

Ahmed, S., Ahmad, M., Swami, B. L., & Ikram, S. (2016). A review on plants extract mediated synthesis of silver nanoparticles for antimicrobial applications: A green expertise. *Journal of Advanced Research, 7*(1), 17–28. doi:10.1016/j.jare.2015.02.007 PMID:26843966

Ashkavand, P., Zarafshar, M., Tabari, M., Mirzaie, J., Nikpour, A., Bordbar, S. K., Struve, D., & Striker, G. G. (2018). Application of SiO2 nanoparticles as pretreatment alleviates the impact of drought on the physiological performance of *Prunus mahaleb* (Rosaceae). *Boletín de la Sociedad Argentina de Botánica, 53*(2), 207–219. doi:10.31055/1851.2372.v53.n2.20578

Ashraf, M. A., Akbar, A., Askari, S. H., Iqbal, M., Rasheed, R., & Hussain, I. (2018). Recent advances in abiotic stress tolerance of plants through chemical priming: An overview. In A. Rakshit & H. Singh (Eds.), *Advances in Seed Priming* (pp. 51–79). Springer. doi:10.1007/978-981-13-0032-5_4

Babaei, K., Sharifi, R. S., Pirzad, A., & Khalilzadeh, R. (2017). Effects of bio fertilizer and nano Zn-Fe oxide on physiological traits, antioxidant enzymes activity and yield of wheat (*Triticum aestivum* L.) under salinity stress. *Journal of Plant Interactions, 12*(1), 381–389. doi:10.1080/17429145.2017.1371798

Bahadar, H., Maqbool, F., Niaz, K., & Abdollahi, M. (2016). Toxicity of Nanoparticles and an Overview of Current Experimental Models. *Iranian Biomedical Journal, 20*(1), 1–11. doi:10.7508/ibj.2016.01.001 PMID:26286636

Bais, H. P., Weir, T. L., Perry, L. G., Gilroy, S., & Vivanco, J. M. (2006). The role of root exudates in rhizosphere interactions with plants and other organisms. *Annual Review of Plant Biology, 57*(1), 233–266. doi:10.1146/annurev.arplant.57.032905.105159 PMID:16669762

Bala, N., Dey, A., Das, S., Basu, R., & Nandy, P. (2014). Effect of hydroxyapatite nanorod on chickpea (*Cicer arietinum*) plant growth and its possible use as nano-fertilizer. *Indian Journal of Plant Physiology / Official Publication of the Indian Society for Plant Physiology, 4*(3), 1061–1069.

Banik, S., & Pérez-de-Luque, A. (2017). In vitro effects of copper nanoparticles on plant pathogens, beneficial microbes and crop plants. *Spanish Journal of Agricultural Research, 15*(2), e1005. doi:10.5424jar/2017152-10305

bin Hussein, M. Z., Zainal, Z., Yahaya, A. H., & Foo, D. W. (2002). Controlled release of a plant growth regulator, alpha-naphthaleneacetate from the lamella of Zn-Al-layered double hydroxide nanocomposite. *Journal of Controlled Release, 82*(2-3), 417–427. doi:10.1016/S0168-3659(02)00172-4

Bindraban, P. S., Dimkpa, C. O., & Pandey, R. (2020). Exploring phosphorus fertilizers and fertilization strategies for improved human and environmental health. *Biology and Fertility of Soils, 56*(3), 299–317. doi:10.100700374-019-01430-2

Boykov, I. N., Shuford, E., & Zhang, B. (2019). Nanoparticle titanium dioxide affects the growth and microRNA expression of switchgrass (*Panicum virgatum*). *Genomics*, *111*(3), 450–456. doi:10.1016/j.ygeno.2018.03.002 PMID:29524463

Broadley, M. R., White, P. J., Hammond, J. P., Zelko, I., & Lux, A. (2007). Zinc in plants. *The New Phytologist*, *173*(4), 677–702. doi:10.1111/j.1469-8137.2007.01996.x PMID:17286818

Carpita, N. C., & Gibeaut, D. M. (1993). Structural models of primary cell walls in flowering plants: Consistency of molecular structure with the physical properties of the walls during growth. *The Plant Journal*, *3*(1), 1–30. doi:10.1111/j.1365-313X.1993.tb00007.x PMID:8401598

Chhipa, H. (2017). Nanofertilizers and nanopesticides for agriculture. *Environmental Chemistry Letters*, *15*(1), 15–22. doi:10.100710311-016-0600-4

Chittora, D., Meena, M., Barupal, T., Swapnil, P., & Sharma, K. (2020). Cyanobacteria as a source of biofertilizers for sustainable agriculture. *Biochemistry and Biophysics Reports*, *22*, 100737. doi:10.1016/j.bbrep.2020.100737 PMID:32083191

Corredor, E., Testillano, P. S., Coronado, M. J., González-Melendi, P., Fernández-Pacheco, R., Marquina, C., Ibarra, M. R., de la Fuente, J. M., Rubiales, D., Pérez-de-Luque, A., & Risueño, M. C. (2009). Nanoparticle penetration and transport in living pumpkin plants: In situ subcellular identification. *BMC Plant Biology*, *9*(1), 45. doi:10.1186/1471-2229-9-45 PMID:19389253

Davarpanah, S., Tehranifar, A., Davarynejad, G., Abadía, J., & Khorasani, R. (2016). Effects of foliar applications of zinc and boron nano-fertilizers on pomegranate (*Punica granatum* cv. *Ardestani*) fruit yield and quality. *Scientia Horticulturae*, *210*, 57–64. doi:10.1016/j.scienta.2016.07.003

Delfani, M., Firouzabadi, M. B., Farrokhi, N., & Makarian, H. (2014). Some physiological responses of black-eyed pea to iron and magnesium nanofertilizers. *Communications in Soil Science and Plant Analysis*, *45*(4), 530–540. doi:10.1080/00103624.2013.863911

Dhuper, S., Panda, D., & Nayak, P. L. (2012). Green synthesis and characterization of zero valent iron nanoparticles from the leaf extract of *Mangifera indica. Nano Trends: A Journal of Nanotechnology and Its Applications, 13*(2), 16–22.

Duhan, J. S., Kumar, R., Kumar, N., Kaur, P., Nehra, K., & Duhan, S. (2017). Nanotechnology: The new perspective in precision agriculture. *Biotechnology Reports (Amsterdam, Netherlands)*, *15*, 11–23. doi:10.1016/j.btre.2017.03.002 PMID:28603692

Ekinci, M., Dursun, A., Yildirim, E., & Parlakova, F. (2014). Effects of nanotechnology liquid fertilizers on the plant growth and yield of cucumber (*Cucumis sativus* L.). *Acta Scientiarum Polonorum. Hortorum Cultus*, *13*(3), 135–141.

El-Saadony, M. T., Lmoshadak, A., Shafi, A. S., Albaqami, M. E., Saad, N. M., El-Tahan, A. M., Desoky, A. M., Elnahal, E. M., Almakas, A., Abd, A., El-Mageed, T. A., Taha, A. E., Elrys, A. S., & Helmy, A. M. (2021). Vital roles of sustainable nano-fertilizers in improving plant quality and quantity-an updated review. *Saudi Journal of Biological Sciences*, *28*(12), 7349–7359. doi:10.1016/j.sjbs.2021.08.032 PMID:34867037

Fatima, F., Bajpai, P., Pathak, N., Singh, S., Priya, S., & Verma, S. R. (2015). Antimicrobial and immunomodulatory efficacy of extracellularly synthesized silver and gold nanoparticles by a novel phosphate solubilizing fungus *Bipolaris tetramera*. *BMC Microbiology*, *15*(1), 52. doi:10.118612866-015-0391-y PMID:25881309

Fatima, F., Hashim, A., & Anees, S. (2021). Efficacy of nanoparticles as nanofertilizer production: A review. *Environmental Science and Pollution Research International*, *28*(2), 1292–1303. doi:10.100711356-020-11218-9 PMID:33070292

Fatima, F., Pathak, N., Verma, S. R., & Bajpai, P. (2018). Toxicity and immunomodulatory efficacy of biosynthesized silver myconanosomes on pathogenic microbes and macrophage cells. *Artificial Cells, Nanomedicine, and Biotechnology*, *46*(8), 1637–1645. doi:10.1080/21691401.2017.1388247 PMID:29022370

Fatima, F., Verma, S. R., Pathak, N., & Bajpai, P. (2016). Extracellular mycosynthesis of silver nanoparticles and their microbicidal activity. *Journal of Global Antimicrobial Resistance*, *7*, 88–92. doi:10.1016/j.jgar.2016.07.013 PMID:27689341

Feizi, H., Rezvani Moghaddam, P., Shahtahmassebi, N., & Fotovat, A. (2012). Impact of bulk and nanosized titanium dioxide (TiO_2) on wheat seed germination and seedling growth. *Biological Trace Element Research*, *146*(1), 101–106. doi:10.100712011-011-9222-7 PMID:21979242

Feregrino-Pérez, A. A., Magaña-López, E., Guzmán, C., & Esquivel, K. (2018). A general overview of the benefits and possible negative effects of the nanotechnology in horticulture. *Scientia Horticulturae*, *238*, 126–137. doi:10.1016/j.scienta.2018.03.060

Fernandez, V., & Eichert, T. (2009). Uptake of hydrophilic solutes through plant leaves: Current state of knowledge and perspectives of foliar fertilization. *Critical Reviews in Plant Sciences*, *28*(1-2), 36–68. doi:10.1080/07352680902743069

Fraceto, L. F., Grillo, R., de Medeiros, G. A., Scognamiglio, V., Rea, G., & Bartolucci, C. (2016). Nanotechnology in agriculture: Which innovation potential does it have? *Frontiers in Environmental Science*, *4*, 20. doi:10.3389/fenvs.2016.00020

Ghafariyan, M. H., Malakouti, M. J., Dadpour, M. R., Stroeve, P., & Mahmoudi, M. (2013). Effects of magnetite nanoparticles on soybean chlorophyll. *Environmental Science & Technology*, *47*(18), 10645–10652. doi:10.1021/es402249b PMID:23951999

Glick, B. R. (2012). Plant growth-promoting bacteria: Mechanisms and applications. *Scientifica*, *963401*, 1–15. Advance online publication. doi:10.6064/2012/963401 PMID:24278762

Gould-Fogerite, S., Mannino, R. J., & Margolis, D. (2003). Cochleate delivery vehicles: Applications to gene therapy. *Drug Delivery Technology*, *3*, 40–47.

Guru, T., Veronica, N., Thatikunta, R., & Reddy, S. N. (2015). Crop nutrition management with nano fertilizers. *International Journal of Environmental Science and Technology*, *1*(1), 4–6.

Hafeez, B., Khanif, M., & Saleem, M. (2013). Role of zinc in plant nutrition: A review. *American Journal of Experimental Agriculture*, *3*(2), 374–391. doi:10.9734/AJEA/2013/2746

Hasanuzzaman, M., Nahar, K., & Fujita, M. (2013). Plant response to salt stress and role of exogenous protectants to mitigate salt induced damages. In P. Ahmad, M. Azooz, & M. Prasad (Eds.), *Ecophysiology and Responses of Plants under Salt Stress* (pp. 25–87). Springer. doi:10.1007/978-1-4614-4747-4_2

Hatami, M. (2017). Toxicity assessment of multi-walled carbon nanotubes on *Cucurbita pepo* L. under well-watered and water-stressed conditions. *Ecotoxicology and Environmental Safety*, *142*, 274–283. doi:10.1016/j.ecoenv.2017.04.018 PMID:28433592

Heidari, M., Salmanpour, I., Ghorbani, H., & Asghari, H. R. (2018). Iron chelate and rhizobacteria changed growth, grain yield, and physiological characteristics in maize. *Scientia Agriculturae Bohemica*, *49*(4), 245–254. doi:10.2478ab-2018-0031

Holt, K. B., & Bard, A. J. (2005). Interaction of silver(I) ions with the respiratory chain of Escherichia coli: An electrochemical and scanning electrochemical microscopy study of the antimicrobial mechanism of micromolar Ag^+. *Biochemistry*, *44*(39), 13214–13223. doi:10.1021/bi0508542 PMID:16185089

Hong, F., Yang, F., Liu, C., Gao, Q., Wan, Z., Gu, F., Wu, C., Ma, Z., Zhou, J., & Yang, P. (2005). Influences of nano-TiO_2 on the chloroplast aging of spinach under light. *Biological Trace Element Research*, *104*(3), 249–260. doi:10.1385/BTER:104:3:249 PMID:15930594

Hossain, Z., Mustafa, G., & Komatsu, S. (2015). Plant Responses to Nanoparticle Stress. *International Journal of Molecular Sciences*, *16*(11), 26644–26653. doi:10.3390/ijms161125980 PMID:26561803

Hussein, M. M., & Abou-Baker, N. H. (2018). The contribution of nano-zinc to alleviate salinity stress on cotton plants. *Royal Society Open Science*, *5*(8), 171809. doi:10.1098/rsos.171809 PMID:30224982

Inada, T., Kasai, Y., Wakabayashi, T., Shibata, H., & Isobe, A. (2007). Ammonium magnesium phosphate containing nanocomposite and its manufacture. Academic Press.

Iravani, S., Korbekandi, H., Mirmohammadi, S. V., & Zolfaghari, B. (2014). Synthesis of silver nanoparticles: Chemical, physical and biological methods. *Research in Pharmaceutical Sciences*, *9*(6), 385–406. PMID:26339255

Jamdagni, P., Rana, J. S., & Khatri, P. (2018). Comparative study of antifungal effect of green and chemically synthesised silver nanoparticles in combination with carbendazim, mancozeb, and thiram. *IET Nanobiotechnology*, *12*(8), 1102–1107. doi:10.1049/iet-nbt.2018.5087 PMID:30964021

Janmohammadi, M., Navid, A., Ebadi Segherloo, A., & Sabaghnia, N. (2016). Impact of nano-chelated micronutrients and biological fertilizers on growth performance and grain yield of maize under deficit irrigation condition. *Biologija (Vilnius, Lithuania)*, *62*(2), 134–147. doi:10.6001/biologija.v62i2.3339

Jeyasubramanian, K., Thoppey, U. U. G., Hikku, G. S., Selvakumar, N., Subramania, A., & Krishnamoorthy, K. (2016). Enhancement in growth rate and productivity of spinach grown in hydroponics with iron oxide nanoparticles. *RSC Advances*, *6*(19), 15451–15459. doi:10.1039/C5RA23425E

Jongeijk, E., & Verheesen, P. (2011). Specific delivery of agrochemicals. Academic Press.

Josef, J., & Katarína, K. (2015). Application of nanotechnology in agriculture and food industry, its prospects and risks. *Ecological Chemistry and Engineering*, *22*(3), 321–361. doi:10.1515/eces-2015-0018

Kah, M. (2015). Nanopesticides and nanofertilizers: Emerging contaminants or opportunities for risk mitigation? *Frontiers in Chemistry, 3*, 64. doi:10.3389/fchem.2015.00064 PMID:26636068

Kah, M., Kookana, R. S., Gogos, A., & Bucheli, T. D. (2018). A critical evaluation of nanopesticides and nanofertilizers against their conventional analogues. *Nature Nanotechnology, 13*(8), 677–684. doi:10.103841565-018-0131-1 PMID:29736032

Kalishwaralal, K., Deepak, V., Ram Kumar Pandian, S., Kottaisamy, M., BarathManiKanth, S., Kartikeyan, B., & Gurunathan, S. (2010). Biosynthesis of silver and gold nanoparticles using *Brevibacterium casei. Colloids and Surfaces. B, Biointerfaces, 77*(2), 257–262. doi:10.1016/j.colsurfb.2010.02.007 PMID:20197229

Karunakaran, G., Suriyaprabha, R., Rajendran, V., & Kannan, N. (2016). Influence of ZrO_2, SiO_2, Al_2O_3 and TiO_2 nanoparticles on maize seed germination under different growth conditions. *IET Nanobiotechnology / IET, 10*(4), 171–177. doi:10.1049/iet-nbt.2015.0007 PMID:27463785

Kasana, R. C., Panwar, N. R., Kaul, R. K., & Kumar, P. (2017). Biosynthesis and effects of copper nanoparticles on plants. *Environmental Chemistry Letters, 15*(2), 233–240. doi:10.100710311-017-0615-5

Kaur, P., Thakur, R., Duhan, J. S., & Chaudhury, A. (2018). Management of wilt disease of chickpea in vivo by silver nanoparticles biosynthesized by rhizospheric microflora of chickpea (*Cicer arietinum*). *Journal of Chemical Technology and Biotechnology (Oxford, Oxfordshire), 93*(11), 3233–3243. doi:10.1002/jctb.5680

Klaine, S. J., Alvarez, P. J., Batley, G. E., Fernandes, T. F., Handy, R. D., Lyon, D. Y., Mahendra, S., McLaughlin, M. J., & Lead, J. R. (2008). Nanomaterials in the environment: Behavior, fate, bioavailability, and effects. *Environmental Toxicology and Chemistry, 27*(9), 1825–1851. doi:10.1897/08-090.1 PMID:19086204

Konate, A., He, X., Zhang, Z., Ma, Y., Zhang, P., Alugongo, G. M., & Rui, Y. (2017). Magnetic (Fe_3O_4) nanoparticles reduce heavy metals uptake and mitigate their toxicity in wheat seedling. *Sustainability, 9*(5), 790. doi:10.3390u9050790

Kottegoda, N., Munaweera, I., Madusanka, N., & Karunaratne, V. (2011). A green slow-release fertilizer composition based on urea-modified hydroxyapatite nanoparticles encapsulated wood. *Current Science, 101*(1), 73–78.

Kottegoda, N., Sandaruwan, C., Priyadarshana, G., Siriwardhana, A., Rathnayake, U. A., Berugoda Arachchige, D. M., Kumarasinghe, A. R., Dahanayake, D., Karunaratne, V., & Amaratunga, G. A. (2017). Urea-hydroxyapatite nanohybrids for slow release of nitrogen. *ACS Nano, 11*(2), 1214–1221. doi:10.1021/acsnano.6b07781 PMID:28121129

Kurepa, J., Paunesku, T., Vogt, S., Arora, H., Rabatic, B. M., Lu, J., Wanzer, M. B., Woloschak, G. E., & Smalle, J. A. (2010). Uptake and distribution of ultrasmall anatase TiO_2 Alizarin red S nanoconjugates in *Arabidopsis thaliana. Nano Letters, 10*(7), 2296–2302. doi:10.1021/nl903518f PMID:20218662

Li, Y., Zhu, N., Liang, X., Bai, X., Zheng, L., Zhao, J., Li, Y.-F., Zhang, Z., & Gao, Y. (2020). Silica nanoparticles alleviate mercury toxicity via immobilization and inactivation of Hg(ii) in soybean (*Glycine max*). *Environmental Science. Nano, 7*(6), 1807–1817. doi:10.1039/D0EN00091D

Liu, M., Feng, S., Ma, Y., Xie, C., He, X., Ding, Y., Zhang, J., Luo, W., Zheng, L., Chen, D., Yang, F., Chai, Z., Zhao, Y., & Zhang, Z. (2019). Influence of surface charge on the phytotoxicity, transformation, and translocation of CeO_2 nanoparticles in cucumber plants. *ACS Applied Materials & Interfaces*, *11*(18), 16905–16913. doi:10.1021/acsami.9b01627 PMID:30993970

Liu, R., & Lal, R. (2015). Potentials of engineered nanoparticles as fertilizers for increasing agronomic productions. *The Science of the Total Environment*, *514*, 131–139. doi:10.1016/j.scitotenv.2015.01.104 PMID:25659311

Liu, X., & Zhang, M. (2005). *Characteristics of nano-subnanocomposites and response of soil and plant nutrition to them* [Ph.D. Dissertation]. Chinese Academy of Agricultural Sciences.

Liu, X., Zhao-bin, F., Fu-dao, Z., & Shu-qing, Z., & Xu-sheng, H. E. (2005). Responses of peanut to nanocalcium carbonate. *Plant Nutrition and Fertitizer Science*, *11*(3), 385–389.

López-Moreno, M. L., Cassé, C., & Correa-Torres, S. N. (2018). Engineered nanomaterials interactions with living plants: Benefits, hazards and regulatory policies. *Current Opinion in Environmental Science & Health*, *6*, 36–41. doi:10.1016/j.coesh.2018.07.013

López-Vargas, E. R., Ortega-Ortíz, H., Cadenas-Pliego, G., De Alba Romenus, K., Cabrera de la Fuente, M., Benavides-Mendoza, A., & Juárez-Maldonado, A. (2018). Foliar application of copper nanoparticles increases the fruit quality and the content of bioactive compounds in tomatoes. *Applied Sciences (Basel, Switzerland)*, *8*(7), 1020. doi:10.3390/app8071020

Lowry, G. V., Gregory, K. B., Apte, S. C., & Lead, J. R. (2012). Transformations of nanomaterials in the environment. *Environmental Science & Technology*, *46*(13), 6893–6899. doi:10.1021/es300839e PMID:22582927

Lyon, D. Y., Fortner, J. D., Sayes, C. M., Colvin, V. L., & Hughe, J. B. (2005). Bacterial cell association and antimicrobial activity of a C60 water suspension. *Environmental Toxicology and Chemistry*, *24*(11), 2757–2762. doi:10.1897/04-649R.1 PMID:16398110

Ma, Y., He, X., Zhang, P., Zhang, Z., Ding, Y., Zhang, J., Wang, G., Xie, C., Luo, W., Zhang, J., Zheng, L., Chai, Z., & Yang, K. (2017). Xylem and phloem based transport of CeO_2 nanoparticles in hydroponic cucumber plants. *Environmental Science & Technology*, *51*(9), 5215–5221. doi:10.1021/acs.est.6b05998 PMID:28383248

Mahmoodzadeh, H., Nabavi, M., & Kashefi, H. (2015). Effect of nanoscale titanium dioxide particles on the germination and growth of canola (*Brassica napus*). *Journal of Ornamental Plants*, *3*(1), 25–32.

Mandal, D., & Lalrinchhani. (2021). Nanofertilizer and its application in horticulture. *Journal of Applied Horticulture*, *23*(1), 70–77. doi:10.37855/jah.2021.v23i01.14

Manikandan, A., & Subramanian, K. S. (2014). Fabrication and characterisation of nanoporous zeolite based N fertilizer. *African Journal of Agricultural Research*, *9*(2), 276–284. doi:10.5897/AJAR2013.8236

Manikandan, A., & Subramanian, K. S. (2016). Evaluation of zeolite-based nitrogen nano-fertilizers on maize growth, yield and quality on inceptisols and alfisols. *International Journal of Plant and Soil Science*, *9*(4), 1–9. doi:10.9734/IJPSS/2016/22103

Marzouk, N. M., Abd-Alrahman, H. A., EL-Tanahy, A. M. M., & Mahmoud, S. H. (2019). Impact of foliar spraying of nano micronutrient fertilizers on the growth, yield, physical quality, and nutritional value of two snap bean cultivars in sandy soils. *Bulletin of the National Research Center*, *43*(1), 84. doi:10.118642269-019-0127-5

Mastronardi, E., Foster, A., Zhang, X., & DeRosa, M. C. (2014). Smart materials based on DNA aptamers: Taking apta sensing to the next level. *Sensors (Basel)*, *14*(2), 3156–3171. doi:10.3390140203156 PMID:24553083

Mastronardi, E., Tsae, P., Zhang, X., Monreal, C., & DeRosa, M. C. (2015). Strategic Role of Nanotechnology in Fertilizers: Potential and Limitations. In M. Rai, C. Ribeiro, L. Mattoso, & N. Duran (Eds.), *Nanotechnologies in Food and Agriculture* (pp. 25–67). Springer International Publishing. doi:10.1007/978-3-319-14024-7_2

Mehrangiz, J. M., Sirous, B., & Ebrahim, A. (2014). Study the effect of foliar application of nano chelate molybdenum fertilizer on the yield and yield components of peanut. *Egyptian Academic Journal of Biological Sciences*, *5*(1), 67–71. doi:10.21608/eajbsh.2014.16829

Mekkdad, A. A. A. (2017). Response of peanut nitrogen fertilizer levels and foliar zinc spraying rates in newly reclaimed sandy soils. *Journal of Plant Production*, *8*(2), 153–159. doi:10.21608/jpp.2017.39240

Meychik, N., Nikolaeva, J., & Yermakov, I. (2005). Ion exchange properties of the root cell walls isolated from the halophyte plants (*Suaeda altissima* L.) grown under conditions of different salinity. *Plant and Soil*, *277*(1-2), 163–174. doi:10.100711104-005-6806-z

Minghetti, M., & Schirmer, K. (2019). Interference of silver nanoparticles with essential metal homeostasis in a novel enterohepatic fish in vitro system. *Environmental Science. Nano*, *6*(6), 1777–1790. doi:10.1039/C9EN00310J

Mirzajani, F., Askari, H., Hamzelou, S., Schober, Y., Römpp, A., Ghassempour, A., & Spengler, B. (2014). Proteomics study of silver nanoparticles toxicity on *Bacillus thuringiensis*. *Ecotoxicology and Environmental Safety*, *100*, 122–130. doi:10.1016/j.ecoenv.2013.10.009 PMID:24290895

Mittal, D., Kaur, G., Singh, P., Yadav, K., & Ali, S. A. (2020). Nanoparticle-based sustainable agriculture and food science: Recent advances and future outlook. *Frontiers in Nanotechnology*, *2*, 10. doi:10.3389/fnano.2020.579954

Mokarram-Kashtiban, S., Hosseini, S. M., Tabari Kouchaksaraei, M., & Younesi, H. (2019). The impact of nanoparticles zero-valent iron (nZVI) and rhizosphere microorganisms on the phytoremediation ability of white willow and its response. *Environmental Science and Pollution Research International*, *26*(11), 10776–10789. doi:10.100711356-019-04411-y PMID:30778927

Morteza, E., Moaveni, P., Farahani, H. A., & Kiyani, M. (2013). Study of photosynthetic pigments changes of maize (*Zea mays* L.) under nano TiO_2 spraying at various growth stages. *SpringerPlus*, *2*(1), 247. doi:10.1186/2193-1801-2-247 PMID:23847752

Mozafari, A. A., Ghadakchi Asl, A., & Ghaderi, N. (2018). Grape response to salinity stress and role of iron nanoparticle and potassium silicate to mitigate salt induced damage under in vitro conditions. *Physiology and Molecular Biology of Plants*, *24*(1), 25–35. doi:10.100712298-017-0488-x PMID:29398836

Mukherjee, R., Kumar, R., Sinha, A., Lama, Y., & Saha, A. K. (2016). A review on synthesis, characterization, and applications of nano zero valent iron (nZVI) for environmental remediation. *Critical Reviews in Environmental Science and Technology, 46*(5), 443–466. doi:10.1080/10643389.2015.1103832

Nair, P., & Chung, I. M. (2017). Regulation of morphological, molecular and nutrient status in Arabidopsis thaliana seedlings in response to ZnO nanoparticles and Zn ion exposure. *The Science of the Total Environment, 575*, 187–198. doi:10.1016/j.scitotenv.2016.10.017 PMID:27741454

Nair, P. M. G. (2018). Toxicological impact of carbon nanomaterials on plants. In K. M. Gothandam, S. Ranjan, N. Dasgupta, C. Ramalingam, & E. Lichtfouse (Eds.), *Nanotechnology, Food Security and Water Treatment* (pp. 163–183). Springer International Publishing. doi:10.1007/978-3-319-70166-0_5

Nair, R., Varghese, S. H., Nair, B. G., Maekawa, T., Yoshida, Y., & Kumar, D. S. (2010). Nanoparticulate material delivery to plants. *Plant Science, 179*(3), 154–163. doi:10.1016/j.plantsci.2010.04.012

Naraghi, L., Negahban, M., Heydari, A., Razavi, M., & Afshari-Azad, H. (2018). Growth inhibition of *Fusarium oxysporum* f. sp. *lycopercisi*, the causal agent of Tomato Fusarium Wilt disease by nano formulations containing *Talaromyces flavus. Ekoloji, 27*(106), 103–112.

Navarro, E., Baun, A., Behra, R., Hartmann, N. B., Filser, J., Miao, A. J., Quigg, A., Santschi, P. H., & Sigg, L. (2008). Environmental behavior and ecotoxicity of engineered nanoparticles to algae, plants, and fungi. *Ecotoxicology (London, England), 17*(5), 372–386. doi:10.100710646-008-0214-0 PMID:18461442

Navarro-León, E., Albacete, A., de la Torre-González, A., Ruiz, J. M., & Blasco, B. (2016). Phytohormone profile in *Lactuca sativa* and *Brassica oleracea* plants grown under Zn deficiency. *Phytochemistry, 130*, 85–89. doi:10.1016/j.phytochem.2016.08.003 PMID:27543253

Navrotsky, A. (2004). Environmental nanoparticles. In S. C. I. Contescu (Ed.), *Dekker encyclopedia of nanoscience and nanotechnology* (Vol. 2, pp. 1147–1155). CRC Press.

Noreen, S., Fatima, Z., Ahmad, S., Athar, H. R., & Ashraf, M. (2018). Foliar application of micronutrients in mitigating abiotic stress in crop plants. In M. Hasanuzzaman, M. Fujita, H. Oku, K. Nahar & B. Hawrylak-Nowak (Eds.), Plant nutrients and abiotic stress tolerance (pp. 95-117). Springer. doi:10.1007/978-981-10-9044-8_3

Odzak, N., Kistler, D., Behra, R., & Sigg, L. (2015). Dissolution of metal and metal oxide nanoparticles under natural freshwater conditions. *Environmental Chemistry, 12*(2), 138–148. doi:10.1071/EN14049

Ombódi, A., & Saigusa, M. (2000). Broadcast application versus band application of polyolefin-coated fertilizer on green peppers grown on andisol. *Journal of Plant Nutrition, 23*(10), 1485–1493. doi:10.1080/01904160009382116

Palmqvist, N., Seisenbaeva, G. A., Svedlindh, P., & Kessler, V. G. (2017). Maghemite nanoparticles acts as nanozymes, improving growth and abiotic stress tolerance in *Brassica napus. Nanoscale Research Letters, 12*(1), 631. doi:10.118611671-017-2404-2 PMID:29260423

Pandey, G. (2018). Challenges and future prospects of agri-nanotechnology for sustainable agriculture in India. *Environmental Technology & Innovation, 11*, 299–307. doi:10.1016/j.eti.2018.06.012

Park, M., Kim, C. Y., Lee, D. H., Choi, C. L., Choi, J., Lee, S. R., & Choy, J. H. (2004). Intercalation of magnesium urea complex into swelling clay. *Journal of Physics and Chemistry of Solids*, *65*(2-3), 409–412. doi:10.1016/j.jpcs.2003.09.011

Parsons, J. G., Lopez, M. L., Gonzalez, C. M., Peralta-Videa, J. R., & Gardea-Torresdey, J. L. (2010). Toxicity and biotransformation of uncoated and coated nickel hydroxide nanoparticles on mesquite plants. *Environmental Toxicology and Chemistry*, *29*(5), 1146–1154. doi:10.1002/etc.146 PMID:20821551

Patra, S., Mishra, P., Mahapatra, S. C., & Mithun, S. K. (2016). Modelling impacts of chemical fertilizer on agricultural production: A case study on Hooghly district, West Bengal, India. *Modeling Earth Systems and Environment*, *2*(4), 1–11. doi:10.100740808-016-0223-6

Pérez-de-Luque, A. (2017). Interaction of nanomaterials with plants: What do we need for real applications in agriculture? *Frontiers in Environmental Science*, *5*, 12. doi:10.3389/fenvs.2017.00012

Pradhan, S., & Mailapalli, D. R. (2017). Interaction of engineered nanoparticles with the agri-environment. *Journal of Agricultural and Food Chemistry*, *65*(38), 8279–8294. doi:10.1021/acs.jafc.7b02528 PMID:28876911

Prasad, R., Bhattacharyya, A., & Nguyen, Q. D. (2017). Nanotechnology in sustainable agriculture: Recent developments, challenges, and perspectives. *Frontiers in Microbiology*, *8*, 1014. doi:10.3389/fmicb.2017.01014 PMID:28676790

Prasad, T. N. V. K. V., Sudhakar, P., Sreenivasulu, Y., Latha, P., Munaswamy, V., Reddy, R. K. T. S., Sreeprasad, P. R., Sajanlal, P. T., & Pradeep, T. (2012). Effect of nanoscale zinc oxide particles on the germination, growth and yield of Peanut. *Journal of Plant Nutrition*, *35*(6), 905–927. doi:10.1080/01904167.2012.663443

Price, R. R., & Wagner, A. (2008). A method for treating agricultural crops using materials associated with tubular carriers. Academic Press.

Qing, Y., Cheng, L., Li, R., Liu, G., Zhang, Y., Tang, X., Wang, J., Liu, H., & Qin, Y. (2018). Potential antibacterial mechanism of silver nanoparticles and the optimization of orthopedic implants by advanced modification technologies. *International Journal of Nanomedicine*, *13*, 3311–3327. doi:10.2147/IJN.S165125 PMID:29892194

Rajonee, A. A., Zaman, S., & Imamul Huq, S. M. (2017). Preparation, characterization and evaluation of efficacy of phosphorus and potassium incorporated nano fertilizer. *Advances in Nanoparticles*, *6*(02), 62–74. doi:10.4236/anp.2017.62006

Rajput, V. D., Minkina, T., Feizi, M., Kumari, A., Khan, M., Mandzhieva, S., Sushkova, S., El-Ramady, H., Verma, K. K., Singh, A., Hullebusch, E., Singh, R. K., Jatav, H. S., & Choudhary, R. (2021). Effects of silicon and silicon-based nanoparticles on rhizosphere microbiome. *Plant Stress and Growth. Biology (Basel)*, *10*(8), 791. doi:10.3390/biology10080791 PMID:34440021

Raliya, R., Nair, R., Chavalmane, S., Wang, W. N., & Biswas, P. (2015). Mechanistic evaluation of translocation and physiological impact of titanium dioxide and zinc oxide nanoparticles on the tomato (*Solanum lycopersicum* L.) plant. *Metallomics*, *7*(12), 1584–1594. doi:10.1039/C5MT00168D PMID:26463441

Rastogi, A., Tripathi, D. K., Yadav, S., Chauhan, D. K., Živčák, M., Ghorbanpour, M., El-Sheery, N. I., & Brestic, M. (2019). Application of silicon nanoparticles in agriculture. *Biotech, 9*(3), 90. doi:10.1007/s13205-019-1626-7

Reda, F. M., El-Saadony, M. T., El-Rayes, T. K., Attia, A. I., El-Sayed, S. A., Ahmed, S. Y., Madkour, M., & Alagawany, M. (2021). Use of biological nano zinc as a feed additive in quail nutrition: Biosynthesis, antimicrobial activity and its effect on growth, feed utilisation, blood metabolites and intestinal microbiota. *Italian Journal of Animal Science, 20*(1), 324–335. doi:10.1080/1828051X.2021.1886001

Reddy Pullagurala, V. L., Adisa, I. O., Rawat, S., Kalagara, S., Hernandez-Viezcas, J. A., Peralta-Videa, J. R., & Gardea-Torresdey, J. L. (2018). ZnO nanoparticles increase photosynthetic pigments and decrease lipid peroxidation in soil grown cilantro (*Coriandrum sativum*). *Plant Physiology and Biochemistry, 132*, 120–127. doi:10.1016/j.plaphy.2018.08.037 PMID:30189415

Rezaei, M., & Abbasi, H. (2014). Foliar application of nanochelate and non-nanochelate of zinc on plant resistance physiological processes in cotton (*Gossipium hirsutum* L.). *Indian Journal of Plant Physiology / Official Publication of the Indian Society for Plant Physiology, 4*(4), 1137–1144.

Rico, C. M., Majumdar, S., Duarte-Gardea, M., Peralta-Videa, J. R., & Gardea-Torresdey, J. L. (2011). Interaction of nanoparticles with edible plants and their possible implications in the food chain. *Journal of Agricultural and Food Chemistry, 59*(8), 3485–3498. doi:10.1021/jf104517j PMID:21405020

Roh, J. Y., Eom, H. J., & Choi, J. (2012). Involvement of *Caenohabditis elegans* MAPK signaling pathways in oxidative stress response induced by silver nanoparticles exposure. *Toxicological Research, 28*(1), 19–24. doi:10.5487/TR.2012.28.1.019 PMID:24278585

Rui, M., Ma, C., Hao, Y., Guo, J., Rui, Y., Tang, X., Zhao, Q., Fan, X., Zhang, Z., Hou, T., & Zhu, S. (2016). Iron oxide nanoparticles as a potential iron fertilizer for peanut (*Arachis hypogaea*). *Frontiers in Plant Science, 7*, 815. doi:10.3389/fpls.2016.00815 PMID:27375665

Safyan, N., Reza, M., Darbaghshahi, N., & Bahari, B. (2012). The effect of microelements spraying on growth, qualitative and quantitative grain corn in Iran. *International Research Journal of Applied and Basic Sciences, 3*, 2780–2784.

Saifullah, M., Shishir, M. R. I., Ferdowsi, R., Rahman, M. R. T., & Van-Vuong, Q. (2019). Micro and nano encapsulation, retention and controlled release of flavor and aroma compounds: A critical review. *Trends in Food Science & Technology, 86*, 230–251. doi:10.1016/j.tifs.2019.02.030

Salama, H. M. H. (2012). Effects of silver nanoparticles in some crop plants, common bean (*Phaseolus vulgaris* L.) and corn (*Zea mays* L.). *International Research Journal of Biotechnology, 3*, 190–197.

Schwab, F., Zhai, G., Kern, M., Turner, A., Schnoor, J. L., & Wiesner, M. R. (2016). Barriers, pathways and processes for uptake, translocation and accumulation of nanomaterials in plants—Critical review. *Nanotoxicology, 10*(3), 257–278. doi:10.3109/17435390.2015.1048326 PMID:26067571

Seghatoleslami, M., Feizi, H., Mousavi, G., & Berahmand, A. (2015). Effect of magnetic field and silver nanoparticles on yield and water use efficiency of *Carum copticum* under water stress conditions. *Polish Journal of Chemical Technology, 17*(1), 110–114. doi:10.1515/pjct-2015-0016

Seleiman, M. F., Almutairi, K. F., Alotaibi, M., Shami, A., Alhammad, B. A., & Battaglia, M. L. (2021). Nano-Fertilization as an Emerging Fertilization Technique: Why Can Modern Agriculture Benefit from Its Use? *Plants*, *10*(1), 2. doi:10.3390/plants10010002 PMID:33375026

Selva Preetha, P., & Balakrishnan, N. (2017). A Review of nano fertilizers and their use and functions in soil. *International Journal of Current Microbiology and Applied Sciences*, *6*(12), 3117–3133. doi:10.20546/ijcmas.2017.612.364

Semida, W. M., Abdelkhalik, A., Mohamed, G. F., Abd El-Mageed, T. A., Abd El-Mageed, S. A., Rady, M. M., & Ali, E. F. (2021). Foliar application of zinc oxide nanoparticles promotes drought stress tolerance in eggplant (*Solanum melongena* L.). *Plants*, *10*(2), 421. doi:10.3390/plants10020421 PMID:33672429

Shang, Y., Hasan, M. K., Ahammed, G. J., Li, M., Yin, H., & Zhou, J. (2019). Applications of nanotechnology in plant growth and crop protection: A review. *Molecules (Basel, Switzerland)*, *24*(14), 2558. doi:10.3390/molecules24142558 PMID:31337070

Sharifi, R., Mohammadi, K., & Rokhzadi, A. (2016). Effect of seed priming and foliar application with micronutrients on quality of forage corn (*Zea mays*). *Environmental and Experimental Biology*, *14*(4), 151–156. doi:10.22364/eeb.14.21

Sharifi, R. S. (2016). Application of biofertilizers and zinc increases yield, nodulation and unsaturated fatty acids of soybean. *Zemdirbyste-Agriculture*, *103*(3), 251–258. doi:10.13080/z-a.2016.103.032

Sharma, P., Sharma, A., Sharma, M., Bhalla, N., Estrela, P., Jain, A., Thakur, P., & Thakur, A. (2017). Nanomaterial fungicides: In vitro and in vivo antimycotic activity of cobalt and nickel nanoferrites on phytopathogenic fungi. *Global Challenges (Hoboken, NJ)*, *1*(9), 1700041. doi:10.1002/gch2.201700041 PMID:31565296

Sharonova, N. L., Yapparov, A. Kh., Khisamutdinov, N. Sh., Ezhkova, A. M., Yapparov, I. A., Ezhkov, V. O., Degtyareva, I. A., & Babynin, E. V. (2015). Nanostructured water-phosphorite suspension is a new promising fertilizer. *Nanotechnologies in Russia*, *10*(7-8), 651–661. doi:10.1134/S1995078015040187

Sheykhbaglou, R., Sedghi, M., Tajbakhsh-Shishevan, M., & Seyed-Sharifi, R. (2010). Effects of nano-iron oxide particles on agronomic traits of soybean. *Notulae Scientia Biologicae*, *2*(2), 112–113. doi:10.15835/nsb224667

Shukla, S., Kumar, R., Mishra, R., Pandey, A., Pathak, A., Zaidi, M., Srivastava, S., & Dikshit, A. (2015). Prediction and validation of gold nanoparticles (GNPs) on plant growth promoting rhizobacteria (PGPR): A step toward development of nano-biofertilizers. *Nanotechnology Reviews*, *4*(5), 439–448. doi:10.1515/ntrev-2015-0036

Siddiqui, M. H., Al-Whaibi, M. H., Faisal, M., & Al Sahli, A. A. (2014). Nano-silicon dioxide mitigates the adverse effects of salt stress on *Cucurbita pepo* L. *Environmental Toxicology and Chemistry*, *33*(11), 2429–2437. doi:10.1002/etc.2697 PMID:25066835

Simarmata, T., Hersanti, Turmuktini, T., Fitriatin, B. N., Setiawati, M. R., & Purwanto. (2018). Application of bioameliorant and biofertilizers to increase the soil health and rice productivity. *Hayati Journal of Biosciences*, *23*(4), 181–184. doi:10.1016/j.hjb.2017.01.001

Singh, M. D. (2015). *Studies on the effect of time of application and concentration of nano zinc sulphide (nZS) on the growth and yield of sunflower (Helianthus annuus L.)* [M. Sc. (Agri.) Thesis]. University of Agricultural Sciences, Dharwad, Karnataka, India.

Sohair, E. E. D., Abdall, A. A., Amany, A. M., Faruque, H. M. D., & Houda, R. A. (2018). Evaluation of nitrogen, phosphorus and potassium nano-fertilizers on yield, yield components and fiber properties of egyptian cotton (*Gossyppium barbadense* L.). *Journal of Plant Sciences and Crop Protection*, *1*(3), 1–10.

Srivastava, A. K., & Malhotra, S. K. (2017). Nutrient use efficiency in perennial fruit crops – A review. *Journal of Plant Nutrition*, *40*(13), 1928–1953. doi:10.1080/01904167.2016.1249798

Subramanian, K. S., Manikandan, A., Thirunavukkarasu, M., & Rahale, C. S. (2015). Nano-fertilizers for Balanced Crop Nutrition. In M. Rai, C. Ribeiro, L. Mattoso, & N. Duran (Eds.), *Nanotechnologies in Food and Agriculture* (pp. 69–80). Springer International Publishing. doi:10.1007/978-3-319-14024-7_3

Suriyaprabha, R., Karunakaran, G., Yuvakkumar, R., Rajendran, V., & Kannan, N. (2012). Silica nanoparticles for increased silica availability in maize (*Zea mays* L) seeds under hydroponic conditions. *Current Nanoscience*, *8*(6), 902–908. doi:10.2174/157341312803989033

Tahmasbi, D., Zarghami, R., Azghandi, A. V., & Chaichi, M. (2011). Effects of nanosilver and nitroxin biofertilizer on yield and yield components of potato minitubers. *International Journal of Agriculture and Biology*, *13*, 986–990.

Tantawy, A. S., Salama, Y. A. M., Abdel-Mawgoud, A. M. R., & Ghoname, A. A. (2014). Comparison of chelated calcium with nano calcium on alleviation of salinity negative effects on tomato plants. *Middle East Journal of Agriculture Research*, *3*(4), 912–916.

Tarafdar, J. C., Raliya, R., Mahawar, H., & Rathore, I. (2014). Development of zinc nanofertilizer to enhance crop production in pearl millet (*Pennisetum americanum*). *Agricultural Research*, *3*(3), 257–262. doi:10.100740003-014-0113-y

Tripathi, D. K., Singh, V. P., Prasad, S. M., Chauhan, D. K., & Dubey, N. K. (2015). Silicon nanoparticles (SiNp) alleviate chromium (VI) phytotoxicity in *Pisum sativum* (L.) seedlings. *Plant Physiology and Biochemistry*, *96*, 189–198. doi:10.1016/j.plaphy.2015.07.026 PMID:26298805

Upadhyaya, H., Shome, S., Tewari, S., Bhattacharya, M. K., & Panda, S. K. (2015). Effect of Zn nanoparticles on growth responses of rice. In B. Singh, A. Kaushik, S. K. Mehta, & S. K. Tripathi (Eds.), *Nanotechnology: Novel Perspectives and Prospects* (pp. 508–512). McGraw Hill Education.

Uzu, G., Sobanska, S., Sarret, G., Muñoz, M., & Dumat, C. (2010). Foliar lead uptake by lettuce exposed to atmospheric fallouts. *Environmental Science & Technology*, *44*(3), 1036–1042. doi:10.1021/es902190u PMID:20063891

Vafa, Z. N., Sirousmehr, A. R., Ghanbari, A., Khammari, I., & Falahi, N. (2015). Effects of nano zinc and humic acid on quantitative and qualitative characteristics of savory (*Satureja hortensis* L.). *International Journal of Biosciences*, *6*(3), 124–136. doi:10.12692/ijb/6.3.124-136

Vannini, C., Domingo, G., Onelli, E., Prinsi, B., Marsoni, M., Espen, L., & Bracale, M. (2013). Morphological and proteomic responses of *Eruca sativa* exposed to silver nanoparticles or silver nitrate. *PLoS One*, *8*(7), e68752. doi:10.1371/journal.pone.0068752 PMID:23874747

Verma, K. K., Song, X. P., Joshi, A., Tian, D. D., Rajput, V. D., Singh, M., Arora, J., Minkina, T., & Li, Y. R. (2022). Recent trends in nano-fertilizers for sustainable agriculture under climate change for global food security. *Nanomaterials (Basel, Switzerland)*, *12*(1), 173. doi:10.3390/nano12010173 PMID:35010126

Wang, F. Y., Wang, L., Shi, Z. Y., Li, Y. J., & Song, Z. M. (2012). Effects of AM inoculation and organic amendment, alone or in combination, on growth, P nutrition, and heavy-metal uptake of tobacco in Pb-Cd-contaminated soil. *Journal of Plant Growth Regulation*, *31*(4), 549–559. doi:10.100700344-012-9265-9

Wang, L., Hu, C., & Shao, L. (2017). The antimicrobial activity of nanoparticles: Present situation and prospects for the future. *International Journal of Nanomedicine*, *12*, 1227–1249. doi:10.2147/IJN.S121956 PMID:28243086

Wang, W. N., Tarafdar, J. C., & Biswas, P. (2013). Nanoparticle synthesis and delivery by an aerosol route for watermelon plant foliar uptake. *Journal of Nanoparticle Research*, *15*(1), 1–13. doi:10.100711051-013-1417-8

Wang, X. P., Li, Q. Q., Pei, Z. M., & Wang, S. C. (2018). Effects of zinc oxide nanoparticles on the growth, photosynthetic traits, and antioxidative enzymes in tomato plants. *Biologia Plantarum*, *62*(4), 801–808. doi:10.100710535-018-0813-4

Wang, Z., Xu, L., Zhao, J., Wang, X., White, J. C., & Xing, B. (2016). CuO nanoparticle interaction with *Arabidopsis thaliana*: Toxicity, parent-progeny transfer, and gene expression. *Environmental Science & Technology*, *50*(11), 6008–6016. doi:10.1021/acs.est.6b01017 PMID:27226046

White, J. C., & Gardea-Torresdey, J. (2018). Achieving food security through the very small. *Nature Nanotechnology*, *13*(8), 627–629. doi:10.103841565-018-0223-y PMID:30082813

Wu, W., & Ma, B. (2015). Integrated nutrient management (INM) for sustaining crop productivity and reducing environmental impact: A review. *The Science of the Total Environment*, *512-513*, 415–427. doi:10.1016/j.scitotenv.2014.12.101 PMID:25644838

Xiao, Y., Vijver, M. G., Chen, G., & Peijnenburg, W. J. (2015). Toxicity and accumulation of Cu and ZnO nanoparticles in *Daphnia magna*. *Environmental Science & Technology*, *49*(7), 4657–4664. doi:10.1021/acs.est.5b00538 PMID:25785366

Yang, L., & Watts, D. J. (2005). Particle surface characteristics may play an important role in phytotoxicity of alumina nanoparticles. *Toxicology Letters*, *158*(2), 122–132. doi:10.1016/j.toxlet.2005.03.003 PMID:16039401

Yang, Y., Wang, J., Xiu, Z., & Alvarez, P. J. (2013). Impacts of silver nanoparticles on cellular and transcriptional activity of nitrogen-cycling bacteria. *Environmental Toxicology and Chemistry*, *32*(7), 1488–1494. doi:10.1002/etc.2230 PMID:23554086

Yassen, A., Abdallah, E., Gaballah, M., & Zaghloul, S. (2017). Role of silicon dioxide Nanofertilizer in mitigating salt stress on growth, yield and chemical composition of cucumber (*Cucumis sativus* L.). *International Journal of Agricultural Research*, *12*(3), 130–135. doi:10.3923/ijar.2017.130.135

Younis, A., Khattab, H., & Emam, M. (2020). Impacts of silicon and silicon nanoparticles on leaf ultrastructure and TaPIP1 and TaNIP2 gene expressions in heat stressed wheat seedlings. *Biologia Plantarum*, *64*, 343–352. doi:10.32615/bp.2020.030

Zhang, X., Davidson, E., Mauzerall, D., Searchinger, T. D., Dumas, P., & Shen, Y. (2015). Managing nitrogen for sustainable development. *Nature*, *528*(7580), 51–59. doi:10.1038/nature15743 PMID:26595273

Zhu, Z. J., Wang, H., Yan, B., Zheng, H., Jiang, Y., Miranda, O. R., Rotello, V. M., Xing, B., & Vachet, R. W. (2012). Effect of surface charge on the uptake and distribution of gold nanoparticles in four plant species. *Environmental Science & Technology*, *46*(22), 12391–12398. doi:10.1021/es301977w PMID:23102049

Zulfiqar, F., Navarro, M., Ashraf, M., Akram, N. A., & Munné-Bosch, S. (2019). Nanofertilizer use for sustainable agriculture: Advantages and limitations. *Plant Science*, *289*, 110270. doi:10.1016/j.plantsci.2019.110270 PMID:31623775

Zuverza-Mena, N., Martínez-Fernández, D., Du, W., Hernandez-Viezcas, J. A., Bonilla-Bird, N., López-Moreno, M. L., Komárek, M., Peralta-Videa, J. R., & Gardea-Torresdey, J. L. (2017). Exposure of engineered nanomaterials to plants: Insights into the physiological and biochemical responses-A review. *Plant Physiology and Biochemistry*, *110*, 236–264. doi:10.1016/j.plaphy.2016.05.037 PMID:27289187

Chapter 7
The Nanotechnological Approaches in Agriculture and Its Allied Fields

Aman Prakash
Jacob Institute of Biotechnology and Bioengineering, Sam Higginbottoms University of Agriculture, Technology, and Sciences, India

Sameer Quazi
https://orcid.org/0000-0002-1258-4088
GenLab Biosolutions Private Limited, Bangalore, India & Department of Biomedical Sciences, Anglia Ruskin University, UK & School of Health Sciences, The University of Manchester, UK & SCAMT Institute, ITMO University, St. Petersburg, Russia

Pragalbh Tiwari
Jacob Institute of Biotechnology and Bioengineering, Sam Higginbottoms University of Agriculture, Technology, and Sciences, India.

ABSTRACT

Nanotechnology is a degree knowledge domain analysis field. In recent efforts, square measures were created to reinforce agricultural yield through a thorough analysis of technology. The revolution resulted in blind usage of pesticides and chemical fertilizers that caused loss of soil selection and developed resistance against pathogens and pests. Nanoparticles are experimenting as anti-infestation as opposed to morbific fungi. Anti-infestation venture nanoparticles of oxide, silver, and titanium oxide have been tested against Macrophomina phaseolina, a big soil-borne infectious agent of pulse and seed crops. A higher antifungal effect was discovered in silver nanoparticles at a decreased level than in flowers of metallic elements and titanic oxide nanoparticles.

DOI: 10.4018/978-1-6684-5533-3.ch007

INTRODUCTION

The study of nanotechnology is an area of degree-level knowledge domain analysis. Square measurements were developed in recent times in an attempt to increase agricultural output by in-depth research of technological developments. Loss of soil selection and the development of resistance to infections and pests were both brought about by the revolution's indiscriminate use of pesticides and artificial fertilisers. Nanoparticles or nanochips may be used to mediate the delivery of materials to plants and to create sophisticated biosensors for precision agriculture. Common fertilisers, insecticides, and herbicides may be improved by being encapsulated in nanoparticles, which then release the nutrients and agrochemicals slowly and steadily, in amounts that are precisely proportional to the plant's needs. The kits based on nanotechnology for detecting infectious illnesses in plants are gaining popularity because of their efficiency in detecting infections quickly and early on. This article discusses the possible benefits and applications of technology in precision agriculture. New tools and procedures based on nanotechnology have the potential to solve the many problems plaguing conventional agriculture and completely transform the industry. As reported by (Mousavi et al., 2011), knowledge-base research in the area of nanotechnology may have some promise. Opportunities in the pharmaceutical industry, medical science, physics, and even farming are all laid bare. A lot of good stuff can be done with and aided by technology. Roughly seven billion people live in the world, with just five hundred residing in Asia. While there is a surplus of food in the industrialised world, a disproportionate number of people in developing nations confront daily food shortages due to environmental consequences or political instability. The goal of agricultural research in developing nations is to create drought- and pest-resistant crops that also enhance output. Health care, textiles, materials, data and communication technology, and the energy industries all stand to benefit greatly from technological advancements. A lot of people these days are also interested in how technology may be used in the food production and farming sectors. The potential benefits of investing in agriculture and food nanotechnologies, such as enhanced food quality and safety, decreased reliance on nonrenewable agricultural inputs, and enhanced efficiency and nutrient density, are giving the sector more clout. Analysis improvements reveal glimpses of possible applications in agricultural, food, and water safety that would have critical effects on rural communities in underdeveloped nations, despite the fact that most investment is established in rich countries. This analysis focuses on cutting-edge strategies for managing water, pesticides, the constraints of using chemical pesticides, and the promise of nanomaterials in agricultural agriculture. (Duhan et.al. 2017)

NANOTECHNOLOGY SCIENCE – DESCRIPTION

Nanotechnology is an interdisciplinary area that has attracted scientists from many other disciplines, including chemists, physicists, botanists, pharmaceutical doctors, and engineers. Nanotechnology refers to the study and development of methods for identifying, manipulating, and living with substances that have a bitsy, fragmentary, or superfragmentary quality. The scope of this field has been loosely defined to include materials, systems, and processes that function on a scale of 100 nm or smaller. There are a billion fewer metres in a nanometer. In general, the term "nano" refers to the range of very small increments of measurement, often between 1 and 100 nanometers (nm). Sunlight, by way of contrast, has a wavelength of somewhere in the range of 400 to 700 nm. At each successively smaller size (a white cell is 10,000 nm in diameter, a microbe is 1,000 to 10,000 nm in diameter, a virus is 75 to 100 nm in diameter, an upper molecule is 5 to 50 nm in diameter, a compound (DNA) is 2 nm in diameter, and an atom is 0.1 nm in diameter), the physical, biological, and chemical characteristics of materials become increasingly distinct from one another, and often exhibit unexpected behaviours. technology takes into account the relative importance of issues with viruses and other diseases. Consequently, offers great promise as a means of identifying and eradicating harmful microorganisms. Nanotechnology has just gained nano-molecules, and there is a danger of operational on nano measuring extent, control, and catalysed actinic reaction of those forms. Nanomaterials are made up of minute particles, and their components have a surprisingly large impact on the material's properties. Nanoparticles may trigger "magic bullets" filled with toxicants and actinic, which will cause particular plant components to secrete their inner content. Nano coverings lessen the likelihood of toxicants successfully penetrating the epidermis, allowing for the controlled, gradual release of the active ingredients. Nanobiotechnology is the integration of nanoscale automation with cytology. The development of nanotechnology necessitates the input of many different professionals, including but not limited to chemists, biologists, physicians, and engineers. The most important nano substances, which have various uses across many fields of study, are nano polymers and nano shells. Nano synthesis consistently yields nano polymers, which are three-dimensional particles, and a nano-shell, which might be a nano-particle with an insulator core and a very thin covering of gold. As reported by (Mousavi et al., 2011)

Biosynthesis of Nanoparticles & Their Use In Agriculture

Since there are now many chemical methods available for combining nanoparticles, many of which involve the use of cyanogenic chemicals, the time has come to switch to safer, more sustainable practises. Scientists are anticipating a wide variety of biotic organisms, such as bacteria, mould, tall plants, actinomycetes, and food for nanoparticles, to reduce the salts that correlate with nanoparticles. The combination of nanoparticles requires botanical origins that are entirely unlike from one another, which is essential in agriculture for achieving pinpoint accuracy. A number of researchers (Duhan et al. 2017, Malik et al., 2020 pharm) have shown that

Applications of Nanotechnology in Precision Agriculture

1) Fertilizers

Extremely large quantities of chemicals, such ammonium ion salts, urea, and nitrate or phosphate compounds, have greatly enhanced food production; nevertheless, they also have various deleterious impacts on the beneficial soil microflora. Most of the fertilisers seem to flow off and pollute the environment rather than being used on plants. These issues may be resolved by using nanomaterial-coated fertilisers. Because of their greater natural phenomena than regular surfaces, nanoparticles in nano-materials have the potential to obfuscate the release of the plant's nourishment. In addition, nano-coatings may shield bigger particles from the environment. Nano-ability coating's to transport nutrients or insecticides is shown schematically. (Duhan et.al. 2017)

Figure 1. Controlled release of pesticides/fungicides Nutrients from Nano-coating.
Source: - Nanotechnology: The new perspective in precision agriculture. Biotechnology Reports, Elsevier.

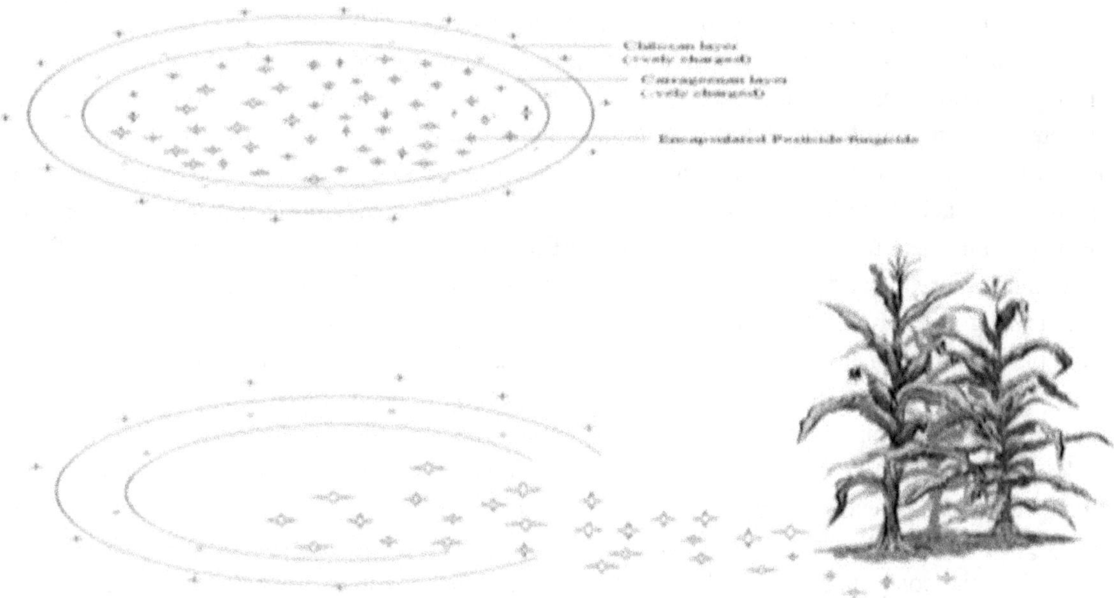

Ø *Chemical Fertilizers*

Since India's nascent technological development, the country's consumption of gas chemicals inside the form of carbamide has grown by a factor of 29. Eighty percent of the increase in regional N2O (a greenhouse gas) that causes amplified air temperature and hence contributes to global warming may be attributed to increased production enabled by more nitrogen software system. To wit: (Park et al., 2012). To compensate for the deficiency of nitrogen (N), phosphorus (P), and potassium (K) in the soil, chemical fertilisers such as urea, diammonium phosphate (DAP), and single superphosphate (SSP) are used in farming. But most of these fertilisers are lost due to run-off or evaporation. As reported by (Trenkel et al., 1997). Forty to seventy percent of the nitrogen, eighty to ninety percent of the phosphorus, and fifty to seventy percent of the atomic number 19 of the dispensed chemical is lost to the environment and cannot be absorbed by the approach means of plant carried out, causing monetary resource loss to the state and environmental pollutants also. This was found by Obboi et al. in the year 2000. Recent methods include using chemical fertilisers such as nanocoated carbamide. Nanocoatings strike a delicate balance between preventing the wasteful application of chemicals and allowing for the gradual, prolonged release of fertiliser that is more easily and effectively taken up by plant roots. Use of slow-release compounds to increase fertiliser use and reduce pollution has recently reemerged as an option. It was found (Wu et al., 2008). Since sulphide satisfaction is most helpful for sulphur-poor soils, fertilisers with a sulphide nano-coating (100 nm surface) (Brady et al., 1999) may work as slow-launch fertilisers. The findings were published in 1995 (Santoso et al. Most Indian soil is deficient in these macronutrients, especially nitrogen, making nano-lined carbamide and phosphate and their continued deployment important for meeting soil and crop needs. To accomplish this gradual release of chemicals, scientists employed a number of flavouring and artificial polymers. Chitosan nanoparticles (78 nm) are a biodegradable substance with excellent results for the controlled release of NPK plant food. This was shown by (Corradini et al., 2010). The slow-release behaviour of fertilisers may also be improved with the help of China stone and microporous bio-compatible nano-particles. In 2008, a group of researchers led by Wilson found that....

To prevent nutrient losses and eliminate unfavourable vitamin contact with microbes, water, and air, nano-fertilizers strike a balance between the release of fertiliser gas and phosphorus and absorption through the plant's approach of methods. To wit: (Emadian et al., 2017). Use of nano-fertilizer has the potential to increase plant nutrient uptake from soil. A nano-fertilizer encased in nano-silica will prevent infections by forming a binary pic at the cellular wall of invading fungi or microorganisms during vitamin absorption, and will go on to promote plant growth at lower earth temperatures and humidity levels while also making plants more resistant to disease. Emadian et al. (2002) cite this study as their source. Fertilizers high in silicon, known as silicon oxide nanoparticles, have long been used to boost plant resilience by encouraging spermatophyte growth and root formation. That was the conclusion reached by researchers (Hutasoit et al., 2013). TiO2 or non-toxic metallic elements might be employed as components in fertilisers to increase meal production. Shaviv et al. (2000) cite this finding. Due to the high solubility of many gas fertilisers and their susceptibility to natural process and denitrification (particularly in the nitrate form), the components in fertilisers can increase water retention. As a result, a wide variety of slow-release fertilisers (SRFs) and managed-launch fertilisers (CRFs) have been developed through the use of artificial or biopolymers. This was shown by Subramanian et al. (2011). Adsorbents like zeolite and more expensive nano-clays like halloysite, montmorillonite, and clay were used to create nitrogen fertilisers with controlled release. To wit: (Sharmila et al., 2010).

The use of poly-methacrylic acid (PMAA) in the production of chitosan nano-particles for incorporation into NPK plant food has already been reported. A combination of gas, atomic number 19, and phosphorus was shown to be optimal for maintaining the chitosan-poly 2-methylpropenoic acid (CS-PMAA) colloid's equilibrium. This was related to the inorganic phosphate's higher ion fee compared to the anion costs of the potassium chloride and urea. Hasaneen et al. (2014) provide evidence for this. The dispersal of CS-PMAA with 500 ppm nitrogen was more stable than with phosphorus. To wit: (Wu et al., 2006) In the presence of harmful services from the carbamide molecules for dispersion with potassium, the addition of more than 500 ppm of nitrogen causes a decrease in top quality costs within the mixture dispersion of CS-PMAA; however, the stability of the solution has been demonstrated with the addition of 400 ppm. For example, (Vasconcelos et al., 2006) This proved the existence of Cl ions (from KCl), which at the highest concentration of 400 ppm were not affecting the dispersion stability of the combination. It has been shown that (Moura et al., 2008)

Biological Amendments

Biofertilizers are beneficial microorganisms that may be added to soil. Mycorrhizal fungi, Rhizobium, Azotobacter, Azospirillum, blue-green algae, the species of bacteria that produce phosphate, and genuine bacteria are all included. This was shown by Wu et al. (2005). Natural deposits are broken down by microorganisms into a simple mixture that boosts plant health, increases crop productivity, and maintains the soil's flavorful environment. Implementing biofertilizers successfully requires careful attention to the processes of preparation, storage, and application. "(Jha et al., 2006)" Temporary shelf life, temperature sensitivity, and garage desiccation concerns are drawbacks to using the product. Possible desiccation-resistant compound nano-particle packages were used to cover biofertilizer arrangements after articulation. One of the ways that the garage stores and disperses its inhabitants is via a combination of water and oil. Reference: (Vandergheynst et al., 2006).

Since the oil acts as a barrier, less water evaporates from the microbe. because of the potential desiccation risk of certain microbes, this is frequently a desirable feature. By incorporating materials into the oil and/or liquid phases, water-in-oil emulsions improve the viability and launch dynamics of all constituent molecules. Yet, one of the most pressing issues is alluviation throughout the whole garage. Due to the limited viability of land and water resources and improvement of farming vegetation of Fabaceae, the need for silver and gold nanoparticles as growth commerce substance is also effective. Hydrophobic silicon oxide nanoparticles shrank molecular sedimentation and advanced molecular viability by thickening the oil section throughout the garage (Vandergheynst et al., 2007). In a 2013 study (Dikshit et al. These nano-particles containing flavorer biofertilizers, which include the bacteria species fluorescens, Bacillus globigii, and Paenibacillus Elgin, have shown excellent underground business activity in a condition analogous to artificial insemination. As a result, a single cubic decimeter of nano-biofertilizers may be used in a wide range of plant areas, despite the fact that this volume is little when compared to the costs of other fertilisers. There are a variety of soil microbes present in the root zone, but the ones that are most suited for plant growth and sales are called plant growth commerce rhizobacteria (PGPR). Fluorescent, true showing subtilis, Paenibacillus Elgin, and Pseudomonas putida were chosen as model bacteria to study the effects of gold nanoparticles on PGPR. The exposure of Pseudomonas putida to gold nanoparticles resulted in no unusual or harmful effects. Gold nanoparticles are also used as nano-biofertilizers since they were shown to promote the development of P. fluorescens, P. Elgin, and B. subtilis. According to a study (Shukla et al., 2015),

2) Supply of Micro-Nutrients in Agriculture Field

It is a proven truth that growth and development need trace minerals including manganese, copper, boron, iron, molybdenum, zinc, and others. The soil's micronutrients, such as zinc, iron, and molybdenum, have been depleted as a result of the revolution and modern agricultural techniques, which have led to a significant rise in crop yields. Incorporating micronutrients at the foliar level improves absorption by the leaves. Micronutrient delivery to plants is another potential use of this technique. To improve soil health and vitality, nanoformulated micronutrients are often sprayed on plants or added to the soil for absorption by roots. Extensive experiments at varying pH and temperatures have been performed on the release behaviour of 1- naphthyl carboxylic acid from chitosan nanoparticles. Slow release of agrochemicals like hormones was shown to be a possibility with the formulation. To meet the needs of plants for micronutrients, several different nanoparticles are tried. Plants mostly grown in high pH and chalky soils may suffer from iron shortage as a common problem. It's possible that using nanoparticle technology to apply iron compounds as a foliar spray may solve the problem. The impact of sprinkling cereal with iron chemical compound nanoparticles on the recovery of height, relaxation, and uniformity has been evaluated. Checks on spike weight, 1,000-grain weight, biological relent, cereal relent, and grain macromolecule content have been performed using concentrations of the iron nano-oxide solution ranging from 0% to 4%. Duhan et al. (2017)

These characteristics were found to be more prevalent in manage two than in manage one. As a result of victimisation by low concentrations of superparamagnetic iron chemical compound nanoparticles, the pigment contents of sub-apical leaves of soybeans increased in a very greenhouse test conducted under aquacultural settings. The results of this research show that iron oxide nanoparticles may be added to soybeans as a source of iron to lessen the health risks associated with iron deficiency anaemia. Peas grown with 500 mg/l iron nanoparticles in the foliage produced 47% more pods per plant, 7 percent more weight per thousand seeds, 34 percent more iron content in the leaves, and 10 percent more colourant than the controls. The use of iron nanoparticles improves batch output even more than would be possible with daily iron salt treatment. Duhan et al. (2017)

Underneath the iron nanoparticles, the framework expands by 28%, 4%, 45%, and 12%, respectively, as compared to when iron salt is used as a serving medium. In addition, the beneficial effect of atomic number 12 nanoparticles employed as nano plant food on black-eyed peas was greatly enhanced by the addition of iron nanoparticles. Most of them need 1-5 mg L1 iron in soil solution for optimal development. Growth of Phaseolus aureus and photosynthesis may be stimulated by Mn nanoparticles. Short-term experiments using zinc-oxide nanoparticles in plant agar media showed improved growth of Vigna radiata and chickpea grass at low concentrations, whereas long-term experiments showed a reduction in growth rates of radicle and sprout at highest concentrations. Duhan et al. (2017)

3) Nano Fungicides

Missteps in the assembly process may be traced back to an infestation infection that occurred during the batch. Although there is a class of fungicides available for commercial use, the causes of their deleterious effects on plants remain unclear. Tissues will be determined in large part by engineering science. Scientists are investigating the potential of nanoparticles as an anti-infestation medium unfriendly to morbific fungus. Testing with Macrophomina phaseolina, a major soil-borne infectious agent of pulse and seed crops, has shown that nanoparticles of oxide, silver, and titanium oxide have antifungal action. Lower quantities of silver nanoparticles were shown to have a more potent anti-infestation impact than higher concentrations of metal oxide or titanic oxide nanoparticles. There is a latent aversion to Phyto-microbe-hostile nano silica in maize's behaviour toward particles in the 20-40 nm range. Comparing Fusarium oxysporum and Aspergillus niger to pure silicon oxide. Extracts of leaves from plants treated with nano silica exhibited increased production of phenolic resin molecules, which are known to be effective against these fungus. These findings demonstrated very high resilience in maize when compared with At 10 and 15 kg/ha, nano silica has been shown to have a positive effect on disease prevention and the production of total phenols, oxidase, and polyphenol oxidase. a plague-like sickness affecting the whole group at once due to a severe drop in numbers during assembly. Duhan et al. (2017)

While there are fungicides on the market, they have negative effects on plants and come at a cost. The field of engineering science will play a crucial part in figuring out this issue. Instead of using nanoparticles to combat pathogenic fungus, researchers are exploring the possibility of using them to combat infestations. Nanoparticles of silver and titanium oxide, as well as oxide, have been explored in an anti-infestation effort against Macrophomina phaseolina. This bacterium is a major soil-borne infectious agent of pulse and seed crops. Silver nanoparticles were shown to have a greater antifungal impact at a lower concentration than metallic element flowers and titanium dioxide nanoparticles. In comparison to bulk silicon oxide, maize served with nano silica (20-40 nm) has been tested for its resistance against Phyto-microbes such Fusarium oxysporum and fungus species niger. The phenolic resin compounds expressed by plants treated with nano silica in collected leaf extracts were enhanced, and in the case of these fungi, the expression of a pressure-reactive enzyme occurred on rare occasions. Duhan et al. (2017)

The illness index and the expression of total phenols, essential amino acid ammonia lyase, oxidase, and polyphenol oxidase all demonstrated that maize treated with Nano silica at 10 and 15 kg/ha was much more resistant than maize treated with bulk. Therefore, in the case of phytopathogens, silica nanoparticles may serve as an additional powerful origin. When compared to other metals, silver's antifungal activity is much stronger. Silver ions may be to blame for this phenomenon because they inactivate thiol teams in the semipermeable membranes of plant plasma membranes, which in turn disrupts the plant's transmembrane, energy metabolism, and lepton transport chain. Other processes include mutations in fungal DNA, dissociation of accelerator complexes critical to the metabolic process chain, decreased membrane permeability, and cell lysis Duhan et al. (2017).

4) Nano-Herbicides

Weeds are the greatest danger in farming because they reduce crop yields by stealing nutrients that would otherwise be used by crop plants. Using time-consuming conventional methods to remove weeds. There is a wide variety of weedkillers on the market. They are effective in eliminating weeds in the fields, but they can cause harm to the crops itself. They are also to blame for soil contamination and the decline of soil fertility. Nano-herbicides will play a crucial role in weed control without leaving behind toxic residues in the soil or surrounding environment. Herbicides that are encased in compound nanoparticles are more safer for the environment. Long-term, improper usage of weed herbicides may leave harmful residues in the soil, harming following plantings. If you keep using the same pesticide on your weeds, eventually they will develop resistance to it. For the purpose of dichlorination of atrazine from soil dust and moisture, the sequential rate of nano valent iron was assessed. Herbicides are loaded onto nanoparticles and sent straight to the weed roots. These chemicals cross the weeds' membranes and enter the cells, where they block the glycolysis-related metabolic processes. The result of this is the plant's demise. The toxicity of poly(-caprolactone) nano-capsules containing ametrine and herbicide against the alga Pseudokirchneriella sub-capital and the water flea similis, which is carried by tiny crustaceans, was evaluated. Because herbicides alone are less lethal to algae and more lethal to micro-crustaceans, herbicides are often encased in poly-nano capsules. Duhan et al. (2017)

Bio-Sensors in Precise Way Agriculture

The long-sought-after aim of precision farming is to enhance crop yields while decreasing the amount of inputs like fertilisers, pesticides, herbicides, and so on. To ascertain whether or not crops are developing at peak efficiency, or to precisely characterise the nature and website of problems, exactitude farming employs computers, sensors, global satellite positioning systems, and remote sensing devices to measure extremely localised environmental conditions. In the end, smart sensors used for precision farming may assist boost agricultural output by giving farmers access to more accurate data. Duhan et al. (2017)

Ø *Nano-biosensors: -*

Nano-biosensors have also shown useful in detecting a wide range of agricultural factors, including plant food, germicides, microbicides, and the pH of moist and soil dust. Property agriculture will benefit greatly from the implementation of biosensor requirements, since this will increase agricultural yields. Accurate farming, aided by smart sensors, has the potential to increase output because of the advantages of modern technology, such as improved fertilisation management, lower input value, and environmental protection. This has been shown by several researchers (Rai et al., 2012). Through precision farming, the very sophisticated transport systems of nanosensors may aid in the environmentally friendly use of fertiliser supplies including water, nutrients, and agrochemicals. (Jones et al. 2006) It is also possible to utilise nano-sensors to detect plant viruses, the concentration of vitamins in the soil, and the presence of crop diseases. As reported by (Brock et al., 2011)

As a result of collecting pressure curves, the weed killer mesopleuron-methyl (an acetolactate synthase inhibitor) was identified using a nano-biosensor based on atomic pressure research tip functionalized with the acetolactate synthase protein. This was shown to be the case (Silva et al., 2013). In addition, the application of bio nano-sensors enables the rapid detection of microscopic organisms and viruses with precise quantification, elevating the safety of the food for the consumer. This has been shown by Olles et al. (2010). Surface functionalized carbon nanotubes tailored with amino firms to regulate inexperienced immobilisation of acetylcholinesterase onto the floor of glassy carbon conductor were utilised to create a biosensor sensitive enough to be used for direct detection of vegetable samples. Several researchers (Yu et al., 2015) came to this conclusion. In order to detect organophosphate microbicides, acetylcholinesterase biosensors based on the stacking of carbon nanotube layers atop cyst bioreactors were developed. In a 2013 study (Yan et al. Using hole gold nano-spheres, a sensitive acetylcholinesterase bio-sensor with detection limits of 0.06 g/dm3 for organophosphate and 0.08 g/dm^3 for carbofuran was developed. Reference: (Sun et al., 2013).

Using graphene-chitosan nano-composite improved electrodes, Zn-Se quantum spots inhibited acetylcholinesterase, an enzyme necessary for testing insect powder insecticides. With an inspection interval of 0.2 nmol/dm3, this biosensor was able to detect insect powder insecticides using methyl group insect powder as a model protein inhibitor. To wit: (Dong et al., 2013). A chemical science biosensor based on acetylcholinesterase immobilised on polyaniline activated on vertically constructed carbon nanotubes and wrapped with ssDNA was developed for the determination of the insecticides methyl parathion and chlorpyrifos. The detection limit of the biosensor for the pesticides was found to be 1 & 10-12 mol/dm3, and this is because the insecticides work by inhibiting the enzyme reaction acetylcholinesterase-acetylcholine, which caused minor changes in near pH inside the surrounding of a conductor surface (Viswanathan et. al., 2009).

For gases such as ammonia, chemical element oxides, gas sulphide, gas, and hazardous organics, electrochemically functionalized single-walled carbon nanotubes (SWCNTs) complete nanosensors with steel/steel oxide nanoparticles or nanotubes have the potential utility in pursuing agricultural pollution and analysing effects on home count or health and in the increase of crop productivity. To wit: (Sekhon et al., 2014). Extremely sensitive (1.0 mol/dm3) screening of organophosphorus retailers using gold nanoparticles and fluorescence qualitative analysis has been proposed.

Ø Quantum Dots (QDs): -

Larger green luminescence, narrow emission spectra, excellent photostability, and robustness proportional to particle size and material composition make quantum dots superior to natural fluorescent dyes in terms of brightness and speed. Due of their broad absorption spectra, QDs may be excited to any colour by using a single-color light source. Pathogens in unusual plant diseases had been sought with the use of quantum dots. According to a study (Warad et al., 2004),

As a result of Phytoplasma aurantifolia, quantum dots-fluorescence resonance electricity transfer-primarily based completely sensors have been developed to detect witches' broom sick of lime. The developed immunosensor was shown to have a high inspection power (5 ca/P, aurantifolia per L) and sensitivity (100 percent). This has been shown to be the case (Rad et al., 2012). Because of beet mortality and yellowish layer virus, rhizomania accounts for practically all detrimental infections in chard. Quantum dots-fluorescence resonance electricity transfer-primarily based completely sensor element successfully identified the transfer of antibiotic medication beta, the most effective known vector of chard harmful yellowish layer, from the soil to the plants. As reported by many researchers (Safarpour et al., 2012)

The optical-electrical device of atomic number 48 chemical compound semiconductor quantum dots enclosed with acetylcholinesterase catalyst through the layer-by-layer meeting approach is used to detect organophosphorus pesticides (OPs) in vegetables and fruits. This resulted in a biosensor with detection limits of 1.05 1011 M for paraoxon and 4.47 1012 M for insect powder, making it far more sensitive than conventional fuel-line activity methods or ampere-metric biosensors (0.5 nM). As reported by (Zheng et al., 2011). Character-stage methyl insect powder detection using the unique optical properties of thiol-stabilized light cadmium telluride (CdTe) energy spots. In a 2010 study (Chouhan et al. Biosensors were needed for the detection of deltamethrin in food processing facilities, specifically for the inspection of soluble CdTe energy spots and particularly fluorescent silicon oxide molecularly imprinted stamp nano-spheres outlining spheres embedded atomic number 48 chemical compound quantum dots. (Ge et al., 2011)

Ø Nano Barcodes

A semi-automated and somewhat accessible technology for electroplating inert metals other than gold, silver, etc. onto templates process particle diameter, and the subsequent release of nanorods from the templates, to create nano barcode trash. When analysing sequence expression in a multiplexed fashion, these micro barcodes serve as identification markers. Improvements in plant tolerance to environmental challenges including drought, salt, and diseases have been seen thanks to engineering scientific power-assisted development in the field of biotechnology. Nanotechnology based on whole gene sequencing is prepared for rapid and cost-effective identification and use of plant gene attribute resources. This means that the agricultural sector may benefit from a variety of engineering science programmes.

Organic Farming &Nano-Technology

The Natural Farming System has been around for a while with the objective of increasing productivity with little input by monitoring environmental factors and implementing targeted action. In order to determine whether or not plants are growing at peak performance, or to precisely pinpoint the individual and location of issues, organic farmers need the use of computers, GPS systems, and remote sensing devices to keep tabs on global environmental conditions. Seeding, fertiliser, chemical, and water consumption may all be optimised by using centralised data to simulate soil conditions and plant growth, lowering production costs and increasing yields for the farmer. Precision farming has the potential to reduce agricultural waste and associated pollution. According to the research (Prasad et al., 2014),

Relation between Agri-business & Nanotechnology

In 2010, the worldwide factory farm business was estimated to be worth anywhere between US$ 20.7 billion to US$ 0.98 trillion, according to economic data from throughout the globe. It is predicted that by 2020, the market would have grown to about US$ 3.4 trillion. The United States is now in the lead because to the National Applied Science Initiative's four-year, $3.70 billion USD investment. The United States is followed by Japan, both of which contribute significantly to the coalition's annual budget (about US$ 750 million and US$ 1.2 billion, respectively) via their respective national governments' budgets. Over 400 companies all over the globe are already engaged in nanotechnology R&D, and this number is expected to rise to well over a thousand over the next decade. According to research (Wang et al., 2021)

Food, Agroforestry & Nanotechnology

New agricultural techniques are created to increase food production via the use of more practicable pesticides and fertilisers in response to the rising global population and the resulting increase in the demand for food. Because of the nature of these technologies, there will be an unchecked release of undesired compounds into the environment, which may damage soil and groundwater. Improved agricultural output and the remediation of polluted areas are two areas where modern applied science shows great promise. Recent uses of nanotechnologies in agro-environmental research are discussed, with an emphasis on the destiny of nanomaterials once introduced in water and soil, as well as their advantages and potential pharmaceutical applications. It has been discovered that using nanoparticles would raise the quality of the environment and make it easier to monitor and clean up unclean areas. Just a narrow band of nanoparticles has undeniable hepatotoxic potential. This unit of square measurement is discussed in detail. According to research (Wang et al., 2021)

Directly and indirectly, agriculture feeds the world's population. The growing human population makes the use of cutting-edge techniques like bio and nanotechnologies in agriculture an absolute requirement. Publicly available technology involving materials, systems, and processes that function at a scale of 100 nanometres (nm) or less has various uses across the board, from cultivation to processing to storage to transportation. Precision farming methods, increased plant absorption of nutrients, more cost-effective and strategically placed inputs, disease detection and management, resistance to environmental stresses, and efficient storage and packaging are just a few examples of how technology will transform agriculture and the food industry. Using nano-particles in animal research may improve the potency of medications. Nanoparticles of silver and iron are used in poultry and uterine diseases therapy. Nano smart mud and gas sensors are often used to swiftly assess levels of establishing pollution area units.

Approaches of Nanotechnology in Allied Fields of Agriculture

1) Applications of Nanotechnology in Animal Science

Future procedures in veterinary medicine and the treatment of pets and livestock may benefit from nanotechnology. Automation has the potential to provide adequate responses for dispensing food products, veterinary services, and prescribed medicines and immunisations for domesticated animals. Using nano tablets to encapsulate and preserve a select group of enzymes and proteins might have significant effects on bovine and poultry feed diets by increasing yield and efficacy in the intended setting. Antibiotics, vaccinations, and probiotics are all examples of helpful drugs that, if administered at the nano level, might prove to be even more effective in curing a wide variety of illnesses, from infections to metabolic and nutritional abnormalities. The research was conducted by (Chakravarthy et al., 2010). The nanoscale application of medication has the triangle properties necessary to overcome organic barriers to improving the efficacy of medicine. The initial benefits of using technology in pharmacological treatments are the increased precision, self-regulation, and adaptability of the drug delivery system. Because of its potent antiseptic (antibacterial and antimicrobial) properties, silver nanoparticles are often utilised as a treatment for animals and birds. In cancer treatment, nanoparticles are applied to cancer cell membrane receptors; once the cells' temperatures reach 55 degrees Celsius as a result of the produced infrared waves, the cancer cells are killed. That was the conclusion reached by researchers (Patil et al., 2009). Additionally, iron nanoparticles combined with increasing magnetic radiation destroy the vast majority of cancer cells. The expense and effort commitment involved in raising bovine for breeding purposes is typical of farms that house other types of livestock. Nano-tubes implanted under the cattle's skin signal the peak of rut and oestrogen hormone, allowing for timely and accurate insemination.

Applications of Nanotechnology in Pests and Plant Diseases Management

Using chemical substances like pesticides, fungicides, and herbicides is the quickest and cheapest way to control pests and diseases today. (Malik et al., 2022) Organic methods of management are also rather costly at the moment. Negative effects on human health, harm to pollinating insects and domestic animals, leaching of insecticides into soil and water, and indirect effects on ecosystems are only a few of the problems caused by unchecked pesticide usage. The clever use of nanoscale chemical substances is also relevant to averting this crisis. These remedies are applied to the damaged parts of plants that have been ravaged by disease or insects. In addition, nanoscale businesses are capable of self-regulation, therefore it is ideal to introduce the necessary dosage of medication into plant tissue. Production of pesticides and chemical fertilisers using nano dust and nano tablets with the ability to regulate or not on time shipping, absorption, and further powerful and environmentally friendly; and production of nano-crystals to growth the performance of insecticides for the utility of insecticides with decrease dose are two examples of how engineering science enables agricultural sciences and reduces environmental pollutants. In the near future, nano detritus for transporting spirit components or medicine molecules might lead the way in curing all pathological afflictions of plants. Beyond compound nano dust, iron compound nano debris, and gold nano debris, there are a plethora of other nano substances that may be carefully manufactured and used as chemical or medication carrying piggybacks. Size, shape, and surface functionalization are all variables that may be used to tailor these nano detritus' pharmacokinetic profiles. They may also be used to alter the kinetic characteristics of drug release, resulting in longer-lasting effects from fewer doses. According to research (Sharon et al., 2010).

Crop yield loss due to diseases is one of the most significant factors. The diagnosis of the exact degree of fore stallion presents a challenge for disease treatment. Most of the time, pesticides are used as a preventive measure, although this may lead to residual toxicity and environmental risks, as well as the repeated use of insecticides if a disease does arrive and cause some crop loss. Infectious agent diseases are the most difficult to control because of the coordinated efforts required to stop the spread of the disease from infected individuals to susceptible ones. When it first starts to develop and show symptoms, though, the chemical application may not be all that helpful. In order to effectively manage diseases, especially viral diseases, it is crucial to determine the real level, which may include the level of infectious agent polymer replication or the synthesis of early viral macromolecule. Multiplexed diagnostic package growth and nano-primarily based, virus-specific diagnostics have gained traction in an effort to pinpoint the exact virus strain and evaluate the efficacy of potential treatments for disease prevention. Biomarker discovery and use, particularly as an indicator of disease severity, is an emerging field of study. activity-dependent macromolecule creation in every healthy and diseased condition ends with the identification of the progression of distinct proteins at some stage of the contamination cycle. These nanotechnology-based, fully-functional diagnostic kits not only increase the detection rate, but also boost the detection power.

2) It is possible that in the future, "smart" agricultural systems may make use of nano-scale devices with unique properties. For instance, farmers may utilise technology to foresee plant health concerns before they become a problem. Devices of this kind may be programmed to react to unusual circumstances by performing the necessary corrective measures. If they do not do so already, they will notify the farmer of a problem. Effectively, this means that smart devices may be used as early warning systems. In the same way that nanomedicine has implications for drug shipping in humans, such devices might be used to deliver chemicals in a controlled and focused manner. Focused delivery (to specific tissues and organs) has shown to be rather effective, and recent developments in nanomedicine are making it possible for the United States to impact rare diseases like cancer in animals with unprecedented specificity. Joseph et al. (2006).

3) Applications of Nanotechnology in the Food Industry

Due to its role in food decomposition and discoloration, oxygen may one day become an integral aspect of food packaging. The development of new polymers for the food packaging industry is one example of the use of engineering science in the food industry. These polymers are manufactured with the help of nanowaste. In order to prevent gas from penetrating the new plastic, nano detritus was deposited zigzagging inside of it. When food is packaged, oxygen molecules have a longer distance to travel before they reach their destination, increasing the risk of spoilage at a later time. Recently, nano-coatings have been developed to cover the fruit's natural ripening pigments and prevent the fruit from losing weight or shrinking. In a 2009 study (Predicala et al. Many businesses have made packaging innovations with the goal of extending the shelf lives of their products. Such packaging systems may be able to repair minor punctures or rips, react to external stimuli, and warn the consumer if the food has been tainted. In 2003, Moraru and colleagues published a study demonstrating that. For instance, engineering science can explain the permeation behaviour of foils, increase barrier properties (mechanical, thermal, chemical, and microbial), improve mechanical and heat-resistance properties, develop energetic antimicrobic and antifungal surfaces, detect and signal changes in microbiology and organic chemistry, and much more. To wit: (Joseph et al., 2006). Enzymes may be protected from environmental degradation by being encased in nanoporous materials, a technique currently under development. As a result, vitamin corruption may be postponed, and their resilience may improve. The most important substance that can be made using nanotechnology is olefine absorbent. Nanomaterials that are "ethylene-absorbent" are designed to soak up the ethylene gas that is naturally generated, allowing fruit to keep for longer. Nanotechnology and nanoprocesses may also be used to evaluate the aesthetic value of agricultural goods. Cornell University researchers are in desperate need of a low-cost, high-throughput, low-mess alternative to UPC for the purpose of disease detection and translation. They developed nano barcodes, or minuscule probes, which may flag several pathogens in a farm and be easily read by any fluorescent-based detection system. This continuing competition typically aims to develop a portable internet detector that may be used by untrained persons on any website. (Li, et al., 2005) With the advent of nanotechnology, nano-primarily based bar codes are now available that may perform the same function as traditional bar codes, helping to monitor and control the quality of the product and delivering all relevant data in a matter of seconds. Several researchers (Prasanna et al., 2007)

An organic biosensor may be anything from a cell to a catalyst to an antibody coupled with a tiny transducer. This transducer supplies electricity (in a variety of forms) to a second device. Although there is often little focus on the tested material, biosensors detect changes in cells and molecules that may be used to quantify and interpret the results. As soon as the drug attaches to the organic component, the electrical gadget displays a symbol whose size is directly related to the amount of the material. Thus, the biosensor may provide a robust sign signalling that the selected meal is unsafe to ingest if there is the enormous attention to microscopic organisms throughout that meal. Mass amounts of food might likewise be easily examined with this technology to ensure customer safety.

Application of Nanotechnology in Agronomy

As a management philosophy for farms, "exactitude agriculture" might represent a paradigm shift. Nano-sensors are installed in every nook and cranny, yet fertiliser and chemical pesticides are still required for every single cargo. As a result, top-quality, risk-free goods may serve as inputs, which boosts productivity and profits. As a result of their sensitivity and small size, nano-sensors are useful tools for maintaining farms that need specialised care and keeping detailed records of the plants' regular needs. The sustainable exploitation of agricultural seasoning supplies like water, vitamins, and chemical compounds by means of precise farming need to be aided by nano-sensors and nano-primarily mainly based complete smart conveyance techniques. Farmers may monitor their crops remotely and look for signs of stress like drought by combining nanomaterials and global positioning systems with television for computer images of fields. Once the cause of an annoyance or drought has been identified, adjustments may be made automatically to chemical programmes or irrigation levels. Nano-sensors dispersed throughout a field can detect plant virus presence and soil nutrient levels. It's possible for plants to take up nano fertilisers unexpectedly and entirely. Another method that has made a comeback to both maintain fertiliser use and reduce pollution is the use of nanoencapsulated, slow-release fertilisers. Nanotechnology-based super water adsorbents play a crucial role in the utility shed, where they may filter out harmful contaminants while also storing potable water for dry areas. This was shown to be the case (DeRosa et al., 2010). Within the field of agricultural instrumentality, engineering has a number of applications, such as a system in machine form and agriculture equipment to increase their resistance towards placed on and corrosion and ultraviolet rays; the generation of long-lasting mechanical additives with the use of nano-coating; the need for bio-sensors in smart machines for mechanical-chemical weed management; the production of nano cowl for bearings to reduce friction; the use of nano-technology in agricultural applications; and so on. In addition, engineering has shown its capacity to improve crop plants' genetic charter, which will aid in the further development of agricultural plant life. In a 2006 study (Jones et al.

Toxicity Issues Due to the Use of Nanotechnology

Precision agriculture is one area where nanotechnology has several potential uses. However, owing to their unique features, toxicity is a major drawback of nano-particles. The phytotoxicity profile of nanoparticles is still in the early stages of development. Many toxicology investigations are required for commercial food crop applications since the impact of nano-particles' distinctive features is not yet recognised. However, the use of nano-particles has neither detrimental nor wholly beneficial impacts on vegetation. To get excellent and efficient results in farming, carbon nano-materials like fullerenes, carbon nano-particles, fuller, and single-walled carbon nanotubes/multiwall carbon nanotubes are needed. Positive results were reported by Tripathi and Sarkar, who found that 150 mg/L (200 L) water-soluble carbon nano-dots accelerated radicle and sprout growth in the batch (10 times) compared to controls after 10 days of exposure. The effects of carbon nanotubes at concentrations of 50, 100, and 200 g/mL on the growth and fertilisation of soybeans, maize, and barley in an alkaloid extract solution were reported by Lahiani. An almost 50% rise in germination rate was seen in Hordeum vulgare and legume after exposure, and a 90% increase in maize compared to untreated controls. Duhan et al. (2017)

The premise lengths of soybean plants increased by up to 26%, while those of liliopsid genus mays increased by 40% and by more than thrice, respectively. It was also anticipated that both Raman spectrographic analysis and transmission microscopy would be able to acquire many layers of carbon nano-tubes. The quantities, growth/exposure settings, and plant species in which deadly carbon nano-materials were detected were the primary points of interest. Kerfahi studied the effects of low to high doses (between 0 and 5,000 mg/kg) of carbon nanotubes on soil microorganisms. They said that over a certain period of time, the soil's microbial community makeup will be contaminated by the high quantities of carbon nanotubes. At the end of the first 8 weeks, there was no discernible difference in the microbial population, other than in the kind of nanotube used. The abrupt, inacidulous quality of carbon nanotubes' many layers causes the pH scale of earth dust to drop at high concentrations of exposure, leading to altered populations of soil microorganisms. Duhan et al. (2017)

Boonyan-it-pong investigated the Phytotoxicity of oxide and titania nanoparticles on rice roots, among other negative consequences. The seed germination package, root length, and root diversity were the three factors studied. Findings revealed that exposure to nanoparticles of zinc and pigment had no negative effect on the germination rate of seeds. However, the primary spermatophyte stage of rice roots was negatively affected by zinc oxide nanoparticles. It was shown that nano-sized zinc chemical compound particles reduce both the depth and the total number of roots. The development of roots was unaffected by the presence of nano-sized oxide particles. This research highlighted the need of environmentally responsible disposal of wastes containing nanoparticles, as well as their increased application in agricultural and environmental systems, by demonstrating that direct exposure to certain kinds of nanoparticles causes important phytotoxicity. For examples, see (Duhan et al., 2017) and (Hajam et al., 2020)

NANO-TECHNOLOGY APPROACHES IN

1) Environment

Natural NPs (NPs) seem to not be the product of any one person's imagination and have instead been consistently present in the wild. Evidence of seasoner NP generation in Cretaceous-Tertiary boundary sediments is provided, for instance, by the detection of NPs in a 10,000-12 month old glacial core floe. Estimates suggest that a few thousand million metric tonnes of component soil are produced annually on a global scale, or maybe with a portion of this as ultrafine particles, this might be infinitely sufficient natural NPs. In contrast to phylogenesis manufacturing, the message to seasoner NPs clearly surpasses any phylogenesis manufacturing if we compare it to making estimations for artificial nanoparticles on the scale of a few thousand lots every year for every fundamental variety of material. According to a study (Qafoku et al., 2010) In both the earth sciences and the biological sciences, NPs may be produced via a wide variety of methods. Chemical weathering, authogenesis/neoformation, and volcanism are all examples of earth science processes. Synthetic NPs are often produced by these geological processes. However, certain creatures may synthesise nano-sized particles of alloys inside their cells, despite their biotic machinery being primarily responsible for the production of natural nano-molecules.

2) Soil

NPs may be imposed in the soil in different techniques for soil correction due to their natural process potential. It is common practise for NPs to interact with polymorphs, minerals, pollutants, and nutrients. Steel and anionic pollutants, as well as flavouring organic matter, natural acids, and important metals, are some of the things they'll be used to absorb. Contamination is sequestered by encapsulating it in the floor complexation or internal interfaces of NP aggregates, which may then be employed for ground-water cleanup and remediation. In order to get rid of radionuclides and other impurities, nano-porous minerals are utilised, whereas compound (CuO) NPs are put to use in the adsorption of elements III and V. (Mura et.al., 2013)

3) Water

The Environmental Protection Agency (EPA) believes that the imposition of these hefty fees will serve as a deterrent to the development and widespread use of clean-up methods and technologies, leading to more efficient, less costly, and quicker site cleanups. Nano-remediation has the potential to not only reduce the cost of cleaning up large-scale contaminated sites, but it also has the potential to reduce the clean-up time, eliminate the need for treatment and disposal of infected dredged soil, reduce some material concentrations to close zero, and is enforced in space. Because of the high price and extended running times of pump-and-treat treatments, regional groundwater treatment technology is on the rise. The first results of this redox-primarily based overall generation are promising, pointing to speedier, more cost-effective results and, eventually, lower standard material levels compared to the conventional techniques of rectification. Iron nano-powders, in particular, are a potent nano-remediation tool for purifying polluted groundwater and soil. (Mura et.al., 2013)

NANO-ETHICS

Potentially significant for our country's future is the field of nanotechnology. Research in this area should have clear societal benefits, such as improved agricultural practises or meal-planning tools. While the enormous capacity boost is welcome, it comes with the responsibility of ensuring the next generation's smart software package is up to snuff. The incontrovertible reality that nanoparticles reveal innovative properties compared to comparable textiles on a larger scale makes them one of the most important aspects of technology from an ethical problem of reading. Using the concept of abuse as a focal point, this moral examination of developing technologies aims at an illusive destination. This threatens the conventional risk-evaluation approach that, in recent years, has portrayed the primary flavouring agent as distinctive in terms of environmental and distinctively enforced ethics, that focus on safety, the environment, and human health. However, before we act, we would need to know whether or not advancing to master technology is morally sound. However, in order to make an ethical assessment, we may desire to know the results, and this knowledge won't be at hand until the horrible investigations that raise the ethical and moral question are completed. The most widely read books on the subject of generation are used to fuel the contentious plans of consequentialists. For example, (Mura et al., 2013)

The optimistic perspective holds that progress in technology is ultimately beneficial. The optimistic perspective additionally presumes that, in the future, the novel technology is proven to be as useful as the majority of other generations in human history, on the presumption that no generation is inherently virtuously bad, that its handiness is only the way in which it is worn inquiry. These arguments frameworks have squarely met the issue of ambiguity and inexperience with the help of maltreatment aiming to lessen the issue of risks and toxicity, which focus on unique morally appropriate elements that may be ahead of or independent of the implications. Minority non-consequentialist viewpoints, however, note that the problem of nano-ethical impacts and risks cannot be solved properly. As a result, one of the primary budgetary goals of future nano-ethics must be to strike a balance between our desire to prevent harm and the need for comprehensive research into expanding technologies and the impacts, risks, and toxicity of doing so. For example, (Mura et al., 2013)

CONCLUSION

Nanotechnology includes the design of equipment with almost endless applications. There may also be an apparent need to test the safety of these novel compounds as soon as they are exposed to the environment and home organisms, preventing the release of hazardous substances and encouraging the development of healthy alternatives. Choices on which of these compounds to use, notwithstanding their adaptability and other advantages, must be made with an eye toward minimising adverse effects on the environment and, by extension, on people. In particular, the astute application of nanomaterials will aid in enhancing environmental friendliness and sustainability by increasing detection and sensing methods for organic and chemical toxins, repairing and harming standard contaminants from water-related soil, and witnessing green industrial techniques that lessen power consumption. Given the identified dangers, it is possible that nanomaterials might have hazardous effects on both humans and the natural world. For example, (Mura et al., 2013)

It should be noted, however, that not all nanomaterials at the present time have harmful effects. Some man-made nanoparticles have been in use for a while and have shown to be rather harmless. Thus, the hypothesis that smaller methods are more reactive and, thus, more toxic, cannot be substantiated by the reported findings. Aside from the obvious health risks, the scale of production and, by extension, the price of prevention, is another major drawback to using nanomaterials more broadly in agriculture. There is a lot of room for manoeuvre in the developing United States of America when it comes to nanotechnologies, and some ground still has to be broken. This is why we are interested in testing and validating the behaviour of these novel drugs. We also need to steer investigations in this way so that we may develop better options and push for much cheaper nanomaterial production and application processes.

REFERENCES

Brady, R. R. (1999). The Nature and Properties of Soils. Prentice Hall.

Branton, D., Deamer, D. W., Marziali, A., Bayley, H., Benner, S. A., Butler, T., Ventra, M. D., Garaj, S., Hibbs, A., Huang, X., Jovanovich, S. B., Krstic, P. S., Lindsay, S., Ling, X. S., Mastrangelo, C. H., Meller, A., Oliver, J. S., Pershin, Y. V., Ramsey, J. M., ... Schloss, J. A. (2008). The potential and challenges of nanopore sequencing. *Nature Biotechnology*, *10*(10), 1146–1153. doi:10.1038/nbt.1495 PMID:18846088

Brock, D. A., Douglas, T. E., Queller, D. C., & Strassmann, J. E. (2011). Primitive agriculture in a social amoeba. *Nature*, *469*(7330), 393–396. doi:10.1038/nature09668 PMID:21248849

Chakravarthy, P. V., & Balaji, N. S. (2010). Applications of nanotechnology in veterinary medicine. *Veterinary World*, *3*, 477–480.

Chouhan, R. S., Vinayaka, A. C., & Thakur, M. S. (2010). Thiol-stabilized luminescent CdTe quantum dot as biological fluorescent probe for sensitive detection of methyl parathion by a fluoro-immune chromatographic technique. *Analytical and Bioanalytical Chemistry*, *397*(4), 1467–1475. doi:10.100700216-009-3433-1 PMID:20101498

Corradini, E., Moura, M. R., & Mattoso, L. H. C. (2010). A preliminary study of the incorporation of NPK fertilizer into chitosan nanoparticles. *Polym. Lett.*, *4*(8), 509–515. doi:10.3144/expresspolymlett.2010.64

Dasary, S. S. R., Rai, U. S., Yu, H. T., Anjaneyulu, Y., Dubey, M., & Ray, P. C. (2008). Gold nanoparticle-based surface enhanced fluorescence for detection of organophosphorus agents. *Chemical Physics Letters*, *460*(1-3), 187–190. doi:10.1016/j.cplett.2008.05.082 PMID:24031096

DeRosa, M. C., Monreal, C., Schnitzer, M., Walsh, R., & Sultan, Y. (2010). Nanotechnology in fertilizers. *Nature Nanotechnology*, *5*(2), 91. doi:10.1038/nnano.2010.2 PMID:20130583

Dikshit, A., Shukla, S. K., & Mishra, R. K. (2013). *Exploring Nanomaterials with PGPR in Current Agricultural Scenario*. Lap Lambert Academic Publishing.

Dong, J., Zhao, H., Qiao, F., Liu, P., Wang, X. D., & Ai, S. Y. (2013). Quantum dot immobilized acetylcholinesterase for the determination of organophosphate pesticides using graphene-chitosan nanocomposite modified electrode. *Analytical Methods*, *5*(11), 2866–2872. doi:10.1039/c3ay26599d

Duhan, J. S., Kumar, R., Kumar, N., Kaur, P., Nehra, K., & Duhan, S. (2017). *Nanotechnology: The new perspective in precision agriculture. In Biotechnology Reports*. Elsevier.

Emadian, S.E. (2017). *Physiological Responses of Loblolly Pine (Finustaeda L.) to Silicon and Water Stress*. Academic Press.

Ge, S., Lu, J., Ge, L., Yan, N., & Yu, J. (2011). Development of a novel deltamethrin sensor based on molecularly imprinted silica nanospheres embedded CdTe quantum dots. *Spectrochimica Acta. Part A: Molecular and Biomolecular Spectroscopy, 79*(5), 1704–1709. doi:10.1016/j.saa.2011.05.040 PMID:21684806

Hajam, Y. A., Rai, S., Kumar, R., Bashir, M., & Malik, J. A. (2020). Phenolic Compounds from Medicinal Herbs: Their Role in Animal Health and Diseases – A New Approach for Sustainable Welfare and Development. In R. Lone, R. Shuab, & A. Kamili (Eds.), *Plant Phenolics in Sustainable Agriculture*. Springer. doi:10.1007/978-981-15-4890-1_10

Hasaneen, M. N. A., Abdel-Aziz, H. M. M., El-Bialy, D. M. A., & Omer, A. M. (2014). Preparation of chitosan nanoparticles for loading with NPK fertilizer. *African Journal of Biotechnology, 13*(31), 3158–3164. doi:10.5897/AJB2014.13699

Hutasoit, S., Suada, J.K., & Susrama, I.G.K. (2013). Antifungal activity test extracts some type of marine life link to Aspergillus flavus and Penicillium sp. *E J. Trop.,* 27–38.

Jha, M. N., & Prasad, A. N. (2006). Efficacy of new inexpensive cyanobacterial biofertilizer including its shelf-life. *World Journal of Microbiology & Biotechnology, 22*(1), 73–79. doi:10.100711274-005-7024-9

Johnson, A. (2005). *Agriculture and nanotechnology.* The University of Wisconsin-Madison.

Jones, P. B. C. (2006). A Nanotech Revolution in Agriculture and the Food Industry. Information Systems for Biotechnology.

Joseph, T., & Morrison, M. (2006). *Nanotechnology in Agriculture and Food.* A Nano forum report, Institute of Nanotechnology.

Li, Y., Hong-Cu, Y. T., & Luo, D. (2005). Multiplexed detection of pathogen DNA with DNA-based fluorescence nano barcodes. *Nature Biotechnology, 23*(7), 885–889. doi:10.1038/nbt1106 PMID:15951805

Malik, J.A., Bhadauria, M., Lone, R., & Hajam, Y.A. (2020). Exploitation of Plant Phenolics in Animal Farming. In *Plant Phenolics in Sustainable Agriculture*. Springer. doi:10.1007/978-981-15-4890-1_3

Malik, J.A., Mishra, G., Hajam, Y.A., Lone, R., & Quazi, S. (2022). Metabolome Analyses in Response to Diverse Abiotic Stress. In *Omics Approach to Manage Abiotic Stress in Cereals*. Springer. doi:10.1007/978-981-19-0140-9_6

Moraru, C. I., Panchapakesan, C. P., Huang, Q., Takhistov, P., Liu, S. J., & Kokini, L. (2003). Nanotechnology: A new frontier in food science. *Food Technology, 57,* 24–29.

Moura, M. R. D., Aouda, F. A., & Mattoso, L. H. C. (2008). Preparation of chitosan nanoparticles using methacrylic acid. *Journal of Colloid and Interface Science, 321*(2), 477–483. doi:10.1016/j.jcis.2008.02.006 PMID:18295778

Mousavi, S.R., & Rezaei, M. (2011). Nanotechnology in Agriculture and Food Production. *Journal of Applied Environmental and Biological Sciences,* 415.

Mura, S., Seddaiu, G., Bacchini, F., Roggero, P.P., & Greppi, G.F. (2013). Advances of nanotechnology in agro-environmental studies. *Italian Journal of Agronomy, 8,* 131-134, 135-136.

Ombodi, A., & Saigusa, M. (2000). Broadcast application versus band application of polyolefin coated fertilizer on green peppers grown on andosol. *Journal of Plant Nutrition, 23*(10), 1485–1493. doi:10.1080/01904160009382116

Otles, S., & Yalcin, B. (2010). Nano-biosensors as new tool for detection of food quality and safety. *Log Forum, 6*, 67–70.

Park, S., Croteau, P., Boering, K. A., Etheridge, D. M., Ferretti, D., Fraser, P. J., Kim, K. R., Krumme, P. B., Langenfelds, R. L., Ommen, T. D. V., Steele, L. P., & Trudinger, C. M. (2012). Trends and seasonal cycles in the isotopic composition of nitrous oxide since 1940. *Nature Geoscience, 5*(4), 261–265. doi:10.1038/ngeo1421

Patil, S. S., Kore, K., & Kumar, P. (2009). Nanotechnology and its applications in veterinary and animal science. *Veterinary World, 2*(1), 475–477. doi:10.5455/vetworld.2009.475-477

Prasad, R., Kumar, V., & Prasad, K. S. (2014). Nanotechnology in sustainable agriculture: Present concerns and future aspects. *African Journal of Biotechnology, 13*, 709.

Prasanna, B.M. (2007). *Nanotechnology in agriculture.* ICAR National Fellow, Division of Genetics, I.A.R.I.

Predicala, B. (2009). *Nanotechnology: Potential for agriculture.* Prairie Swine Centre Inc., University of Saskatchewan.

Qafoku, N. P. (2010). Terrestrial nanoparticles and their controls on soil geo-processes and reactions. *Advances in Agronomy, 107*, 33–91. doi:10.1016/S0065-2113(10)07002-1

Rad, F., Mohsenifar, A., Tabatabaei, M., Safarnejad, M. R., Shahryari, F., Safarpour, H., Foroutan, A., Mardi, M., Davoudi, D., & Fotokian, M. (2012). Detection of Candidatus Phytoplasma aurantifolia with a quantum dots fret-based biosensor. *Journal of Plant Pathology, 94*, 525–534.

Rai, V., Acharya, S., & Dey, N. (2012). Implications of nano biosensors in agriculture. *J Biomatter Nano Biotechnol, 3*, 315–324.

Safarpour, H., Safarnejad, M. R., Tabatabaei, M., Mohsenifar, A., Rad, F., Basra, M., Shahryari, F., & Hasanzadeh, F. (2012). Development of a quantum dots FRET-based biosensor for efficient detection of Polymyxa betae. *Canadian Journal of Plant Pathology, 34*(4), 507–515. doi:10.1080/07060661.2012.709885

Santoso, D., Lefroy, R. D. B., & Blair, G. J. (1995). Sulphur and phosphorus dynamics in an acid soil/crop system. *Australian Journal of Soil Research, 33*(1), 113–124. doi:10.1071/SR9950113

Scott, N. R. (2005). Nanotechnology and animal health. Revue scient fique et technique. International Office of Epizootics. doi:10.20506/rst.24.1.1579

Sekhon, B. S. (2014). Nanotechnology in agri-food production: An overview. *Nanotechnology, Science and Applications, 7*, 31–53. doi:10.2147/NSA.S39406 PMID:24966671

Sharmila, R. C. (2010). *Nutrient Release Pattern of Nano-Fertilizer Formulations.* Tamil Nadu Agricultural University.

Sharon, M., Choudhary, A. K., & Kumar, R. (2010). Nanotechnology in agricultural diseases and food safety. *Journal of Phytology*, *2*, 83–92.

Shaviv, A. (2000). Advances in controlled release of fertilizers. *Advances in Agronomy*, *71*, 1–49.

Shukla, S. K., Kumar, R., Mishra, R. K., Pandey, A., Pathak, A., Zaidi, M., Srivastava, S. K., & Dikshit, A. (2015). Prediction and validation of gold nanoparticles (GNPs) on plant growth promoting rhizobacteria (PGPR): A step toward the development of nano-biofertilizers. *Nano Reviews*, *4*, 439–448.

Silva, A. C. N., Deda, D. K., da Róz, A. L., Prado, R. A., Carvalho, C. C., Viviani, V., & Leite, F. L. (2013). Nanobiosensors based on chemically modified AFM probes: A useful tool for metsulfuron-methyl detection. *Sensors (Basel)*, *13*(2), 1477–1489. doi:10.3390130201477 PMID:23348034

Subramanian, K. S., & Tarafdar, J. C. (2011). Prospects of nanotechnology in Indian farming. *International Journal of Agricultural Sciences*, *81*, 887–893.

Sun, X., Zhai, C., & Wang, X. Y. (2013). A novel and highly sensitive acetyl-cholinesterase biosensor modified with hollow gold nanospheres. *Bioprocess and Biosystems Engineering*, *36*(3), 273–283. doi:10.100700449-012-0782-5 PMID:22829062

Trenkel, M. E. (1997). *Controlled-release and stabilized fertilizers in agriculture. Int. Fert.* Ind. Assoc., 234–318.

Vandergheynst, J. S., Scher, H., & Yun, G. H. (2006). Design of formulations for improved biological control agent viability and sequestration during storage. *Industrial Biotechnology (New Rochelle, N.Y.)*, *2*(3), 213–219. doi:10.1089/ind.2006.2.213

Vandergheynst, J. S., Scher, H. B., Gou, H. Y., & Schultz, D. L. (2007). Water in oil emulsions that improve the storage and delivery of the biolarvacide Lagenidium giganteum. *Biological Control*, *52*, 207–229.

Vasconcelos, C. L. D., Bezerril, P. M., Santos, D. E. S. D., Dantas, T. N. C., Pereira, M. R., & Fonseca, J. L. C. (2006). Effect of molecular weight and ionic strength on the formation of polyelectrolyte complexes based on poly (methacrylic acid) and chitosan. *Biomacromolecules*, *7*(4), 1245–1252. doi:10.1021/bm050963w PMID:16602745

Viswanathan, S., Radecka, H., & Radecki, J. (2009). Electrochemical biosensor for pesticides based on acetylcholinesterase immobilized on polyaniline deposited on vertically assembled carbon nanotubes wrapped with s-DNA. *Biosensors & Bioelectronics*, *24*(9), 2772–2777. doi:10.1016/j.bios.2009.01.044 PMID:19269805

Wang, D., Behura, S., Karim, F., Kazi, A., Wangila, G., & Wang, S. (2021). *Nanoscience and Nanotechnology for Food and Agroforestry*. Academic Press.

Wang, L. J., Wang, Y. H., Li, M., Fan, M. S., Zhang, F. S., Wu, X. M., Yang, W. S., & Li, T. J. (2002). Synthesis of ordered biosilica materials. *Chinese Journal of Chemistry*, 107–110.

Warad, H. C., Ghosh, S. C., Thanachayanont, C., Dutta, J., & Hilborn, J. G. (2004). Highly luminescent manganese doped ZnS quantum dots for biological labelling. *Proc. Int. Conf. on Smart Materials/Intelligent Materials*, 203–206.

Wilson, M. A., Tran, N. H., Milev, A. S., Kannangara, G. S. K., Volk, H., & Lu, G. Q. M. (2008). Nanomaterials in soils. *Geoderma*, *146*(1-2), 291–302. doi:10.1016/j.geoderma.2008.06.004

Wu, L., & Liu, M. (2008). Preparation and properties of chitosan-coated NPK compound fertilizer with controlled release and water-retention. *Carbohydrate Polymers*, *72*(2), 240–247. doi:10.1016/j.carbpol.2007.08.020

Wu, S. C., Cao, Z. H., Li, Z. G., Cheung, K. C., & Wong, M. H. (2005). Effects of biofertilizer containing N-fixer, P and K solubilizers and AM fungi on maize growth: A greenhouse trial. *Geoderma*, *125*(1-2), 155–166. doi:10.1016/j.geoderma.2004.07.003

Wu, Y., Guo, J., Yang, W., Wang, C., & Fu, S. (2006). Preparation and characterization of chitosan-poly (acrylic acid) polymer magnetic microspheres. *Polymer*, *47*(15), 5287–5294. doi:10.1016/j.polymer.2006.05.017

Yan, J. X., Guan, H. N., Yu, J., & Chi, D. F. (2013). Acetylcholinesterase biosensor based on assembly of multiwall carbon nanotubes onto liposome bioreactors for detection of organophosphates pesticides. *Pesticide Biochemistry and Physiology*, *105*(3), 197–202. doi:10.1016/j.pestbp.2013.02.003

Yu, G. X., Wu, W. X., Zhao, Q., Wei, X. Y., & Lu, Q. (2015). Efficient immobilization of acetylcholinesterase onto amino-functionalized carbon nanotubes for the fabrication of highly sensitive organophosphorus pesticides biosensors. *Biosensors & Bioelectronics*, *68*, 288–294. doi:10.1016/j.bios.2015.01.005 PMID:25594160

Zheng, Z., Zhou, Y., Li, X., Liu, S., & Tang, Z. (2011). Highly-sensitive organo phosphorous pesticide biosensors based on nanostructured films of acetylcholinesterase and CdTe quantum dots. *Biosensors & Bioelectronics*, *26*(6), 3081–3085. doi:10.1016/j.bios.2010.12.021 PMID:21196108

Chapter 8
The Impact of Nanomaterials on the Environment and Perspectives for Sustainable Energy Generation

Aline Varella Rodrigues
Federal University of Technology Paraná, Brazil

Fernanda da Costa Romeiro
https://orcid.org/0000-0002-9632-4290
São Paulo State University, Brazil

Caroline Varella Rodrigues
University of São Paulo, Brazil

ABSTRACT

In this chapter, the eco-friendly applications, disposal, and the risks of inorganic nanoparticles to human health and to the environment are discussed, including the advancements in the green applications of green nanomaterials for their sustainable and competitive energy systems. The first section summarizes an introduction with the importance of the nanomaterials in scientific and industrial applications. The second and third sections present green nanoparticles applications for energy generation (green electricity generation by wind and solar energies, luminescent solar concentrators, and photovoltaics and bioenergy production). In the fourth section, the effects of nanotechnology on the human health and ecosystems and also regulations to work with nanomaterials are presented. This chapter illustrates the main principles and progress of nanotechnology and their positive and negative impacts into the environment and human health.

DOI: 10.4018/978-1-6684-5533-3.ch008

INTRODUCTION

Nanoscience and nanotechnology are considered relevant topics, which involves materials and devices with novel properties due to the arrangement of atoms and formation of materials from 1 to 100 nanometers (nm). The term nanoscience, includes the study and processes to obtain nanomaterials by physics, materials science and biology areas, in particular, focusing on the unique, size-dependent properties of solid-state materials; while the nanotechnology term is ascribed to the ability to measure, manipulate, assemble, control, and manufacture matter or devices at the nanometer scale (Hannah & Thompson, 2008). This subject has attracted great interest because of their scientific and industrial applications. Several areas have applied nanomaterials including; chemistry, biology, materials science, physics, medicine, and engineering (Lemos et al., 2022; Patra et al., 2018; Romeiro et al., 2022; Zhang et al., 2020). In recent years, the worldwide society have observed the use of these tiny compounds to human health, such as for cancer treatment (Bayda et al., 2019; Hannah & Thompson, 2008). The benefits of using nanomaterials have been reported in recent years, especially after the recognition of the famous lecture "There is plenty of room at the bottom" by Richard Feynman (1959), where the concept of manipulating individual atoms was reported.

Nanomaterials can be obtained mainly by two synthetic routes: top-down and bottom-up approaches. The top-down methodology is described as a process to break down bulk material to get nanoparticles. The bottom-up approach, such as self-assembly or chemical processes atom-by-atom or molecule-by-molecule can be manipulated to build more advanced ordered materials. In the bottom-up approach, chemicals as solvents and precursors are employed to obtain nano-sized materials. These nanomaterials need to be disposed properly into the environment, otherwise they can damage the nature and human health (Papp et al., 2008). Therefore, the purpose of this chapter is to present the current state of knowledge of the eco-friendly applications of nanomaterials and also present the effects of nanomaterials with regard to human health and to the environment. The ecofriendly and sustainable applications of nanoparticles, such as electricity by wind and solar energies and for biological processes to produce bioenergy will be discussed. This chapter also brings information of the direct exposure or the use of commercial products made of nanomaterials and recommendations on safe handling to prevent adverse implications.

BACKGROUND

The Benefits and Importance of Nanotechnology

Since the 2000s, the development of nanotechnology has led to an enormous investment in nanomaterials production. The nanotechnology encompasses the understanding and control of individual atoms and molecules, leading to revolutions in several technological areas that benefit society. Indeed, the nanotechnology have been receiving significant funding support from developed economies including USA, Japan, Western Europe, Canada, Australia, and South Korea (Hannah & Thompson, 2008). There are several applications of nanotechnological products: waste reduction, energy production, biological and medical areas. For medical applications, nanomaterials have been used to target-oriented drug delivery, where noncompounds release the drug in a desirable body location (Patra et al., 2018). In addition, tissue regeneration nanotechnology has presented progress by the ability to mimic the of human tissues and organs at the nanoscale (Alarçin et al., 2016). Therefore, its significant role in the advancement of medicine/drug formulations, and in controlled drug release could be an option to solve these critical medical issues.

To renew agriculture, the use of nanotechnology-based systems has been introduced, for instance, in the innovative-nanopesticides obtainment. These engineered nanomaterials minimize application losses, improve coverage on the leaf, and can reduce the quantities of formulation's chemical reagents. The use of nanopesticides can increase water solubility and can control weeds and insects in the crops (Chaud et al., 2021). However, the cytotoxicity and genotoxicity of these nanocompounds nanomaterial cannot be neglected. Other important application of nanomaterials can be the cleaning of the environment and for efficient energy solutions, such as nanomaterials based solar cells or CO_2 conversion to fuels (Lemos et al., 2017; Romeiro et al., 2021). Besides, by the nanotechnology, indoors and outdoors toxic gases in the air can be detected at the molecular level using precise sensors based on nanomaterials (Bai et al., 2016; Suman et al., 2015).

The same excitement achieved with the development of nanotechnology also brings uncertainties about its risks and challenges of this new technology. Up to date, few studies have been performed with the aim of discovering the effects of nanomaterials. Studies in this research topic can provide the risk issues concerning nanoscience. For instance, TiO_2 nanoparticles are extensively used in textiles, plastics, paper and cosmetics. These commercial applications contribute to the overall consumption of TiO_2, increasing the amount of nanomaterial in the environment. According to (Rashid et al., 2021) risk management strategies are crucial and should consider the short or long-term exposure, as well as the frequency of exposure to nanomaterials. in this way, regulations and legalization are important to control the adequate use of nanomaterials and to guarantee their sustainable use.

NANOMATERIALS ON ENERGY SYSTEMS TO PRODUCE ELECTRICITY

Considering the rising energy demand and the environmental impact of fossil fuels, alternative approaches to obtain green nanomaterials to be used in sustainable energy generation have been attracted attention from scientific community. The application of nanomaterials has shown positive results for the advancement of human health, energy security, and environmental protection. As an interesting energy production alternative, the sunlight can be converted to electricity without heat and with little maintenance of the photovoltaic (PV) system, which has the capacity of working with outputs from microwatts to megawatts. Thus, sunlight has been demonstrated to be the most promising solutions to the world crisis. The PV facilitates a clean, reliable, scalable, and affordable electricity, converting direct current to alternating current electricity, even storing the energy which is used later. However, there is a spectral mismatch which causes energy losses, and consequently, a degradation of the PV system and its efficiency. In order to overcome that energy losses, new technologies have been developed. Connected with PV panels, these technologies mitigate or eliminate the mismatch effects, even though difference irradiance, temperature, dirt, or object shadow (Al-Ezzi & Ansari, 2022; Romeiro & Rodrigues, 2021).

PVs which combine luminescent properties, nanoscaled sizes, polymers, and ionic liquids (ILs)' functionalities are named as smart materials (Figure 1). Sensors, actuators, batteries, fuel cells, environmental and biological applications are also included. With the goal of improving the PVs' performance, smart materials have the capability of minimizing or mitigating self-absorption energy losses from luminophores, therefore, they have been taken into account, e.g. nanoparticles (NPs), quantum dots (QDs). From adding IL in a specific solution, the IL-based materials are developed, whose enhancement of LSC-PV system performance, with LSC as luminescent solar concentrator, is achieved by collecting the radiation from the down-shifting layer, and then interacting cations and anions from the ILs with the polymer matrix. Therefore, the IL-based material is the optically active center in the LSC, which improves the LSCs' features, such as their chemical/thermal stability, optical transmittance, and luminescence intensity. The LSCs are obtained from doping a transparent matrix with species with good optical properties, e.g. a poly(methyl methacrylate) (PMMA) doped with lanthanides or metals. Moreover, semiconductor QDs with variable diameter, e.g. organic dyes in PMMA usually exhibit great stability and tunable emission wavelength, which minimize the overlap occurred with absorption and emission spectra in LSC, and then lead the producing energy with great quantum, conversion, and power conversion efficiencies, thus the PCE enhancement, and consequently the conversion of solar energy to electricity, with lower energy price. Accordingly, the power conversion efficiency (PCE) depends therefore the irradiance of the photoluminescent (PL) collected in the LSC and the quantum efficiency of the PV system. Also exhibiting a good QY and a specific geometrical factor, semiconductor core/shell QDs with heavy metals should be used as luminophores into the IL-based materials (Rodrigues et al. 2022).

Figure 1. The combination of luminescent (1), nanoscaled size (2), polymer (3), and IL (4) properties, which generate Smart materials, which have the capability of minimizing or mitigating self-absorption energy losses, producing improvement of LSC-PV system performance

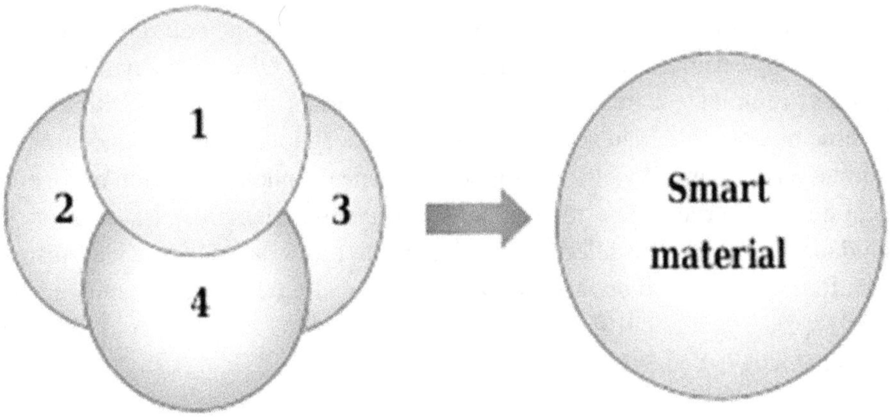

Wind and Solar Energies

Wind turbines and solar PV systems are part of the most useful systems to generate electricity with reducing carbon dioxide emissions into the Earth planet. Even as renewable energy sources, the PVs have some advantages over the wind systems, e.g. the PVs have constant solar radiation in a wide area which supplies more power, and they are more economic than wind systems, with lower electricity prices and recovery of primary investment capital (Alic and Sarewitz 2016; Kassem, Al Zoubi, and Gökçekuş 2019). In Europe, the renewable energy sources as wave, wind, and solar PVs produce 30% of the total energy consumption, which should reduce CO_2 emissions by the year 2030 (Brouwer et al. 2014; Gul, Kotak, and Muneer 2016; Malinauskaite et al. 2017). China, Japan, USA, Germany, and UK were the countries which installed the most PV systems in 2014. The PVs' production was led sequentially by China, EU, USA, and Japan in 2017, as well as the winds' capacity by China, EU, USA, and India in the same year (El Bakari and Kling 2010; Zsiborács et al. 2019). Recent researches reported that renewable sources used to produce electricity in the worldwide supply are 26.5%, with 5.6% from the wind power, and 1.9% of PV systems (Rodrigues et al. 2022; Sampaio and González 2017; Zsiborács et al. 2019).

The reasons which shed light on the rising development and application of wind and solar PV systems in the middle of this century are especially political, economic, and environmental. Moreover, their systems' designs and climatology studies influence the performance of both renewable energy systems (Al-Dousari, Al-Nassar, and Ahmed 2020; Bergin et al. 2017; Leitão, Torres, and Fernandes 2020).

Luminescent Solar Concentrators and Photovoltaics: Nanomaterials and Improvement of Energy Conversion Systems

The transition to a sustainable energy generation is a big challenge for scientists in the 21st century. In this way, the synthesis and application of environmentally friendly nanomaterials with chemical and optical stabilities are extremally necessary to no longer harm the environment. Taking that into account, sensitized environmentally friendly lanthanides, perovskites QD semiconductors, and doped QDs semiconductors with small extinction coefficient and large Stokes shift should be chosen with the deal of minimizing or mitigating the luminophore self-absorption losses. This self-absorption elimination occurs through the excitation of states within the band gap of the semiconductor by the insertion of dopants, transforming high into low-energy photons, and then downshifting the emission to the semiconductor absorption (Figure 2). Sensitized inorganic and colloidal doped QDs with small dimensions also use their great features to eliminate scattering effects. These materials have dopants with low phonon-energy efficient emission, tunable shift, and strong broad band absorption. Minimal concentration quenching, high photoluminescence quantum yield, and higher resistance to photo-oxidation make doped QDs high photo/chemical stable materials. Thus, re-emissions to longer wavelength emissions, which facilitate the emissions' guidance to the PVs installed at the edge of the LSCs, are due to wider absorption spectra, large Stokes shift, core shells and dopants of QDs (Jalali et al. 2021; Coropceanu and Bawendi 2014; Creutz et al. 2017; Daorta et al. 2014; Han et al. 2014; Liu et al. 2018; Lu et al. 2023; Sharma et al. 2017; Waldron et al. 2017; You et al. 2019).

Figure 2. Sunlight absorption, phosphor excitation, and wavelength emission from the host semiconductor edge
Source: Adapted from (Rodrigues et al., 2022)

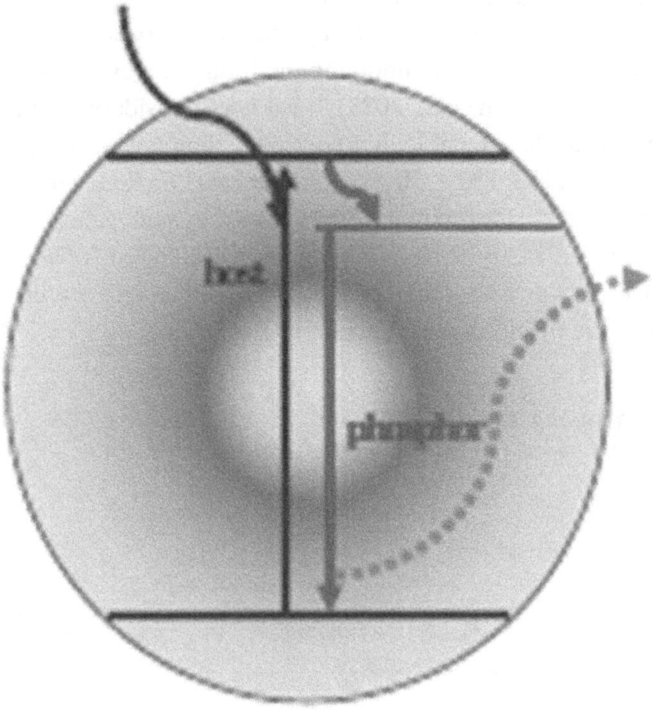

NANOMATERIALS ON BIOLOGICAL PROCESS TO PRODUCE BIOENERGY

Anaerobic Digestion (AD) for Biogas Production

The depletion of fossil fuel resource and energy shortages have gained worldwide concern (Xiaomei et al., 2022). Irreversible effects on climate change could be expected in the next years, due to double of CO_2 emissions and the rise of greenhouse gases (GHG). In addition, the threat to environment and human life could be given by improper management and disposal of solid waste and wastewater by the discharge of untreated effluents with contaminants or by the inefficiency of wastewater treatment plants (Fazzino et al., 2021). One of the biggest concerns in the world regarding industrial activities in the last 30 years has been wastewater with potential toxic agents. In addition, water consumption in industries comprises a large drain, corresponding to up to about a third of the fresh water in the world. Thus, the need to recover these wastewaters becomes an important factor for the environmental issue (Saravanan et al., 2023).

Due to the above-mentioned, it is necessary to investigate different kinds of sustainable energy sources, originating from solid waste or wastewaters, as renewable feedstocks for the bioenergy generation (Fazzino et al., 2021). Anaerobic digestion (AD) corresponds to in an adequate process to treat residues, converting biodegradable organic matter into biogas by many microorganisms, being pure or mixed cultures, under anaerobic conditions (Liu et al., 2021) (Kumar et al., 2021).

The generation of biogas from AD presents a superior energy balance compared to other biofuels, being able to cover up to 25% of the consumption in the future energy demand, being of great interest for the future society (Chozhavendhan et al., 2023). It has been considered as the best substitute to fossil fuel, because it is renewable, biodegradable and non-toxic component, being composed of especially by methane (CH_4) (50 ~ 70%), CO_2 (30 ~ 50%) and traces of H_2S (Rodrigues, Santana, et al., 2020).

The process of AD comprises in four steps namely, hydrolysis, acidogenesis, acetogenesis, and methanogenesis (Figure 3). The volatile fatty acids (VFAs) generated at the end of acetogenesis step are broken down into acetic acid and H_2, which they are consumed by methanogenic archaea (methanogenesis) to produce biogas (CH_4) (Figure 3) (S, Vasa, & Sabumon, 2022).

Figure 3. Stages of anaerobic digeston (AD)
Source: Adapted from (Sá, Cammarota, & Ferreira-Leitão (2014) and Ajay et al., 2020)

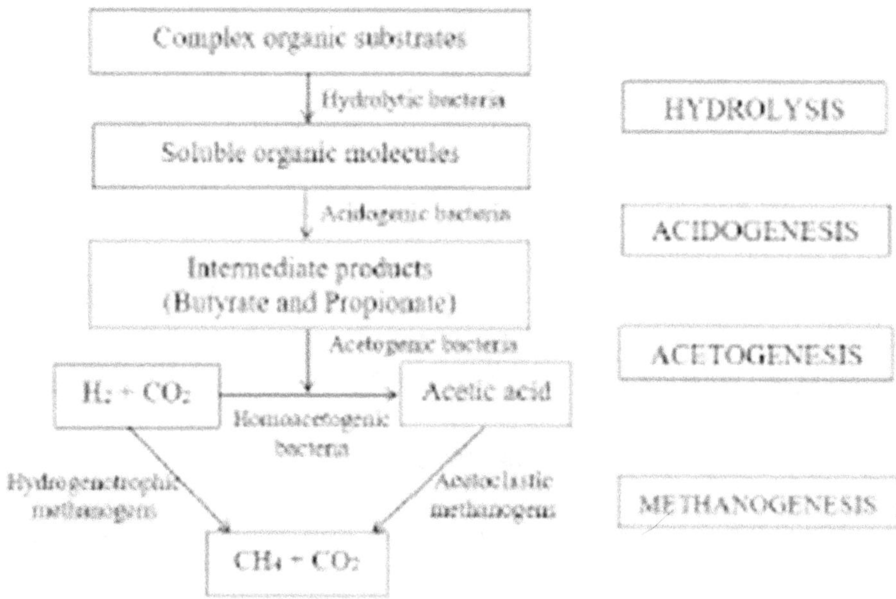

In fact, biogas from AD can be generated by different types of organic substrates coming from sewage sludge (Díaz et al., 2020; Zakaria & Dhar, 2022), animal excreta (Ihoeghian et al., 2022), wastes from agro-industrial process (Pachiega et al., 2019; Rodrigues, Alcaraz, et al., 2020; Rodrigues, Rios Alcaraz, et al., 2020) and food wastes (Le Pera et al., 2022), for example. However, with this current process, the efficiency of biogas generation is still low. Therefore, some techniques have been studied to increase biogas production like the use of nanomaterials (Aguilar-Moreno et al., 2020).

Studies have shown that nanoscale nutrient supplementation positively impact the generation of biogas, implying a greater methane production when compared to microscale supplementation, due to the increase in the surface area of these materials making them more reactive and available to react in anaerobic processes. Nanoparticles have a size of less than 100 μm in at least one dimension (0D, 1D, 2D, or 3D), and can be grouped into nanoparticles based on metals, ceramics, carbon, polymers, semiconductors and lipids. Thus, the use of these nanomaterials has represented a promising alternative for AD, promoting the stability of the processes, increasing the generation of biogas in addition to improving the quality of the effluent with a greater removal of organic matter and toxic components (Rocha-Meneses et al., 2022).

Direct Interspecies Electron Transfer (DIET) and Nanoparticles in AD

AD processes involving the degradation of organic matter depend exclusively on syntrophic interactions between bacteria and methanogenic archaea in order to overcome the energy dilemma, with the effective transport and diffusion of electrons between these microorganisms being a vital factor for the performance of anaerobic digestion (Wu et al., 2023). This interaction is facilitated by direct interspecies electron transfer (DIET) among the species (Ajay et al., 2020; S, Vasa, & Sabumon, 2022) (Figure 4).

Figure 4. DIET mechanism between fermentative bacteria and methanogen by nanomaterials
Source: Adapted from (Ajay et al., 2020)

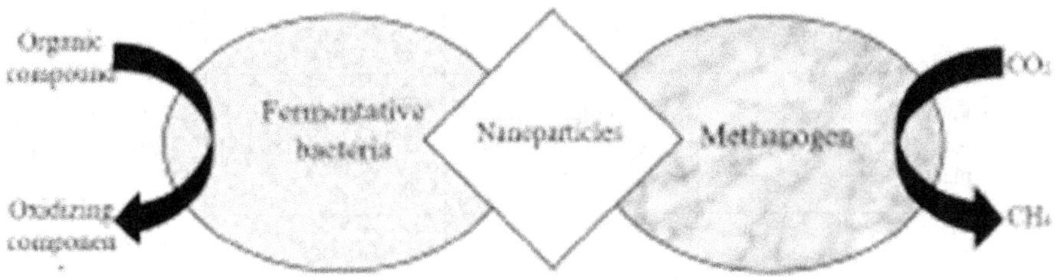

The hydrolysis step (Figure 3) is known to be the determining step during the anaerobic digestion, being enhanced by several chemical, physical, and biological pre-treatment methods, co-digestion, and nanoparticle additions. Among these methods, addition of nanoparticles has been gaining great attention (S, Vasa, & Sabumon, 2022).

The mobility of nanomaterials in an anaerobic environment influences the DIET, promoting stable reaction conditions and thereby enhances biogas production (Ajay et al., 2020; Jadhav et al., 2022). Nanomaterials can break cell membranes, promoting electron transformation through biological barriers, making them useful in several biological applications (Jadhav et al., 2022).

The use of nanomaterials is an alternative route to electron transfer in AD, which facilitate the DIET between the microorganism, without the interference of an electron carrier, enhancing the quality and quantity of biogas generated. In that way, the addition of nanoparticles is very beneficial. Many of nanoparticles can be used in the AD process, being efficiently in terms of biogas generated, such as metallic, conductive and carbon based, for example (S, Vasa, & Sabumon, 2022).

The carbon-based materials can be 3D graphite, 2D graphene, 1D nanotubes, and 0D fullerenes. The addition of carbon-based in AD causes a damage in cell membrane, releasing the intracellular substances to solution, considered as a type of mechanical pretreatment for enhanced hydrolysis. In addition, carbon-based materials can enhance the acid formation processes, while the excess of H_2 can benefit the hydrogenotrophic methanogenesis. However, a range of concentration for certain types of carbon-based materials should be investigated due to the negative impact on the biological activities, being toxic to aquatic environment. The enhancement effects of carbon-based materials added to AD should overcome their yielded cytotoxicity on the incorporated microorganisms, being studied the development of safe and effective methods for its use (Lu et al., 2020).

Iron based nanoparticles hold great promise as an additive during the AD, due to their high surface area, chemical stability, improving the microorganism's interactions through DIET, in addition to buffering the system even at high concentrations of volatile fatty acids (Juntupally et al., 2022; S, Vasa, & Sabumon, 2022). It is a fundamental micro nutrient for syntrophic interactions between bacteria and archaea, being responsible to reduce the lag phase and digestion time and it has many other unique important properties as an additive for AD (Ajay et al., 2020; S, Vasa, & Sabumon, 2022). Three forms of Iron nanoparticles can be commonly added to AD, namely Zero Valent Iron (Fe^0), Hematite (Fe_2O_3), and Magnetite (Fe_3O_4) (Juntupally et al., 2022). Several authors have been reported the use of iron-based nanoparticles to increase the methane production, as can be seen in Table 1.

Table 1. Studies with iron-based nanoparticles in AD

NP type	Substrate	Inoculum	Operational condition	CH_4 yield	Ref.
Magnet-Fe_3O_4 composite	From a municipal wastewater treatment plant	Mixed culture (from a long-term stable semi-continuous reactor)	Anaerobic batch reactors, at 37°C, with agitation at 100 rpm	78.6% of methane (~320 mL CH_4 g^{-1} SV)	(Li et al., 2022)
Fe_3O_4	Crude glycerol	Mixed culture (anaerobic granular sludge)	Anaerobic batch reactors, at 37°C	868.8 mL CH_4 g^{-1} TVS (49.8% increase on CH_4 production in reactors with 0.2 g L^{-1} of Fe_3O_4)	(Boscaro et al., 2022)
Fe_3O_4@PANI	Nutrient solution under the oxygen-free condition	Mixed culture (anaerobic sludge)	Anaerobic batch reactors, at 35°C	CH_4 production rate increased by 26.98%	(Huang et al., 2022)
Fe_3O_4	Chicken litter (animal waste, food waste and bedding material containing wood chips)	MeMixed culture (mesophilic digester fed for 3 years with chicken litter)	Anaerobic batch reactors, at 36°C	137.23 mL CH_4•g_{vsf}^{-1} (20 mg L^{-1} of Fe_3O_4 increases CH_4 production by 73.9%)	(Aguilar-Moreno et al., 2020)
Nanoscale zero valente iron (NZVI)	Dewatered sludge	Mixed culture (waste activated sludge)	Continuous stirred-tank reactors (CSTR), at 35 °C	Biogas production increased by 21.66%	(Xiang et al., 2019)

Source: the References are mentioned above

The addition of Fe^0 nanoparticles in AD process increase the enzyme activities during the hydrolysis and acidification steps even in the presence of recalcitrant substances and antibiotics; providing electrons directly or producing hydrogen through the iron anaerobic oxidation, being responsible for the growth of hydrogenotrophic methanogens (Barrena et al., 2021). In addition, this nanoparticle was also reported to be able to generate ferrous ions, serving as extracellular electron donors during AD (Xiang et al., 2019). However, adverse effects on the methane generation were also reported at high doses, representing a negative effect (Barrena et al., 2021).

Fe_3O_4 nanoparticles can gradually releasing Fe^{2+} and Fe^{3+} ions, when in solution, which represents an important role in microorganisms vital functions such as DNA replication, being responsible by energy generation. Besides that, magnetite nanoparticles contribute to the synthesis and activity of enzymes, coenzymes, and other essential molecules to AD processes (Boscaro et al., 2022). This nanoparticle presents the least toxicity effect, being applied as a supplement for microbial growth in AD, resulting in high methane generation (Ajayi-Banji & Rahman, 2021).

As well as the use of other iron nanoparticles, Fe_2O_3 also improved DIET and methane production. Anchoring Fe_2O_3 nanoparticles with small size and good dispersion in the conductive matrices can provide an excellent electron transfer in the methanogenic step (Yang et al., 2022).

The use of nanomaterials in the bionanotechnological sector for the sustainable development in bioenergy has been widely investigated (Pandey, 2022). The addition of nanoparticles exhibits positive effects in terms of biogas/methane yields, in addition to decrease the duration of lag phase during the AD (Grosser et al., 2021). However, the biogas generation depends on the type and quantity of nanoparticle used in the process and its cost may have an impact on the overall economics of biogas production (Ajay et al., 2020). Although the use of nanoparticles is promising, the economic and environmental feasibility of their use in anaerobic digestion needs to be explored in the future.

INDUSTRY AND ENVIRONMENTAL REGULATIONS

The Effects of Nanotechnology on the Human Health and Ecosystems

All the environmentally friendly applications of the nanomaterials have helped the society in many advances. However, the scientific community need to keep in mind that the environmental exposure to nanomaterials is inevitable, and in this way, nanotoxicity research has attracted attention to assess the risks of nanomaterials on the human health and ecosystems (Hannah & Thompson, 2008; Patra et al., 2018; Najahi-Missaoui, et al., 2021; Ray et al., 2009; Toropova et al., 2022). In a short period of time, investments in nanotechnology has gained tremendous public attraction in research and development. The global nanotechnology market is expected to reach $290.93 billion investment in 2028, and this increasing growth can be ascribed to the reduce in manufacturing costs, offer products at more cost-effective prices, lighter and more functional devices that require less energy. In this way, several governments are working on strategies to promote the development and application of nanotechnologies, and simultaneously taking into consideration its associated risks.

Each day about 2-3 million tons of waste from agriculture and industrial have been released into the environment. Indeed, contaminants such as trace metals and nanopesticides can be found in aquatic ecosystems from numerous sources, such as waste water effluents from domestic and industrial sources or from agricultural sources (Forero Lopez, et al., 2022). These primary sources of metals can release compounds such as mercury (Hg), cadmium (Cd), copper (Cu), nickel (Ni), cobalt (Co), zinc (Zn) and palladium (Pb) into nature. Trace metal contamination, considered as one global environmental problem, influences plants and aquatic organisms by the process of biological accumulation and magnification (Kumar et al., 2019; Uddin et al., 2020). For instance, the plants-based human and animals' diet can offer a risk when considering that vegetables accumulate metals above the recommended limits. The bioaccumulation involves nanoparticle accumulation in the organisms and then, the biomagnification follows in the food chain (Uddin et al., 2020). Therefore, various interactions of engineered nanomaterials/ metals compounds with the environment include dissolution, sedimentation, agglomeration or chemical transformations, followed by the uptake and transfer to the living organisms. Table 2 summarizes the impact of different nanocompounds on several organisms.

Table 2. Organisms exposed to nanoparticles and metal traces reported in the literature

Organism	Contaminants	Environment	Reference
Aquatic organisms	Carbon based nanomaterials	Aquatic System	(Freixa et al., 2018)
Lolium perenne, bocage variety plant	Ag nanoparticles	Soil from organic agriculture	(Courtois et al., 2021)
Cantareus aspersus, mollusk	Ag nanoparticles	Soil from organic agriculture	(Courtois et al., 2021)
Aquatic *Daphnia magna* (zooplankton) and fish	Plastic nanoparticles	Aquatic System	(Mattsson et al., 2017)
Oncorhynchus mykiss, fish	Copper nanoparticles	Aquatic System	(Shaw et al., 2012)
Tigriopus fulvus and *Corophium insidiosum* marine crustacean species	ZnO nanoparticles	Aquatic System	(Vimercati et al., 2020)
Crustacean *Gammarus fossarum*	Au nanoparticles	-	(Baudrimont et al., 2018)
Crustacean *Daphnia magna*, duckweed *Lemna minor* and bacteria *Vibrio fscheri*	Au nanoparticles	Aquatic System	(Hlavkova et al., 2020)
Algae (*Microcystis aeruginosa* and *Scenedesmus obliquus*	TiO_2 nanoparticles	Aquatic System	(Luo et al., 2018)
Bacteria	Copper nanoparticles and herbicide atrazine	Soil-plant system	(Parada et al., 2019)

Source: the References are mentioned above

Besides the use and release of nanoparticles into the environment by agriculture, industries and society, the use of nanoparticles in food need to be assessed. Nanoparticles in food includes several applications such as on food packaging, food additives to improve color, and improve compositional stability. For instance, packaging plays an important role, protecting the products from external conditions and keeping the food quality. However, the exposure and safety evaluations of the presence of nanoparticles in packaging food are still under debate for decades. The consumers can ingest nanomaterials on food, and any nanoparticle migration from packaging can occur and exposure human beings to the nanoparticles (Wang et al., 2021).

Nanoparticles such as nano-titanium dioxide (TiO_2) is normally used to improve the appearance and sensory properties of candy products such as chewing gum and chocolate (Weir et al., 2012). According to The Food and Drug Administration (FDA) of the United States, the content of TiO_2 in food must not be higher than 1 wt%, although the content of TiO_2 nanoparticles in sweets is high, reaching 2.5 mg Ti/g of food (Wang et al., 2021). SiO_2 nanoparticles have been extensively used as food additives in flour, coffee and milk essence. The content no higher than 2% by weight of SiO_2 in food was determined by FDA (Wang et al., 2021). In addition to the use of TiO_2 and SiO_2 nanoparticles in food market, silver (Ag) nanoparticles are increasingly incorporated into personal care goods, clothing and food containers (Rogers et al., 2018). Because of the anti-microbial properties of nano-Ag is employed to maintain food fresh for a long period of time and to avoid any contamination (Besha et al., 2020).

The invention of new materials provides new opportunities for their potential use, however the ethics regarding the sustainable use of nanomaterials need to be evaluated. Considerations on whether the use of nanoparticles is sustainable, without any damage to humans, animals and to the ecosystems can justify the good environmental and ecological outcomes through their applications. The social and ecological pros and cons associated with these considerations provides approaches to confront the challenges associated with the nanomaterials. At the same time, reports in the literature show the potential negative effects of nanoparticles on the environment and public health, such as ability of nanoparticles penetrating into cells, tissues, and organs, because of their tiny dimensions (Besha et al., 2020; Tirumala, et al., 2021). In this way, the sustainable use of nanomaterials should allow us to maintain a healthy environment for all lifeforms, in which the present and future generations can benefit from it.

Regulatory Approaches to Workers in Nanotechnology Industry and Universities

Precautionary measures need to be taken when handling the nanotechnology. The scientific institutions and the industrial sector must implement preventative and protective measures to minimize the risk of nanomaterials. Institutions, for instance, should provide mitigation approaches for the use of toxic and hazardous nanoparticles, and these procedures should be included in their internal regulations. Employees and researchers need to know about the physics, biological and chemical risks of substances they are exposed to in the work environment. Responsible entities such as the Ecole Polytechnique Fédérale de Lausanne (EPFL) developed measures to minimize the risk of dealing with nanotechnology by the 'user-friendly' approach for a safety and health management of nanomaterials. According with their proposal, nanomaterials can be classified into three hazard types: the highest hazard (Nano 3) to lowest hazard (Nano 1) (Groso et al., 2010). It is recommended to other institutions to follow these mitigation approaches in their internal regulations.

Equipment and procedures must be applied when working with nanomaterials: Laboratory fume hoods can significantly reduce workers' exposure to nanoparticles. Other good practice can be related to use of efficient ventilation systems. Personal protective equipment (PPE) is highly encouraged to be used when the workers are dealing with nano-scaled materials. They can be used to complement other control measures, and should be used as part of a comprehensive workplace program. Some PPE to be used as a barrier to nanoparticles have been largely employed, such as respirators, protective clothing, gloves and safety glasses for eye protection.

European union (EU) legislations apply to imported or manufactured chemical substances, in which nanomaterials should be addressed and mentioned within its sector-specific regulations. For cosmetic products (Regulation (EC) 1223/2009), the EU should be informed about cosmetic products containing nanoparticles before their selling or marketing. The toxicological and safety information should be clarified. For food additives (Regulation (EC) 1333/2008), any change in specification shall be provided before its sale. In Regulation (EU) 528/2012, the risk assessment is necessary for biocidal products containing engineered nanoparticles and product label should inform that it comprises nanomaterials (Besha et al., 2020; Pavlicek, et al., 2021). Unfortunately, the effectiveness of legislation for controlling ENPs are under debate. Even if nanomaterials are covered by existing legislations, it is still unsure if the current legislations are applicable considering the several applications of nanomaterials. The disposal of nanomaterials or products containing nanoparticles require special processes to minimize potential risks. Therefore, effective legislations and regulations for the production, sale, application, and disposal of different nanomaterials highly needed and involves ethical dilemmas to overcome.

FUTURE RESEARCH DIRECTIONS

This chapter has been exposed the most factors that influence the choice of materials, whose application is based on at least the minimum prejudice to the environment, from their production processes to their lifetime and disuse. Commercial products and recommendations on safe prevent adverse implications, always attempting the eco-friendly, cost-effective, and simple nanomaterials synthesis methods, which reduce the effects of nanomaterials directly to the human health and environment.

CONCLUSION

A better life is always accompanied with increasing population, technology, and economic development. Good results from the past to now, e.g. safety activities which considerate toxicological effects and regulatory of nanomaterials, renewable energy technologies instead of fuel fossils that generate electricity, and high added value products as biogas through a green energy generation must make increasingly harmonic interaction with environment.

ACKNOWLEDGMENT

The authors acknowledge the São Paulo State University-UNESP and USP, as well as the *Federal Technological University of Paraná (UTFPR).*

REFERENCES

Aguilar-Moreno, G. S., Navarro-Cerón, E., Velázquez-Hernández, A., Hernández-Eugenio, G., Aguilar-Méndez, M. Á., & Espinosa-Solares, T. (2020). Enhancing methane yield of chicken litter in anaerobic digestion using magnetite nanoparticles. *Renewable Energy*, *147*, 204–213. doi:10.1016/j.renene.2019.08.111

Ajay, C. M., Mohan, S., Dinesha, P., & Rosen, M. A. (2020). Review of impact of nanoparticle additives on anaerobic digestion and methane generation. *Fuel*, *277*, 118234. doi:10.1016/j.fuel.2020.118234

Ajayi-Banji, A. A., & Rahman, S. (2021). Efficacy of magnetite (Fe3O4) nanoparticles for enhancing solid-state anaerobic co-digestion: Focus on reactor performance and retention time. *Bioresource Technology*, *324*, 124670. doi:10.1016/j.biortech.2021.124670 PMID:33453521

Al-Dousari, A., Al-Nassar, W., & Ahmed, M. (2020). Photovoltaic and Wind Energy: Challenges and Solutions in Desert Regions. *E3S Web of Conferences*, *166*, 1–6. doi:10.1051/e3sconf/202016604003

Al-Ezzi, A. S., & Ansari, M. N. M. (2022). Photovoltaic Solar Cells: A Review. *Applied System Innovation*, *5*(67), 1–17. doi:10.3390/asi5040067

Alarçin, E., Guan, X., Kashaf, S. S., Elbaradie, K., Yang, H., Jang, H. L., & Khademhosseini, A. (2016). Recreating composition, structure, functionalities of tissues at nanoscale for regenerative medicine. *Regenerative Medicine*, *11*(8), 849–858. doi:10.2217/rme-2016-0120 PMID:27885900

Alic, J. A., & Sarewitz, D. (2016). Rethinking Innovation for Decarbonizing Energy Systems. *Energy Research & Social Science*, *21*, 212–221. doi:10.1016/j.erss.2016.08.005

Bai, S., Guo, W., Sun, J., Li, J., Tian, Y., Chen, A., Luo, R., & Li, D. (2016). Synthesis of SnO2–CuO heterojunction using electrospinning and application in detecting of CO. *Sensors and Actuators. B, Chemical*, *226*, 96–103. doi:10.1016/j.snb.2015.11.028

Barrena, R., Vargas-García, M. del C., Capell, G., Barańska, M., Puntes, V., Moral-Vico, J., Sánchez, A., & Font, X. (2021). Sustained effect of zero-valent iron nanoparticles under semi-continuous anaerobic digestion of sewage sludge: Evolution of nanoparticles and microbial community dynamics. *The Science of the Total Environment*, *777*, 145969. Advance online publication. doi:10.1016/j.scitotenv.2021.145969 PMID:33676214

Baudrimont, M., Andrei, J., Mornet, S., Gonzalez, P., Mesmer-Dudons, N., Gourves, P.-Y., Jaffal, A., Dedourge-Geffard, O., Geffard, A., Geffard, O., Garric, J., & Feurtet-Mazel, A. (2018). Trophic transfer and effects of gold nanoparticles (AuNPs) in Gammarus fossarum from contaminated periphytic biofilm. *Environmental Science and Pollution Research International*, *25*(12), 11181–11191. doi:10.100711356-017-8400-3 PMID:28091991

Bayda, S., Adeel, M., Tuccinardi, T., Cordani, M., & Rizzolio, F. (2019). The History of Nanoscience and Nanotechnology: From Chemical-Physical Applications to Nanomedicine. *Molecules (Basel, Switzerland)*, *25*(1), 112. doi:10.3390/molecules25010112 PMID:31892180

Bergin, M. H., Ghoroi, C., Dixit, D., Schauer, J. J., & Shindell, D. T. (2017). Large Reductions in Solar Energy Production Due to Dust and Particulate Air Pollution. *Environmental Science & Technology Letters*, *4*(8), 339–344. doi:10.1021/acs.estlett.7b00197

Besha, A. T., Liu, Y., Bekele, D. N., Dong, Z., Naidu, R., & Gebremariam, G. N. (2020). Sustainability and environmental ethics for the application of engineered nanoparticles. *Environmental Science & Policy*, *103*, 85–98. doi:10.1016/j.envsci.2019.10.013

Boscaro, M. E., Marin, D. F. C., da Silva, D. C., & Maintinguer, S. I. (2022). Biomass and Bioenergy Effect of Fe 3 O 4 nanoparticles on microbial diversity and biogas production in anaerobic digestion of crude glycerol. *Biomass and Bioenergy*, *160*, 106439. doi:10.1016/j.biombioe.2022.106439

Brouwer, A. S., van Den Broek, M., Seebregts, A., & Faaij, A. (2014). Impacts of Large-Scale Intermittent Renewable Energy Sources on Electricity Systems, and How These Can Be Modeled. *Renewable & Sustainable Energy Reviews*, *33*, 443–466. doi:10.1016/j.rser.2014.01.076

Chaud, M., Souto, E. B., Zielinska, A., Severino, P., Batain, F., Oliveira-Junior, J., & Alves, T. (2021). Nanopesticides in Agriculture: Benefits and Challenge in Agricultural Productivity, Toxicological Risks to Human Health and Environment. *Toxics*, *9*(6), 131. Advance online publication. doi:10.3390/toxics9060131 PMID:34199739

Chozhavendhan, S., Karthigadevi, G., Bharathiraja, B., Kumar, R. P., Abo, L. D., Prabhu, S. V., Balachandar, R., & Jayakumar, M. (2023). Current and prognostic overview on the strategic exploitation of anaerobic digestion and digestate: A review. *Environmental Research*, *216*, 114526. doi:10.1016/j.envres.2022.114526 PMID:36252837

Coropceanu, I., & Bawendi, M. G. (2014). Core/Shell Quantum Dot Based Luminescent Solar Concentrators with Reduced Reabsorption and Enhanced Efficiency. *Nano Letters*, *14*(7), 4097–4101. doi:10.1021/nl501627e PMID:24902615

Courtois, P., de Vaufleury, A., Grosser, A., Lors, C., & Vandenbulcke, F. (2021). Transfer of sulfidized silver from silver nanoparticles, in sewage sludge, to plants and primary consumers in agricultural soil environment. *The Science of the Total Environment*, *777*, 145900. doi:10.1016/j.scitotenv.2021.145900 PMID:33676211

Creutz, S. E., Fainblat, R., Kim, Y., De Siena, M. C., & Gamelin, D. R. (2017). A Selective Cation Exchange Strategy for the Synthesis of Colloidal Yb3+-Doped Chalcogenide Nanocrystals with Strong Broadband Visible Absorption and Long-Lived Near-Infrared Emission. *Journal of the American Chemical Society*, *139*(34), 11814–11824. doi:10.1021/jacs.7b04938 PMID:28750510

Daorta, S. F., Proto, A., Fusco, R., Andreani, L. C., & Liscidini, M. (2014). Cascade Luminescent Solar Concentrators. *Applied Physics Letters*, *104*(15), 10–14. doi:10.1063/1.4871481

Díaz, I., Fdz-Polanco, F., Mutsvene, B., & Fdz-Polanco, M. (2020). Effect of operating pressure on direct biomethane production from carbon dioxide and exogenous hydrogen in the anaerobic digestion of sewage sludge. *Applied Energy*, *280*, 115915. doi:10.1016/j.apenergy.2020.115915

El Bakari, K., & Kling, W. L. (2010). Virtual Power Plants: An Answer to Increasing Distributed Generation. *IEEE PES Innovative Smart Grid Technologies Conference Europe,* 1–6. doi:10.1109/ISGTEUROPE.2010.5638984

Fazzino, F., Folino, A., Mauriello, F., Pedullà, A., & Calabrò, P. S. (2021). Biofuel production from fruit and vegetable market waste and mature landfill leachate by an active filter-anaerobic digestion integrated system. *Energy Conversion and Management: X, 12*, 100130. doi:10.1016/j.ecmx.2021.100130

Forero López, A. D., Fabiani, M., Lassalle, V. L., Spetter, C. V., & Fernandez Severini, M. D. (2022). Critical review of the characteristics, interactions, and toxicity of micro/nanomaterials pollutants in aquatic environments. *Marine Pollution Bulletin, 174*, 113276. doi:10.1016/j.marpolbul.2021.113276 PMID:35090270

Freixa, A., Acuña, V., Sanchís, J., Farré, M., Barceló, D., & Sabater, S. (2018). Ecotoxicological effects of carbon based nanomaterials in aquatic organisms. *The Science of the Total Environment, 619–620*, 328–337. doi:10.1016/j.scitotenv.2017.11.095 PMID:29154051

Groso, A., Petri-Fink, A., Magrez, A., Riediker, M., & Meyer, T. (2010). Management of nanomaterials safety in research environment. *Particle and Fibre Toxicology, 7*(1), 40. doi:10.1186/1743-8977-7-40 PMID:21143952

Grosser, A., Grobelak, A., Rorat, A., Courtois, P., Vandenbulcke, F., Lemière, S., Guyoneaud, R., Attard, E., & Celary, P. (2021). Effects of silver nanoparticles on performance of anaerobic digestion of sewage sludge and associated microbial communities. *Renewable Energy, 171*, 1014–1025. doi:10.1016/j.renene.2021.02.127

Gul, M., Kotak, Y., & Muneer, T. (2016). Review on Recent Trend of Solar Photovoltaic Technology. *Energy Exploration and Exploitation, 34*(4), 485–526. doi:10.1177/0144598716650552

Han, H.-V., Lin, C.-C., Tsai, Y.-L., Chen, H.-C., Chen, K.-J., Yeh, Y.-L., Lin, W.-Y., Kuo, H.-C., & Yu, P. (2014). A Highly Efficient Hybrid GaAs Solar Cell Based on Colloidal-Quantum-Dot- Sensitization. *Scientific Reports, 4*(1), 1–9. doi:10.1038rep05734 PMID:25034623

Hannah, W., & Thompson, P. B. (2008). Nanotechnology risk and the environment: A review. *Journal of Environmental Monitoring, 10*(3), 291–300. doi:10.1039/b718127m PMID:18392270

Hlavkova, D., Caloudova, H., Palikova, P., Kopel, P., Plhalova, L., Beklova, M., & Havelkova, B. (2020). Effect of Gold Nanoparticles and Ions Exposure on the Aquatic Organisms. *Bulletin of Environmental Contamination and Toxicology, 105*(4), 530–537. doi:10.100700128-020-02988-6 PMID:32940716

Huang, W., Zhou, J., Hu, Q., Qiu, B., Huang, M., Murugadoss, V., & Guo, Z. (2022). Improved methanogenesis in anaerobic wastewater treatment by magnetite @ polyaniline (Fe 3 O 4 @ PANI) composites. *Chemosphere, 296*, 133953. doi:10.1016/j.chemosphere.2022.133953 PMID:35157884

Ihoeghian, N. A., Amenaghawon, A. N., Ajieh, M. U., Oshoma, C. E., Ogofure, A., Erhunmwunse, N. O., Edosa, V. I. O., Tongo, I., Obuekwe, I. S., Isagba, E. S., Emokaro, C., Ezemonye, L. I. N., Lag-Brotons, A. J., Semple, K. T., & Martin, A. D. (2022). Anaerobic co-digestion of cattle rumen content and food waste for biogas production: Establishment of co-digestion ratios and kinetic studies. *Bioresource Technology Reports, 18*, 101033. doi:10.1016/j.biteb.2022.101033

Jadhav, P., Khalid, Z. B., Zularisam, A. W., Krishnan, S., & Nasrullah, M. (2022). The role of iron-based nanoparticles (Fe-NPs) on methanogenesis in anaerobic digestion (AD) performance. *Environmental Research, 204*(PB), 112043. doi:10.1016/j.envres.2021.112043

Jalali, H. B., Sadeghi, S., Baylam, I., & Han, M. (2021). Exciton Recycling via InP Quantum Dot Funnels for Luminescent Solar Concentrators. *Nano Research, 14*(5), 1488–1494. doi:10.100712274-020-3207-9

Juntupally, S., Begum, S., Arelli, V., Mamindlapelli, N. K., Srinivasan, S., & Anupoju, G. R. (2022). Evaluating the impact of Iron Oxide nanoparticles (IO-NPs) and IO-NPs doped granular activated carbon on the anaerobic digestion of food waste at mesophilic and thermophilic temperature. *Journal of Environmental Chemical Engineering, 10*(3), 107388. Advance online publication. doi:10.1016/j.jece.2022.107388

Kassem, Y., Zoubi, R. A., & Hüseyin, G. (2019). The Possibility of Generating Electricity Using Small-Scale Wind Turbines and Solar Photovoltaic Systems for Households in Northern Cyprus: A Comparative Study. *Environments – MDPI, 6*(4), 1–22. doi:10.3390/environments6040047

Kumar, S., Prasad, S., Yadav, K. K., Shrivastava, M., Gupta, N., Nagar, S., Bach, Q.-V., Kamyab, H., Khan, S. A., Yadav, S., & Malav, L. C. (2019). Hazardous heavy metals contamination of vegetables and food chain: Role of sustainable remediation approaches - A review. *Environmental Research, 179*, 108792. doi:10.1016/j.envres.2019.108792 PMID:31610391

Kumar, S. S., Ghosh, P., Kataria, N., Kumar, D., Thakur, S., Pathania, D., Kumar, V., Nasrullah, M., & Singh, L. (2021). The role of conductive nanoparticles in anaerobic digestion: Mechanism, current status and future perspectives. *Chemosphere, 280*, 130601. doi:10.1016/j.chemosphere.2021.130601 PMID:33945900

Le Pera, A., Sellaro, M., Bencivenni, E., & D'Amico, F. (2022). Environmental sustainability of an integrate anaerobic digestion-composting treatment of food waste: Analysis of an Italian plant in the circular bioeconomy strategy. *Waste Management, 139*(July), 341–351. doi:10.1016/j.wasman.2021.12.042

Leitão, D., Torres, J. P. N., & Fernandes, J. F. P. (2020). Spectral Irradiance Influence on Solar Cells Efficiency. *Energies, 13*(19), 1–18. doi:10.3390/en13195017

Lemos, S. C. S., Rezende, T. K. L., Assis, M., Romeiro, F. C., Peixoto, D. A., Gomes, E. O., Jacobsen, G. M., Teodoro, M. D., Gracia, L., Ferrari, J. L., Longo, E., Andrés, J., & de Lima, R. C. (2022). Efficient Ni and Fe doping process in ZnO with enhanced photocatalytic activity: A theoretical and experimental investigation. *Materials Research Bulletin, 111849*. Advance online publication. doi:10.1016/j.materresbull.2022.111849

Lemos, S. C. S., Romeiro, F. C., de Paula, L. F., Goncalves, R. F., de Moura, A. P., Ferrer, M. M., Longo, E., Patrocinio, A. O. T., & Lima, R. C. (2017). Effect of Er3+ ions on the phase formation and properties of In2O3 nanostructures crystallized upon microwave heating. *Journal of Solid State Chemistry, 249*, 58–63. doi:10.1016/j.jssc.2017.02.011

Li, L., Liu, H., Chen, Y., Yang, D., Cai, C., Yuan, S., & Dai, X. (2022). Effect of Magnet-Fe3O4 composite structure on methane production during anaerobic sludge digestion: Establishment of direct interspecies electron transfer. *Renewable Energy, 188*, 52–60. doi:10.1016/j.renene.2022.01.101

Liu, H., Li, S., Chen, W., Wang, D., Li, C., Wu, C., Hao, J., Zhou, Z., Wang, X., & Wang, K. (2018). Scattering Enhanced Quantum Dots Based Luminescent Solar Concentrators by Silica Microparticles. *Solar Energy Materials and Solar Cells, 179*, 380–385. doi:10.1016/j.solmat.2018.01.029

Liu, Y., Xiao, Q., Ye, X., Wang, C., Jia, Z., Du, J., Kong, X., & Xi, Y. (2021). Effect of different charged Fe 3 O 4 nanoparticles on methane production for anaerobic digestion of wheat straw. *Journal of Cleaner Production*, *328*, 129655. doi:10.1016/j.jclepro.2021.129655

Lu, C., Li, X., Guo, X., Fu, S., Zhang, W., Yuan, H., & Fang, J. (2023). Efficient inverted CsPbI3 perovskite solar cells fabricated in common air. *Chemical Engineering Journal*, *452*, 139495. doi:10.1016/j.cej.2022.139495

Lu, J. S., Chang, J. S., & Lee, D. J. (2020). Adding carbon-based materials on anaerobic digestion performance: A mini-review. *Bioresource Technology*, *300*(November), 122696. doi:10.1016/j.biortech.2019.122696

Luo, Z., Wang, Z., Yan, Y., Li, J., Yan, C., & Xing, B. (2018). Titanium dioxide nanoparticles enhance inorganic arsenic bioavailability and methylation in two freshwater algae species. *Environmental Pollution*, *238*, 631–637. doi:10.1016/j.envpol.2018.03.070 PMID:29614472

Malinauskaite, J., Jouhara, H., Czajczyńska, D., Stanchev, P., Katsou, E., Rostkowski, P., Thorne, R. J., Colón, J., Ponsá, S., Al-Mansour, F., Anguilano, L., Krzyżyńska, R., López, I. C., Vlasopoulos, A., & Spencer, N. (2017). Municipal Solid Waste Management and Waste-to-Energy in the Context of a Circular Economy and Energy Recycling in Europe. *Energy*, *141*, 2013–2044. doi:10.1016/j.energy.2017.11.128

Mattsson, K., Johnson, E. V., Malmendal, A., Linse, S., Hansson, L.-A., & Cedervall, T. (2017). Brain damage and behavioural disorders in fish induced by plastic nanoparticles delivered through the food chain. *Scientific Reports*, *7*(1), 11452. doi:10.103841598-017-10813-0 PMID:28904346

Najahi-Missaoui, W., Arnold, R. D., & Cummings, B. S. (2021). Safe Nanoparticles: Are We There Yet? *International Journal of Material Science*, *22*(1), 385. doi:10.3390/ijms22010385 PMID:33396561

Pachiega, R., Rodrigues, M. F., Rodrigues, C. V., Sakamoto, I. K., de Oliveira, J. E., Varesche, M. B. A., & Maintinguer, S. I. (2019). Hydrogen bioproduction with anaerobic bacteria consortium from brewery wastewater. *International Journal of Hydrogen Energy*, *44*(1), 155–163. doi:10.1016/j.ijhydene.2018.02.107

Pandey, M. D. (2022). Perspective of nanomaterials for sustainable biofuel and bioenergy production. *Materials Letters*, *313*, 131686. doi:10.1016/j.matlet.2022.131686

Papp, T., Schiffmann, D., Weiss, D., Castranova, V., Vallyathan, D. V., & Rahman, Q. (2008). Human health implications of nanomaterial exposure. *Nanotoxicology*, *2*(1), 9–27. doi:10.1080/17435390701847935

Parada, J., Rubilar, O., Sousa, D. Z., Martínez, M., Fernández-Baldo, M. A., & Tortella, G. R. (2019). Short term changes in the abundance of nitrifying microorganisms in a soil-plant system simultaneously exposed to copper nanoparticles and atrazine. *The Science of the Total Environment*, *670*, 1068–1074. doi:10.1016/j.scitotenv.2019.03.221 PMID:31018422

Patra, J. K., Das, G., Fraceto, L. F., Campos, E. V. R., Rodriguez-Torres, M. del P., Acosta-Torres, L. S., Diaz-Torres, L. A., Grillo, R., Swamy, M. K., Sharma, S., Habtemariam, S., & Shin, H.-S. (2018). Nano based drug delivery systems: Recent developments and future prospects. *Journal of Nanobiotechnology*, *16*(1), 71. doi:10.118612951-018-0392-8 PMID:30231877

Pavlicek, A., Part, F., Rose, G., Praetorius, A., Miernicki, M., Gazsó, A., & Huber-Humer, M. (2021). A European nano-registry as a reliable database for quantitative risk assessment of nanomaterials? A comparison of national approaches. *NanoImpact, 21*, 100276. doi:10.1016/j.impact.2020.100276 PMID:35559789

Rashid, M. M., Forte Tavčer, P., & Tomšič, B. (2021). Influence of Titanium Dioxide Nanoparticles on Human Health and the Environment. *Nanomaterials (Basel, Switzerland), 11*(9), 2354. doi:10.3390/nano11092354 PMID:34578667

Ray, P. C., Yu, H., & Fu, P. P. (2009). Toxicity and environmental risks of nanomaterials: Challenges and future needs. *Journal of Environmental Science and Health. Part C, Environmental Carcinogenesis & Ecotoxicology Reviews, 27*(1), 1–35. doi:10.1080/10590500802708267 PMID:19204862

Rocha-Meneses, L., Hari, A., Inayat, A., Shanableh, A., Abdallah, M., Ghenai, C., Shanmugam, S., & Kikas, T. (2022). Application of nanomaterials in anaerobic digestion processes: A new strategy towards sustainable methane production. *Biochemical Engineering Journal, 188*, 108694. doi:10.1016/j.bej.2022.108694

Rodrigues, A. V., de Souza, D. A. R., Garcia, F. D. R., & Ribeiro, S. J. L. (2022). Renewable Energy for a Green Future: Electricity Produced from Efficient Luminescent Solar Concentrators. *Solar Energy Advances, 2*, 1–13. doi:10.1016/j.seja.2022.100013

Rodrigues, C. V., Alcaraz, F. A. R., Nespeca, M. G., Rodrigues, A. V., Motteran, F., Adorno, M. A. T., Varesche, M. B. A., & Maintinguer, S. I. (2020). Biohydrogen production in an integrated biosystem using crude glycerol from waste cooking oils. *Renewable Energy, 162*, 701–711. doi:10.1016/j.renene.2020.08.061

Rodrigues, C. V., Santana, K. O., Nespeca, M. G., Rodrigues, A. V., Pires, L. O., & Maintinguer, S. I. (2020). Energy valorization of crude glycerol and sanitary sewage in hydrogen generation by biological processes. *International Journal of Hydrogen Energy, 45*(21), 11943–11953. doi:10.1016/j.ijhydene.2020.02.168

Rogers, K. R., Navratilova, J., Stefaniak, A., Bowers, L., Knepp, A. K., Al-Abed, S. R., Potter, P., Gitipour, A., Radwan, I., Nelson, C., & Bradham, K. D. (2018). Characterization of engineered nanoparticles in commercially available spray disinfectant products advertised to contain colloidal silver. *The Science of the Total Environment, 619–620*, 1375–1384. doi:10.1016/j.scitotenv.2017.11.195 PMID:29723948

Romeiro, F. C., Castro, N. S., Scolfaro, L., Borges, P. D., & Lima, R. C. (2022). Theoretical and experimental study of effects of Co2+ doping on structural and electronic properties of ZnO. *Journal of Physics and Chemistry of Solids, 162*, 110501. doi:10.1016/j.jpcs.2021.110501

Romeiro, F. C., & Rodrigues, A. V. (2022). Green Synthesis of Semiconductors and Environmental Applications: From Single to High Added Value Products. Handbook of Research on Green Synthesis and Applications of Nanomaterials, 27–50. doi:10.4018/978-1-7998-8936-6.ch002

Romeiro, F. C., Silva, B. C., Martins, A. S., Zanoni, M. V. B., & Orlandi, M. O. (2021). Superior performance of rGO-tin oxide nanocomposite for selective reduction of CO2 to methanol. *Journal of CO2 Utilization, 46*. doi:10.1016/j.jcou.2021.101460

S, R., Vasa, T. N., & Sabumon, P. C. (2022). A mini review on effect of nano particles of Fe in the anaerobic digestion of waste activated sludge. *Materials Today: Proceedings, 51*, 1482–1488. doi:10.1016/j.matpr.2021.10.265

Sá, L. R. V., Cammarota, M. C., & Ferreira-Leitão, V. S. (2014). Produção de Hidrogênio via Fermentação anaeróbia - aspectos gerais e possibilidade de utilização de resíduos agroindustriais brasileiros. *Quimica Nova, 37*(5), 857–867. doi:10.5935/0100-4042.20140138

Sampaio, P. G. V., & González, M. O. A. (2017). Photovoltaic Solar Energy: Conceptual Framework. *Renewable & Sustainable Energy Reviews, 74*, 590–601. doi:10.1016/j.rser.2017.02.081

Saravanan, A., Kumar, P. S., Rangasamy, G., Hariharan, R., Hemavathy, R. V., Deepika, P. D., Anand, K., & Karthika, S. (2023). Strategies for enhancing the efficacy of anaerobic digestion of food industry wastewater: An insight into bioreactor types, challenges, and future scope. *Chemosphere, 310*, 136856. doi:10.1016/j.chemosphere.2022.136856 PMID:36243094

Sharma, M., Gungor, K., Yeltik, A., Olutas, M., Guzelturk, B., Kelestemur, Y., Erdem, T., Delikanli, S., McBride, J. R., & Demir, H. V. (2017). Near-Unity Emitting Copper-Doped Colloidal Semiconductor Quantum Wells for Luminescent Solar Concentrators. *Advanced Materials, 29*(30), 1–10. doi:10.1002/adma.201700821 PMID:28605062

Shaw, B. J., Al-Bairuty, G., & Handy, R. D. (2012). Effects of waterborne copper nanoparticles and copper sulphate on rainbow trout, (Oncorhynchus mykiss): Physiology and accumulation. *Aquatic Toxicology (Amsterdam, Netherlands), 116–117*, 90–101. doi:10.1016/j.aquatox.2012.02.032 PMID:22480992

Suman, P. H., Felix, A. A., Tuller, H. L., Varela, J. A., & Orlandi, M. O. (2015). Comparative gas sensor response of SnO2, SnO and Sn3O4 nanobelts to NO2 and potential interferents. *Sensors and Actuators. B, Chemical, 208*, 122–127. doi:10.1016/j.snb.2014.10.119

Tirumala, M. G., Anchi, P., Raja, S., Rachamalla, M., & Godugu, C. (2021). Novel Methods and Approaches for Safety Evaluation of Nanoparticle Formulations: A Focus Towards In Vitro Models and Adverse Outcome Pathways. *Frontiers in Pharmacology, 12*, 1–12. doi:10.3389/fphar.2021.612659 PMID:34566630

Toropova, A. P., & Toropov, A. A. (2022). Nanomaterials: Quasi-SMILES as a flexible basis for regulation and environmental risk assessment. *The Science of the Total Environment, 823*, 153747. doi:10.1016/j.scitotenv.2022.153747 PMID:35149067

Uddin, M. N., Desai, F., & Asmatulu, E. (2020). Engineered nanomaterials in the environment: Bioaccumulation, biomagnification and biotransformation. *Environmental Chemistry Letters, 18*(4), 1073–1083. doi:10.100710311-019-00947-0

Vimercati, L., Cavone, D., Caputi, A., De Maria, L., Tria, M., Prato, E., & Ferri, G. M. (2020). Nanoparticles: An Experimental Study of Zinc Nanoparticles Toxicity on Marine Crustaceans. General Overview on the Health Implications in Humans. *Frontiers in Public Health, 8*, 192. Advance online publication. doi:10.3389/fpubh.2020.00192 PMID:32509719

Waldron, D. L., Preske, A., Zawodny, J. M., Krauss, T. D., & Gupta, M. C. (2017). PbSe Quantum Dot Based Luminescent Solar Concentrators. *Nanotechnology*, *28*(9), 095205. Advance online publication. doi:10.1088/1361-6528/aa577f PMID:28060769

Wang, M., Li, S., Chen, Z., Zhu, J., Hao, W., Jia, G., Chen, W., Zheng, Y., Qu, W., & Liu, Y. (2021). Safety assessment of nanoparticles in food: Current status and prospective. *Nano Today*, *39*, 101169. doi:10.1016/j.nantod.2021.101169

Weir, A., Westerhoff, P., Fabricius, L., Hristovski, K., & von Goetz, N. (2012). Titanium Dioxide Nanoparticles in Food and Personal Care Products. *Environmental Science & Technology*, *46*(4), 2242–2250. doi:10.1021/es204168d PMID:22260395

Wu, L., Jin, T., Chen, H., Shen, Z., & Zhou, Y. (2023). Conductive materials as fantastic toolkits to stimulate direct interspecies electron transfer in anaerobic digestion: New insights into methanogenesis contribution, characterization technology, and downstream treatment. *Journal of Environmental Management*, *326*(PB), 116732. doi:10.1016/j.jenvman.2022.116732

Xiang, Y., Yang, Z., Zhang, Y., Xu, R., Zheng, Y., Hu, J., Li, X., Jia, M., Xiong, W., & Cao, J. (2019). Influence of nanoscale zero-valent iron and magnetite nanoparticles on anaerobic digestion performance and macrolide, aminoglycoside, β-lactam resistance genes reduction. *Bioresource Technology*, *294*, 122139. doi:10.1016/j.biortech.2019.122139 PMID:31525586

Xiaomei, Z., Rujing, L., Jun, X., Yingying, H., Xinying, Z., & Li, X. (2022). Enhanced methane production by bimetallic metal – organic frameworks (MOFs) as cathode in an anaerobic digestion microbial electrolysis cell. *Chemical Engineering Journal*, *440*, 135799. Advance online publication. doi:10.1016/j.cej.2022.135799

Yang, Z. M., Guo, R. B., & Dong, X. H. (2022). Promoting biomethane production from propionate with Fe2O3@carbon nanotubes composites. *The Science of the Total Environment*, *818*, 151762. doi:10.1016/j.scitotenv.2021.151762 PMID:34800454

You, Y., Tong, X., Wang, W., Sun, J., Yu, P., Ji, H., Niu, X., & Wang, Z. M. (2019). Eco-Friendly Colloidal Quantum Dot-Based Luminescent Solar Concentrators. *Advancement of Science*, *6*(9), 1801967. Advance online publication. doi:10.1002/advs.201801967 PMID:31065522

Zakaria, B. S., & Dhar, B. R. (2022). Intermittent energization catalyzes direct interspecies electron transfer in electro-anaerobic digestion of sewage sludge. *Chemical Engineering Journal*, *136177*, 136177. Advance online publication. doi:10.1016/j.cej.2022.136177

Zhang, D., Ma, X., Gu, Y., Huang, H., & Zhang, G. (2020). Green Synthesis of Metallic Nanoparticles and Their Potential Applications to Treat Cancer. *Frontiers in Chemistry*, *8*, 799. doi:10.3389/fchem.2020.00799 PMID:33195027

Zsiborács, H., Baranyai, N. H., Vincze, A., Zentkó, L., Birknei, Z., Máté, K., & Pintér, G. (2019). Intermittent Renewable Energy Sources: The Role of Energy Storage in the European Power System of 2040. *Electronics (Switzerland)*, *8*(7), 1–18. doi:10.3390/electronics8070729

ADDITIONAL READING

Krasniewska, K., Galus, S., & Gniewosz, M. (2020). Biopolymers-Based Materials Containing silver Nanoparticles as Active Packaging for Food Applications–A Review. *International Journal of Molecular Sciences*, *21*(3), 1–18. doi:10.3390/ijms21030698 PMID:31973105

Pavlicek, A., Part, F., Rose, G., Praetorius, A., Miernicki, M., Gazso, A., & Huber-Humer, M. (2021). Current Legislative Framework for Nanomaterials & Introduction to the Impact Assessment on Transparency Measures. *NanoImpact*, *21*, 1–10.

KEY TERMS AND DEFINITIONS

Anaerobic Digestion: Corresponds to in an adequate bioprocess to treat residues, converting biodegradable organic matter into biogas by microorganisms.

Biofuel: Can be generated during the anaerobic digestion, been manly composite of CO_2, CH_4 and H_2S.

DIET: Direct interspecies electron transfer among the microorganisms of bacteria and archaea.

Nanotechnology: Involves the research, development, manipulation of materials in the range of 0.1–100 nanometers.

Photoluminescence: Property of emitting luminescent radiation after an external excitation source.

Photovoltaic: Technology with the capacity of producing produce electricity using solar energy.

Toxicology: Study of the adverse effects of chemical compounds organisms and involves the diagnosing and treatment of toxicants.

Chapter 9
Toxicity Analysis of Different Types of Two-Dimensional Nanomaterials Used in Biomedical Applications

Rafael Vargas-Bernal
https://orcid.org/0000-0003-4865-4575
Instituto Tecnológico Superior de Irapuato, Mexico

ABSTRACT

The study of biological interactions with two-dimensional nanomaterials provides the scientific foundation for identifying and managing development risks in both biomedical and non-biomedical technologies. The purpose of this chapter is to carry out an analysis of the toxicity of two-dimensional materials in biomedical applications or in other applications where they could have contact with biological entities. The toxicity of graphene, graphdiyne, transition metal dichalcogenides (TMDs), MXenes, hexagonal boron nitride (h-BN), graphitic phase C^3N^4, phosphorene (BP), exfoliated pnictogens (antimonene, arsenene, bismuthene), Xenes (borophene), and layered metal oxides (LMOs) are analyzed. The different mechanisms involved in the interactions were identified to propose solutions to reduce or eliminate the adverse effects of their cytotoxicity. Future research topics that should be developed to ensure the safe use of two-dimensional nanomaterials in biomedical applications, before their use in commercial applications, are discussed.

INTRODUCTION

Due to its amazing and unique features, two-dimensional nanomaterials have stimulated the interest of academic and industrial researchers since the initial isolation of graphene from graphite in 2004 (Chia, 2018). These layered materials have the potential to be used in a variety of applications, ranging from everyday items to water purification and biomedical applications (Fojtů, 2017). Since their positive application potential is still expanding, the degradation of such materials, their biocompatibility, and their

DOI: 10.4018/978-1-6684-5533-3.ch009

effect on living creatures have begun to attract considerable attention. In short, the study of aspects such as biocompatibility, nanotoxicity, and biodistribution of each of them must be carried out. Graphene and its derivatives (Bazina, 2021; Cao, 2021; Chowdhury, 2014; Fahmi, 2017; Farahani, 2020; Li, 2014; Wen, 2015; Yang, 2013), transition metal dichalcogenides (TMDs) (Latiff, 2017; Teo, 2014), black phosphorus (Latiff, 2015; Wu, 2021), MXenes (Gu, 2021), exfoliated pnictogens (Chia, 2019), and other two-dimensional nanomaterials have all been described thus far. Transition metal dichalcogenides are less harmful than organic equivalents like graphene oxide and halogenated graphene so far (Teo, 2014). The greatest impediment to the commercialization of two-dimensional materials is their little-researched toxicity and the absence of studies defining their behavior under real-world environmental and physiological conditions, highlighting the critical need for more study in this area. As a result, it is critical to recognize the various qualities that may influence the material's safety as well as the potential issues associated with its use to take advantage of the extraordinary properties it possesses. Currently, the most studied two-dimensional material concerning toxicity is graphene and its derivatives, which coincides with having been the first to be introduced more extensively (Bazina, 2021; Cao, 2021; Chowdhury, 2014; Fahmi, 2017; Farahani, 2020; Li, 2014; Wen, 2015; Yang, 2013). However, some other two-dimensional materials are being studied by different research groups (Chia, 2019; Gu, 2021; Latiff, 2015; Latiff, 2017; Teo, 2014; Wu, 2021), which has made it possible to carry out in vitro studies even in human cells (Bazina, 2021; Cao, 2021; Gu, 2021). A parallel effort is underway to better understand the biological and environmental interactions of synthetic nanosheets, both to enable biomedical advances and to ensure human health and safety in all sectors of application. Many two-dimensional materials are projected to undergo reactive dissolution processes, which will be crucial to understanding their behaviors and interpreting biological response data (Wang, 2016). Furthermore, because the elucidation and understanding of the physicochemical properties of this novel class of nanomaterials are still in their early stages, data on biocompatibility and nanotoxicity in vitro and in vivo is few and poorly understood (Kenry, 2016).

Toxicity is a two-edged sword: it can be poisonous to cells and animals, but it can also be advantageous when employed as a medicine, a drug carrier, or even an artificial skin, for example, or inserted into food packaging by increasing the toxic effect and hence the killing effectiveness of target cells (antibacterial effect, cancer cells, etc.) (Tan, 2019). Toxicity must be assessed at a meaningful dosage for use in normal and worst-case circumstances, with and without suitable precautions. However, the risk that these materials pose as manufacturing and scale expand must not be overlooked and must be thoroughly researched before mass commercialization and manufacture. There must also be adequate regulation of the synthesis and testing of these additives to compare them throughout the literature, as well as methodologies for hazard assessment in likely exposure scenarios.

To evaluate whether exposure to this type of nanomaterials harms our well-being, an in vitro toxicity assessment of two-dimensional materials must be done for them to be commercialized in the future. Despite all the advances made so far, research on the toxicity of two-dimensional materials is still in its infancy, since the number of them is expanding every day as an endless range of possibilities, however, some mechanisms of toxicity have already could be established. The purpose of this chapter is to review the mechanisms involved in the toxicity that the different types of two-dimensional materials present to determine the possibility of reducing it and to expand the potential applications with the greatest safety for human beings.

TWO-DIMENSIONAL NANOMATERIALS

Over the last decade, a lot of work has gone into understanding and exploring different forms of two-dimensional (2D) nanomaterials. Optoelectronics, photonics, energy storage and conversion, and biomedicine are just a few of the applications where 2D nanomaterials show promise. These monolayers and/or multilayer 2D nanomaterials are appealing due to their different structure, and when layered in numerous layers, they make use of van der Waals forces. The distinctive features of 2D materials stem from their parent material's crystalline structure, which is made up of vertically stacked layers. This allows them to be thinner mostly on the atomic layer level while somehow modifying their physicochemical features (Fojtů, 2017). The researchers are continuing to assess the various allotropes to determine which are the most stable.

2D nanomaterials have attracted a lot of interest in disease detection and biomedicine because of their unique structure and excellent properties. The main two-dimensional nanomaterials that can be used in biomedical applications are illustrated in Figure 1. The key distinction between nanoparticles and bulk materials is the difference in surface area per unit mass, which has a significant impact on reactive areas and toxicity (Tan, 2019). However, as the manufacture of nanomaterials for a wide range of applications grows, so does the availability of man-made nanomaterials like those produced for specific use. In the following subsections, the basic physicochemical properties of the different two-dimensional materials that have been most applied to various medical applications will be described.

Figure 1. Main two-dimensional nanomaterials used in biomedical applications

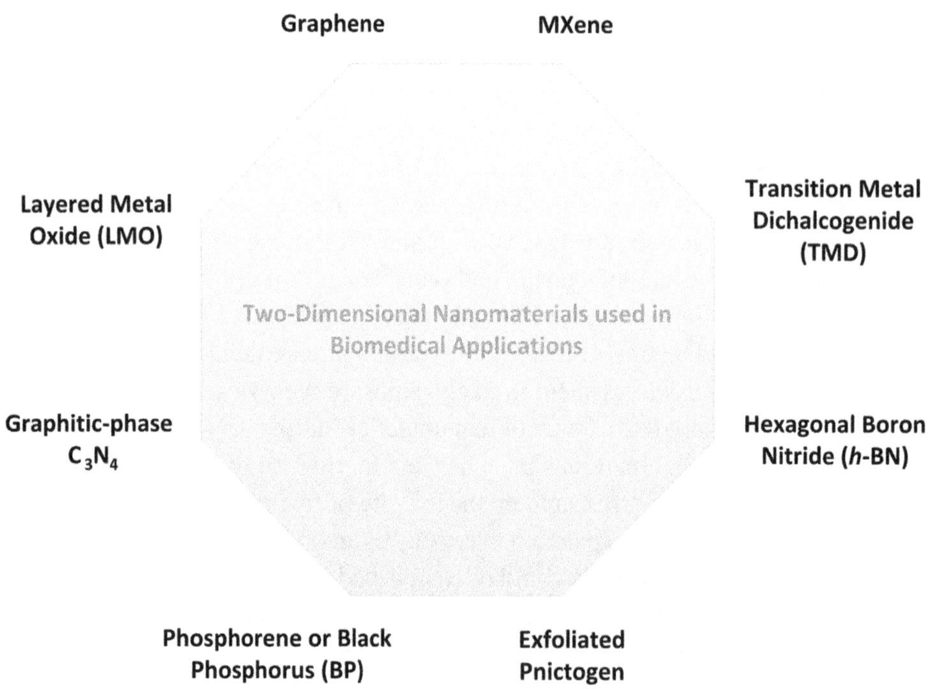

Introduction to Graphene

A monolayer of carbon atoms arranged into a sandwich structure lattice structure made up of sp^2 interconnected carbons in a one-atom-thick sheet, creating a two-dimensional array of carbon atoms in a hexagonal configuration, is commonly termed graphene (Tan, 2019). The functionalization of graphene exploded, as a result, enhancing its present qualities, and expanding its applications. Many graphene derivatives, including the boron and nitrogen-doped graphene, reduced graphene oxide (rGO), graphene oxide (GO), functionalized graphene with metal nanoparticles, polymer, and biomolecules, have been produced because of the functionalization of pure graphene, and have been employed in a variety of key areas. Because of its exceptional electrical properties and biocompatibility, graphene has gained considerable interest for possible uses in biotechnology (Li, 2014). Since 2004, graphene has been a materials science superstar, holding enormous promise in a variety of sectors such as power storage, nanoscale devices, batteries, transparent conductors, healthcare, and plenty more (Yang, 2013). Graphene and its functionalized analogs have lately emerged as fascinating nanomaterials with promising biomedical applications (Wen, 2015). In addition, graphene seems to have low toxicity to humans, which aids in achieving the antibacterial goal by restricting toxicity (Liu, Y. 2021).

A monolayer of sp^2-hybridized carbon atoms organized in a hexagonal lattice makes up graphene (Yu, 2021). High electron mobility, a large specific surface area (2630 m^2g^{-1}), and extremely strong mechanical stiffness (exhibiting Young's modulus of approximately 1 TPa and fracture strength of approximately 125 GPa) are just a few of the intriguing properties that graphene and its derivatives possess. Graphene and its derivatives have undergone substantial research and have been used in a variety of fields since their discovery 18 years ago. Polymers can be combined with graphene to improve their barrier, thermal, and mechanical properties for use in food packaging. The structure between graphene and the polymer matrix has been explained by two different simultaneous intermolecular interactions. One interaction takes place between the polymer fragments themselves, and the other is between the polymer substrates and the graphene. Furthermore, due to its propensity to penetrate cell membranes and absorb significant quantities of phospholipids, graphene, like many other 2D materials, demonstrates a high antibacterial activity. As a result, some polymer-graphene nanocomposites also exhibit strong antibacterial properties.

Introduction to Graphdiyne (GDY)

Graphdiyne (GDY) is a novel member of the family of carbon-based 2D nanomaterials with *sp*- and sp^2-hybridized carbons (Cao, 2021). GDY shares many characteristics with GO but also has some distinguishing characteristics, such as the presence of a quantifiable quantity of *sp* carbon in its molecular structure, which makes GDY acceptable for applications in domains where GO is used. As a result, since the initial synthesis of GDY in 2010, the potential of GDY has piqued the curiosity of numerous researchers.

Introduction to Transition Metal Dichalcogenides (TMDs)

TMDs are 2D materials of the future that rival graphene (Tan, 2019). TMDs nanosheets contain an abundance of elemental compositions, allowing for more exact control of their physicochemical characteristics, in addition to numerous fascinating qualities related to those of graphene. TMDs have the empirical formulae MX_2, where M is a transition metal from Group IV, V, or VI, and X is a chalcogen species such as S, Se, or Te. Distinct combinations of these elements produce around 60 different TMDs, with two-thirds of these compounds exhibiting multilayer architectures (Chia, 2018). It has been claimed that the bonds inside each layer are covalent, however, the interactions between two separate MX_2 layers are normally kept together through van der Waals interactions of contact, permitting TMD exfoliation down to individual layers. In addition, unsaturated *d*-orbitals of transition metals produce band structures, which give TMDs materials their unique characteristics. TMDs have unique and desirable features such as wide surface area, variable band gaps, photo corrosion resistance, and low shear resistance. Electrocatalytic hydrogen evolution, high-performance electrochemical supercapacitors, biosensors, field-effect transistors (FETs), photodetectors, heterojunctions, photovoltaics, and lubricant additives are examples of such uses.

TMDs are semiconductors of the MX_2 class. The letters *M* and *X* stand for the chalcogen element (S, Se, and Te) and the transition metal (Nb, V, Mo, W, and Ta), respectively (Yu, 2021). The TMD molybdenum disulfide (MoS_2) in its monolayer, where the Mo atom layer is in the center and two S atoms are on either side, has received the greatest research attention to far. Strong covalent connections between the Mo and S atoms in each monolayer hold them together. Weak van der Waals forces link adjacent monolayers. MoS_2 is a well-known star in optoelectronics, biomedicine, and other fields due to its remarkable adsorption capacity and other fascinating features. Researchers have begun to investigate using MoS_2 with polymers like poly (vinyl alcohol) and polyacrylonitrile in food packaging. The thermal stability, barrier property, and mechanical performance of these composite films are noticeably enhanced by the hydrogen bonding interactions between MoS_2 and the polymer matrix. Additionally, some TMDs like MoS_2 and tungsten disulfide (WS_2) have been used to create sensors and smart labels for intelligent packaging because of their intriguing optoelectronic and catalytic capabilities.

Introduction to MXenes

MXenes are a class of atomically thin, two-dimensional (2D) transition metal carbides and carbonitrides with a variety of appealing features (Rasool, 2016). Two-dimensional $Ti_3C_2T_x$ (MXene) has recently been investigated for use in water desalination/purification membranes. MXenes have been exploited as novel nanomaterials for a variety of applications since their discovery by Gogotsi et al. in 2011, including energy storage, electrodes, and environmental catalysts (Gu, 2021). MXene-based nanomaterials have lately been used in biomedical applications. MXenes will become quantum dots exhibiting tunable fluorescence when turned from 2D to 0D (lateral dimension 20 nm) and so could be exploited in biosensing and bio-imaging.

The 2D materials category now includes MXenes. In the ternary-layered intermetallic complex Ti_3AlC_2, the metal Al layer was selectively etched away to successfully prepare Ti_3C_2 MXene for the first time (Yu, 2021). $M_{n+1}X_nT_z$ is a chemical formula that describes MXenes. The third or fourth main group of chemical elements is represented by *X*, while *M* stands for the early transition metal elements. *T* denotes a C or N atom. MXenes have been used as fillers to enhance the mechanical property and thermal stability of polymers because of their strong mechanical strength and thermal conductivity. MXene surfaces are typically terminated with O, OH, and/or F groups, resulting in MXenes with high hydrophilicity. Due to their strong hydrophilicity and abundance of hydroxyl groups in repeating molecular units, some polymers used in food packaging, such as polyvinyl alcohol (PVA), are compatible with them.

Introduction to Hexagonal Boron Nitride (*h*-BN)

Hexagonal boron nitride (*h*-BN) monolayers are made up of alternating covalently bound boron and nitrogen atoms (Kenry, 2016). Indeed, the stacked 2D hexagonal boron nitride nanomaterial has received a lot of interest due to its beneficial biological features. Even though it is still in its early stages and has not been fully proven, numerous reports have demonstrated that it has superior biological qualities to graphene, including biocompatibility. As a result, *h*-BN is an attractive nanomaterial for a wide range of biological and medicinal applications.

h-BN, often known as "white graphite," exhibits stacked 2D layer structures resembling graphene. (Yu, 2021). Equal numbers of nitrogen and boron atoms are positioned in each layer in a hexagonal arrangement and joined by strong covalent bonds. Weak van der Waals forces join adjacent layers. Due to its graphene-like structure, *h*-BN shares several characteristics with the material, such as a very high in-plane elastic modulus and a very large surface area. Because the planar networks of boron nitride hexagons are consistently stacked, *h*-BN has been used as a filler to boost the barrier characteristics of polymers. Additionally, the strong hydrogen-bond interactions between *h*-BN nanosheets and polymer chains will allow for the creation of polymers with extraordinary mechanical properties when *h*-BN is combined with polymers like polyvinyl alcohol.

Introduction to Graphitic-Phase g-C_3N_4 Nanosheets

With a weak vdW force between CN layers, graphitic-phase C_3N_4 (g-C_3N_4) exhibits a graphite-like structure (Tan, 2019). Since the first report, g-C_3N_4 has attracted tremendous attention due to its promising practical applications and theoretical value, especially in electrocatalysis, electronic devices, and photoreactivity.

g-C_3N_4 is a typical polymer semiconductor with 2D layer structures like graphene (Yu, 2021). In g-C_3N_4, C and N atoms sp^2 hybridize to form a delocalized conjugated system. Because of its many advantages, including strong visible-light absorption, ease of preparation, non-toxicity, and high physicochemical stability, g-C_3N_4 is widely used in many fields, including photocatalysis and environmental remediation. Because of its adjustable electron band structure and high photocatalytic activity, g-C_3N_4 has been widely used for antibacterial purposes. When exposed to visible light, g-C_3N_4 generates many electron-hole pairs, which produce reactive oxygen species (ROS) and kill bacteria. Several studies have used g-C_3N_4 as a co-catalyst to improve the antimicrobial properties of food packaging materials.

Introduction to Black Phosphorus or Phosphorene (BP)

The only component of black phosphorus (BP) is phosphorus, which is bound together by van der Waals contacts in the form of sheets of six-membered rings (Tan, 2019). As with graphite, every single BP sheet is split into two chemically bonded twofold layers, rather than being flat. Black phosphorus (BP) is a monolayer and/or multilayer bi-dimensional nanomaterial (2D) that has lately emerged as one of the most appealing 2D nanomaterials due to its intriguing physicochemical and optoelectronic features (Song, 2018). In comparison to other 2D materials, black phosphorus (BP) possesses an adjustable bandgap (0.3–2 eV) from bulk to monolayer form (Xiong, 2018). Although their intrinsic toxicity has not been determined, BP in layers could have intriguing applications in biomedical sectors including drug delivery, photodynamic/photothermal treatment, and bioimaging. BP's recent study has revealed interesting features like variable electronic bandgap and high electron mobility, which could be useful in nanoelectronics, battery materials, biosensing, photocatalysis, and medicinal applications (Tan, 2019). As a result, it is critical to understand the toxicological consequences of layered black phosphorus before using it in medicinal applications.

Compounds containing phosphorus have been used extensively in the biomedical industry. Numerous phosphorus-based nanomaterials have been created for biomedical applications as nanotechnology advances (Zhu, 2021). Two-dimensional black phosphorus (2D BP), also known as phosphorene, has recently been found to have great biological properties, including low toxicity and biodegradability, making it particularly outstanding for use in biosensors for the identification of diseases. Reactive oxygen species (ROS) or heat can be produced to treat the disease upon the application of light or ultrasound to the phosphorene. Additionally, phosphorene is a clever and effective medication delivery system. This makes it appropriate for many different singles or combined therapeutic techniques. As a byproduct of BP breakdown, the minor quantity of phosphoric acid coated on BP may also contribute to antimicrobial action. BP can treat the bacterial infection while also repairing bone deficiency. BP is an intriguing substance for potential bone regeneration, to put it simply.

Introduction to Exfoliated Pnictogens

The experimental success of phosphorene, which has superior electrical, physical, and optical properties, has attracted attention in all other Group 15 elements 2D nanomaterials including arsenene, antimonene, and bismuthene (Chia, 2019). Because of the unique and valuable properties of pnictogen nanosheets, intensive research efforts have resulted in the discovery of a wide range of potential applications, including optical Kerr material, optoelectronic devices, pnictogen-decorated fibers, high-speed field-effect transistors, and flexible 2D devices.

Arsenene, bismuthene, and antimonene are projected to exhibit ferroelectric (and anti-ferroelectric in bismuthene) features along Curie temperatures above room temperature (RT), a development that, if understood practically, might change how 2D-Xenes are used (Khan, 2021). Meanwhile, buckle-shaped antimonene complexes were shown to be inert chemically and air-stable. Additionally, different levels of mechanical anisotropy were suggested for antimonene crystal formations. Black, gray, and yellow As are the three arsenene allotropes. The most characteristic and stable form of As is gray, which has a layered rhombohedral structure. The three allotrope kinds of bulk Sb are well-known (for example, black, gray, and explosive Sb). Rhombohedral crystals make up the gray-Sb because gray As is a relatively stable color. Like bulk gray As, gray-Sb is semimetallic. Bulk-Bi exists in one stable form, which is a rhombohedral structure.

Introduction to Xenes

Xenes are graphene-like monoelemental 2D materials with intra-layer covalent bonds and weak van der Waals forces between layers (Han, 2020). The Xenes are two-dimensional materials based on the elements of groups IIIA, IVA, VA, and VIA. Borophene has arisen as a two-dimensional monoelemental nanomaterial with high drug-loading capacity and photothermal characteristics. Borophene has recently sparked widespread interest in the sectors of energy, environment, chemistry, and biomedicine due to its superior morphological characteristics and physicochemical properties. In comparison to graphene and phosphorene, borophene has demonstrated more biomedical potential because of its unique tractability and biocompatibility.

The development of new two-dimensional materials, particularly monoelemental materials (Xenes), for use in a variety of technological domains has demonstrated the potential of nature in addition to fundamental research, which is addressing the new findings (Khan, 2021). Due to their exceptional chemical, physical, optical, and electronic properties, as well as in more than a few other new bio-uses, the second-generation materials known as Xenes, such as Group-IIIA (borophene, gallenene, and aluminene), Group IVA (silicene, germanene, stanene), Group VA (phosphorene, arsenene, antimonene, and bismuthene) and Group VIA (tellurene and selenene) for synthetic exploration, are chemically tractable materials that are considered capable mediators for biomedical applications. The 2D-Xenes are viewed as skilled biological theranostic mediators, particularly in the biomedicines, due to their exceptional optical and electronic nature (for analysis the photo-acoustic (PA) imaging (PAI), photo-thermal imaging (PTI), tomography (CT) imaging is computed by X-ray, photoluminescence lifetime imaging (PLTI), fluorescence imaging (FI), photo-thermal therapy (PTT), and optical-therapeutic strategies photodynamic therapy (PDT). The 2D-Xenes' ultrahigh surface-to-volume ratio, which is made possible by their therapeutic nature and extensive surface interactions, is unavailable in conventional nanosized particle-based drug delivery platforms, which are mainly notable for their extremely high load capability.

Introduction to Layered Metal Oxides (LMOs)

Because of their outstanding electrical and photochemical properties, layered metal oxides (LMOs) are also another prominent family of 2D nanomaterials that are rapidly being explored for biological and medical applications (Kenry, 2016). Even though layered metal oxide nanomaterials have indeed been actively investigated for decades, most of these studies have centered almost entirely on their 0D, 1D, and 3D counterparts due to the difficult nature of the production of 2D metal oxide nanomaterials. Only recently has significant progress been made in the effective delamination of 2D layered metal oxide nanosheets which include TiO_2, MnO_2, ZnO, Fe_3O_4, and WO_3 nanosheets.

MoO_3 and WO_3 are two examples of layered metal oxides that have been extensively investigated for use in biological applications due to their outstanding biocompatibility, low toxicity, and ease of manufacture (Zhou, 2021). Defective metal oxide nanoparticles, including SiO_x, TiO_{2-x}, and MoO_x, have also been investigated as photothermal agents for PTT in the NIR-II window due to their great biocompatibility, low toxicity, and ease of synthesis. The investigations have shown that structural engineering of layered metal oxides is a potent method to adjust their characteristics and hence improve their efficacy in photothermal cancer therapy. Engineering LMOs for additional biomedical applications, such as photodynamic treatment, chemodynamic therapy, and sonodynamic therapy, might be done using an even more straightforward and effective technique.

TOXICITY IN TWO-DIMENSIONAL MATERIALS

Bacteria are essential to the ecosystem. Furthermore, because bacteria are single-cell organisms, they are good models for assessing the dangers of chemicals to the ecological environment (Xiong, 2018). The bacterial toxicity of 2D materials can provide important information regarding their impact once discharged into the environment. A few mechanisms are proposed to explain the bactericidal activity of

2D materials in general. First, physical damage to the bacterial membrane caused by interaction with the sharp edges of 2D materials, known as the nano-knives effect, can cause intracellular substance leakage and, finally, cell death (Pham, 2015). Furthermore, oxidative stress caused by the generation of reactive oxygen species (ROS) (Yang, 2014) or charge transfer activity between the bacterial membrane and 2D materials (Li, 2014) can interfere with bacterial metabolism, disrupt cellular processes, and eventually render the bacterial cells inactive. Additionally, by wrapping or trapping bacterial membranes, the unique 2D lateral structure can form a flexible barrier to segregate bacteria from their surrounding nutrients and environment (Carpio, 2012). Other mechanisms, including lipid bilayer extraction, protein-protein interaction interference, and the "self-killing" effect, also contribute to the antibacterial activity of 2D materials (Luan, 2015; Tu, 2013). Given the unique properties and enormous potential of BP nanosheets, it is vital to examine their environmental concerns by researching bacterial toxicity and the underlying mechanisms.

An essay presented a paradigm based on fundamental materials physics and chemistry for a more systematic exploration of the biological behavior of 2D materials (Wang, 2016). This approach considers three key interaction modes: (i) chemical interactions and phase transitions, (ii) electronic and interfacial electrochemical activities, and (iii) physical and mechanical interactions specific to near-atomically thin, high-aspect-ratio materials.

The cytotoxicity of 2D nanomaterials is heavily determined by their contact with biological systems, which in turn can be regulated by nanomaterial physicochemical features such as form, size, and chemical composition (Fojtů, 2017). As a result, toxicity studies on well-characterized nanomaterials should be conducted systematically so that an adequate correlation between a material's physicochemical variations and its potential impacts on biological systems can also be established.

Surprisingly, biological reactions to 2D nanomaterials vary greatly due to their extraordinary physical and chemical diversity, with some findings demonstrating low toxicity for several nanomaterials (e.g., MoS_2 and *h*-BN) but higher toxicity toward others (e.g., some selenides and tellurides) (Kenry, 2016). While these differences present several potentials for additional investigation, they also present a significant difficulty for thorough assessments of their biological impact. This chapter illustrates for the first time to the best of the author's knowledge a graph that compares the toxicity of different two-dimensional materials. The toxicity levels of the different two-dimensional nanomaterials that can be used in biomedical applications are shown in Figure 2.

Figure 2. Toxicity levels reached by the different two-dimensional nanomaterials

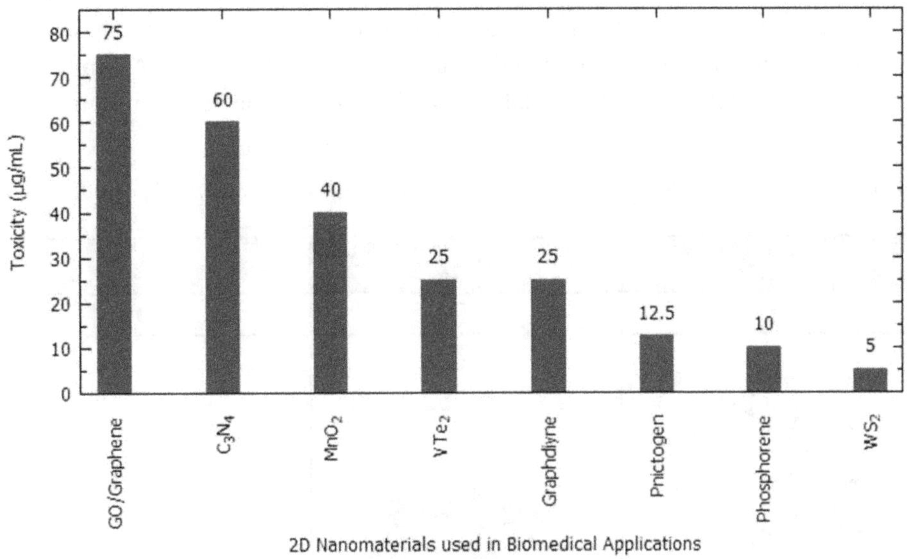

The emergence of multi-drug resistance pathogenic microbial species is a serious public health concern in the twenty-first century (Shaw, 2021). Recent technology is creating numerous options for 2D nanomaterials to be used as possible next-generation antimicrobials. Because of their unique physicochemical properties, 2D nanomaterials have shown antibacterial activity against a broad of pathogenic fungal and bacterial cells. Based on their dimensionality, nanomaterials are classified as zero-dimensional (0D), one-dimensional (1D), or two-dimensional (2D). Among the nanomaterials that have shown antibacterial action across many dimensions are 0D black phosphorous (BP), 1D zinc oxide (ZnO), and 2D molybdenum disulfide (MoS_2). Pertinently, it has been proposed that nanomaterials can control the great potential of microbial resistance while also allowing control over antipathogenic mechanisms by tuning the materials' physicochemical properties, such as size, shape, and composition, which include heterostructures and the inclusion of functional groups. For simplicity, the authors evaluated the currently described antibacterial mechanisms of nanomaterials and classified them into seven major groups: (i) *Basal plane effects*: The basal plane of nanomaterials adds to antimicrobial activity by inhibiting nutrient/waste exchange and/or causing perturbations in the cell membrane via hydrophobic interactions or electrical charge transfer effects due to their large surface area and amenability to chemical functionalization, (ii) *Edge effects*: The edges of nanomaterials have unique physicochemical factors that contribute to the antibacterial action by puncturing microbial cell walls upon contact, resulting in intracellular component release and cell lysis, (iii) *Extraction of phospholipids from the membrane*: Interaction with the edge of nanomaterials might result in "nanoscale de-wetting" due to strong hydrophobic interactions between the basal plane and the phospholipids, eventually culminating in membrane collapse, (iv) *Oxidative stress*: Reactive oxygen species (ROS) in aqueous media can damage biological components such as proteins, nucleic acids, and phospholipids, leading to cell lysis, (v) *Light-driven mechanisms*: First, photocatalytic activity promotes the generation of reactive oxygen species (ROS), which causes cell lysis, and second, photothermal activity generates temperature rise, which inhibits the cellular activity through cellular damage and enzyme denaturation, (vi) *Synergistic effect*: Combining nanomaterials can offer better stability or antibacterial activity, by releasing ions or chemical species or by using zeta potential, and (vii) *Other physical interactions*: When 2D nanomaterials come into contact with microbial cells, they cause bodily harm the membrane through electrical and mechanical interactions by wrapping around, encapsulating, or entrapping the cell. More details can be found in Shaw, 2021. A summary of the main antipathogenic properties of nanomaterials related to their toxicity is depicted in Figure 3.

Figure 3. Antipathogenic properties found in different nanomaterials (adapted from Shaw, 2021)

Anisotropic		Isotropic
2D	1D	0D
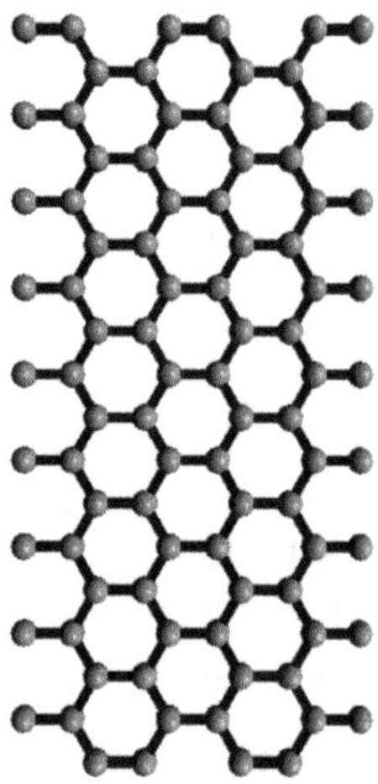	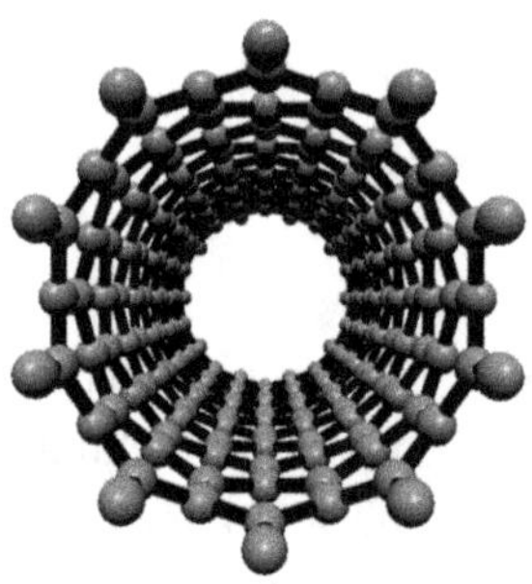	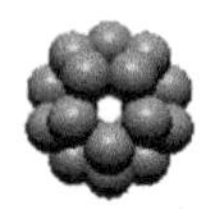
Thickness < 100 nm Sheets, Platelets, Films	Diameter: 10-100 nm Wires, Rods, Tubes	Diameter < 10 nm Spheres, Clusters, Dots
Antipathogenic Mechanisms in General		
• Cell Membrane Damage • Phospholipids Disruption • Charge Transfer • Reactive Oxygen Species (ROS) Production • Oxidative Stress	• Physical Entrapment • Membrane Disruption • Electronic Interactions • Reactive Oxygen Species (ROS) Production • Oxidative Stress	• Membrane Destruction • Free Radical Production • Genetic Material Binding • Metabolic Inhibition • Oxidative Stress

Even though 2D materials have a wide range of applications and a lot of potentials, their toxicological effects may have a significant negative impact on people's health (Naikoo, 2022). The precise size or concentration of 2D materials that play a vital role in exhibiting certain toxicological features is unknown, even though most of their toxicity is recognized.

In the following subsections, the toxicity of the different two-dimensional materials that have been most applied to various medical applications will be described.

Toxicity of Graphene

Understanding how nanoparticles interact with cell membranes is linked to how they produce cytotoxicity and is thus crucial for developing safer biomedical applications (Tu, 2013). Graphene has recently been proven to have antibacterial activity against *Escherichia coli*, however, the underlying molecular pathways are unknown. The scientists demonstrated experimentally and conceptually that pristine graphene and graphene oxide nanosheets can cause the inner and outer cell membranes of *Escherichia coli* to degrade, reducing their viability. Because of the strong dispersion interactions between graphene and lipid molecules, graphene nanosheets may penetrate and remove substantial volumes of phospholipids from cell membranes. This disruptive extraction provides a new technique for understanding the chemical basis of graphene's cytotoxicity and antibacterial activity.

The rising use of graphene for biological and material science purposes needs a detailed assessment of its potential toxicity under both purposeful and unintentional exposure situations (Chowdhury, 2014). When compared to bath sonicated or unsonicated suspensions, the results showed that low concentration (20 μg/mL) suspensions of graphene sheets prepared by as little as 1 minute of sonicating can cause a significant reduction in the overall metabolic function of cells in vitro and elevated embryo/larval mortality in vivo. According to a structural study, probe sonication causes graphene structure breakdown and the formation of tiny carbonaceous debris, and this may be the origin of the observed toxicity. As a result, structural disruption raises the toxicity of graphene nanoribbons.

The stimulation of DNA fragmentation caused by endogenous cellular nucleases is one of the most severe and permanent cytotoxic properties of carbon nanomaterials (Fahmi, 2014). This study found that pure graphene, when exposed to cultivated kidney tubular epithelial cells, can induce DNA fragmentation as determined by the terminal deoxynucleotidyl transferase dUTP nick finished labeling (TUNEL) assay, which would be normally linked with cell death. Low doses of graphene, up to 25 μg/mL, were found to not affect DNA fragmentation when compared to vehicle control samples, and hence were not considered cytotoxic. However, beginning at 50 μg/mL, the increase in TUNEL above the vehicle control showed statistically significant, indicating cytotoxicity.

The antibacterial properties of large-area monolayer graphene nanosheets over conductivity Cu, semiconductor Ge, and insulator SiO_2 were studied using Gram-positive *Staphylococcus aureus* (*S. aureus*) as well as Gram-negative *Escherichia coli* (*E. coli*) (Li, 2014). The findings reveal that graphene nanosheets on Cu and Ge can suppress the growth of both bacteria, particularly the former. The graphene nanosheet atop SiO_2 cannot, however, appreciably limit the multiplication of either bacteria. The morphology of *S. aureus* and *E. coli* on graphene nanosheets reveals that direct interaction of both bacteria with graphene on Cu and Ge can cause membrane damage and ruin membrane integrity, however, graphene on SiO_2 causes no visible membrane destruction. A reasonable method for explaining this occurrence is proposed here from the standpoint of charge transfer. The surface electron states of *E. coli* cells and *S. aureus* cells differ, which may contribute to *E. coli* cells having more electrical resistance against direct contact interaction with graphene nanosheet on Cu, Ge, or SiO_2 substrate than *S. aureus* cells, accounting for the observed difference in antimicrobial activity against Gram-negative *E. coli* and Gram-positive

S. aureus. This research could lead to a better understanding of graphene film's antibacterial properties and the development of graphene-based antibiotics or other biological uses.

The antimicrobial behavior of graphene sheet surfaces allows for the assessment of the existing understanding of graphene cytotoxicity (Pham, 2015). Graphene was produced with varying edge lengths and angles of orientation of the sheets. When these substrates were exposed to *Pseudomonas aeruginosa* and *Staphylococcus aureus* bacteria, it was discovered that they have varying bactericidal effectiveness against these two harmful bacteria. It was shown that the density of graphene edges was one of the primary characteristics that led to the antibacterial action of graphene nanosheets. The work offered experimental and theoretical evidence that the antibacterial action of graphene nanosheets is caused by the creation of pores in the bacterial cell wall, resulting in osmotic imbalance and cell death.

Researchers used large-scale all-atom molecular dynamics simulations to examine the possible toxicity of graphene to a living cell at the molecular level (Luan, 2015). The simulation results reveal that a graphene nanosheet can disrupt the hydrophobic protein-protein interaction (or recognition) that is required for biological processes. Because graphene is hydrophobic, it is energetically advantageous for a graphene nanosheet to enter the hydrophobic interface of two interacting proteins, such as a dimer. The forced separation of two functioning proteins can impair the cell's metabolism and potentially cause it to die.

The genomic omnibus database was used to obtain the patterns of gene expression of graphene-exposed cells (Farahani, 2020). The route, protein-protein interaction (PPI) structure, and module analyses were then used to analyze differentially expressed proteins and their functional roles. Through a systems toxicology assessment, the study revealed the potentially harmful effects of graphene-based products. The functional examination of important genes revealed that these graphene–gene interactions might be very useful for future research.

The toxicity of graphene depends on the particle size, particle state, oxygen content, and surface charge (Liu, Y. 2021). Therefore, its toxicity can be reduced by induced degradation, the use of biocompatible coatings, electrical charge treatment, and surface passivation. Halogenated graphene induces significant toxicity, while fluorinated graphene reduces the toxicity 2-3 times of graphene. The cytotoxicity of reduced graphene oxide is 5 times higher than that of graphene oxide. The toxicity of graphene-based nanocomposites depends on the exposure time and applied dose. For the effective removal of bacteria and viruses, graphene can be prepared in colloidal solutions, coatings, plural gels, and composite fibers.

Toxicity of Graphdiyne (GDY)

The toxicity of graphene oxide (GO) and graphdiyne (GDY) on human umbilical vein endothelial cells (HUVECs) has been investigated (Cao, 2021). Cytotoxicity was caused by up to 100 μg/mL GDY and GO exposure, however, there was no statistical difference between the two materials. At noncytotoxic concentrations of 25 μg/mL GDY or GO, nanomaterials were internalized, typically in the cytoplasm but not in the nucleus. The results indicated that GDY and GO may be absorbed by HUVECs and cause cytotoxicity.

Toxicity of Transition Metal Dichalcogenide (TMDs)

Yang et al. (2014) discovered that two-dimensional (2D) chemically exfoliated MoS_2 nanosheets have antibacterial activity. This activity of chemical exfoliated MoS_2 (MoS_2) nanosheets was significantly higher than that of raw MoS_2 powders used in the manufacture of MoS_2 nanosheets, perhaps due to the 2D MoS_2's planar structure (high specific surface area) and better electrical conductivity. The MoS_2 nanosheets could create reactive oxygen species (ROS). The oxidation capacity of the MoS_2 nanosheets toward glutathione oxidation displayed a time and concentration-dependent trend, which is consistent with the MoS_2 nanosheets' antibacterial behavior. The findings imply that antimicrobial actions are caused by both membrane and oxidative stress. MoS_2–bacteria interaction produced membrane stress, superoxide anion (O_2^-) driven ROS generation by MoS_2, and subsequent superoxide anion-independent oxidation are among the antibacterial mechanisms. This research suggests that altering the dimensions of nanomaterials as well as their electrical characteristics could influence antibacterial activity.

A proof-of-concept study was proposed to assess the potential of functionalized two-dimensional chemically exfoliated MoS_2 (MoS_2) toward inhibitory and bactericidal activity against two representative ESKAPE pathogenic strains—a Gram-positive *Staphylococcus aureus* (MRSA) and a Gram-negative *Pseudomonas aeruginosa* (Pandit, 2016). More importantly, the mechanistic investigation demonstrates a distinct degree of oxidative stress as well as fast membrane depolarization in association with MoS_2 ligands of varying charge and hydrophobicity. The implications of their findings are examined considering planktonic bacteria's lack of survival and biofilm breakdown in vitro. When compared to commonly used small compounds and other nanomaterial-based therapies, 2D MoS_2 outperforms them as a novel class of antibiotics.

Transition metal Group V vanadium dichalcogenides have demonstrated interesting features in a variety of applications, including batteries, capacitors, and electrocatalysts for hydrogen production, among others (Latiff, 2017). The cytotoxicity of exfoliated VS_2, VSe_2, and VTe_2 was examined by combining different concentrations of the materials with human lung cancer (A549) cells for 24 hours and determining the remaining cell viability after the treatment. They discovered that vanadium dichalcogenides are more hazardous than Group VI transition metal dichalcogenides (TMDs), such as MoS_2, WS_2, and WSe_2. VS_2 had the highest cell viability readings among the three vanadium dichalcogenides at all doses tested. VSe_2 and VTe_2, on the other hand, have profiles with relatively equal degrees of cell viability loss. This implies that VSe_2 and VTe_2 have similar toxicity. To demonstrate, after a 50 µg/mL treatment with the exfoliated vanadium dichalcogenides, cell viability for VS_2 drops to around 43 percent, whereas cell viability for VSe_2 and VTe_2 drops to around 9 percent and 8 percent, respectively.

Using the WST-8 assay, researchers investigated the cytotoxicity of Group 5 ditellurides TMDs on A549 cells (Chia, 2018). Aside from investigating the cytotoxic effects of Group 5 transition metal tellurides, the toxicological data acquired might be used to conduct comparative research with other TMDs to gain a better knowledge of their relative toxicities. They discovered also that Group 5 transition metal tellurides are usually more toxic than the Group 6 TMDs (particularly MoS_2, WS_2, and WSe_2) under similar settings (such as experiencing a 24 h exposure of materials with A549 cells). The exception was WSe_2, which was shown to be more toxic than the Group 5 transition metal tellurides, with only VTe_2 being much more toxic than WSe_2. VTe_2 was found to be cytotoxic to A549 cells, with more than half of the cells dying even at low concentrations of 25 µg/mL and less than ten percent of the cells dying at the highest dosage of 200 µg/mL. This finding is consistent with prior toxicity tests on exfoliated VTe_2, which also revealed that VTe_2 was extremely toxic.

The current study examines the biotransformations and *in vitro* cytotoxicity of seven 2D nanomaterials in the gastrointestinal tract: graphene, graphene oxide (GO), partially reduced graphene oxide (*pr*GO), reduced graphene oxide (*r*GO), hexagonal boron nitride (*h*-BN), molybdenum disulfide (MoS_2), and tungsten disulfide (WS_2) (Bazina, 2021). A triculture model of the small intestine epithelium was utilized to investigate the effects of the digested products (digestas) on epithelial layer integrity, oxidative stress, viability, cytotoxicity, and the start of apoptosis. Based on the toxicity endpoints assessed, the majority of ingested 2D nanomaterials exhibited no acute cytotoxicity at 1 μg/mL and 5 μg/mL, with only WS_2 inducing substantial cell death at the highest dose.

According to a study, the administration of 3.52 mg/L of MoS_2 particles alone resulted in 10% of cell death, indicating the particles' high level of toxicity (Naikoo, 2022). The chalcogenides break down and release free Te or Se elements, which have extremely harmful effects on both the human body and the environment. In addition to the material's chemical composition, the exfoliation procedure can significantly alter the behavior of TMD materials *in vitro* and *in vivo* as well as their potential for toxicity. The surface modification of two-dimensional nanosheets can dramatically lessen their toxicity, according to *in vitro* data, allowing for their safe usage in biomedical applications.

Toxicity of MXenes

The resistance to biofouling is a key success indicator for every water treatment membrane. The scientists tested the antibacterial capabilities of single- and few-layer $Ti_3C_2T_x$ MXene flakes in colloidal solution to validate this and better understand the health and environmental consequences of the new 2D carbides (Rasool, 2016). $Ti_3C_2T_x$ was investigated for antibacterial activity against *Escherichia coli* (*E. coli*) and *Bacillus subtilis* (*B. subtilis*) utilizing bacterial growth curves based on optical densities (OD) and colony growth on agar nutritive plates. $Ti_3C_2T_x$ outperforms graphene oxide (GO), which has been frequently described as an antibacterial agent, in terms of antibacterial efficacy against both Gram-negative *E. coli* and Gram-positive *B. subtilis*. Antibacterial activity was concentration-dependent, with more than 98 percent bacterial cell viability loss reported at 200 μg/mL $Ti_3C_2T_x$ for both bacterial cells within 4 h of treatment, as confirmed by the colony-forming unit (CFU) and regrowth curve. $Ti_3C_2T_x$ induced reactive oxygen species (ROS) dependent and independent stress in two distinct abiotic experiments. MXenes are believed to be biofouling resistant and to have antibacterial effects.

A study investigated the likely mechanism of quantum dots (QDs) cytotoxicity to human umbilical vein endothelial cells (HUVECs) (Gu, 2021). For 24 hours, exposure to up to 100 μg/mL Ti_3C_2 QDs, but not Nb_2C QDs, significantly triggered cytotoxicity. At the same mass densities, Nb_2C QDs proved to become more biocompatible to HUVECs than Ti_3C_2 QDs, implying that the proportion of MXene QDs influences their cytotoxic activity endothelial cells.

MXenes' cytotoxicity and genotoxicity effects at the cellular level via the signaling pathway have not been thoroughly studied (Lim, 2021). MXene has the potential to reach organs via normal physiological barriers such as the blood-air barrier, blood-testis barrier, blood-brain barrier, blood-placental barrier, and skin barrier. It is necessary to determine the time-dependent degradability of the biopolymers encapsulating MXenes. Mxene's biocompatibility is currently being investigated *in vitro* and *in vivo* in lower-order animal models such as mice and zebra fish. This is a major concern because MXene nanoparticles have the potential to bioaccumulate in the organs.

MXenes exhibit a variety of chemistries due to their varied chemical elements, therefore while some of the chemistries containing these elements might be safe, others might not be (Naikoo, 2022). However, only a few numbers of groups have been investigated in terms of their toxic potential and consequences on the environment, including $Ti_3C_2T_x$, $Nb_4C_3T_x$, $Mo_2Ti_2C_3T_x$, and a few others. No research has so far been able to confirm that MXenes have any negative effects or toxic potential on human or animal cells. MXenes have the greatest application potential among 2D materials and the lowest hazardous potential.

Toxicity of Graphitic-phase C_3N_4 Nanosheets

g-C_3N_4 has recently sparked interest as a potential option for medical and biomedical imaging applications due to its high photoluminescence intensity (Tan, 2019). An MTT assay was used to assess the biocompatibility of the atomically thin *g*-C_3N_4 nanosheets after they were incubated with *HeLa* cells for 48 hours. According to the findings, even at concentrations of incubated *g*-C_3N_4 nanosheets as large as 600 μg/mL, no obvious decrease in cell viability was seen, demonstrating the ultrathin *g*-C_3N_4 nanosheets' exceptional biocompatibility and minimal nanotoxicity.

Because of the high charge transfer and rapid separation of photogenerated electron-hole pairs in the nanosheets structure, the Bi_2S_3-*x*@*g*-C_3N_4 nanosheets/light system has improved photocatalytic properties compared to *g*-C_3N_4 alone (Abdel-Moniem, 2021). In the dark, the antimicrobial activity of the prepared materials was investigated. After 240 minutes in the dark, Bi_2S_3-10@*g*-C_3N_4 nanosheets demonstrated excellent deactivation of *E. coli, S. aureus*, and *Candida albicans* at a dose of 5 ppm. The antibacterial efficacy of bare *g*-C_3N_4 against *E. coli, S. aureus*, and *Candida albicans* is 78%, 13%, and 12%, respectively. The underlying mechanism is that reactive oxygen species (ROS) can weaken the bacterial membrane. The Bi_2S_3@*g*-C_3N_4 nanosheets show promise for environmental decontamination and disinfection, including the sterilization of water and public facilities.

Toxicity of Black Phosphorus or Phosphorene (BP)

Several contradictory findings of black phosphorus toxicity have been recorded, implying that a thorough understanding of the cytotoxicity and underlying processes of black phosphorus is critical (Latiff, 2015; Mu, 2017; Zhang, 2017; Song, 2018; Xiong, 2018; Eke, 2020). Like other 2D nanomaterials, the toxicity of layered black phosphorus varies greatly depending on its quantity, size, shape, surface chemistry, and exposure period.

The toxicity of black phosphorus was unknown, and it is unclear whether it is as harmful as white phosphorous (Latiff, 2015). The cytotoxicity of black phosphorus to epithelial cells of human lung cancer (A549) was researched in response to the prospect of using it in commercial products and biomedical devices. Cell survival was assessed using water-soluble salt tetrazolium (WST-8) and methyl-thiazolyl diphenyltetrazolium bromide (MTT) assays after cells were exposed to varied black phosphorus concentrations for 24 hours. The toxicological effects of black phosphorus were shown to be dosage dependent, with 50 μg/mL exposure lowering cell viability by 48 percent (WST-8) and 34 percent (MTT). This toxicity was found to be somewhere between that of graphene oxides and that of exfoliated transition-metal dichalcogenides (MoS_2, WS_2, WSe_2). Because of its low toxicity, black phosphorus can be used.

Black phosphorus (BP) has sparked a lot of interest in cancer therapy as a potential successor to two-dimensional layered materials. In an *in vivo* mouse model, it was discovered that while black phosphorus can cause cell death and necrosis due to transient intracellular reactive oxygen species (ROS)-mediated oxidative stress, the induced oxidative stress can be gradually restored to normal levels without causing long-term inflammation or obvious damage (Mu, 2017). Black phosphorous quantum dots at high concentrations (200 μg/mL) had strong apoptotic effects on *HeLa* cells, according to in vitro studies. Black phosphorus quantum dots (BPQDs) can produce acute toxicity because of oxidative stress responses, although inflammatory reactions can take up to 30 days to recover. As a result, they are unable to produce detectable long-term toxicological reactions.

A label-free, real-time cell analysis technique was used to examine the cytotoxicity of layered black phosphorus as it was cleaved, revealing a cell type, concentration, and size-dependent response (Zhang, 2017). The multilayer structure has the potential to disrupt and destroy cell membranes, causing intracellular contents to leak out and cell death. Aggregates were seen adhering to the cell surface after a 4 h treatment with 10 μg/mL layered black phosphorus. Black phosphorus induces cytotoxicity in two ways, both of which have been proven: (1) A sensitive ROS probe is used to detect intracellular reactive oxygen species (ROS) using inverted fluorescence microscopy and flow cytometry; (2) A dissipative quartz crystal microbalance is used to examine the interaction of layered black phosphorus and the model cell membrane, demonstrating the disruption of cell membrane integrity.

Song et al. (2018) used cytotoxicity experiments with multiple endpoints, such as cell metabolic activity, membrane integrity, and intracellular ROS production, to determine the cytotoxicity of layered black phosphorus on fibroblast cells based on its concentration and exposure period. The cytotoxicity of black phosphorus is proportional to its concentration and exposure period, which is influenced by oxidative effort-mediated enzyme activity reduction and membrane rupture (Song, 2018). Layered phosphorene, on the other hand, did not show any cytotoxicity at doses lower than 4 μg/mL. As a result, it's thought that layered phosphorene could be useful as medicinal delivery and imaging vehicles.

Even though BP's biological effects have been investigated, its environmental implications have received less attention (Xiong, 2018). For the first time, the bacterial toxicity of exfoliated BP nanosheets was assessed against two model bacteria strains, Gram-negative *Escherichia coli* (*E. coli*) and Gram-positive *Bacillus subtilis* (*B. subtilis*). The bacterial toxicity of BP nanosheets was investigated by monitoring the bacterial growth curve and colony counting. After 6 hours of treatment, Gram-negative *E. coli* generated greater toxicity than Gram-positive *B. subtilis*, which was reversed at 12 hours due to *E. coli* membrane self-healing. Bacterial toxicity was time- and concentration-dependent, with a maximum bactericidal efficiency of 91.65 percent for *E. coli* and 99.69 percent for *B. subtilis* after 12 hours of exposure.

Because phosphorene is a novel material, toxicity studies with a model nematode, *Caenorhabditis elegans*, were conducted to gain insight into its biocompatibility and safety (Eke, 2020). The results showed that phosphorene-modified membranes had higher protein rejection but lower flux values. Furthermore, data showed that phosphorene loss was negligible within the membrane matrix regardless of pH. Phosphorene toxicity was observed in nematodes in their free form, but not in membrane permeates. The main issue with phosphorene is its rapid degradation under ambient conditions because of reactive oxygen species formation. However, research has shown that incorporating phosphorene into polymers preserves its structure and properties.

Toxicity of Exfoliated Pnictogens

The cytotoxicity of pnictogen nanosheets (As, Sb, and Bi) was investigated after 24 hours of incubation with various test materials and A549 lung cancer cells (Chia, 2019). Following the treatment time, the remaining cell viability was determined by absorbance measurements with the WST-8 and MTT tests. These findings show that the toxicity of pnictogen nanosheets reduces along with Group 15, with arsenic nanosheets being the most toxic and bismuth nanosheets causing low cytotoxicity. Shear exfoliated arsenene displayed a dosages trend and is the most cytotoxic of the three pnictogens employed, whereas antimony and bismuth nanosheets generate significantly lesser cytotoxicity. When subjected to arsenic nanosheets, cells exhibited nearly 50% cell death at a low dosage of 12.5 μg/mL, and only 12.8 percent of cells remained viable at the greatest test concentration of 200 μg/mL. Cell viability percentages, on the other hand, remained relatively high after 24 hours of treatment with antimony and bismuth nanosheets, at 57.2 and 71.5 percent, respectively.

As^{III} reacts with closely spaced protein thiols, which is primarily responsible for arsenic toxicity but also serves as a therapeutic mechanism (Liu, C. 2021). One of the most encouraging stories in the library of drug rediscovery and clinical oncology is the "re-branding" of arsenical compounds—from historical accounts of poisoning agents to clinical therapeutic agents. However, recent research indicates that layered pnictogens have a diverse range of toxicological profiles. Bismuth, despite being a heavy metal, is relatively harmless to humans while exhibiting high selectivity and toxicity toward pathogens. The main reason for its low toxicity is thought to be its low uptake by the human gastrointestinal tract. Bismuth is primarily metabolized in the human body into bismuth-sulfur particles, which are only found in certain organs. Layered pnictogen-based photo-radiotherapy has many advantages, including minimal invasiveness and high therapeutic efficiency, implying that semimetal nanomaterials have the potential to be effective nanomedicines. Despite the biomedical therapeutic efficacy studies on layered pnictogens that have already been conducted, it must be continued the work to understand their actual biological safety, including selective cytotoxicity, biodistribution, biodegradability, and elimination pathways in physiological systems.

Toxicity of Xenes

According to research on borophene, graphene, and phosphorene, the 2D monoelemental nanosheets share comparable properties by exhibiting similar surface changes in the circulatory system (Han, 2020). The protein contents adsorbed by these nanosheets were linked to plasma protein abundance. Most types of plasma proteins were adsorbed by phosphorene, most of the three nanosheets (involving immune proteins). Furthermore, these 2D monoelemental nanosheets with identical properties may exhibit the same cellular ingestion pattern. Graphene and borophene exhibited immunoregulation solely in the presence of plasma corona, but phosphorene exhibited higher immunoregulation in both the absence and presence of corona.

Toxicity of Layered Metal Oxides (LMOs)

2D LMO nanosheets have also shown concentration-dependent nanotoxicity of stacked 2D nanomaterials (Kenry, 2016). Ultrathin MnO_2 nanosheets, for example, have been claimed to exhibit low cytotoxicity and great biocompatibility. Throughout this study, 2D MnO_2 nanosheets with an average lateral size of about 200 nm were created by reducing $KMnO_4$. The MTT assay was used to assess the in vitro cytotoxic effects of these nanosheets on MCF-7 breast cancer cells. MCF-7 cells maintained a high survival with no loss after 24 hours of exposure to MnO_2 nanosheets at varied concentrations below 40 µg/mL. Nonetheless, cellular viability began to decline to around 80%, with a concomitant increase in MnO_2 nanosheet concentration to 50 µg/mL.

Numerous 2D oxides are poisonous and extremely reactive, particularly when present as nanomaterials (Naikoo, 2022). This is because when the size of the 2D metal oxide reduces, they exhibit improved catalytic function and high dissolving rates. For instance, V_2O_5, a 2D metal oxide with significant hazardous potential and a well-known extremely poisonous substance, is a good example. Like this, the 2D oxide MoO_3 exhibits extremely hazardous characteristics. Although other 2D materials, such as MoS_2, are developed using this 2D material, it is also known to be harmful to both human and animal life. MoO_3 does not cause any fatalities while it is present in gaseous form, and this toxicity is only seen when it is present in solid form. Studies have also shown that oxides, such as copper oxide nanoparticles, exhibit hazardous characteristics in a variety of cell lines. The diameter, shape, functionalization, and concentration of CuO NPs are a few of the variables that affect their toxicity.

SOLUTIONS AND RECOMMENDATIONS TO REDUCE THE TOXICITY

The toxicological research available aids in the understanding of the safety and possible toxicity of 2D nanomaterials including various transition metals and surface changes. It is necessary to pay close attention to the methods of action of two-dimensional nanomaterials as well as their potential biocompatibility (Szuplewska, 2022). Different etching processes, surface functionalization techniques, and intercalation with a variety of organic and inorganic chemicals can all be used to modulate the bioactive characteristics. Only the finished 2D nanomaterial layout may be obtained by manipulating the synthesis conditions. It does, however, have a significant impact on their biological features, which are critical for their toxicity and medical uses. Until now, the main strategies to reduce the toxicity of two-dimensional materials consist of functionalizing them or including them in biocompatible polymeric matrices to form composite materials, as illustrated in Figure 4.

Figure 4. Strategies to reduce the cytotoxicity of two-dimensional nanomaterials

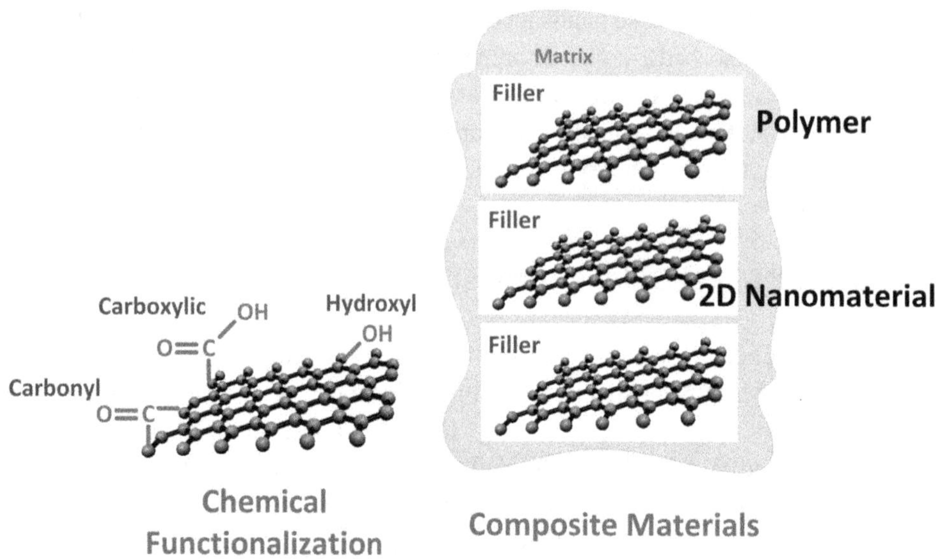

In the following subsections, the strategies to reduce or eliminate the toxicity of the different two-dimensional materials that have been most applied to various medical applications will be described.

Graphene Toxicity Reduction Strategies

It is vital to produce highly effective antimicrobial drugs that are not hazardous to humans and have no negative environmental consequences (Carpio, 2012). Even though antimicrobial investigations of graphene-based nanomaterials are currently restricted, several researchers have focused on such nanocomposites as prospective candidates for the next generation of antimicrobial treatments. The polyvinyl-*N*-carbazole (PVK)–graphene oxide (GO) nanocomposite (PVK–GO), which includes only 3 wt% GO well-dispersed in a 97 wt% PVK matrix, exhibits outstanding antibacterial capabilities while exhibiting no appreciable cytotoxicity to mammalian cells. The high polymer content of this nanocomposite enables future large-scale material synthesis in a high-yield adiabatic bulk polymerization technique. The toxicity of PVK–GO was investigated using planktonic microbial cells, biofilms, and NIH 3T3 fibroblast cells in this work. Two Gram-negative bacteria, *Escherichia coli* and *Cupriavidus metallidurans*, and two Gram-positive bacteria, *Bacillus subtilis,* and *Rhodococcus opacus*, were tested for antibacterial activity. The results reveal that the PVK–GO nanocomposite outperforms the pure GO in terms of antibacterial activity. The efficiency of the PVK–GO nanocomposite in solution was demonstrated when the nanocomposite "encapsulated" the bacterial cells, resulting in decreased microbial metabolic activity and cell death. The fact that the PVK–GO did not cause considerable cytotoxicity in fibroblast cells opens a wide range of possible applications in biomedical and environmental sectors.

Graphene-based platforms could be used to detect a variety of biological molecules and cancer indicators. Graphene oxide (GO) and other graphene-related materials have also been employed as innovative carriers for gene transfer, cell imaging, and cancer chemo- and photothermal treatments. Graphene has also been used as a medium to determine the rate, distinction, and activities of specific cell types, such as stem cells, as a single-atom thin sheet (Tan, 2019). The toxicity of pristine graphene and GO to mice was shown to be dose-dependent, limiting their biological applicability. *In vitro* and *in vivo* toxicity of graphene was lowered after functionalization (e.g., by biocompatible polymer coating).

The cytotoxicity of graphene oxide (GO) and graphene in blood was discovered to be dependent on the functional 2D nanomaterial composition as well as the exposure conditions (Ou, 2016). To illustrate the benefits of strategies to modify the toxicity of two-dimensional materials, the values reported in the graph shown in Figure 5 have been compiled from different references included in this chapter.

Figure 5. Toxicity levels reached by graphene-based materials, whether natural, functionalized, or used in composite materials

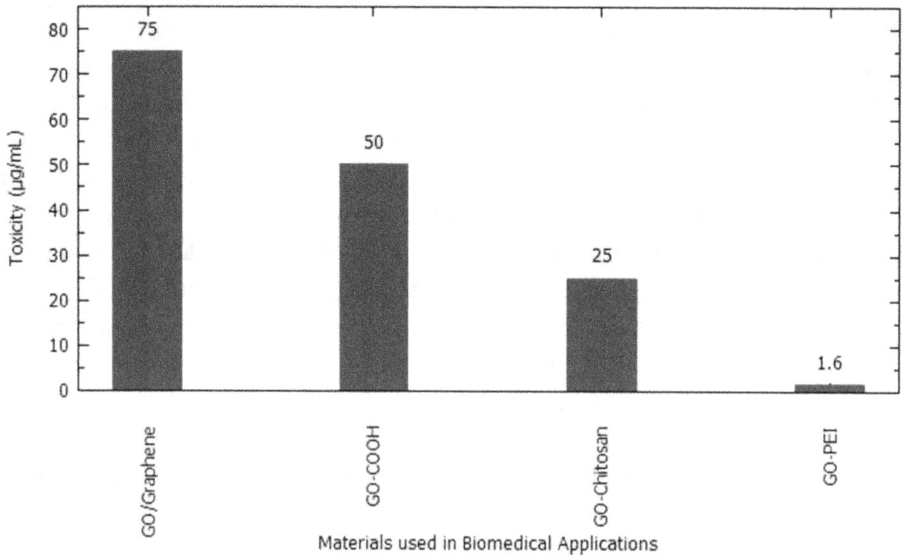

The cytotoxicity of the modified graphene showed likewise significantly lower than that of the unmodified particles (Fojtů, 2017). For example, graphene oxides can exist in a variety of layers and sizes, with diverse functional groups like carbonyl, carboxyl, and hydroxyl, resulting in considerable variances in their reported toxicity. This emphasizes the significance of surface properties and particle coating, which have a significant impact on environmental safety.

MXenes Toxicity Reduction Strategies

The essentials for MXenes' surface changes in modern biotechnology and nanomedicine were discussed in a paper (Szuplewska, 2022). They focus on the overarching issues of introducing MXenes within nanostructured organic/inorganic bioactive structures, putting the material-structure-chemistry-property link into perspective. Controlling MXene production and surface modifications for lower toxicity and improved colloidal stability with longer in vivo circulation duration could lead to commercialization in biomedical applications. Surface chemistry will influence the metabolic processes and toxicity of a substance from a fundamental standpoint. It can also offer novel ways to fine-tune the ensuing biological activity to get the desired results. Different synthetic processes and/or modification strategies can be used to design surface chemistry.

Researchers around the world are being inspired by the development of novel methods for the preparation of multifunctional MXene-based materials and to promote the practical biomedical applications of MXenes using surface modification to reduce their toxicity (Huang, 2020). In most biomedical applications, for example, hydrophilic polymers are commonly used to modify MXene-based materials to improve water dispersity via non-covalent and physical adsorption interactions. However, in some extreme circumstances, non-covalent and physisorption interactions are not stable enough.

Black Phosphorus or Phosphorene Toxicity Reduction Strategies

To learn more about phosphorene's biocompatibility and safety, researchers conducted toxicity tests on a model nematode, *Caenorhabditis elegans* (Eke, 2020). The overall goal was to create phosphorene-containing polymeric membranes that were both stable and non-toxic. The free form of phosphorene was hazardous to nematodes, while the membrane constructed of polymer-phosphorene composite material was not toxic. Fouling is reduced and the photocatalytic and electrical characteristics of polymeric membranes based on a blend of polysulfone and sulfonated polyether ether ketone are improved when phosphorene is added. Because BP has a high level of biocompatibility and minimal cytotoxicity, it can be used in biomedical applications.

When exogenous nanoparticles interface with living biological specimens, nano-bio interactions occur very quickly, causing the nanomaterials to react at the molecular and cellular levels. Because of its potential applications in imaging and tailored medication administration, phosphorene has piqued the interest of the bioengineering community. Many research groups have reported on the use of functionalized phosphorene in biomedical applications. Nevertheless, untreated black phosphorus is unstable, and if used biologically, it may produce negative nano-bio interactions and have long-term impacts on the biological system. The toxicity of phosphorene *in vitro* and *in vivo* can be minimized by selectively functionalizing its surface with biomaterials such as hydrogels, polymers, and fibrin (Wu, 2021).

FUTURE RESEARCH DIRECTIONS

The creation of novel materials must be balanced with risk evaluations of newer materials. Because the properties of a bulk material change when it is nanosized, it should be registered separately rather than deposited under its bulk attachment counterpart. Increasing public information may also aid in bringing the public on board with scientists and companies who recognize the positive advantages of these nanomaterials rather than the unknown risks propagated by the nonscientific sector. This, contrary to the context of genetic modification in Europe, can lead to improved public acceptance and suitable laws and can speed up the commercialization of items using these novel nanomaterials. Certainly, further short- and long-term research is needed to completely understand their pathogenicity. Toxicity can develop when the action of cell cycle progression is disrupted. The findings highlighted the importance of 2D nanomaterials-based materials for the development of research into cell cycle regulator targeting. More research into the toxicity of various forms of 2D materials in vitro and in vivo is needed (Lim, 2021).

The interaction between biological media and two-dimensional materials requires that several high-priority research topics continue to be developed (Wang, 2016). The list of proposed topics that involve this interaction includes studies based on the simulation, design, and experimentation of (i) release, dissolution, and exposure of these materials in biological media, (ii) phase transition in complex media, (iii) physical-chemical kinetics of the biological-material system, (iv) biodegradation and possible corrosion of materials, (v) physical-mechanical interaction between biological entities and materials, (vi) chemical hazards and safety possibly caused, (vii) development of methods analytical and biochemical screening tools, and (viii) diseases, abnormalities or allergies involving phagocytosis, lung clearance, translocation, inflammatory activity or otherwise.

More attention and effort should be directed in the future toward the development of novel and effective methods for surface modification of 2D nanomaterials to meet the requirements for biomedical applications, as well as an in-depth investigation of the medical application and toxicity of 2D nanomaterials and their composites. The cell absorption activity of 2D nanomaterials-based materials, as well as the impact of surface functionalization and 2D nanomaterials-based material features on biocompatibility, are still completely unexplored. There is a dearth of data on the distribution, accumulation, and clearance behavior, *in vivo* toxicity, and long-term toxicity. As a result, further surface modification strategies based on covalent approaches and the production of dynamic bonds for conjugating polymers but also controlled release drug delivery systems in biomedical applications are likely to be developed.

CONCLUSION

Understanding the interactions of 2D nanomaterials with biological systems, as well as the toxic effects of 2D nanomaterials *in vitro* and *in vivo*, is critical for the further development of 2D nanomaterials-based biomedicine, as is providing safety guidance to all scientists working with these novel nanomaterials. The toxicity of 2D nanomaterials has been examined in this chapter by explaining the behavior of these and their derivative in microbes, cells, and animals. Despite some discrepancies in several in-depth experimental results and hypotheses of toxicity mechanisms, multiple reports conclude that physicochemical properties of graphene, such as functional groups on the surface, charges, coatings, sizes, and structural defects, may affect theirs *in vitro/in vivo* activity as well as its toxicity to living organisms. 2D nanomaterials, with their ultra-small sizes, biocompatible protective coatings, superior dispersibility, and durability in physiological conditions, appear to be substantially less damaging to cells and animals *in vitro* and *in vivo*. The toxicity of two-dimensional materials is greatly reduced with the use of functionalization or the incorporation of these in nanocomposite materials with biocompatible polymers. However, how to eliminate toxicity and hasten excretion or even destruction of functionalized graphene in living organisms is a challenging task for scientific investigations. More research into the toxicity of various forms of 2D materials *in vitro* and *in vivo* is needed. Moreover, more systematic studies are still needed to properly understand the biological implications and solve safety concerns well before 2D nanomaterials-based materials may be used in medicine.

ACKNOWLEDGMENT

The author is grateful for the support of the Instituto Tecnológico Superior de Irapuato (ITESI) for the development of this scientific research. The author wants to thank his wife and son for their support during the time to conduct this scientific research.

REFERENCES

Abdel-Moniem, S. M., El-Liethy, M. A., Ibrahim, H. S., & Ali, M. E. M. (2021). Innovative Green/Non-Toxic Bi_2S_3@g-C_3N_4 Nanosheet for Dark Antimicrobial Activity and Photocatalytic Depollution: Turnover Assessment. *Ecotoxicology and Environmental Safety*, *226*, 112808. doi:10.1016/j.ecoenv.2021.112808 PMID:34600290

Bazina, L., Bitounis, D., Cao, X., DeLoid, G. M., Parviz, D., Strano, M. S., Greg Lin, H.-Y., Bell, D. C., Thrall, B. D., & Demkritou, P. (2021). Biotransformations and Cytotoxicity of Graphene and Inorganic Two-Dimensional Nanomaterials using Simulated Digestions coupled with a Triculture in vitro Model of the Human Gastrointestinal Epithelium. *Environmental Science. Nano*, *8*(11), 3233–3249. doi:10.1039/D1EN00594D

Cao, Y., Xiao, W., Li, S., & Qiu, D. (2021). A Comparative Study of Graphdiyne and Graphene Oxide to Human Umbilical Vein Endothelial Cells. *Journal of Applied Toxicology*, *41*(12), 2021–2030. doi:10.1002/jat.4182 PMID:33973267

Carpio, I. E. M., Santos, C. M., Wei, X., & Rodrigues, D. F. (2012). Toxicity of a Polymer-Graphene Oxide Composite against Bacterial Planktonic Cells, Biofilms, and Mammalian Cells. *Nanoscale*, *4*(15), 4746–4756. doi:10.1039/c2nr30774j PMID:22751735

Chia, H. L., Latiff, N. M., Gusmão, R., Sofer, Z., & Pumera, M. (2019). Cytotoxicity of Shear Exfoliated Pnictogen (As, Sb, Bi) Nanosheets. *Chemistry (Weinheim an der Bergstrasse, Germany)*, *25*(9), 2242–2249. doi:10.1002/chem.201804336 PMID:30637810

Chia, H. L., Latiff, N. M., Sofer, Z., & Pumera, M. (2018). Cytotoxicity of Group 5 Transition Metal Ditellurides (MTe_2; M = V, Nb, Ta). *Chemistry (Weinheim an der Bergstrasse, Germany)*, *24*(1), 206–211. doi:10.1002/chem.201704316 PMID:29027716

Chowdhury, S. M., Dasgupta, S., McElroy, A. E., & Sitharaman, B. (2014). Structural Disruption increases Toxicity of Graphene Nanoribbons. *Journal of Applied Toxicology*, *34*(11), 1235–1246. doi:10.1002/jat.3066 PMID:25224919

Eke, J., Mills, P. A., Page, J. R., Wrigth, G. P., Tsyusko, O. V., & Escobar, I. C. (2020). Nanohybrid Membrane Synthesis with Phosphorene Nanoparticles: A Study of the Addition, Stability and Toxicity. *Polymers*, *12*(7), 1555. doi:10.3390/polym12071555 PMID:32674304

Fahmi, T., Branch, L. D., Nima, Z. A., Jang, D. S., Savenka, A. V., Biris, A. S., & Basnakian, A. G. (2017). Mechanism of Graphene-induced Cytotoxicity: Role of Endonucleases. *Journal of Applied Toxicology*, *37*(11), 1325–1332. doi:10.1002/jat.3462 PMID:28543094

Farahani, M., Rezaei-Tavirani, M., Zali, H., Hatamie, S., & Ghasemi, N. (2020). Systems Toxicology Assessment Revealed the Impact of Graphene-based Materials on Cell Cycle Regulators. *Journal of Biomedical Materials Research. Part A*, *108*(7), 1520–1533. doi:10.1002/jbm.a.36923 PMID:32170903

Fojtů, M., Teo, W. Z., & Pumera, M. (2017). Environmental Impact and Potential Health Risks of 2D Nanomaterials. *Environmental Science. Nano*, *4*(8), 1617–1633. doi:10.1039/C7EN00401J

Gu, M., Dai, Z., Yan, X., Yan, X., Ma, J., Niu, Y., Lan, W., Wang, X., & Xu, Q. (2021). Comparison of Toxicity of Ti_3C_2 and Nb_2C Mxene Quantum dots (QDs) to Human Umbilical Vein Endothelial Cells. *Journal of Applied Toxicology*, *41*(5), 745–754. doi:10.1002/jat.4085 PMID:33048420

Han, M., Zhu, L., Mo, J., Wei, W., Yuan, B., Zhao, J., & Cao, C. (2020). Protein Corona and Immune Responses of Borophene: A Comparison of Nanosheet-Plasma Interface with Graphene and Phosphorene. *ACS Applied Bio Materials*, *3*(7), 4220–4229. doi:10.1021/acsabm.0c00306 PMID:35025423

Huang, H., Jiang, R., Feng, Y., Ouyang, H., Zhou, N., Zhang, X., & Wei, Y. (2020). Recent Development and Prospects of Surface Modification and Biomedical Applications of MXenes. *Nanoscale*, *12*(3), 1325–1338. doi:10.1039/C9NR07616F PMID:31872839

Kenry, & Lim, C.T. (2016). Biocompatibility and Nanotoxicity of Layered Two-Dimensional Nanomaterials. *ChemNanoMat*, *3*(1), 5-16. doi:10.1002/cnma.201600290

Khan, K., Tareen, A. K., Iqbal, M., Wang, L., Ma, C., Shi, Z., Ye, Z., Ahmad, W., Sagar, R. U. R., Shams, S. S., Sophia, P. J., Ullah, Z., Xie, Z., & Zhang, H. (2021). Navigating Recent Advances in Monoelemental Materials (Xenes) - Fundamental to Biomedical Applications. *Progress in Solid State Chemistry*, *63*, 100326. doi:10.1016/j.progsolidstchem.2021.100326

Latiff, N. M., Sofer, Z., Fisher, A. C., & Pumera, M. (2017). Cytotoxicity of Exfoliated Layered Vanadium Dichalcogenides. *Chemistry (Weinheim an der Bergstrasse, Germany)*, *23*(3), 684–690. doi:10.1002/chem.201604430 PMID:27781318

Latiff, N. M., Teo, W. Z., Sofer, Z., Fisher, A. C., & Pumera, M. (2015). The Cytotoxicity of Layered Black Phosphorus. *Chemistry (Weinheim an der Bergstrasse, Germany)*, *21*(40), 13991–13995. doi:10.1002/chem.201502006 PMID:26291565

Li, Y., Feng, L., Shi, X., Wang, X., Yang, Y., Yang, K., Liu, T., Yang, G., & Liu, Z. (2014). Surface Coating-dependent Cytotoxicity and Degradation of Graphene Derivatives: Towards the Design of Non-Toxic, Degradable Nano-Graphene. *Small*, *10*(8), 1544–1554. doi:10.1002mll.201303234 PMID:24376215

Lim, G. P., Soon, C. F., Ma, N. L., Morsin, M., Nayan, N., Ahmad, M. K., & Tee, K. S. (2021). Cytotoxicity of MXene-based Nanomaterials for Biomedical Applications: A Mini Review. *Environmental Research*, *201*, 111592. doi:10.1016/j.envres.2021.111592 PMID:34175291

Liu, C., Shin, J., Son, S., Choe, Y., Farokhzad, N., Tang, Z., Xiao, Y., Kong, N., Kim, J. S., & Tao, W. (2021). Pnictogens in Medical Chemistry: Evolution from Erstwhile Drugs to Emerging Layered Photonic Nanomedicine. *Chemical Society Reviews*, *50*(4), 2260–2279. doi:10.1039/D0CS01175D PMID:33367452

Liu, Y., Ding, J., Wang, Q.-Q., Wen, M.-L., Tang, T.-T., Liu, Y., Yuan, R., Li, Y.-F., & An, M.-W. (2021). Research Progress on the Biomedical Uses of Graphene and Its Derivatives. *New Carbon Materials*, *36*(4), 779–793. doi:10.1016/S1872-5805(21)60073-2

Luan, B., Huynh, T., Zhao, L., & Zhou, R. (2015). Potential Toxicity of Graphene to Cell Functions via Disrupting Protein-Protein Interactions. *ACS Nano*, *9*(1), 663–669. doi:10.1021/nn506011j PMID:25494677

Mu, X., Wang, J.-Y., Bai, X., Xu, F., Liu, H., Yang, J., Jing, Y., Liu, L., Xue, X., Dai, H., Liu, Q., Sun, Y.-M., Liu, C., & Zhang, X.-D. (2017). Black Phosphorus Quantum Dot induced Oxidative Stress and Toxicity in Living Cells and Mice. *ACS Applied Materials & Interfaces*, *9*(24), 20399–20409. doi:10.1021/acsami.7b02900 PMID:28553710

Naikoo, G. A., Arshad, F., Almas, M., Hassan, I. U., Pedram, M. Z., Aljabali, A. A. A., Mishra, V., Serrano-Aroca, A., Birkett, M., Charbe, N. B., Goyal, R., Negi, P., El-Tanani, M., & Tambuwala, M. M. (2022). 2D Materials, Synthesis, Characterization and Toxicity: A Critical Review. *Chemico-Biological Interactions*, *365*, 110081. doi:10.1016/j.cbi.2022.110081 PMID:35948135

Ou, L., Song, B., Liang, H., Liu, J., Feng, X., Deng, B., Sun, T., & Shao, L. (2016). Toxicity of Graphene-Family Nanoparticles: A General Review of the Origins and Mechanisms. *Particle and Fibre Toxicology*, *13*(1), 57. doi:10.118612989-016-0168-y PMID:27799056

Pandit, S., Karunakaran, S., Boda, S. K., Basu, B., & De, M. (2016). High Antibacterial Activity of Functionalized Chemically Exfoliated MoS_2. *ACS Applied Materials & Interfaces*, *8*(46), 31567–31573. doi:10.1021/acsami.6b10916 PMID:27933975

Pham, V. T. H., Truong, V. K., Quinn, M. D. J., Notley, S. M., Guo, Y., Baulin, V. A., Al Kobaisi, M., Crawford, R. J., & Ivanova, E. P. (2015). Graphene induces Formation of Pores that Kill Spherical and Rod-shaped Bacteria. *ACS Nano*, *9*(8), 8458–8467. doi:10.1021/acsnano.5b03368 PMID:26166486

Rasool, K., Helal, M., Ali, A., Ren, C. E., Gogotsi, Y., & Mahmoud, K. A. (2016). Antibacterial Activity of $Ti_3C_2T_x$ Mxene. *ACS Nano*, *10*(3), 3674–3684. doi:10.1021/acsnano.6b00181 PMID:26909865

Shaw, Z. L., Kuriakose, S., Cheeseman, S., Dickey, M. D., Genzer, J., Christofferson, A. J., Crawford, R. J., McConville, C. F., Chapman, J., Truong, V. K., Elbourne, A., & Walia, S. (2021). Antipathogenic Properties and Applications of Low-Dimensional Materials. *Nature Communications*, *12*(1), 3897. doi:10.103841467-021-23278-7 PMID:34162835

Song, S.-J., Shin, Y. C., Lee, H. U., Kim, B., Han, D.-W., & Lim, D. (2018). Dose- and Time-Dependent Cytotoxicity of Layered Black Phosphorus in Fibroblastic Cells. *Nanomaterials (Basel, Switzerland)*, *8*(6), 408. doi:10.3390/nano8060408 PMID:29882820

Szuplewska, A., Kulpińska, D., Jakubczak, M., Dybko, A., Chudy, M., Olszyna, A., Brzózka, Z., & Hastrzębska, A. M. (2022). The 10th Anniversary of MXenes: Challenges and Prospects for Their Surface Modification toward Future Biotechnological Applications. *Advanced Drug Delivery Reviews*, *182*, 114099. doi:10.1016/j.addr.2021.114099 PMID:34990793

Tan, E., Li, B. L., Ariga, K., Lim, C.-T., Garaj, S., & Leong, D. T. (2019). Toxicity of Two-Dimensional Layered Materials and Their Heterostructures. *Bioconjugate Chemistry*, *30*(9), 2287–2299. doi:10.1021/acs.bioconjchem.9b00502 PMID:31381854

Teo, W. Z., Chng, E. L. K., Sofer, Z., & Pumera, M. (2014). Cytotoxicity of Exfoliated Transition-Metal Dichalcogenides (MoS_2, WS_2, and WSe_2) is Lower Than that of Graphene and Its Analogues. *Chemistry (Weinheim an der Bergstrasse, Germany)*, *20*(31), 9627–9632. doi:10.1002/chem.201402680 PMID:24976159

Tu, Y., Lv, M., Xiu, P., Huynh, T., Zhang, M., Castelli, M., Liu, Z., Huang, Q., Fan, C., Fang, H., & Zhou, R. (2013). Destructive extraction of Phospholipids from *Escherichia coli* Membranes by Graphene Nanosheets. *Nature Nanotechnology*, *8*(8), 594–601. doi:10.1038/nnano.2013.125 PMID:23832191

Wang, Z., Zhu, W., Qiu, Y., Yi, X., von dem Bussche, A., Kane, A., Gao, H., Koski, K., & Hurt, R. (2016). Biological and Environmental Interactions of Emerging Two-Dimensional Nanomaterials. *Chemical Society Reviews*, *45*(6), 1750–1780. doi:10.1039/C5CS00914F PMID:26923057

Wen, K.-P., Chen, Y.-C., Chuang, C.-H., Chang, H.-Y., Lee, C.-Y., & Tai, N.-H. (2015). Accumulation and Toxicity of Intravenously-Injected Functionalized Graphene Oxide in Mice. *Journal of Applied Toxicology*, *35*(10), 1211–1218. doi:10.1002/jat.3187 PMID:26099253

Wu, N., Wang, X., Das, C. M., Ma, M., Qiao, N., Fan, T., Zhang, H., Xu, G., & Yong, K.-T. (2021). Bioengineering Applications of Black Phosphorus and Their Toxicity Assessment. *Environmental Science. Nano*, *8*(12), 3452–3477. doi:10.1039/D1EN00273B

Xiong, Z., Zhang, X., Zhang, S., Lei, L., Ma, W., Li, D., Wang, W., Zhao, Q., & Xing, B. (2018). Bacterial Toxicity of Exfoliated Black Phosphorous Nanosheets. *Ecotoxicology and Environmental Safety*, *161*, 507–514. doi:10.1016/j.ecoenv.2018.06.008 PMID:29913419

Yang, K., Li, Y., Tan, X., Peng, R., & Liu, Z. (2013). Behavior and Toxicity of Graphene and Its Functionalized Derivatives in Biological Systems. *Small*, *9*(9-10), 1492–1503. doi:10.1002mll.201201417 PMID:22987582

Yu, Y., Zheng, J., Li, J., Lu, L., Yan, J., Zhang, L., & Wang, L. (2021). Applications of Two-Dimensional Materials in Food Packaging. *Trends in Food Science & Technology*, *110*, 443–457. doi:10.1016/j.tifs.2021.02.021

Zhang, X., Zhang, Z., Zhang, S., Li, D., Ma, W., Ma, C., Wu, F., Zhao, Q., Yan, Q., & Xing, B. (2017). Size Effect on the Cytotoxicity of Layered Black Phosphorus and Underlying Mechanisms. *Small*, *13*(32), 1701210. doi:10.1002mll.201701210 PMID:28696584

Zhou, Z., Wang, X., Zhang, H., Huang, H., Sun, L., Ma, L., Du, Y., Pei, C., Zhang, Q., Li, H., Ma, L., Gu, L., Liu, Z., Cheng, L., & Tan, C. (2021). Activating Layered Metal Oxide Nanomaterials via Structural Engineering as Biodegradable Nanoagents for Photothermal Cancer Therapy. *Small*, *17*(12), 2007486. doi:10.1002mll.202007486 PMID:33590671

Zhu, Y., Xie, Z., Li, J., Liu, Y., Li, C., Liang, W., Huang, W., Kang, J., Cheng, F., Kang, L., Al-Hartomy, O. A., Al-Ghamdi, A., Wageh, S., Xu, J., Li, D., & Zhang, H. (2021). From Phosphorus to Phosphorene: Applications in Disease Theranostics. *Coordination Chemistry Reviews*, *446*, 214110. doi:10.1016/j.ccr.2021.214110

ADDITIONAL READING

Alhussain, H., Augustine, R., Hussein, E. A., Gupta, I., Hasan, A., Al Moustafa, A.-E., & Elzatahry, A. (2020). MXene Nanosheets May Induce Toxic Effect on the Early Stage of Embryogenesis. *Journal of Biomedical Nanotechnology*, *16*(3), 364–372. doi:10.1166/jbn.2020.2894 PMID:32493546

Gurunathan, S., & Kim, J.-H. (2016). Synthesis, Toxicity, Biocompatibility, and Biomedical Applications of Graphene and Graphene-related Materials. *International Journal of Nanomedicine*, *11*, 1927–1945. doi:10.2147/IJN.S105264 PMID:27226713

Hu, T., Mei, X., Wang, Y., Weng, X., Liang, R., & Wei, M. (2019). Two-Dimensional Nanomaterials: Fascinating Materials in Biomedical Field. *Science Bulletin*, *64*(22), 1707–1727. doi:10.1016/j.scib.2019.09.021 PMID:36659785

Ihsanullah, I., & Ali, H. (2020). Technological Challenges in the Environmental Applications of MXenes and Future Outlook. *Case Studies in Chemical and Environmental Engineering*, *2*, 100034. doi:10.1016/j.cscee.2020.100034

Markovic, M., Kumar, A., Andjelkovic, I., Lath, S., Kirby, J. K., Losic, D., Batley, G. E., & McLaughlin, M. J. (2018). Ecotoxicology of Manufactured Graphene Oxide Nanomaterials and Derivation of Preliminary Guideline Values for Freshwater Environments. *Environmental Toxicology and Chemistry, 37*(5), 1340–1348. doi:10.1002/etc.4074 PMID:29314166

Nel, A., Xia, T., Mädler, L., & Li, N. (2006). Toxic Potential of Materials at the Nanolevel. *Science, 311*(5761), 622–627. doi:10.1126cience.1114397 PMID:16456071

Ramanathan, A. (2019). Toxicity of Nanoparticles: Challenges and Opportunities. *Applied Microscopy, 49*(1), 2. doi:10.100742649-019-0004-6 PMID:33580298

Walker, N. J., & Bucher, J. R. (2009). A 21st Century Paradigm for Evaluating the Health Hazards of Nanoscale Materials? *Toxicological Sciences, 110*(2), 251–254. doi:10.1093/toxsci/kfp106 PMID:19468057

KEY TERMS AND DEFINITIONS

Cytoxicity: The ability of some cells to be toxic to other cells that have been changed.

Graphene: Allotrope or carbon formed by a single layer of carbon atoms located in a hexagonal lattice.

MXene: Two-dimensional inorganic compound with thickness layers of few atoms based on transition metal carbides, nitrides, or carbonitrides.

Nanosheet: A two-dimensional basic structural element with a thickness of 1 to 100 nanometers.

Phosphorene: Stable orthorhombic form of phosphorus with two-dimensional structure at room temperature and pressure of entangled rings with six members where each atom is linked to three other atoms.

Pnictogen: An element that belongs to the nitrogen group (nitrogen, phosphorus, arsenic, antimony, bismuth, and ununpentium), Group 15 of the chemical elements (formerly numbered as Group V or Group VA).

Transition Metal Dichalcogenide (TMD): Chemical compound consisting of two chalcogen anions (sulfur, selenium, tellurium) and at least one transition metal (molybdenum, tungsten, cobalt, etc.) with two-dimensional structure.

Two-Dimensional (2D) Materials: Crystalline materials made up of single or few layers of atoms that have substantially stronger in-plane interactions than those along the stacking direction.

Chapter 10
Recent Advances in Green-Synthesis of Silver Nanoparticles and Their Biomedical Applications:
A Future Perspective

Richa Arora
Department of Microbiology, Punjab Agricultural University, Ludhiana, India

Vijay K. Bharti
Defence Institute of High Altitude Research, DRDO, India

ABSTRACT

Various researchers have been focused on the synthesis of silver nanoparticles (SNPs) owing to their widespread benefits in various fields including material sciences, biotechnology, medical sciences, engineering, etc. Biological applications and toxicity of SNPs are greatly dependent on their method of synthesis, experimental model species, duration and dose regime, etc. Green-synthesis is a commonly used methodology for synthesis medical grade SNPs that has minimal to nil toxicity and good biocompatibility in various animal models and human beings. Therefore, a number of SNPs technologies and products have been licensed and commercialized in the consumer markets for human application and treatment of various medical problems. Hence, this review has discussed the recent advances and updates on various approaches of green synthesis of SNPs, their biomedical applications, and their toxicological effects. This will help researchers and readers to investigate further techniques of synthesis of SNPs and industrial applications.

DOI: 10.4018/978-1-6684-5533-3.ch010

1. INTRODUCTION

Nanotechnology is synthesis and application of materials at nanoscale with control over shape and size that affects their properties in physical and biological environment. As compared to larger particles, nanoparticles (NPs) exhibit superior characteristics and bioavailability like optical properties, catalytic properties, antimicrobial properties, magnetic properties, at specific size (1-100 nm), structure, shape, chemical nature and surface area to volume ratio. Therefore, since few years, research has been more focused towards the synthesis of various kinds of nanoparticles (NPs) owing to their applications in various fields using different metals, biological molecules, capping agents, drugs molecules, etc. (Rafique et al., 2017). Among the most widely exploited NPs, metal based NPs are very common viz. zinc oxide (Cao et al., 2019; Deekala et al., 2019), cadmium sulphide (Solanki and Rajaram, 2017), gold (Suchomel et al., 2018), silver (Gondil et al., 2019), aluminium (Krause et al., 2020), copper (Kamikoriyama et al., 2019), iron (Saif et al., 2016), nickel (Ahghari et al., 2020), etc. However, noble metals like silver and gold exhibit superior characteristics due to their electrical properties, optical properties and thermal conductivity (Radzuan et al., 2017). Hence, there is a growing interest among the researchers for the synthesis of silver nanoparticles (SNPs) as they have wide range of applications like biological sensors, bio-imaging (Tan et al., 2020), antimicrobials (Yin et al., 2020), microbial contamination control (Simbine et al., 2019), energy generation (Huff et al., 2020), and other biomedical applications (Hussein and Abdullah, 2020).

Therefore, various methods for the synthesis of silver nanoparticles have been reported, including, ball milling (Yadav et al., 2012), pulse laser ablation (Machmudah et al., 2012), evaporation-condensation (Harra et al., 2012), tube furnace synthesis (Baby and Sundara, 2013), etc. However, these methods suffer from several lacunae like non-ecofriendly, more space requirement, surface structure defects, requirement of high temperature, costly equipments, poor-bio-compatability, toxicity, etc. (Iravani et al., 2014), which limits their biological applications. There are various reducing agents including ascorbic acid (Malassis et al., 2016), sodium borohydride (Vu et al., 2018), sodium citrate (Quintero-Quiroz et al., 2019), polyethylene glycol (Fleitas-Salazar et al., 2017), etc. have been reported in the literature. However, considering the toxic effect of these chemicals on the environment and generation of by-products, the researchers shifted their interest towards green synthesis of SNPs by exploiting eco-friendly and biological materials for chemical reduction. Hence, currently efforts are being made for the green synthesis of NPs using biological materials viz. microorganisms, plant extracts, virus DNA, enzymes, etc. (Rafique et al., 2017).

Apart from being eco-friendly, green synthesis of nanoparticles offers other benefits including easy scale up, economical, high reaction rate, less energy-dependent and no requirement of harsh chemicals (Ahmad et al., 2019). Keeping in view the advantages of green synthesis of silver nanoparticles along with their applications as antimicrobials and antioxidant agents, the present review explores the recent advances in the biosynthesis of silver nanoparticles (SNPs) and their bio-medical applications.

2. SILVER NANOPARTICLES: AN OVERVIEW

Silver is a soft and lustrous metal with high thermal and electrical conductivity. Since ancient times, it has been exploited for its therapeutic and medical benefits (Sim et al., 2018). The mechanism of action of silver nanoparticles on the microbial cells is reported through several theories, among them one of the important mechanism is SNPs adhere to the cell wall of bacteria and gradually penetrate in it, thereby, resulting in structural changes in the cell membrane and finally causing the cell death (Mikhailova, 2020). Prabhu and Poulose (2012) proposed the possibility of release of silver ions from the nanoparticles, which subsequently, inactivates the bacterial enzymes. In another theory, Ahmad et al. (2020) hypothesized the interaction of silver NPs with phosphorus and sulphur present in bacterial DNA, thereby, hindering the replication of bacterial DNA and finally causing bacterial cell death.

2.1. Green Synthesis

Conventional methods used for synthesis of SNPs are non-eco-friendly, utilize harsh chemicals and costly. So, to overcome these issues, the researchers have started investigating natural materials, for example, bacteria (Saravanan et al., 2018), fungi (Guilger-Casagrande and Lima, 2019), extracts from plants (Castillo-Henriquez et al., 2020) etc., thereby, undergoing green synthesis of NPs. Bacteria produce several extracellular and intracellular inorganic materials which can be exploited for the synthesis of NPs. Several researchers have reported bacterial strains like *Pseudomonas stutzeri* (Rajora et al., 2016), *Bacillus licheniformis* (Elbeshehy et al., 2015), *Proteus mirabilis* (Parveen et al., 2018), *Klebsiella pneumonia* (Sayyid and Zghair, 2021), *Escherichia coli* (Viet et al., 2018) etc. for synthesis of SNPs. The utilization of bacteria for NP synthesis makes the process flexible, and appropriate for scale-up. However, a major constraint in using bacteria for NP synthesis is slow rate of synthesis and limited range of size and shape of silver NPs (Ahmad et al., 2019).

Apart from bacteria, fungi have also been explored for the synthesis of NPs owing to their high metal bioaccumulation potential with high binding capacity and secretion of several enzymes for reduction of silver in ionic state (Guilger-Casagrande and Lima, 2019). Therefore, several fungi have been reported for SNPs synthesis, e.g. *Cladosporium cladosporioides* (Hulikere al 2019), *Fusarium oxysporum* (Khan et al, 2017), *Penicillium duclauxii* (Almaary et al., 2020), *Aspergillus fumigatus* (Othman et al, 2019), *Aspergillus flavus* (Sulaiman et al., 2015), *Aspergillus terreus* (Kumari et al, 2020), *Trichoderma reesei* (Guilger-Casagrande et al., 2019) etc., with size ranging from 1-50 nm. However, unlike bacteria, the fungus-based NPs are formed on the mycelium surface rather than in solution. Therefore, SNPs get adsorbed on cell surface owing to electrostatic interaction of silver ions with carboxylate groups of enzymes and get reduced by the cell wall enzymes, forming silver NP (Rafique et al., 2017). In comparison to bacteria, fungus-based NPs offer more benefits in terms of cost as more amount of enzymes are secreted, thereby, increasing the reaction rate and the process can be easily scaled up (Ahmad et al., 2019). However, their application can be enhanced once these enzymes are explored further biological value.

Plant-based extracts are also explored by many researchers for green synthesis of Ag NPs using leaves, stem, seeds, latex, roots, as these plant parts contain several bioactive compounds viz. phenolics, terpenoids, alkaloids, polysaccharides, proteins etc. that helps in stabilization and reduction of NPs (Vijayaraghavan and Ashokkumar, 2017; Kumari and Sangal, 2020). Therefore, many researchers has reported successful SNPs synthesis methodology using various parts of plants, viz. alfa-alfa sprouts (Park et al., 2016), peanut shell extract (Velmurugan et al., 2015), pineapple juice (Saod et al., 2020), *Argemonemexicana* (Chandrasekhar and Vinay, 2017), neem extracts (Chand et al, 2019), *domestica* (Anand et al., 2019), papaya extract (Singh et al., 2021), *Ocimum canum* leaf extract (Tailor et al, 2020), *Cassia auriculata* flower extract (Muthu and Priya, 2017), *Olea europaea* (Matteis et al., 2019), etc. Plant-based SNPs are more advantageous over microorganism-based NPs as they can be synthesized in short period of time at room temperature, stable NPs, and are easily available and safer to handle (Singh et al., 2020).

3. BIOLOGICAL APPLICATIONS

SNPs have been reported to have broad range of applications (Table 1). From the last few years, silver has been used in the form of nanomaterials owing to its attractive physicochemical, antimicrobial, anti-inflammatory and anti-cancerous properties (Ahmad et al., 2019). These applications are described in detail below.

3.1. Antimicrobial

Owing to the antimicrobial properties, SNPs have been incorporated in various cosmetic products, toothpastes, oral care gels, anti-dandruff shampoos etc. (Effiong et al., 2020; Ong and Nyam, 2022). Kedziora et al (2018) reported that the nanoparticles form of silver possesses more antimicrobial activity as compared to the silver in ionic form. Antimicrobial activity of SNPs have been explored in various biomedical applications, e.g. wound dressings (Paladini and Pollini, 2019), silicon heart valves (Burdusel et al., 2018); reducing biofilms on catheters (Gondil et al., 2019; Yassin et al., 2019), dental composites (Barot et al., 2020), orthopedic implants (Brennan et al., 2019), surgical gloves (Nakamura et al., 2019), etc. Nanoparticles based antimicrobials unveil various advantages including augmented bioavailability, targeted drug release and negligible side effects (Patra et al., 2018). In this context, researchers are exploring the possibility of overcoming multidrug resistant (MDR) bacteria using silver NPs at global level as they are known to provide a broad spectrum of antimicrobial activity against MDR bacteria (Alharbi and Alarfaj, 2020; Liao et al., 2019).

Lohani et al (2014) reported the use of silver and other metal oxide nanoparticles in nail paints to for the treatment of fungal infections in nails. Guerra et al (2020) reported 55% reduction in fungal growth with SNPs. Further, silver nanoparticles have a great potential to replace conventional antibiotics for the treatment of nosocomial infections (Singh et al., 2018). A number of conjugating agents like polymers, drugs, amino acids, fatty acids etc. can be incorporated with SNPs for their wide applications as biotherapeutics (Zhao et al., 2019; Anwar et al., 2019). However, the efficacy of the silver nanoparticles is highly dependent on the shape and size of NPs. Smaller size is reported to have higher antibacterial

Table 1. Recent reports on various applications of silver nanoparticles

Applications	Size (nm)	Main findings	References
Antimicrobial activity	10-50	Antimicrobial activity against *Campylobacter jejuni* Minimum Bactericidal Concentration: Between 9.85 and 39.4μg/mL Minimum Bacteriostatic Concentration: Between 1.23 and 4.92 μg/mL	Silvan et al, 2018
	20-30	Antimicrobial activity against *Escherichia coli* and *Salmonella typhimurium* Minimum Bactericidal Concentration: 9.5μg/mL (*E. coli* and *S. typhimurium*) Minimum Bacteriostatic Concentration: 4.75 μg/mL (*E. coli* and *S. typhimurium*)	Chandhru et al, 2019
	47.3	Antimicrobial activity against *Vibrio parahaemolyticus*and *Staphylococcus aureus* Minimum Bactericidal Concentration: 12.5 μg/mL (*V. parahaemolyticus*); 100 μg/mL (*S. aureus*) Minimum Bacteriostatic Concentration: 6.25 μg/mL (*V. parahaemolyticus*); 50 μg/mL (*S. aureus*)	Du et al, 2019
	5-15	Antimicrobial activity against *Escherichia coli* and methicillin resistant *Staphylococcus aureus* Minimum Bactericidal Concentration: 7.5 μg/mL (*E. coli*) Minimum Bactericidal Concentration: >30 μg/mL (methicillin resistant *Staphylococcus aureus*)	Navarro Gallon et al, 2019
Cytotoxic effect on cell lines	6	Cytotoxic effect on human colon epithelial cells (HT-28 and HCT-116) 50% cytotoxicity at a 4 μg/mL	Gurunathan et al, 2018
	10-31.4	Cytotoxic effect on Human breast epithelial cells (MCF-7) 70% Cytotoxicity occurred at a concentration between 60 μg/mL Less cytotoxic against non-cancerous cell line	Khorrami et al, 2018
In vivo models	10, 40 and 100	Silver nanoparticles with size 10 and 100 nm caused developmental effects in the muscles and intestine of embryo of zebrafish and nanoparticles with 40 nm caused lethal effects	Liu et al, 2019
	<100	Growth inhibition and reproductive toxicity in *Caenorhabditiselegans* at a concentration of ³5 mg kg^{-1}	Moon et al, 2019
	3.44	Toxic dose 10 nM for *Drosophila melanogaster* Depigmentation, intracellular ROS and DNA damage was observed	Alaraby et al, 2019
Abiotic stress tolerance	-	Improved chilling stress tolerance in *Phaseolus vulgaris* with increased seedling height, fresh and dry weight and net photosynthesis	Prazak et al, 2020
Anticancer activity	27	Anticancerous effect of silver aqueous pollen extract of *Phoenix dactylifera* (Male Date Palm) flowers against MCF-7 cell line with apoptotic cell death IC50: 9.06 μg/mL	Banu et al, 2018

Continued on following page

Table 1. Continued

Applications	Size (nm)	Main findings	References
	21.65–41.05	Anticancerous effect of silver *Phoenix dactylifera* (Aqueous Root Extract) against MCF-7 cell line with apoptotic cell death, necrosis and cell growth inhibition IC50: 29.6 μg/mL	Oves et al, 2018
	30	Anticancerous effect of silver *Punicagranatum* (polysaccharide (PSP001) extracted from fruit peel) against A549 with Caspase mediated programmed cell death IC50: 54.5 μg/mL	Padinjarathil et al. (2018)
	2-50	Anticancerous effect of silver *Calotropis gigantean* (Ethanolic extract of latex) against Jurkat with upregulation of caspase-3 indicating activation of apoptotic pathways IC50: 11.99 μg/mL	Maity et al, 2018

properties due to more surface area, thereby, effectively disrupting the cell membrane of the microorganisms (Ong and Nyam, 2022). In another study, truncated triangular shaped SNPs were found to better inhibit the gram negative bacteria than rod or spherical shaped NPs (Dakal et al., 2016).

3.2. Anti-Inflammatory

The anti-inflammatory effect of SNPs is well reported in recent studies. SNPs are reported to disrupt vascular endothelial growth factor pathway (VEGF), thereby, disrupting the inflammation (Ong and Nyam, 2022). Fehaid et al (2020) reported anti-inflammatory effect of SNPs in NCI-H292 cell line. Owing to these properties, they are well exploited in skin ointments, especially for treatment of acnes, sunscreens, formulations for UV damaged skin etc. (Patil and Muthusamy, 2020).

3.3. Anti-Cancer

Metal NPs can be conjugated with several anticancerous drugs for cancer treatment (Carvalho et al., 2019). Karuppaiah et al (2020) reported targeted delivery of drugs conjugated with SNPS to the cancer cells, thereby, decreasing the cytotoxic effects to the normal cells. Shejawal et al (2021) reported anticancerous properties of lycopene conjugated SNPs against colon cancer cell lines. Yuan et al (2018) reported enhanced anti-cancerous activity of SNPs conjugated with camptothecin against Hela cancer cells. In a similar study, Gomathi et al (2020) investigated the efficacy of SNPs prepared from *Tamarindus indica* against breast cancer cell lines.

3.4. Other Applications

Apart from the above mentioned applications, SNPs can be exploited in several other fields. SNPs have also been explored as biosensors with nanoprobes (Loiseau et al., 2019) and in agricultural science as nanofertilizers (Zulfiqar et al., 2019) and nanopesticides (Konappa et al., 2021). Some of the other applications of SNPs are their incorporation in water detoxification (Mehwish et al., 2021), hand wash

(Ahmed et al., 2019), deodorants (Ballarin et al., 2015), food packaging (Carbone et al., 2016), dye effluent treatment (Marimuthu et al., 2020), etc. Yin et al (2020) reported the application of SNPs for the preparation of dental resins and coatings. Deshmukh et al (2019) reported the blending of SNPs with fabrics for disinfection and better hygiene. Further, the are also exploited as components of automobile interiors for durability (Shafique and Luo, 2019). Gangadoo et al (2020) supplemented broiler feed with nanoparticles and reported improved quality of feed with no toxicity. Moreover, the application of SNPs in crops helps to enhance the abiotic stress tolerance and boosts their antioxidant defence mechanisms (Alabdallah and and Hasan, 2021).

These applications are an indicative of wider acceptability of SNPs and therefore, more research should be focused on developing methodology using other capping and binding agents. It is further necessitate the research on toxicity and bio-compatability evaluation using various animal model and dose regime to enhance SNPs application for human welfare.

4. TOXICOLOGICAL IMPACT

Although silver nanoparticles have been exploited at a large scale for the production of various nanotechnological products, but their unjudicious use can pose toxicological effects (Glinski et al., 2021). Their increased use in food and medicine has increased the exposure via oral routes (Sahoo et al., 2021). Nanoparticles are also reported to interact with the biomolecules present in food molecules, thereby, reducing their neutraceutical and nutritional value (Xie et al., 2021). Apart from food, nanoparticles have also been reported to pose the toxic effect on animal models like zebrafish (Mahjoubian et al., 2021); mice (Ramirez-Garcia et al., 2021); guinea pigs (Zheng et al., 2020) etc. Therefore, further studies are required for targeted and remote control of nanoparticles at the site of action. A better comprehension of mechanism of action of nanoparticles on the cells and tissues can help to alleviate their adverse effects. Moreover, hybridized nanomaterials may be exploited for biological systems to reduce their toxic effects.

5. CONCLUSION

In conclusion, silver nanoparticles can be synthesized using green chemistry approach having low toxicity, better bio-compatability, eco-friendly, and economical. Further, this green-synthesis method are easy to produce and therapeutic biomolecules can be capped or bounded over it would give synergistic and additive therapeutic effect, including better stability, which can be explored for commercial purposes. This review further explored many biomedical applications, among them microbial control is important. Therefore, more investigations are required to synthesize more potent SNPs to control bacterial resistance and development of alternative antimicrobials.

Conflict of interest

The authors declare no conflict of interest.

REFERENCES

Ahghari, M. R., Soltaninejad, V., & Maleki, A. (2020). Synthesis of nickel nanoparticles by a green and convenient method as a magnetic mirror with antibacterial activities. *Scientific Reports*, *10*(1), 12627. doi:10.103841598-020-69679-4 PMID:32724123

Ahmad, S., Munir, S., Zeb, N., Ullah, A., Khan, B., Ali, J., Bilal, M., Omer, M., Alamzeb, M., Salman, S. M., & Ali, S. (2019). Green nanotechnology: A review on green synthesis of silver nanoparticles-an ecofriendly approach. *International Journal of Nanomedicine*, *14*, 5087–5107. doi:10.2147/IJN.S200254 PMID:31371949

Ahmad, S. A., Das, S. S., Khatoon, A., Ansari, M. T., Afzal, M., Hasnain, M. S., & Nayak, A. K. (2020). Bactericidal activity of silver nanoparticles: A mechanistic review. *Materials Science for Energy Technologies*, *3*, 756–769. doi:10.1016/j.mset.2020.09.002

Ahmed, M. M., Fatima, F., Anwer, M. K., Alshahrani, S. M., Alalaiwe, A., & Katakam, P. (2019). Biosynthesis, characterization and anti-microbial activity of silver nanoparticle based gel hand wash. *Green Processing and Synthesis*, *8*(1), 577–583. doi:10.1515/gps-2019-0027

Alabdallah, N. M., & Hasan, M. (2021). Plant based green synthesis of silver nanoparticles and its effective role in abiotic stress tolerance in crop plants. *Saudi Journal of Biological Sciences*, *28*(10), 5631–5639. doi:10.1016/j.sjbs.2021.05.081 PMID:34588874

Alaraby, M., Romero, S., Hernandez, A., & Marcos, R. (2019). Toxic and genotoxic effects of silver nanoparticles in Drosophila. *Environmental and Molecular Mutagenesis*, *60*(3), 277–285. doi:10.1002/em.22262 PMID:30353950

Alharbi, F. A., & Alarfaj, A. A. (2020). Green synthesis of silver nanoparticles from *Neuradaprocumbens* and its antibacterial activity against multi-drug resistant microbial pathogens. *Journal of King Saud University. Science*, *32*(2), 1346–1352. doi:10.1016/j.jksus.2019.11.026

Almaary, K. S., Sayed, S. R. M., Abd-Elkader, O., Dawoud, T. M., El-Orabi, N. F., & Elgorban, A. M. (2020). Complete green synthesis of silver-nanoparticles applying seed-borne *Penicillium duclauxii. Saudi Journal of Biological Sciences*, *27*(5), 1333–1339. doi:10.1016/j.sjbs.2019.12.022 PMID:32346343

Anand, M. A. V., Vinayagam, R., Vijayakumar, S., Balupillai, A., Herbert, F. J., Kumar, S., Ghidan, A. Y., Al-Antary, T., & David, E. (2019). Green synthesis, characterization and antibacterial activity of silver nanoparticles by *Malus domestica* and its cytotoxic effect on (MCF-7) cell line. *Microbial Pathogenesis*, *135*, 103609. doi:10.1016/j.micpath.2019.103609 PMID:31247255

Anwar, A., Abdalla, S. A. O., Aslam, Z., Shah, M. R., Siddiqui, R., & Khan, N. A. (2019). Oleic acid-conjugated silver nanoparticles as efficient antiamoebic agent against *Acanthamoeba castellanii. Parasitology Research*, *118*(7), 2295–2304. doi:10.100700436-019-06329-3 PMID:31093751

Baby, T. T., & Sundara, R. (2013). Synthesis of silver nanoparticle decorated multiwalled carbon nanotubes-graphene mixture and its heat transfer studies in nanofluid. *AIP Advances*, *3*(1), 012111. doi:10.1063/1.4789404

Ballarin, B., Mignani, A., Mogavero, F., Gabbanini, S., & Morigi, M. (2015). Hybrid material based on ZnAl hydrolcite and silver nanoparticles for deodorant formulation. *Applied Clay Science*, *114*, 303–308. doi:10.1016/j.clay.2015.06.014

Banu, H., Renuka, N., Faheem, S. M., Ismail, R., Singh, V., Saadatmand, Z., Khan, S. S., Narayanan, K., Raheem, A., Premkumar, K., & Vasanthakumar, G. (2018). Gold and silver NPs biomimetically synthesized using date palm pollen extract-induce apoptosis and regulate p53 and Bcl-2 expression in human breast adenocarcinoma cells. *Biological Trace Element Research*, *186*(1), 122–134. doi:10.100712011-018-1287-0 PMID:29552710

Barot, T., Rawtani, D., & Kulkarni, P. (2020). Physicochemical and biological assessment of silver nanoparticles immobilized Halloysite nanotubes-based resin composite for dental applications. *Heliyon*, *6*(3, e03601), 1–10. doi:10.1016/j.heliyon.2020.e03601 PMID:32195405

Brennan, S. A., Fhoghlu, C. N., Devitt, B. M., Mahony, F. J. O., Brabazon, D., & Walsh, A. (2019). Review synthesis and application of silver nanoparticles (Ag NPs) for the prevention of infection in healthcare workers. *International Journal of Molecular Sciences*, *20*(15), 3620. doi:10.3390/ijms20153620 PMID:31344881

Burdusel, A. C., Gherasim, O., Grumezescu, A. M., Mogoanta, L., Ficai, A., & Andronescu, E. (2018). Biomedical applications of silver nanoparticles: An Up-to-Date overview. *Nanomaterials (Basel, Switzerland)*, *8*(9), 681. doi:10.3390/nano8090681 PMID:30200373

Cao, D., Gong, S., Shu, X., Zhu, D., & Liang, L. (2019). Preparation of ZnO nanoparticles with high dispersibility based on Oriented Attachment (OA) Process. *Nanoscale Research Letters*, *14*(1), 210. doi:10.118611671-019-3038-3 PMID:31222635

Carbone, M., Donia, D. T., Sabbatella, G., & Antiochia, R. (2016). Silver nanoparticles in polymeric matrices for fresh food packaging. *Journal of King Saud University. Science*, *28*(4), 273–279. doi:10.1016/j.jksus.2016.05.004

Carvalho, S. M., Leonel, A. G., Mansur, A. A. P., Carvalho, I. C., Krambrock, K., & Mansur, H. S. (2019). Bifunctional magnetopolymersomes of iron oxide nanoparticles and carboxymethylcellulose conjugated with doxorubicin for hyperthermochemotherapy of brain cancer cells. *Biomaterials Science*, *7*(5), 2102–2122. doi:10.1039/C8BM01528G PMID:30869664

Castillo-Henriquez, L., Alfaro-Aguilar, K., Ugalde-Alvarez, J., Vega-Fernandez, L., de Oca-Vasquez, G. M., & Vega-Baudrit, J. R. (2020). Green synthesis of gold and silver nanoparticles from plant extracts and their possible applications as antimicrobial agents in the agricultural area. *Nanomaterials (Basel, Switzerland)*, *10*(9), 1763. doi:10.3390/nano10091763 PMID:32906575

Chand, K., Ishaque Abro, M., Aftab, U., Shah, A. H., Lakhan, M. N., Cao, D., Mehdi, G., & Mohamed, A. M. A. (2019). Green synthesis characterization and antimicrobial activity against *Staphylococcus aureus* of silver nanoparticles using extracts of neem, onion and tomato. *RSC Advances*, *30*(30), 17002–17015. doi:10.1039/C9RA01407A PMID:35519862

Chandhru, M., Logesh, R., Rani, S. K., Ahmed, N., & Vasimalai, N. (2019). One-pot green route synthesis of silver nanoparticles from jack fruit seeds and their antibacterial activities with *Escherichia coli* and *Salmonella* bacteria. *Biocatalysis and Agricultural Biotechnology*, *20*, 101241. doi:10.1016/j.bcab.2019.101241

Chandrasekhar, N., & Vinay, S. P. (2017). Yellow colored blooms of *Argemone mexicana* and *Turnera ulmifolia* mediated synthesis of silver nanoparticles and study of their antibacterial and antioxidant activity. *Applied Nanoscience*, *7*(8), 851–861. doi:10.100713204-017-0624-5

Dakal, T. C., Kumar, A., Majumdar, R. S., & Yadav, V. (2016). Mechanistic basis of antimicrobial actions of silver nanoparticles. *Frontiers in Microbiology*, *7*, 1–17. doi:10.3389/fmicb.2016.01831 PMID:27899918

Deekala, V., Babu, B. K., & Rudraraju, R. (2019). Pharmacological studies of zinc oxide nanoparticles. *Indian Journal of Biochemistry & Biophysics*, *56*, 500–505.

Deshmukh, S. P., Patil, S. M., Mullani, S. B., & Delekar, S. D. (2019). Silver nanoparticles as an effective disinfectant: A review. *Materials Science and Engineering C*, *97*, 954–965. doi:10.1016/j.msec.2018.12.102 PMID:30678983

Du, J., Hu, Z., Yu, Z., Li, H., Pan, J., Zhao, D., & Bai, Y. (2019). Antibacterial activity of a novel *Forsythia suspensa* fruit mediated green silver nanoparticles against food-borne pathogens and mechanisms investigation. *Materials Science and Engineering C*, *102*, 247–253. doi:10.1016/j.msec.2019.04.031 PMID:31146997

Effiong, D., Uwah, T., Jumbo, E., & Akpabio, A. (2020). Nanotechnology in cosmetics: Basics, Current Trends and Safety Concerns- A Review. *Advances in Nanoparticles*, *9*, 1–22.

Elbeshehy, E. K. F., Elazzazy, A. M., & Aggelis, G. (2015). [REMOVED HYPERLINK FIELD]Silver nanoparticles synthesis mediated by new isolates of *Bacillus* spp., nanoparticle characterization and their activity against Bean Yellow Mosaic Virus and human pathogens. *Frontiers in Microbiology*, *6*, 453. doi:10.3389/fmicb.2015.00453 PMID:26029190

Fehaid, A., Fujii, R., Sato, T., & Taniguchi, A. (2020). Silver nanoparticles affect the inflammatory response in a lung epithelial cell line. *The Open Biotechnology Journal*, *14*(1), 113–123. doi:10.2174/1874070702014010113

Fleitas-Salazar, N., Silva-Campa, E., Pedroso-Santana, S., Tanori, J., Pedroza-Montero, M. R., & Riera, R. (2017). Effect of temperature on the synthesis of silver nanoparticles with polyethylene glycol: New insights into the reduction mechanism. *Journal of Nanoparticle Research*, *19*(3), 113. doi:10.100711051-017-3780-3

Gangadoo, S., Dinev, I., Willson, N.-L., Moore, R. J., Chapman, J., & Stanley, D. (2020). Nanoparticles of selenium as high bioavailable and non-toxic supplement alternatives for broiler chickens. *Environmental Science and Pollution Research International*, *27*(14), 16159–16166. doi:10.100711356-020-07962-7 PMID:32107689

Glinski, A., de Souza, T. L., da Luz, J. Z., Junior, A. G. B., de Oliveira, C. C., Ribeiro, C. A. O., & Neto, F. F. (2021). Toxicological effects of silver nanoparticles and cadmium chloride in macrophage cell line (RAW 264.7): An in vitro approach. *Journal of Trace Elements in Medicine and Biology*, *68*, 126854. doi:10.1016/j.jtemb.2021.126854 PMID:34488184

Gomathi, A. C., Xavier Rajarathinam, S. R., Mohammed Sadiq, A., & Rajeshkumar, S. (2020). Anticancer activity of silver nanoparticles synthesized using aqueous fruit shell extract of *Tamarindus indica* on MCF-7 human breast cancer cell line. *Journal of Drug Delivery Science and Technology*, *55*, 101376. doi:10.1016/j.jddst.2019.101376

Gondil, V. K., Kalaiyarasan, T., Bharti, V. K., & Chhibber, S. (2019). Antibiofilm potential of Seabuckthorn silver nanoparticles (SBT@AgNPs) against *Pseudomonas aeruginosa. 3Biotech*, *9*, 402.

Guerra, J. D., Sandoval, G., Avalos-Borja, M., Pestryakov, A., Garibo, D., Susarrey-Arce, A., & Bogdanchikova, N. (2020). Selective antifungal activity of silver nanoparticles: A comparative study between *Candida tropicalis* and *Saccharomyces boulardii. Colloid and Interface Science Communications*, *37*, 100280. doi:10.1016/j.colcom.2020.100280

Guilger-Casagrande, M., Germano-Costa, T., Pasquoto-Stigliani, T., Fraceto, L., & Lima, R. (2019). Biosynthesis of silver nanoparticles employing *Trichoderma harzianum* with enzymatic stimulation for the control of *Sclerotinia sclerotiorum. Scientific Reports*, *9*(1), 14351. doi:10.103841598-019-50871-0 PMID:31586116

Guilger-Casagrande, M., & Lima, R. (2019). Synthesis of Silver Nanoparticles Mediated by Fungi: A Review. *Frontiers in Bioengineering and Biotechnology*, *7*, 287. doi:10.3389/fbioe.2019.00287 PMID:31696113

Gurunathan, S., Qasim, M., Park, C., Yoo, H., Kim, J. H., & Hong, K. (2018). Cytotoxic potential and molecular pathway analysis of silver nanoparticles in human colon cancer cells HCT116. *International Journal of Molecular Sciences*, *19*(8), 2269. doi:10.3390/ijms19082269 PMID:30072642

Harra, J., Makitalo, J., Siikanen, R., Virkki, M., Genty, G., Kobayashi, T., Kobayashi, T., Kauranen, M., & Makela, J. M. (2012). Size-controlled aerosol synthesis of silver nanoparticles for plasmonic materials. *Journal of Nanoparticle Research*, *14*(6), 870. doi:10.100711051-012-0870-0 PMID:22844206

Huff, C., Long, J. M., & Abdel-Fattah, T. M. (2020). Beta-cyclodextrin-assisted synthesis of silver nanoparticle network and its application in a hydrogen generation reaction. *Catalysts*, *10*(9), 1014. doi:10.3390/catal10091014

Hulikere, M., Chandrashekhar, M., & Joshi, G. (2019). Characterization, antioxidant and antimicrobial activity of silver nanoparticles synthesized using marine endophytic fungus-*Cladosporium cladosporioides. Process Biochemistry*, *82*, 199–204. doi:10.1016/j.procbio.2019.04.011

Hussein, H. A., & Abdullah, M. A. (2020). Biosynthesis, mechanisms, and biomedical applications of silver nanoparticles. In D. Thangadurai, J. Sangeetha, & R. Prasad (Eds.), *Functional Bionanomaterials. Nanotechnology in the Life Sciences*. Springer. doi:10.1007/978-3-030-41464-1_14

Iravani, S., Korbekandi, H., Mirmohammadi, S. V., & Zolfaghari, B. (2014). Synthesis of silver nanoparticles: Chemical, physical and biological methods. *Research in Pharmaceutical Sciences*, *9*, 385–406. PMID:26339255

Kamikoriyama, Y., Imamura, H., Muramatsu, A., & Kanie, K. (2019). Ambient aqueous-phase synthesis of copper nanoparticles and nanopastes with low-temperature sintering and ultra-high bonding abilities. *Scientific Reports*, *9*(1), 899. doi:10.103841598-018-38422-5 PMID:30692589

Karuppaiah, A., Rajan, R., Hariharan, S., Balasubramaniam, D. K., Gregory, M., & Sankar, V. (2020). Synthesis and characterization of folic acid conjugated gemcitabine tethered silver nanoparticles (FA-GEM-AgNPs) for targeted delivery. *Current Pharmaceutical Design*, *26*(26), 3141–3146. doi:10.2174/1381612826666200316143239 PMID:32175835

Kedziora, A., Speruda, M., Krzyzewska, K., Rybka, J., Lukowiak, A., & Bugla-Ploskonska, G. (2018). Similarities and differences between silver ions and silver in nanoforms as antibacterial agents. *International Journal of Molecular Sciences*, *19*(2), 444. doi:10.3390/ijms19020444 PMID:29393866

Khan, N. T., Jameel, M., & Jameel, J. (2017). Silver nanoparticles biosynthesis by *Fusarium oxysporum* and determination of its antimicrobial potency. *Journal of Nanomedicine & Biotherapeutic Discovery*, *7*(01), 1. doi:10.4172/2155-983X.1000145

Khorrami, S., Zarrabi, A., Khaleghi, M., Danaei, M., & Mozafari, M. R. (2018). Selective cytotoxicity of green synthesized silver nanoparticles against the MCF-7 tumor cell line and their enhanced antioxidant and antimicrobial properties. *International Journal of Nanomedicine*, *2018*(13), 8013–8024. doi:10.2147/IJN.S189295 PMID:30568442

Konappa, N., Krishnamurthy, S., Arakere, U. C., Chowdappa, S., Akbarbasha, R., & Ramachandrappab, N. S. (2021). Nanofertilizers and nanopesticides: Recent trends, future prospects in agriculture. Woodhead Publishing.

Krause, B. C., Kriegel, F. L., Rosenkranz, D., Dreiack, N., Tentschert, J., Jungnickel, H., Jalili, P., Fessard, V., Laux, P., & Luch, A. (2020). Aluminum and aluminum oxide nanomaterials uptake after oral exposure - a comparative study. *Scientific Reports*, *10*(1), 2698. doi:10.103841598-020-59710-z PMID:32060369

Kumari, R. M., Kumar, V., Kumar, M., Pareek, N., & Nimesh, S. (2020). Assessment of antibacterial and anticancer capability of silver nanoparticles extracellularly biosynthesized using *Aspergillus terreus*. *Nano Express*, *1*(3), 030011. doi:10.1088/2632-959X/abc2e4

Kumari, V., & Sangal, A. (2020). Antimicrobial study of *Arjuna Terminalia* loaded PLGA nanoparticle. *Indian Journal of Biochemistry & Biophysics*, *58*, 253–260.

Liao, S., Zhang, Y., Pan, X., Zhu, F., Jiang, C., Liu, Q., Cheng, Z., Dai, G., Wu, G., Wang, L., & Chen, L. (2019). Antibacterial activity and mechanism of silver nanoparticles against multidrug-resistant *Pseudomonas aeruginosa*. *International Journal of Nanomedicine*, *2019*, 1469–1487. doi:10.2147/IJN.S191340 PMID:30880959

Liu, H., Wang, X., Wu, Y., Hou, J., Zhang, S., Zhou, N., & Wang, X. (2019). (Toxicity responses of dierent organs of zebrafish (*Daniorerio*) to silver nanoparticles with different particle sizes and surface coatings. *Environmental Pollution*, *246*, 414–422. doi:10.1016/j.envpol.2018.12.034 PMID:30579210

Lohani, A., Verma, A., Joshi, H., Yadav, N., & Karki, N. (2014). Nanotechnology-based cosmeceuticals. *ISRN Dermatology, 2014*, 1–14. doi:10.1155/2014/843687 PMID:24963412

Loiseau, A., Asila, V., Boitel-Aullen, G., Lam, M., Salmainm, M., & Boujday, S. (2019). Silver-based plasmonic nanoparticles for and their use in biosensing. *Biosensors (Basel), 9*(2), 78. doi:10.3390/bios9020078 PMID:31185689

Machmudah, S., Sato, T., Wahyudiono, Sasaki, M., & Goto, M. (2012). Silver nanoparticles generated by pulsed laser ablation in supercritical CO_2 medium. *High Pressure Research, 32*(1), 60–66. doi:10.1080/08957959.2011.649277

Mahjoubian, M., Naeemi, A. S., & Sheykhan, M. (2021). Toxicological effects of Ag_2O and Ag_2CO_3 doped TiO_2 nanoparticles and pure TiO_2 particles on zebrafish (*Danio rerio*). *Chemosphere, 263*, 128182. doi:10.1016/j.chemosphere.2020.128182 PMID:33297149

Maity, P., Bepari, M., Pradhan, A., Baral, R., Roy, S., & Choudhury, S. M. (2018). Synthesis and characterization of biogenic metal NPs and its cytotoxicity and anti-neoplasticity through the induction of oxidative stress, mitochondrial dysfunction and apoptosis. *Colloids and Surfaces. B, Biointerfaces, 161*, 111–120. doi:10.1016/j.colsurfb.2017.10.040 PMID:29055863

Malassis, L., Dreyfus, R., Murphy, R. J., Hough, L. A., Donnio, B., & Murray, C. B. (2016). One-step green synthesis of gold and silver nanoparticles with ascorbic acid and their versatile surface post-functionalization. *RSC Advances, 6*(39), 33092–33100. doi:10.1039/C6RA00194G

Marimuthu, S., Antonisamy, A. J., Malayandi, S., Rajendran, K., Pugazhendhi, A., & Ponnusamy, V. K. (2020). Silver nanoparticles in dye effluent treatment: A review on synthesis, treatment methods, mechanisms, photocatalytic degradation, toxic effects and mitigation of toxicity. *Journal of Photochemistry and Photobiology. B, Biology, 205*, 111823. doi:10.1016/j.jphotobiol.2020.111823 PMID:32120184

Matteis, V. D., Rizzello, L., Ingrosso, C., Liatsi-Douvitsa, E., Giorgi, M. L. D., Matteis, G. D., & Rinaldi, R. (2019). Cultivar-dependent anticancer and antibacterial properties of silver nanoparticles synthesized using leaves of different *Olea Europaea* trees. *Nanomaterials (Basel, Switzerland), 2019*(9), 1544. doi:10.3390/nano9111544 PMID:31671618

Mehwish, H. M., Rajoka, M. S. R., Xiong, Y., Cai, H., Aadil, R. M., Mahmood, Q., He, Z., & Zhu, Q. (2021). Green synthesis of a silver nanoparticle using *Moringa oleifera* seed and its applications for antimicrobial and sun-light mediated photocatalytic water detoxification. *Journal of Environmental Chemical Engineering, 9*(4), 105290. doi:10.1016/j.jece.2021.105290

Mikhailova, E. O. (2020). Silver nanoparticles: Mechanism of action and probable bio-application. *Journal of Functional Biomaterials, 11*(4), 84. doi:10.3390/jfb11040084 PMID:33255874

Moon, J., Kwak, J. I., & An, Y. J. (2019). The effects of silver nanomaterial shape and size on toxicity to *Caenorhabditis elegans* in soil media. *Chemosphere, 215*, 50–56. doi:10.1016/j.chemosphere.2018.09.177 PMID:30312916

Muthu, K., & Priya, S. (2017). Green synthesis, characterization and catalytic activity of silver nanoparticles using *Cassia auriculata* flower extract separated fraction. *Spectrochimica Acta A. Molecular & Biomolecular, 15*(179), 66–72. doi:10.1016/j.saa.2017.02.024

Nakamura, S., Sato, M., Sato, Y., Ando, N., Takayama, T., Fujita, M., & Ishihara, M. (2019). Synthesis and application of silver nanoparticles (Ag NPs) for the prevention of infection in healthcare workers. *International Journal of Molecular Sciences*, *20*(15), 3620. doi:10.3390/ijms20153620 PMID:31344881

Navarro Gallon, S. M., Alpaslan, E., Wang, M., Larese-Casanova, P., Londono, M. E., Atehortua, L., Pavon, J. J., & Webster, T. J. (2019). Characterization and study of the antibacterial mechanisms of silver nanoparticles prepared with microalgal exopolysaccharides. *Materials Science and Engineering C*, *99*, 685–695. doi:10.1016/j.msec.2019.01.134 PMID:30889742

Ong, W. T. J., & Nyam, L. L. (2022). Evaluation of silver nanoparticles in cosmeceutical and potential biosafety complications. *Saudi Journal of Biological Sciences*, *29*(4), 2085–2094. doi:10.1016/j.sjbs.2022.01.035 PMID:35531241

Othman, A. M., Elsayed, M. A., Al-Balakocy, N. G., Hassan, M. M., & Elshafei, A. M. (2019). Biosynthesis and characterization of silver nanoparticles induced by fungal proteins and its application in different biological activities. *Journal of Genetic Engineering and Biotechnology*, *17*(1), 8. doi:10.118643141-019-0008-1 PMID:31673864

Oves, M., Aslam, M., Rauf, M. A., Qayyum, S., Qari, H. A., Khan, M. S., Alam, M. Z., Tabrez, S., Pugazhendhi, A., & Ismail, I. M. (2018). Antimicrobial and anticancer activities of silver NPs synthesized from the root hair extract of *Phoenix dactylifera*. *Materials Science and Engineering C*, *89*, 429–443. doi:10.1016/j.msec.2018.03.035 PMID:29752116

Padinjarathil, H., Joseph, M. M., Unnikrishnan, B. S., Preethi, G. U., Shiji, R., Archana, M. G., Maya, S., Syama, H. P., & Sreelekha, T. T. (2018). Galactomannan endowed biogenic silver NPs exposed enhanced cancer cytotoxicity with excellent biocompatibility. *International Journal of Biological Macromolecules*, *118*, 1174–1182. doi:10.1016/j.ijbiomac.2018.06.194 PMID:30001604

Paladini, F., & Pollini, M. (2019). Antimicrobial silver nanoparticles for wound healing application: Progress and future trends. *Materials (Basel)*, *12*(16), 2540. doi:10.3390/ma12162540 PMID:31404974

Park, S., Sung, H. K., & Kim, Y. (2016). Green synthesis of metal nanoparticles using sprout plants: Pros and cons. *Journal of Nanoscience and Nanotechnology*, *6*(5), 4444–4449. doi:10.1166/jnn.2016.10970 PMID:27483771

Parveen, A., Yalagatti, M. S., Abbaraju, V., & Deshpande, R. (2018). Emphasized mechanistic antimicrobial study of biofunctionalized silver nanoparticles on model *Proteus mirabilis*. *Journal of Drug Delivery*, *2018*, 3850139. doi:10.1155/2018/3850139 PMID:29951316

Patil, S., & Muthusamy, O. (2020). A bio-inspired approach of formulation and evaluation of *Aegle marmelos* fruit extract mediated silver nanoparticle gel and comparison of its antibacterial activity with antiseptic cream. *European Journal of Integrative Medicine*, *33*, 101025. doi:10.1016/j.eujim.2019.101025

Patra, J. K., Das, G., Fraceto, L. F., Campos, E. V. R., Rodrigeuz-Torres, M. P., Acosta-Torres, L. S., Diaz-Torres, L. A., Grillo, R., Swamy, M. K., Sharma, S., Habtemariam, S., & Shin, H.-S. (2018). Nano based drug delivery systems: Recent developments and future prospects. *Journal of Nanobiotechnology*, *16*(1), 71. doi:10.118612951-018-0392-8 PMID:30231877

Prabhu, S., & Poulose, E. K. (2012). Silver nanoparticles: Mechanism of antimicrobial action, synthesis, medical applications, and toxicity effects. *International Nano Letters, 2*(1), 32. doi:10.1186/2228-5326-2-32

Prazak, R., Święciło, A., Krzepiłko, A., Michałek, S., & Arczewska, M. (2020). Impact of Ag nanoparticles on seed germination and seedling growth of green beans in normal and chill temperatures. *Agriculture, 10*(8), 312. doi:10.3390/agriculture10080312

Quintero-Quiroz, C., Acevedo, N., Zapata-Giraldo, J., Botero, L. E., Quintero, J., Zarate-Trivino, D., Saldarriaga, J., & Perez, V. Z. (2019). Optimization of silver nanoparticle synthesis by chemical reduction and evaluation of its antimicrobial and toxic activity. *Biomaterials Research, 23*(1), 27. doi:10.118640824-019-0173-y PMID:31890269

Radzuan, N. A., Sulong, A. B., & Sahari, J. (2017). A review of electrical conductivity models for conductive polymer composite. *International Journal of Hydrogen Energy, 42*(14), 9262–9273. doi:10.1016/j.ijhydene.2016.03.045

Rafique, M., Sadaf, I., Rafique, M. S., & Tahir, M. B. (2017). A review on green synthesis of silver nanoparticles and their applications. *Artificial Cells, Nanomedicine, and Biotechnology, 45*(7), 1272–1291. doi:10.1080/21691401.2016.1241792 PMID:27825269

Rajora, N., Kaushik, S., Jyoti, A., & Kothari, S. L. (2016). Rapid synthesis of silver nanoparticles by *Pseudomonas stutzeri* isolated from textile soil under optimised conditions and evaluation of their antimicrobial and cytotoxicity properties. *IET Nanobiotechnology / IET, 10*(6), 367–373. doi:10.1049/iet-nbt.2015.0107 PMID:27906136

Ramirez-Garcia, G., Gutierrez-Granados, S., Gallegos-Corona, M. A., Palma-Tirado, L., Orlye, F., Varenne, A., Mignet, N., Richard, C., & Martinez-Alfaro, M. (2017). Long-term toxicological effects of persistent luminescence nanoparticles after intravenous injection in mice. *International Journal of Pharmaceutics, 532*(2), 686–695. doi:10.1016/j.ijpharm.2017.07.015 PMID:28705622

Sahoo, M., Vishwakarma, S., Panigrahi, C., & Kumar, J. (2021). Nanotechnology: Current applications and future scope in food. *Food Frontiers, 2*(1), 3–22. doi:10.1002/fft2.58

Saif, S., Tahir, A., & Chen, Y. (2016). Green synthesis of iron nanoparticles and their environmental applications and implications. *Nanomaterials (Basel, Switzerland), 6*(11), 209. doi:10.3390/nano6110209 PMID:28335338

Saod, W.M., Ghafoor, D. A., Najeeb, L. M., Al-Taee, M. J. M. (2020). Study of analytical characterization and antibacterial activity of silver nanoparticles synthesized by using pineapple juice as a reducing agent. *Indian Journal of Forensic Medicine & Toxicology*, 14.

Saravanan, M., Barik, S. K., MubarakAli, D., Prakash, P., & Pugazhendhi, A. (2018). Synthesis of silver nanoparticles from *Bacillus brevis* (NCIM 2533) and their antibacterial activity against pathogenic bacteria. *Microbial Pathogenesis, 116*, 221–226. doi:10.1016/j.micpath.2018.01.038 PMID:29407231

Sayyid, N. H., & Zghair, Z. R. (2021). Biosynthesis of silver nanoparticles produced by *Klebsiella pneumonia. Materials Today: Proceedings, 42*, 2045–2049. doi:10.1016/j.matpr.2020.12.257

Shafique, M., & Luo, X. (2019). Nanotechnology in transportation vehicles: An overview of its applications, environmental, health and safety concerns. *Materials (Basel), 12*(15), 2493. doi:10.3390/ma12152493 PMID:31390752

Shejawal, K. P., Randive, D. S., Bhinge, S. D., Bhutkar, M. A., Todkar, S. S., Mulla, A. S., & Jadhav, N. R. (2021). Green synthesis of silver, iron and gold nanoparticles of lycopene extracted from tomato: Their characterization and cytotoxicity against COLO320DM, HT29 and Hella cell. *Journal of Materials Science. Materials in Medicine, 32*(2), 1–12. doi:10.100710856-021-06489-8 PMID:33576907

Silvan, J. M., Zorraquin-Pena, I., Gonzalez de Llano, D., Moreno-Arribas, M. V., & Martinez-Rodriguez, A. J. (2018). Antibacterial activity of glutathione-stabilized silver nanoparticles against *Campylobacter* multidrug-resistant strains. *Frontiers in Microbiology, 9*, 458. doi:10.3389/fmicb.2018.00458 PMID:29615993

Sim, W., Barnard, R. T., Blaskovich, M. A. T, & Ziora, Z. M. (2018). Antimicrobial silver in medicinal and consumer applications: A patent review of the past decade (2007–2017). *Antibiotics (Basel), 7*(4), 93.

Simbine, E. O., Rodrigues, L. C., Lapa-guimaraes, J., Kamimura, E. S., Corassin, C. H., & Oliveira, C. A. F. (2019). Application of silver nanoparticles in food packages: A review. *Food Science and Technology (Campinas), 39*(4), 793–802. doi:10.1590/fst.36318

Singh, A., Gautam, P. K., Verma, A., Singh, V., Shivapriya, P. M., Shivalkar, S., Sahoo, A. K., & Samanta, S. K. (2020). Green synthesis of metallic nanoparticles as effective alternatives to treat antibiotics resistant bacterial infections: A review. *Biotechnology Reports (Amsterdam, Netherlands), 25*, e00427. doi:10.1016/j.btre.2020.e00427 PMID:32055457

Singh, R., Vora, I., Nadhe, S. B., Wadhwani, S. A., Shedbalkar, U. U., & Chopade, B. A. (2018). Antibacterial activities of bacteriagenic silver nanoparticles against nosocomial *Acinetobacter baumannii. Journal of Nanoscience and Nanotechnology, 18*(6), 3806–3815. doi:10.1166/jnn.2018.15013 PMID:29442713

Singh, S. P., Mishra, A., Shyanti, R. K., Singh, R. P., & Acharya, A. (2021). Silver nanoparticles synthesized using *Carica papaya* leaf extract (AgNPs-PLE) causes cell cycle arrest and apoptosis in human prostate (DU145) cancer cells. *Biological Trace Element Research, 199*(4), 1316–1331. doi:10.100712011-020-02255-z PMID:32557113

Solanki, R. G., & Rajaram, P. (2017). Structural, optical and morphological properties of CdS nanoparticles synthesized using hydrazine hydrate as a complexing agent. *Nano-Structures & Nano-Objects, 12*, 157–165. doi:10.1016/j.nanoso.2017.10.003

Suchomel, P., Kvitek, L., Prucek, R., Panacek, A., Halder, A., Vajda, S., & Zboril, R. (2018). Simple size-controlled synthesis of Au nanoparticles and their size-dependent catalytic activity. *Scientific Reports, 8*(1), 4589. doi:10.103841598-018-22976-5 PMID:29545580

Sulaiman, G. M., Hussien, H. T., & Saleem, M. M. N. M. (2015). Biosynthesis of silver nanoparticles synthesized by *Aspergillus flavus* and their antioxidant, antimicrobial and cytotoxicity properties. *Bulletin of Materials Science, 38*(3), 639–644. doi:10.100712034-015-0905-0

Tailor, G., Yadav, B. L., Chaudhary, J., Joshi, M., & Suvalka, C. (2020). Green synthesis of silver nanoparticles using *Ocimum canum* and their anti-bacterial activity. *Biochemistry and Biophysics Reports, 24*, 100848. doi:10.1016/j.bbrep.2020.100848 PMID:33305022

Tan, P., Li, H. S., Wang, J., & Gopinath, S. C. B. (2020). Silver nanoparticle in biosensor and bioimaging: Clinical perspectives. *Biotechnology and Applied Biochemistry*, 1–7. doi:10.1002/bab.2045 PMID:33043496

Velmurugan, P., Sivakumar, S., Song Young-Chae, S., Seong-Ho, J., Pyoung-In, Y., Jeong-Min, S., & Sung-Chul, H. (2015). Synthesis and characterization comparison of peanut shell extract silver nanoparticles with commercial silver nanoparticles and their antifungal activity. *Journal of Industrial and Engineering Chemistry, 31*, 51–54. doi:10.1016/j.jiec.2015.06.031

Viet, P. V., Sang, T. T., Ngoc Bich, N. H., & Thi, C. M. (2018). An improved green synthesis method and *Escherichia coli* antibacterial activity of silver nanoparticles. *Journal of Photochemistry and Photobiology. B, Biology, 182*, 108–114. doi:10.1016/j.jphotobiol.2018.04.002 PMID:29656219

Vijayaraghavan, K., & Ashokkumar, T. (2017). Plant-mediated biosynthesis of metallic nanoparticles: A review of literature, factors affecting synthesis, characterization techniques and applications. *Journal of Environmental Chemical Engineering, 5*(5), 4866–4883. doi:10.1016/j.jece.2017.09.026

Vu, X. H., Duong, T. T. T., Pham, T. T. H., Trinh, D. K., Nguyen, X. H., & Dang, V.-S. (2018). Synthesis and study of silver nanoparticles for antibacterial activity against *Escherichia coli* and *Staphylococcus aureus*. *Advances in Natural Sciences: Nanoscience and Nanotechnology, 9*(2), 025019. doi:10.1088/2043-6254/aac58f

Xie, Y., Wang, H., & He, Z. (2021). Recent advances in polyphenols improving vascular endothelial dysfunction induced by endogenous toxicity. *Journal of Applied Toxicology, 41*(5), 701–712. doi:10.1002/jat.4123 PMID:33251608

Yadav, T. P., Yadav, R. M., & Singh, D. P. (2012). Mechanical Milling: A top down approach for the synthesis of nanomaterials and nanocomposites. *Nanoscience and Nanotechnology, 2*(3), 22–48. doi:10.5923/j.nn.20120203.01

Yassin, M. A., Elkhooly, T. A., Elsherbiny, S. M., Reicha, F. M., & Shokeir, A. A. (2019). Facile coating of urinary catheter with bio–inspired antibacterial coating. *Heliyon, 5*(12), e02986. doi:10.1016/j.heliyon.2019.e02986 PMID:31886428

Yin, I. X., Zhang, J., Zhao, I. S., Mei, M. L., Li, Q., & Chu, C. H. (2020). The antibacterial mechanism of silver nanoparticles and its application in dentistry. *International Journal of Nanomedicine, 15*, 2555–2562. doi:10.2147/IJN.S246764 PMID:32368040

Yuan, Y. G., Zhang, S., Hwang, J. Y., & Kong, I. K. (2018). Silver nanoparticles potentiates cytotoxicity and apoptotic potential of camptothecin in human cervical cancer cells. *Oxidative Medicine and Cellular Longevity, 2018*, 6121328. doi:10.1155/2018/6121328 PMID:30647812

Zhao, J., Liu, P., Ma, J., Li, D., Yang, H., Chen, W., & Jiang, Y. (2019). Enhancement of radiosensitization by silver nanoparticles functionalized with polyethylene glycol and aptamer As1411 for glioma irradiation therapy. *International Journal of Nanomedicine*, *14*, 9483–9496. doi:10.2147/IJN.S224160 PMID:31819445

Zheng, H., Pan, L., Lv, J., Zhang, Z., Wang, Y., Hu, W., Liu, X., Zhou, P., Wang, Y., & Zhang, Y. (2020). Comparison of immune responses in guinea pigs by intranasal delivery with different nanoparticles-loaded FMDV DNA vaccine. *Microbial Pathogenesis*, *142*, 104061. doi:10.1016/j.micpath.2020.104061 PMID:32061916

Zulfiqar, F., Navarro, M., Ashraf, M., Akram, N. A., & Munne-Bosch, S. (2019). Nanofertilizer use for sustainable agriculture: Advantages and limitations. *Plant Science*, *289*, 110270. doi:10.1016/j.plantsci.2019.110270 PMID:31623775

Chapter 11
Impact of Heavy Metal-Based Nanomaterials on Environment and Health:
Small Things Having Big Impacts!

Somi N. Sharma
https://orcid.org/0000-0002-6557-9661
Biochemistry Department, Biological Sciences Division, ICMR-National Institute of Occupational Health (NIOH), Ahmedabad, India

Uzma Kadri
https://orcid.org/0000-0001-8773-467X
Biochemistry Department, Biological Sciences Division, ICMR-National Institute of Occupational Health (NIOH), Ahmedabad, India

Nibedita Naha
https://orcid.org/0000-
Biochemistry Department, Biological Sciences Division, ICMR-National Institute of Occupational Health (NIOH), Ahmedabad, India

ABSTRACT

Nanotechnology possesses a wide array of materials in the form of nano-tubes, fibres, plates, chips, particles, sensors, and so on; the application of which has dramatically increased over the last few years due to their unique physicochemical characteristics (i.e., catalytic activity as well as optical, electronic, antibacterial, and magnetic properties). However, the exposure to nanoparticles, and consequent threat to human health and the environment are valid concerns because of the magnitude of nanotechnology-based applications, mainly in the pharmaceutical industries during drug designing, targeted drug delivery and disease diagnostics, and manufacturing of commercial products. For this purpose, the majority of nanomaterials are synthesized either from carbon (organic) and/or heavy metals, metal oxides (inorganic). Accordingly, lead, zinc, copper, chromium, iron, cadmium, and their oxides are important inorganic nanomaterials that are discussed in the chapter.

DOI: 10.4018/978-1-6684-5533-3.ch011

INTRODUCTION

In the 20th century, the word "nano" is equivalent to modernization, nanotechnology is the 'apple-of-eye' of every person working in the field of Science and Innovation on the basis of its promising and beneficial effects in every field of human endeavors such as, industry, agriculture, business, medicine and public health. According to the estimation at global level, it has been found that the nanomaterial market will be worth of USD 7.9 billion in 2020 and USD 12.1 billion by 2026, with 9.7% compound annual growth rate (Barhoum et al., 2022). As per the report of Barhoum (2022), the nanomaterial market size substantially expands during the period of 2021-2027 due to its implementation in the industrial sectors as shown in **(Figure 1A and 1B)**.

Nanomaterials are not something we are naive to; from the little activities of the beginning of our day such as, brushing our teeth using toothpaste to the natural phenomenon occurring at the end of our day like watching beautiful sunset are the result of nanomaterials (Hobson et al., 2016). Even in the ancient times, the existence of the nanomaterials in various events such as Lycurgus cup from the British museum, stained glass windows by Romans, swords by Damascans, etc. are recorded (Hochella et al., 2019). With the advancement of science and the possibility of the exploration of benefits of nanomaterials, day-by-day, there is an increase in the consumer and industrial products that ultimately result in a new branch of science, named "Nanotechnology".

Nanotechnology is the manipulation of matter at the atomic level so that the final product exhibits diversified and unique physical, chemical and biological properties, leading to several applications. But, it is well-experienced that the manipulations are not always easily accepted by the nature; henceforth, though positive impacts possess by these "nano" products, one cannot ignore the big negative impact of these products that may have on human and environmental health in order to maintain sustainability.

Further, the susceptibility to nanoparticles (NPs), and resulting risks to human health and environment, are subjects of concern because of the vast applications of nanotechnology in different industries during drug designing, targeted drug delivery and disease diagnostics, as well as wide spectra of manufacturing the commercial products like toothpaste, sunblock creams and skin care, sports supplies, and textile paints, semiconductors, electric devices, beverages, edibles and agricultural items, building materials, etc. For this purpose, a large proportion of nanomaterials are produced either from carbon-based organic and/or heavy metals and metal oxides-based inorganic materials. Lead (Pb), cadmium (Cd), iron (Fe), zinc (Zn), copper (Cu), and their oxides like lead oxide (PbO), zinc oxide (ZnO), iron oxide (Fe_3O_4), and others are important in this respect. Due to their enormous use and toxic effects to not only the manufacturers but also the consumers and the environment, heavy metal-based nanomaterials impact a major concern throughout the world. For example, a profound negative impact at both cellular and tissue levels including ultra-structural changes in primary and secondary organs due to sub-chronic exposure of PbO-NPs through inhalation (Dumkova et al., 2017).

Although the properties can be different on nanoscale level for heavy metal-based nanomaterials due to the changes in magnetic properties, physical and chemical natures, electrical conductivity, etc., people working with nanotechnology in a wide horizon raises global concern because of their biodegradability, bioleaching, environmental exposure, etc., which causes negative impacts on health especially, in front-line workers of different occupations (Kurwadkar et al., 2015a; Colvin, 2003; Balali-Mood et al., 2021). Thus, to answer the above-mentioned concerns, nanotoxicology comes into action, which defines what products can expose threat to environment and public health, and to what extent that ultimately responsible for further burden to our society. As the modernization is essential part of our lives, it is expected that the production of different nanomaterials in various industrial processes are increasing day-by-day.

Therefore, in the above perspective, the current book chapter deals with the hazards cause by heavy metal-based nanomaterials, and their possible health and environmental impacts, risk factors as well as safety measures in the concerned field targeting specific occupational groups. For this purpose, data is searched through PubMed, Medline, Toxline, Research Gate and Google using numerous combinations of terms pertaining to several aspects of heavy metal-based nanomaterials, their uses in different industries and effects on health and environment.

NANOMATERIALS IN INDUSTRY

Nanomaterials have extensive future applications in a majority of modern sectors comprising construction, aviation, textile, pharmacy, water, agri-food, environment, beauty, fuel, computers, etc. Material science has recently experienced a revolution, particularly in the fields of mechanics, optics, electronics, medicine, aerospace, plastics and many more (Subhan et al., 2021). Due to their bigger surface area-to-volume than bulk counterparts, nanoscale substances' enhance the reactivity, better and size-adjustable magnetic, visual, and electrical capabilities (Baig et al., 2021). Moreover, nanomaterials are frequently employed in a variety of industries throughout the world not only to create new products but also improve the product standard considering the high demands of the consumers. Numerous applications of the nanomaterials in various industrial areas are observed worldwide (Subhan et al., 2021).

Types of Nanomaterials

Nanomaterial is a structure with a size <100 nm, or between 1 nm to 100 nm. The Greek word "nanos," which means "dwarf," is the source of the prefix "nano." Nanotechnology exploits a material's key features at a level where the material's properties are surprisingly different. Nanomaterials provide interesting opportunities to control material's distinctive properties and utilize their capacity for numerous applications, which would be tough to achieve with their macrocounterparts. Henceforth, nanomaterials are divided on the basis of their origin, size, shape and content, etc. as discussed below. The following **Figure 1C** depicts the various types of existing nanomaterials: -

Content-Based Classification

This segment discusses the dimension- and material-based classification:

Carbon-Based Nanomaterials containing carbon, and are found in hollow tubes, ellipsoids, or spheres like structures. Carbon-containing nanomaterials include fullerenes (C60), carbon nanotubes (CNTs), carbon nanofibers, graphene (Gr), carbon black, and carbon onions. CNTs possess cylindrical nanostructure and are allotropes of carbon. CNTs are built with length-to-diameter ratio of upto 2,80, and 00,000:1 that is found to be significantly larger. CNTs possess extraordinary strength and unique electrical properties, which make them substantially useful in various applications in electronics, nanotechnology, optics, materials science, as well as in architectural fields. However, the final usage of carbon-based nanomaterials is limited by their potential toxicity as reported by Majhi and Yadav (2021).

Inorganic Compounds-Based Nanomaterials mainly comprise metals and their oxide-based NPs. These nanomaterials can be produced as metal nanoparticles (silver (Ag) and gold (Au), metal oxide nanoparticles (ZnO-NPs, titanium oxide (TiO_2), semiconductor nanoparticles (Silicon (Si), and ceramic NPs (Chavali and Nikolova, 2019).

Metal-Based Nanomaterials are utilized as thermoelectric, semiconductors and electrical luminescent materials. In addition, these NPs are utilized in drug delivery as "antibacterial" function to the reach areas that are recently difficult to access by regular medicines in the field of biomedicines. However, as Pumera (2010) showed the hazardous effects of the various metal and metal-oxide NPs on cells like, oxidation and DNA fragmentation, mutation, change of characteristics, decrease viability, enhance apoptosis as well as necrosis and reduce proliferation; hence, whether the original features of these metals and metal oxide-based NPs cause negative impact on the environment and health, needs to be studied in-depth.

Metal-Oxide Nanomaterials are widely used as catalysts in industrial sectors. However, according to several animal model studies, TiO_2 NPs may interfere with the insulin response in Fao cells and cause problem in pregnancy (Akcan et al., 2020). Other metal-oxide nanoparticles have been shown to negatively impact on reproductive and newborn development.

Organic-Based Nanomaterials include nanomaterials mainly synthesize from organic components except carbon-based, or inorganic-based nanomaterials. The involvement of weak non-covalent exchanges for the self-organization and molecular assembly aids to convert the organic nanomaterials into desired structures such as dendrimers, liposomes, and polymer NPs (Patra et al., 2018).

Composite-Based Nanomaterials are polyphase NPs with nanoscale at one phase can combine with either NPs, or bulk-type materials, or more complex structures such as, a organo-metallic framework (Jeevanandam et al., 2018). The composite-based nanomaterials may combine with metal-based, carbon-based, or organic-based nanomaterials through various types of metals, or polymer bulk materials (Majhi and Yadav, 2021).

ORIGIN-BASED CLASSIFICATION

Apart from dimension and material-based classifications, based on their origin, the classification of nanomaterials may be natural, or synthetic: -

Natural Nanomaterials are produced in nature mostly by two sources, i.e., natural species, or through climatic activities. Now-a-days, natural resources may easily be utilized today to produce artificial surfaces having unique micro and nanoscale arrangement, and features for technological uses. Nanomaterials are found naturally all over the spheres of the earth such as, hydrosphere (oceans, rivers, lakes, ground waters, hydrothermal vents), atmosphere (whole troposphere), lithosphere (rock, soil, lava, magna) and even in the biosphere (micro- and higher- organisms including human) regardless of human activities (Jeevanandam et al., 2018).

Synthetic Nanomaterials are produced mechanically such as, through grinding, exhaust gases, and smoke, or by physicochemical, biological, or hybrid techniques (Patra and Baek, 2014; Al-Hakkani et al., 2020). The topic of risk assessment approaches has recently come to light due to the rising development and subsequent dissemination of engineered nanomaterials, as well as their use in several consumer items and industrial applications (Barhoum et al., 2022). Overall, NPs are mainly classified as incidental and engineered. Incidental NPs are the byproducts of man-made activities and generally have awfully controlled sizes and shapes, and may be a blend of several elements, whereas engineered (man-made)

NPs with multimeric-organic and inorganic structures are specifically moulded and precisely synthesized to meet the requirements of the consumers and the markets (Singh, 2017).

Figure 1. Nanomaterials and nanotechnology: Demands vs. exposures in every sector of daily lives

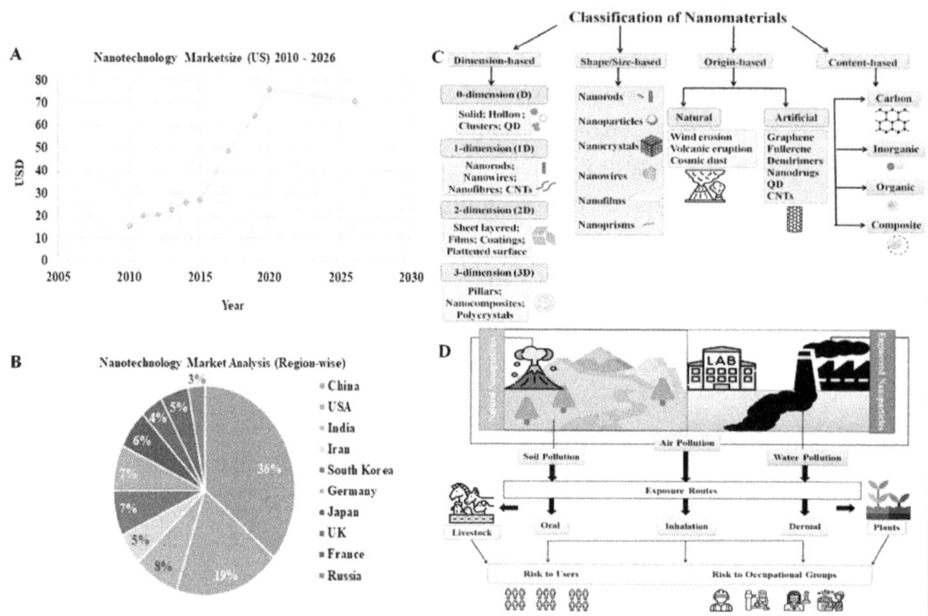

TYPES OF INDUSTRIES INVOLVED: PRODUCTS AND APPLICATIONS OF NANOMATERIALS

By creating high-volume, commercial nanomaterials such as, fullerenes, quantum dots (QDs), CNTs, and metal-oxide NPs, nanotechnology has grown into a multibillion-dollar business around the world. Moreover, it continues to play a critical role in a wide range of human activities in almost every sector of our daily lives like health, food and nutrition, water treatment, manufacturing and engineering, and many more as shown above (**Figure 1D**). However, at the same time, these nanoproducts are foreign to environment and have certain unfavorable effects causing particle toxicity upon release to environment, or through the contact with nanobased products depending on the parameters of NPs; for example, size, shape, absorption, surface area, physical and chemical properties, etc. (Contado, 2015). Further, as this field is highly multidisciplinary and extend application in the field of power and energy, electrical appliances, skin care, textiles, pharmacy, agriculture, food and environment, the industrial uses of nanomaterials are increasing day-by-day inspite of their toxicological impacts (Mody and Choi, 2013; Jane Cypriyana et al., 2021).

Currently, 5224 nanomaterial-based products are found in the database (The Nanodatabase: https://nanodb.dk/) for commercial usage such as, fillers, catalysts, opacifying agents, water filtration, microelectronics, semiconductors, cosmetics, etc., leading to intended and unintended exposures to environment and human, which are briefly described below.

Nanomaterial's Chemistry

Nanotechnology plays a pivoting role during chemical catalysis and filtration techniques. The synthesis imparts innovative materials with customized characteristics and chemical properties; like NPs with a different chemical atmosphere (ligand), or particular optical features. NPs, in particular, due to exceptionally high surface-to-volume ratio, have a wide range of applications in catalysis especially in the chemical industry, from fuel cells to catalytic converters to photocatalytic devices. Recently, due to high surface area, platinum NPs are evaluated for use in the next generation of vehicle catalytic converters, which could minimize the requirement of elemental platinum (Navarro-Espinoza et al., 2021).

Agriculture Sector

Nanotechnology in the form of rising technology, has the capability to revolutionize agriculture and food industry. With more than 60% of the population depending on agriculture for their livelihood (Brock et al., 2011), most developing countries have developed better systems for monitoring environmental conditions, supplying fertilisers, or pesticides as necessary (Mittal et al., 2020). Nanotechnology can also help us to understand the biology of different crops, which could increase yields, or improve nutrition. It can also open doors to higher-value crops, or environmental cleaning (Kumar and Arora, 2020). With the advancement of nanotechnology, the integrated sensing, monitoring, and regulating system find the signs of illness of the crops even before actual occurrence, alert the farmers, and turn on bioactive systems including medications, insecticides, nutrition, probiotics, nutraceuticals, and implantable cell bioreactors (Ranjha et al., 2022). In order to achieve the highest possible biological efficacy without risk of overdosing, nano-fertilizers and nano-pesticides that target nano-formulations of conventional active ingredients, or inorganic nanomaterials allow a manageable release of agrochemicals (Omara et al., 2019). Further, nanoscale sensors and nano-remediation techniques detect and destroy ecological contaminants in large-scale use of the agricultural sector (Baruah and Dutta, 2009).

Food and Bioprocess Engineering Industries

In the food and bioprocess engineering industries, nanotechnology encounters a variety of engineering and scientific challenges such as, creating high-quality, safe food in an effective and sustainable way. Innovative nanotechnology applications in the food industry include biosensors for germ detection and food quality monitoring, active, intelligent and smart systems for food packaging, safety management, and nano-encapsulation of bioactive food ingredients (Neethirajan and Jayas, 2011). Using a nanocomposite coating technique, antimicrobial compounds are applied directly to the surface of the coated film's, which enhancing safer food packaging (Chawla et al., 2021). Nanocomposites could also alter the gas permeability of various fillers to suit the needs of varied commodities (Vasile, 2018). Additionally, they improve the mechanical and thermal resistance of the packaging material of food products (Honarvar et al., 2016).

Medicines and Diagnostics Sector

Apart from the usage of nanomaterials in consumer products, biomedical applications including the role as medication delivery agents, biosensors, and imaging contrast agents, etc. are notable that uses the distinctive features of nanomaterials in the biological and medical fields; for example, the contrast agents for cell imaging and cancer treatment. According to experts, nanopores, QD and other devices for therapeutic detection as well as diagnostics, etc. could be accessible for medical usage in 5-15 years (Zrazhevskiy et al., 2010). NPs are used in site-specific delivery of medications to specific cells; dendrimers and nanoporous materials (Hughes, 2005; Wanunu et al., 2010), block co-polymers in encapsulation to produce micelles, etc. (Cabral et al., 2018). The best merit of such kind of commercial applications to carry little medicine molecules, ensuring optimum success in treatment procedures; for example, iron NPs, or gold shells in cancer treatment (Wu et al., 2013).

Overall, targeted, or customized therapy reduces drug consumption and treatment expenses, providing a social benefit by lowering the costs of the public health system. Due to their first-order kinetics, implanted delivery devices are typically preferred over injectable medications. Nanotechnology is expanding the possibilities in implanted delivery devices. In other words, blood concentration rises quickly but falls exponentially with time (Duggal, 2011). However, this quick increase could lead to NP toxicity, and medication efficacy might decline if drug concentrations drop below the target range (Mitchell et al., 2021).

Filtration Technologies

Due to the scarcity of traditional energy sources, photochemistry is expected to have a significant impact on air purification, waste-water treatment and energy accumulating devices in the near future. With the development of nanotechnology, effective filtration procedures can be created via mechanical, or chemical approaches. The principle behind one kind of filtration method is the employment of membranes with the correct pore sizes through which the liquid is forced. NTs- based nanoporous membranes are suitable for mechanical filtration, or nanofiltration with pores as tiny as 10 nm (Wang et al., 2022). Nanofiltration is primarily used to remove ions, or separate different fluids. Ultrafiltration in kidney dialysis is an important application of nanotechnology (Ma et al., 2020). Additionally, employing magnetic segregation methods, magnetic NPs offer a reliable and efficient way to remove heavy metal pollutants from waste water (Liosis et al., 2021).

Cosmetic Manufacturing Industry

Nanosized ingredients are used by cosmetic manufacturers to improve protection from harmful UV rays, skin absorbance and penetration, colour, aroma release, finish quality, age prevention effect, and a variety of other properties. Mostly inorganic nanomaterials such as, TiO_2-, ZnO-, Au- and Ag-NPs, carbon black, fullerenes, dendrimers, etc. are using in the manufacturing of sunscreens, lipstick, mascara, facemasks, anti-ageing creams and tooth-paste (Santos et al., 2019).

Textile Industry

NPs can offer excellent durability and better affinity for processed fabrics because to their huge surface area and high surface energy toward the intended textile function. Nanotechnology enables textiles to become multipurpose, versatile and produce fabrics with unique properties such as, antibacterial, UV protection, water and stain resistant, easy cleaning, and odour-free (Patra and Gouda, 2013). The use of engineered nanofibers makes clothing water- and stain-repellent as well as wrinkle-free; thus, reducing the need for frequent washing at lower temperatures. Nanotechnology, which incorporates membrane made of carbon NPs, provides the consumer with whole-surface sheilding from electrostatic charges. Now-a-days, 'smart textile' is a new emerging technology in the textile industry, which can be changed as per the surrounding atmosphere (Shah et al., 2022). Smart textiles save us from wearing sweaters by keeping warm and can even convert the body heat to energy that can be stored in the body during winter season. This fascinating technology is called "energy harvesting" (Dolez, 2021), yet another big surprise by the small things that matters so much! Coating processes, which have specific uses in the textile sector, produce different organic and inorganic compounds as particles on the scale of nano-size. Examples are polyacetylene-, polypyrrole-, polyaniline-, Au- (Velmurugan et al., 2016), Ag- (Vigneshwaran et al., 2007), Palladium (Pd)- (Yaqoob et al., 2020), Cu- (Sun et al., 2011), Silica (Si)-, Copper oxide (CuO)-, ZnO- (Tania and Ali, 2021), TiO^2- (Tudu et al., 2020), and Gr-oxide-NPs (Bhattacharjee et al., 2020). Furthermore, nanomaterial-modified textiles have noteworthy applications in wound healing, air refining, drug delivery, skin care, renewable energy generation, and electronic applications such as, the fabrication and characterization of on-body diodes, circuitry and transistors (Shah et al., 2022).

Optics

The study of light behaviour at the nanoscale scale, as well as how nanometer-scale objects interact with light, is known as nanophotonics. Laser and visible light are only capable of passing through submicron structures like nanopores, nanoneedles and nanoslits (Robinson and Nakkeeran, 2013). Photonic crystals and QD are two important examples of nano-optics. Photonic crystals have periodic distinction in refractive index and a lattice parameter half the light wavelength (Chen et al., 2011). They feature a semiconductor for light, or photons rather than electrons, and offer a preferred band gap for the wavelength propagation (Nihei et al., 2009). QDs are the nanometer-scale substances that are primarily used in the fabrication of lasers as well as other applications such as, nanoscale sensors, tetrahertz (THz) imaging technology, Time-of-flight (TOF) camera applications and medical imaging for diagnosis (Liu et al., 2021).

Computers and IT

Nanotechnology has revolutionized the application of computer from the use of CNTs as semiconductor to transistor in computer chips and development of qubits-based quantum computers, computational nanotechnology, DNA computation, etc., where researchers design single DNA strands as representation of bits of binary data (Passian and Imam, 2019; Zhang et al., 2022). Moreover, the development of next generation nanomachines in the form of nanobots are so high-end that can be able to sense and adapt to external stimuli such as, heat, light, sounds, surface textures and substances; to move, communicate and collaborate; to conduct molecular assembly; and to some extent repair, or even replicate themselves (Zhou et al., 2021).

The following table briefly represents different nanomaterials-based products and their uses in various industrial and commercial sectors (**Table 1**). "

Table 1. Products based on nanomaterials in various industries

Industry	Nanomaterials	Products/Uses	References
Construction Industry	TiO_2-NPs; CNT; MgO-NPs; Nano-smectite-based clay	Increase photocatalytic properties of cement sample; Reinforce cement composite	Papadaki et al., 2018
Textile Industry	TiO_2-NPs; ZnO-NPs; CNT-reinforced polymer composite fiber	UV protection; Wrinkle-free; Improve strength, durability; Weight reduction	Shah et al., 2022; Dolez, 2021
Water Treatment Industry	CNTs	Biological treatment of industrial wastewater	Schirmer & Auffan, 2015; Abbas et al., 2020
Agriculture Industry	CNTs; Nano-oxides of Cu, Ag, Mn, Mo, Zn, Fe, Si & Ti	Develop nanopesticides & nanofertilizer; Nanoformulations	Mittal et al., 2020; Kumar & Arora, 2020
Food Industry	ZnO-NPs with polystyrene film; Ag-NPs; Nanocomposites	Food packaging; Antibacterial agents for food safety	Neethirajan & Jayas, 2011; Honarvar et al., 2016
Aeronautics Industry	Fe_3O_4 nano-powders	Manufacture multifunctional structural composites	Viscardi et al., 2018
Environmental Industry	TiO_2-NPs; ZnO-NPs	Degradation of organic pollutants in water & air	Liosis et al., 2021; Ma et al., 2020
Cosmetics Industry	TiO_2-NPs; ZnO-NPs; Polycaprolactone (PCL) nanocapsules; Ag-NPs	Sunscreen; Non-prescription drug products; Encapsulation of elements in cosmetic items; Preservative agent	Santos et al., 2019
Computer Industry	NTs/CNTs	Manufacture smaller transistors, or electronic devices; DNA computation	Passian & Imam, 2019
Oil & Gas Industry	Air-suspended Ni-Fe-NPs	Inject in hydrate formation for deep penetration into hydrate reservoir by passing through the cavities	Jeyaraj et al., 2019
Medicinal Industry	Nanodrugs QD; Dendrimers	Drug delivery agent for site-specific target(s)	Wanunu et al., 2010; Duggal, 2011

NANOMATERIALS: INDUSTRY TO ENVIRONMENT

Environment surrounds us, which is harshly influenced by man-made activities and ultimately affects not only one species but also several residing in the same environment. Environmental hazards of nanotechnology are typically distinguished based on their effect, which finally causes risk of human and animal health. Thus, the sustainable growth of nanotechnology for a better structure should comprehend the influence of these nanosized materials on health aspects (Falsini et al., 2018). If the coin is flipped, along with the various benefits of NPs, it causes adverse outcomes and pollution in the environment due to inadequate handling and disposal of nanomaterials and nanomaterial-based products (Taghavi et al., 2013). Prompt advances in nanotechnology have therefore raised questions about the occurrence, distribution, fate and movement of nanomaterials in the environment in recent years; hence, suggesting proper treatment of industrial nanowastes before disposal such as, chemical, thermal, physical, or biochemical processing with an aim to deactivate, or lowering the hazardous effects of nanowaste (Abd Elkodous et al., 2021). Currently, three ways are developed for disposal of nanowastes, i.e., landfill, recycling (Zahra et al., 2022) and heating through thermal treatment. The conditions to which product matrices incorporating nanomaterials are exposed vary slightly since the recycling process depends on the type of material. Furthermore, recycling may have three negative effects: (1) health effects due to occupational exposure from recycling processes; (2) environmental influence from the treatment of recycling remanats that ends up in burning, landfill, or sewage management, and (3) the entry of residing nanomaterials into reprocessed items (Gupta and Xie, 2018).

As nanomaterials shift from research laboratories to our everyday lives as machineries of several consumer products as mentioned above, and to various industrial procedures such as, remediation items, or health diagnostics, catalysts; thus from slight, immediate effects to long-term exposure effects, one should be extremely concerned about their toxicity and impact on organisms (Pumera, 2010). The overall production of metal/metal oxide-NPs and CNTs is expected to exceed 1.0 million tons per year in no time, leading to increase in the exposure of population to nanomaterials like never before (Perez-Hernandez et al., 2021).

Furthermore, NPs enter the environment through a variety of channels, including deliberate release to clean up the environment, inadvertent release, and improper handling, or disposal of consumer goods (Hochella et al., 2019), where the routes of exposure of human and animals are soils, plants, water and atmosphere that uptake NPs through various pathways (Smita et al., 2012). The transport of NPs in the above mentioned processes is determined by particle characteristics such as, charge, size, diffusion, agglomeration, deposition, solubility and gravitational attraction (Akter et al., 2018), particle-wise varies significantly. NPs have been shown to biodegrade and be bioavailable once released into the environment. The potential for bioaccumulation, deposition into the environment and exposure are determined based on the characteristics of the particular particle (Bundschuh et al., 2018).

Exposure to Air

Nanomaterials especially NPs are significant determinants of air pollution from both natural and anthropogenic sources. Because of their ultrasmall sizes, they can remain suspended in the atmosphere for longer duration and enable them to travel greater distances, posing a number of health risks upon exposure. Unintentional sources such as, incomplete combustion (automobiles and other industries), biomass fire and non-biodegradable waste incineration exacerbate the situation (Bundschuh et al., 2018). Intentional contributors include the widespread use of pesticides and chemical fertilisers, which generate a large amount of NPs (Sonwani et al., 2021). Surface mining and mine shaft excavation can also directly produce NPs; indirect production through decanting, sedimenting, filtering and floating processes (Strambeanu et al., 2015). Further, inhalation of metal fumes especially, Cu and Zn as well as dust from the industrial exposure of Nickel (Ni), Chromium (Cr) and Cobalt (Co) seeds pulmonary fibrosis and, in acute cases, leading to cancer, attributing to 15% of occupational hazards due to exposure of elements in the air (Buzea et al., 2007).

Exposure to Soil

Soil is distinguished by the existence of terrestrial ecosystems, the most important element contributing to biodiversity (Thakur et. al., 2020). The introduction of nanomaterials to soil is of highly vital in nature due to their submicron and nanoscale dimension (Khan et al., 2019b). Once NPs reach the soil, they undergo biotransformation through a step-wise procedure that defines their value of bioavailability and toxicity, leading to several hazards in the environment, ecosystem including plants, animals and human via transferring through food chain (Maharramov et al., 2019). The mechanisms of NPs into soil define their entry, acquisition and movement in the surrounding soil; for example, NPs that can access soil via atmospheric precipitation, dust and aerosols provides a way for their sedimetation, as well as direct assimilation in the form of gaseous compounds, leaf abscission, or anthropogenic activity (Khan et al., 2019b; Rajput et al., 2020). Generally, anthropogenic sources of releasing NPs are categorized to point (factories, refineries, sewage treatment plants, etc.) and non-point (fertilizers, herbicides, oil and grease from industries) bases (Hochella et al., 2019). Through direct contact with the environment, or through NP degradation, free NPs may disperse into the ecosystem. The first phase in the bioaccumulation of NPs is believed to be their chemisorption by plant roots from the soil (Nair et al., 2010). For example, CuO-NPs are adsorbed on root surface, partially which cannot be desorbed further during competition for ions (Lespes et al., 2020). Overall, accumulation in plants via adsorption through roots and shoots, most of the heavy metal-based NPs are distributed in different plant tissues by dissolution of crystal phase, biotransformation and bioaccumulation (Hannah and Thompson, 2008).

Exposure to Water

NPs in natural waterbodies from various industrial sources, including wastewater treatment plants causes enormous health hazards to environment in terms of the extent and the form of nanopollution, which need to be assessed in different real-life conditions. The most critical components influencing the fate of NPs in aquatic systems are aggregation and agglomeration, dissolution, transformation and redox reactions of larger particles into new solid phases, which leads to sedimentation (Hedberg et al., 2019). NPs are moderately bound by agglomeration, whereas strong chemical bonds between particles are formed upon aggregation (Drobne, 2007). Bioavailability of NPs and their dissolution behaviour rises remarkably for significantly bigger agglomerated particles, which is derived by the medium composition, primarily ionic strength, pH and natural organic matter concentration (Abbas et al., 2020).

NANOPARTICLES: ENVIRONMENT TO HUMAN HEALTH

The entry of NPs to human health system at several points from environment are multifaced, i.e., point of origin (synthesis of nanomaterials), point of uses, point of disposal as well as point of re- entry of NPs in case of improper disposal. From environment, mainly through inhalation of the airborne NPs and through dermal exposure in case of spillage, or direct contact, the particular occupational groups are exposed (Oberbek et al., 2019). However, research data is still lacking on the existing levels of exposure to the workers specifically to engineered heavy metal-based NPs in the developing and underdeveloped countries though some data are available regarding the exposure of artificial NPs to the general population (Lespes et al., 2020). Also, it is difficult to determine the precise amounts of NPs exposure; researcher's claims that measurable quantities exist in the environment (Jane Cypriyana et al., 2021).

ROUTES OF EXPOSURE

Inhalation

After inhalation, NPs deposit throughout the upper and lower respiratory tracts (Qiao et al., 2015), leading to adverse effect on the immune system due to modulatory role of NPs on intrinsic defensive function of macrophages during their reactivity to infections (Liu et al., 2017). Inhalation of NPs increases the expression of a few virus-related receptors, making macrophages hyper-reactive to pathological infections, causing extreme inflammation (for example, ZnO-, Cu-· TiO_2-NPs); on the other hand, exposure to SiO_2- and TiO_2-NPs (Sonwani et al., 2021) reduces the expression of certain receptors of virus and bacteria such as, Toll-like receptor 9 (TRP9) leading to lowering of resistance (Ferdous and Nemmar, 2020).

Oral

Ingestion is another route of entry of various NPs into the body, though majority of the NP toxicity studies have focused primarily on respiratory tract exposures, with limited studies emphasizing the gastrointestinal tract (GI) toxicity (Peixe at al., 2015), which is typically caused by unintentional transfer from hand to mouth. Furthermore, it could also occurs during handling of NPs containing materials in workplaces as well as in case of oral nano-medicines. After oral exposure, NPs are distributed to kidney, liver, spleen, lungs, brain via crossing blood-brain- barrier (B-B-B) and GI tract (Bergin and Witzmann, 2013)· Some NPs like TiO_2- and SiO_2-NPs can enter orally, and are quickly removed in faeces and urine, signifying their permeability through GI walls and in to the circulation (Brandelli, 2020). On the contrary, NPs such as, Cu-, ZnO-, Ag-NPs accumulate in liver during first-pass metabolism where usually drugs metabolism occurs, which reduces its adversity as compared to the original NPs (Izci et al., 2021).

Dermal

With the advent of introduction of nanotechnology in cosmetics, skin is the primary target of NPs (Yildirimer et al., 2011). Apart from dermal interface with the environment, skin also protects the body against external interferences and acts as an insulator. However, alteration of such barrier during dermatitis, lesion, or scrapes might acts as easy exposure route to enter the body for NPs, which in most of the cases are generally overlooked. Very recently people are aware about the dermal exposure of nanomaterials, especially during occupational/accidental exposure of TiO_2-NPs into cosmetic and drug products, and ZnO-NPs in sunscreen (Musial et al., 2020)· Ultimately, nanomaterials are more likely to penetrate (absorb) through the epidermal layer of skin than macro-sized particles (Bahadar et al., 2016). NPs can also penetrate to dermis and even stratum corneum by passing through, or amidst epithelial cells, or via adnexa of skin, the best example of which is cosmetics containing NPs reported by both *in vitro* and *in vivo* experiments (Bahadar et al., 2016).

EFFECTS OF NANOTOXICOLOGY: OCCUPATIONAL STATISTICS IN INDIA

India has a lot of promise with nanotechnology, and a multi-pronged approach ensures its realization. Long-term financing that accommodate coherent research programmers with high- impact outcomes, may be applicable in this regard (Singh, 2017). Various research centers across the country are recently collaborating to provide yield with superior results through streamlined research and large number of skilled manpower in the field of nanotechnology (Ghobadi et al., 2020); Peixe et al., 2015); hence, the frequency of occupational hazards in this field should be taken into account during risk factor assessment. Moreover, in-depth study involving both cross-sectional and longitudinal aspects through proper surveillance, are the current medical needs of the country (Piexe et al., 2015).

HEAVY METAL-BASED NANOMATERIALS

Heavy metals are defined as elements with an atomic number >20 and a relatively high, atomic specific gravity of > 5 g/cm 3 with metal-like characteristics (Fergusson, 1990). Heavy metals such as, Pb, mercury (Hg), Cd and arsenic (As) have no biological role and accumulate in various organs over time; thus, exert cumulative toxic effect on body even at ppb levels due to their non-biodegradable nature. The widespread uses of these heavy metals in various anthropogenic activities result in extensive environmental contamination, human exposure and other terrestrial animal exposure that ultimately raise significant public health concern with huge global burden of disease (WHO, 2011; Ericson et al., 2016). Apart from nanomaterials synthesized from organic (carbon-based) materials, there is a vast array of products that involve the use of the above mentioned heavy metals and their oxides using nanotechnology, which are typically encase in an organic, or inorganic substance, or metal oxide shell (Liu et al., 2016). These particles are commonly regarded as QD, nanogold, nanosilver and oxides with metal bases, which are briefly described below.

Involvement of Industries

The application of heavy metal-based nanomaterials has dramatically increased over the last few years due to unique physicochemical characteristics as discussed above, especially in biomedical sciences and pharmaceutical industries due to their unique chemical binding and conjugated properties that enable them to be altered with various chemistry of the functional groups to ensure further conjugation with immune bodies, drugs of interest and ligands (Khan et al., 2019a; Del Solar and Contel, 2019; Attarilar et al., 2020). This opens an extensive spectrum of potential applications in drug designing and targeted drug delivery, magnetic separation, preconcentration of target analytes (Egbuna et al.,2021; Mody et al., 2010), gene vehicles as well as diagnostic imaging like Surface-enhanced Raman Spectroscopy (SERS), Magnetic Resonance Imaging (MRI), Computed Tomography (CT), Positron Emission Tomography (PET), ultrasound, and many more (Mody et al., 2010; Yu et al., 2018), Nanomaterials also find their application in wide spectrum of profitable routine products' manufacturing for instance, sunscreen gels and cosmetics, toothpaste, textiles paints, sports equipment, semiconductors, electronic products, construction materials, agricultural and food products, etc. (Kurwadkar et. al., 2015b; Srivastava et al., 2015). However, as the enormous use of heavy metal-based nanomaterials is in almost every aspect of modern life; hence, there is an urgent need of evaluating the threat to human health and environment by exposure and effect's analysis as well as the trends and magnitude of the problem in the respective field.

Commercial Use, Exposures and Effects of Nanotoxicity

Though it is well established that heavy metals like Pb, Zn, Cu, Cd, Fe, As, etc. are well known toxicants, there is huge negligence regarding their nanoparticle toxicity due to enormous industrial application based on their commercial aspects and growing market demands. In this book chapter, only six heavy metal-based NPs and their oxides are discussed considering "no safe level" in human body. However, other metal oxides-NPs like TiO_2, silicon dioxide (SiO_2), cerium dioxide (CeO_2), aluminum oxide (Al_2O_3), cuprous oxide (Cu_2O), CuO, nickel oxide (NiO), magnesium oxide (MgO), indium oxide (In_2O_3), zirconium dioxide (ZrO_2), lanthanum oxide (La_2O_3), etc. are also used (Rajput et al., 2018) that may be later discussed somewhere.

The following table (**Table 2**) summarizes the applications and unique properties of heavy metals and their oxide-based NPs as discuss below.

Table 2. Heavy metals & their oxides: Physicochemical properties, exposures and commercial applications

Heavy Metals & Oxides	Physicochemical Properties	Exposure & Sources	Commercial Applications	References
Pb Pb_2O_3, Pb_3O_4, Pb_9O_{12}, PbS	Magnetic resonance; Electronic, mechanical & optical properties	Mining; Smelting; Burning fossil fuels; Manufacturing/recycling of batteries, e-waste; Contaminated drinking water; dust; Pipes, paint & pigments; Ammunition; Solder; Stained glass & glassware; Toys; Cosmetics; Glazed, or -soldered food container; Ceramics; Pottery; Spices	Luminescent glassy materials; Paints; Pigments; Gas sensors; Storage batteries; Optics; Nanoscale electronic devices; Stock piling gadgets; Modifiers; Harmful protectant in oxide glasses	Mathew & Krishnamurthy, 2018 Nafees et al., 2017 Karami et al., 2008 Lafront et al., 2010 Martos et al., 2001
As As_2O_3	Physical stability; Favorable Size; Moderate release rate; Cellular internalisation	Ground water contamination due to tectonic disturbances; Contaminated in water in food preparation & irrigation; Industrial processes; Tobacco Smoking	Alloying agents in transistors, lasers & semiconductors; Cancer treatment	Madhavi et al., 2022 Song et al., 2018
Cd CdS	Thermal constancy repellent; Large bandgap-energy; Metal Photochemical properties; Solar panels optical properties; Unique Melting point; Crystal configuration; Optoelectronic absorption spectra; High stability	Dirt; Cigarette smoke; Cadmium-contaminated food; Smelting; & Catalytic refining; Battery manufacturing; High fluorescence & plastics; Coatings; Solar panels	Encapsulated coatings of semiconductor QDs; Pigments; Paints; Engineered plastic materials; Optoelectronics (photocells, LED); Photonics; Photodetectors; Optical filters; Optical switches; Photovoltaics & Photo catalysis; Diagnostics & Imaging (cancer); Soft tissue drug delivery (photodynamic cancer therapy); Air-water purifier; H_2 production	Gupta, 2019

Continued on following page

Table 2. Continued

Heavy Metals & Oxides	Physicochemical Properties	Exposure & Sources	Commercial Applications	References
Cu CuO, CuS	High electrical conductivity; Stability; Spin dynamics; Electron correlation; High-temperature superconductivity	Contaminated air, food, drinking water & soil (skin contact); Asphalt making; Electroplating; Wallpaper; Embosses; Solders; Mines; Smelters (ore)	Plant growth promoter; Pesticides; Fungicides; Antimicrobial reagents; Fertilizers; Integrated coatings; Plastics; Textiles; Antimicrobial, antioxidant, Antidiabetic & anti-inflammatory agents; Biosensors; Electrochemical sensors; Lubricants; Nanocomposite coating; Inks; Filler; Heat transfer fluids; E-sensitized solar cells; Lithium ion batteries; Gas sensing; Heterogeneous catalysis; Anticancer agent; Bioimaging	Aruoja et al., 2009 Siddiqi and Husen, 2020 Perlman et al., 2015 El-Trass et al., 2012 Attarilar et al., 2020
Zn ZnO	UV emission & absorption capacity; Conductivity; Piezoelectricity	Smelter; Welding; Fume Generation during thermal/ chemical processes; Contaminated water & food	Paints; Wave filters; Sensors; UV detectors; UV-protective materials (textiles); Gas sensors; Biosensors; Electronic sensors; Personal care products (sunscreens); Enzymes; Medical & healthcare products; Food products (additives, supplements); Containers; Packaging; Fuels; Photographic catalysts; Electronics; Semiconductors; Air filter; Solar photovoltaics; Transducers; Pharmaceutical; Biomedical Engineering	Schulte et al., 2018 Schmalz et al., 2017 Kolodziejczak- Radzimska & Jesionowski, 2014 Ma et al., 2013 Hoffmann et al., 1995 Attarilar et al., 2020
Fe FeO, Fe_2O_3, Fe_3O_4	Magnetic properties; Ultrafine size; Biocompatibility; SPION	Fumes; Arc welding; Smelting; Coal gas purification	MRI; Targeted drug delivery/imaging; Gene therapy; Cancer, inflammation, diabetes & atherosclerosis; Memory chip/tape; Resonance imaging; Stem cell tracking; Molecular/ cellular tracking; Magnetic separation/ imaging (rapid DNA sequencing; Glass; Ceramics; Plastics; Textiles; Environmental remediation; Nanowires coatings; Alloy; Catalyst	Mody et al., 2010 Bahadar et al., 2016 Attarilar et al., 2020

Lead Oxide Nanoparticles

Pb having many oxides in numerous stoichiometric arrangements, e.g., PbO, PbO_2, Pb_2O_3, Pb_3O_4 and Pb_9O_{12} depending on lead-to-oxygen ratio (Nafees et al., 2017). PbO finds its conceivable applications in nanodevices and functionalized materials (Xi et al., 2004) owing to its unique electronic, mechanical and optical properties. Engineered PbO-NPs are known to have special assets and extensive variety of utilizations in science and technology due to magnetic resonance, data storage and imaging aspects (Mathew and Krishnamurthy, 2018). Lead sulfide (PbS)-NPs and nano-wires have wide-ranging applications such as, network modifiers in luminescent glassy materials, optics, pigments, gas sensors, ammonia gas-sensing agents, paints, Pb acid storage batteries, valve that regulate Pb acid batteries and lithium secondary batteries, nanoscale electronic devices, modifiers, stockpiling devices, and UV protectant oxide glasses (Karami et al., 2008; Lafront et al., 2010; Martos et al., 2001). Apart from these formats, PbO in different forms use as formation of nanoplates, nanostars, nanodendrites, nanorods, nanopowders, nanosheets and nanotubes (Bratovcic et al., 2020). It has been known that PbO_2^- NPs form inside the lead-coated pipes, or lead comprising faucets in consuming water arrangements that discharge toxic Pb^{2+}, producing adulteration (Minigaliyeva et al., 2020). Although there has been extensive research in the area of adverse effects of lead in occupational and environmental setups, there is paucity of data, which emphasizes about the use of lead in NPs formation and its toxicity on humans (Oszlanczi et al., 2011; Shaikh et al., 2015; Amiri et al., 2016; Al Naggar, 2016; Al Naggar et al., 2020; Miri et al., 2018; Aleksiichuk et al., 2018; Dumkova et al., 2017; Khalil et al., 2020). Both type of PbO_2 particles, nanoscale and microscale bulk reductively dissolve into Pb^{2+} and accumulate in the GI tract as primary uptake route and gills of the medaka fish, *Oryzias latipes*; however, nanoscale Pb particles show enhanced accumulation in liver and brain while comparing with the microscale counterpart (Ng et al., 2019). This *in vivo* research highlights the increase danger of exposure to lead dissolution from PbO_2 particles in the digestive tract by drinking water, which may increase the bioavailability of lead uptake and toxicity in people. Sub-chronic exposure to PbO-NPs causes profound negative effects at both cellular and tissue levels in rats along with significant ultramicroscopic changes in primary and secondary organs upon inhalation of heavy metal-based NPs (Dumkova et al., 2017). Overall, the health risks due to Pb-containing NPs are associated not only with their persistent production and usage, but also with their inevitable by-production as toxic air pollutants as they are regarded as "ultrafine particles", which can be dispersed as bioaerosols in different operations of metallic Pb, its alloys either alone, or in conjugation with Cu (Minigaliyeva et al., 2020).

Copper Oxide Nanoparticles

Engineered Cu-based NPs possess exceptional characteristics like high temperature superconductivity, spin dynamics, and electron correlation effects (El-Trass et al., 2012) and are chiefly used in manufacturing of semiconductors, heat transfer fluids, intrauterine contraceptive devices and antimicrobial reagents (Aruoja et al., 2009). These NPs are of remarkable importance owing to their easy availability, low cost and high electrical conductivity (Umer et al., 2012; Jain et al., 2015; Tiwari et al., 2016) as well as their applications in pesticides, fungicides and fertilizers (Borkow and Gabbay, 2009; Zheng et al., 2000). These NPs upon integration in coatings, plastics and textiles have antioxidant, antimicrobial, antidiabetic, anti-inflammatory and antifouling functions, as well as biocide and plant growth promoter (Apostolov et al., 2014; Thiruvengadam et al., 2019; Lee et al., 2008, Siddiqi and Husen, 2020). There are further reports of the use of copper, or copper oxide nanoparticles, in devices, lithium ion batteries, gas sensors, e-sensitized solar cells, heterogeneous catalysis and bioimaging systems (Rubilar et al., 2013; Waser et al., 2013; Sharma et al., 2015; Wang et al., 2016; Joshi et al., 2019; Perlman et al., 2015). The various *in vitro* and *in vivo* toxicity studies on Cu-based NPs such as, hepatic and renal toxicity are also reported (Lei et al., 2008). Studies of experimental animals reveal severe impairment in liver, kidney and spleen exposed to nano-Cu (Chen et al., 2006; Meng et al., 2007). It has been reported that highly reactive ionic copper (Cu^{2+}) following oral administration accumulate in the kidney of the exposed animals post interaction with gastric juice. CuO-NPs of 50 nm are known to disturb cell membrane integrity and inducing oxidative stress, and are genotoxic and cytotoxic as revealed by an *in vitro* study. Furthermore, compared to other nanosized metal oxides and CNTs, both single-wall and multi-wall CuO-NPs are particularly hazardous to human lung cells by producing oxidative lesions and damaging DNA (Karlsson et al., 2008). The possible mechanism of toxicity is attributed to disintegration of Cu into its ionic states, which subsequently leads to accumulation of metal ions in the cells as the pre-requisite of cellular damage (Wongrakpanich et al., 2016). According to a study evaluating the toxicity of CuO-

NPs towards astrocytes in brain, the NPs quickly accumulate in astrocytes through endocytosis; thereby increasing cellular Cu content and producing reactive oxygen species (ROS), reducing cell viability, and causing diseases by metabolic disturbances in brain-Cu balance (Bulcke and Dringen, 2016; Bulcke et al., 2017). As per studies on specific cell lines, the unique characteristic of Cu- and CuO-NPs is that the disintegrating NPs in cells bind to organelles and demonstrate a Trojan horse-like mechanism, resulting in blockage of cellular development as a whole (Studer et al., 2010; Bulcke et al., 2017).

Iron Oxide Nanoparticles

Iron (III) oxide (Fe_2O_3), the main oxide of Fe besides FeO and Fe_3O_4, is a reddish brown, paramagnetic inorganic compound (Mody et al., 2010). These superparamagnetic iron oxide NPs (SPION) are promising candidates for a wide range of biomedical applications, including high-resolution contrast agents for MRI, targeted drug delivery and imaging, hyperthermia, gene therapy, stem cell tracking, molecular/cellular tracking, and magnetic separation technologies like rapid DNA sequencing and early detection for inflammation, cancer, diabetes and atherosclerosis (Morales et al., 2003; Babes et al., 1999; Yigit et al., 2008; Peng et al., 2008; Chertok et al., 2008; Gonzales-Weimuller et al., 2009; Wei et al., 2006; Elias and Tsourkas, 2009; Moore et al., 2000; Kooi et al., 2003) owing to their magnetic properties, ultrafine size and biocompatibility. However, these NPs are accumulated in liver and other reticuloendothelial system mostly upon inhalation (Albukhaty et al., 2013). Additionally, these NPs remain in cellular organelles especially in endosomes/lysosomes of spleen and lungs, and release into cytoplasm after decomposing; thereby, contribute to cellular Fe pool (Rana, 2020; Zhu et al, 2008). These NPs disturb blood coagulation system and causing cell lysis and inflammation. Moreover, SPION also show the ability to cross B-B-B (Liu et al., 2013). *In vitro* studies reveal diminished cell viability as the most common lethal effect of iron oxide NPs, the proposed mechanism of action is thought to be unwarranted excessive production of ROS, which further elicits DNA damage in addition to lipid peroxidation.

Zinc Oxide Nanoparticles

Engineered Zn-based NPs are used in paints, wave filters, gas sensors, UV detectors, enzymes, sunscreens, and many individual care products in addition to vigor sector as oils and chemical agents; electronics, air filter, semiconductors, food products as additives, containers and packaging, supplements, drinking water as well as in the pharmaceutical industry and biomedical engineering (Huang et al., 2006 and 2010, Schulte et al., 2018; Schmalz et al., 2017). Owing to their conductivity, piezoelectricity and UV emission capacity, these ZnO-NPs have numerous applications in photovoltaics and transducers, electronic sensors, solar, and photographic catalyst with exceptional UV absorption and reflection properties (Hoffmann et al., 1995); hence, human exposure to ZnO-NPs is imminent (Ma et al., 2013). The most common toxicity induced by Zn-based nanomaterials includes cytotoxicity, cell membrane damage, increase oxidative stress as observed in several bacteria and mammalian cells (Huang et al., 2010). Several *in vitro* studies in human hepatocytes and embryonic kidney cells have reported altered cell morphology and mitochondrial activity along with DNA damage (Guan et al., 2012). Besides cytotoxicity, the genotoxic potential of ZnO-NPs has also been described in both *in vivo* and *in vitro* toxicity studies (Brunner et al 2006) Cell death on exposure to high concentration of ZnO-NPs and a decrease in human dermal fibroblast cell viability is reported by Meyer et al. (2011). The genotoxicity of Zno-NPs exerts via DNA damaging in HEp-2 cell line by (Osman et al., 2010). Bahadar et al., (2016) reveal that chronic exposure of

ZnO-NPs (300 mg/kg) results in oxidative DNA damage along with enzymatic alteration mostly in the liver. A study exploring acute toxicity of ZnO-NPs on marine crustaceans shows an increase in average mortality relative frequency in a dose-dependent manner (Vimercati et al., 2020).

Arsenic Oxide Nanoparticles

As is considered one of the worst pollutants in bulk form as it is highly soluble in natural waters due to geological processes i.e., predominant sedimentary deposits, and industrial effluents such as, smelting and mining (USEPA, 2001, Gonzalez et al., 2006) that causes dermal lesions and skin, bladder and lung cancers (WHO, 2011). Arsenicals are primarily used as alloying agents in manufacturing of transistors, lasers and semiconductors (Madhavi et al., 2022). As-NPs are mainly present as Arsenic (III) trioxide, and are widely used in breast cancer treatment due to cytotoxic potentiality (Subastri et al., 2018). Acute promyelocytic leukaemia and PTEN/AKT- mediated apoptosis are reported to be inhibited by As_2O_3-NPs. As_2O_3-NPs induce apoptosis and cytotoxicity in NB4 tumor cells by inhibiting Bcl-2 expression (Wang et al., 2018). In both the cases, as itself contributes a lot on genotoxicity and carcinogenicity (Dong et al., 2015). Mitochondrial damage, DNA fragmentation, reduction in protein content, defects in enzyme activities, ROS elevation, and induction of apoptosis are the most common mechanism of As- and AsO-NPs-induced toxicity in human health (Das et al., 2012; Pazhanisamy and Indra 2007; Humtsoe et al., 2007; Amsath et al.,2017; Luh et al.,1973). Further, high concentrations of As-NPs are fatal, and kidney, gills and liver of aquatic species are majorly affected through imbalance of protein, and induction of oxidative stress, DNA damage and necrosis. As-NPs exposure-mediated apoptotic lesions and organelle damage in hepatocytes occur through lysosomal mechanism (Jahangirnejad et al., 2020). Though all these evidences suggest higher toxicity of As- and AsO-NPs than the bulk metal, still there is a paucity of literature regarding nanotoxicity to human (occupational groups and consumers), which needs to be addressed as it is a matter of global concern.

Cadmium Sulfide Nanoparticles

Cd is a naturally occurring metal that can be found in water, soil, minerals, and the salt forms of sulphide, sulphate, carbonate and hydroxide. Ingestion of contaminated food can also cause major exposure to Cd (Balali-Mood et al., 2021). Exposure of Cd to environment is primarily through manufacturing of batteries, steel, plastics, fertilizers and mining, which causes substantial human exposure through contaminated water, air and soil. Cd is known to accumulate in kidneys with a long biological life of 10-35 years in humans and is classified as a carcinogen. CdS-NPs endorse distinctive physical, chemical and structural properties due to their exceptional crystal configuration, bandgap energy, melting point, ease of preparation and handling, optoelectronic absorption spectra, and high stability and availability (Zhu et al., 2009). In addition to the surface-to-volume ratio, the atomic distribution over the NPs shell is crucial in semiconductivity (Jamakala and Rani, 2014). Because of their high thermal stability and dirt repellency, CdS-NPs have a wide range of applications as pigments in paints and engineered plastics (Acharya, 2009; Zhu et al., 2009). At room temperature, CdS-NPs have a large bandgap-energy of 2.42 eV, which makes them ideal for optoelectronics (photocells), LED photonics (photo detectors, sensors, optical filters, optical switches, photovoltaics), and photo catalysis (field-effect transistors, decoders, lasers) as reported by Lin et al. (2002), Li et al. (2010), Duan et al. (2014) and Ma et al (2007). CdS-NPs are used as air-water purifier and molecular hydrogen production due to their photochemical and catalytic properties (Song et al., 2011). In biomedical fields, Cd-based NPs are used via cellular internalisation in cancer cells, which can be additionally visualized for chemotherapy and in drug delivery to several soft tissues for diagnosis and treatment of various cancer types (Pandian et al., 2011; El-Kemary et al., 2009; Kozhevnikova et al., 2010). The concentration-dependent decrease during hatching using mercaptopropionic acid-coated Cd-Se-NPs, and QD in zebrafish with degenerated morphology and embryonic toxicity are also reported (Li et al., 2009; Zhang et al., 2012). When CdS-NPs are exposed to chick embryos coated with maltodextrin, they cause a dose- dependent increase in developmental anomalies. However, less amount of CdS-NPs exposures found to remain nontoxic and thus, are recommended for bioimaging applications (Rodriguez-Fragoso et al., 2012). The cytotoxicity of CdS-NPs may be due to the release of CdS, which results in intracellular production of ROS and depletion of GSH in the cell (Trabelsi et al., 2013). Cd itself is xenotoxic to central and peripheral nervous system as cross the B-B-B (Wang and Du, 2013), which is manifested by neuropathy, reduce vasomotor function, Parkinson-like symptoms, concentration deficit, learning inability and decrease balancing (Cao et al., 2009; Pesch et al., 2000; Kim et al., 2005; Monroe and Halvorsen, 2006). Further, Cd-based NPs also cause austere malignancy of liver, pancreas, stomach, lungs, kidney, hematopoietic system, prostate and bladder during occupational exposure that seems to be aggravated toxicity in nanoscale form (Waalkes, 2000; Joseph et al., 2001; Lopez et al., 2003).

The following table (**Table 3**) summarizes the *in vitro* and *in vivo* toxicity studies of different heavy metals and their oxide-based NPs.

Table 3. Toxicity assessment of metal oxide NPs in different in vitro and in vivo models. ↓ indicates decrease and ↑ indicates increase

Base Metal for Metal/ Metal Oxide NPs	Study Type	Study Models	Reported NPs Toxicity	References
Pb	*In vitro*	Human fibroblast	Cytotoxicity	Bushueva et al., 2019
		Neuro2A cancer cell line	Cytotoxicity (>30 μg/ml)	Miri et al., 2018
	In vivo	Mice	↓ Inflammatory response; Sustained pathological changes in target organs	Dumkova et al., 2020
		Rats	Significant microscopic, ultramicroscopic changes in primary & secondary organs	Dumkova et al., 2017
		Adult medaka fish (*Oryzias latipes*)	Hepatic bioaccumulation & toxicity	Ng et al., 2019
		Wistar Rats	Neurotoxicity	Oszlanczi et al., 2011
		Mice	Bioaccumulation; Behavior, body weight & organ index changes	Shaikh et al., 2015
		Honey Bees (*Apis millefera L.*)	↑ α-tocopherol content	Al Naggar, 2016
			Neurotoxicity; ↑ Catalase; ↓ AChE activity	Al Naggar et al., 2020
		Rats	↑ Bioaccumulation; ↓ Hb, RBC & platelet counts	Amiri et al., 2016
		Rats	Hepatotoxicity; Biochemical changes in blood & serum	Aleksiichuk et al., 2018
		Rats	Bioaccumulation; ↑ Reticulocytes proportion; ↑ δ-ALA-excretion; ↑ Arterial hypertension	Sutunkova et al., 2020
		Mice	Undesired immune system stimulation; Inflammation; ↑ Oxidative stress in gut; Pro-fibrotic response in kidneys; Massive accumulation in bones	Tulinska et al., 2022

Continued on following page

Table 3. Continued

Base Metal for Metal/ Metal Oxide NPs	Study Type	Study Models	Reported NPs Toxicity	References
Cu	*In vitro*	Airway epithelial cells	Cytotoxic; Oxidative stress	Fahmy & Cormier, 2009
		Human A549 lung epithelial cell line	Oxidative lesions; DNA damage	Karlsson et al., 2008
		Chinese hamster oocytes; HeLa cells	Cytotoxicity via Trojan horse-type mechanism	Studer et al., 2010
		Human skin keratinocytes	Cytotoxicity; Genotoxicity	Alarifi et al., 2013
		Cardiac microvascular endothelial cells	Cytotoxicity; Permeability impairment; Inflammation	Sun et al., 2011
		HepG2 cells	Cytotoxicity	Cohen et al., 2013
		Human skin organ culture		
		Lung epithelial cells	ROS induction via disintegration into ionic state	Wongrakpanich et al., 2016
		Human lung epithelial cells	Dose-dependent genotoxicity by ROS	Akhtar et al., 2016
		Chinook salmon cells	Cytotoxicity; Abnormal morphology; Oxidative stress	Srikanth et al., 2016
		Caco-2 intestinal epithelial cells	Concentration-dependent cytotoxicity; IL-8 production	Ude et al., 2017
		Brain astrocytes	Endocytosis-mediated accumulation; ↑ Cellular Cu content; ROS production; ↓ Cell viability; Metabolic disturbances in brain-Cu balance	Bulcke & Dringen, 2016
		Human lung epithelial cells	Genotoxicity; Cytotoxicity; Oxidative stress; Alter cell membrane integrity	Ahamed et al., 2010
		Breast cancer MCF-7 cells	Morphological changes; Cell cycle arrest; Apoptosis	Laha et al., 2014
		Lung epithelial cells	Cytotoxicity; Oxidative stress	Jing et al., 2015
		Caco-2 intestinal epithelial cells	Cytotoxicity	Bannunah et al., 2014
		A549THP-1 cell line		Lanone et al., 2009
		A549SAEC cell line	Cell cycle arrest; Inhibit cell proliferation genes; Apoptosis	Hanagata et al., 2011
		HUVECs, RAW & HeLa cell lines	↓ Cell viability; Cytotoxicity	Feng et al., 2015
	In vivo	Mouse	Epigenetic changes	Lu et al., 2016
		Rats	Hepatotoxicity; Nephrotoxicity	Lei et al, 2008
		Sprague-Dawley rats	Alter blood cells number; Lymphocyte subpopulation in spleen; Antibody production; Histopathological changes	Zhou et al., 2019
		Rats	Bioaccumulation; Liver, kidney & spleen toxicity	Chen et al., 2006 Meng et al., 2007
		Zebrafish embryos & larvae	Neurotoxicity; Hepatotoxicity	Li et al., 2016
		Raphanus sativus, Lolium perenne & Lolium rigidum	DNA Damage	Atha et al., 2012
		Rats	↑ Alanine amino-transferase (ALT)	Lee et al., 2016

Continued on following page

Table 3. Continued

Base Metal for Metal/ Metal Oxide NPs	Study Type	Study Models	Reported NPs Toxicity	References
Fe	*In vitro*	Human lymphoblastoid TK6 cells; Primary human blood cells	Alter cellular behavior uptake; Dose-dependent cytotoxicity; Genotoxicity via DNA damage	Magdolenova et al., 2015
		Murine macrophage cells	↓ Cell viability; Bioaccumulation in liver & reticuloendothelial system	Naqvi et al., 2010
		A549 human lung epithelial cell line	Oxidative DNA lesions	Karlson et al., 2009
		Neural stem cells	Bioaccumulation in liver & reticuloendothelial system	Albukhaty et al., 2013
		Murine C17.2 neural progenitor cells; Human blood outgrowth endothelial cells	Inhibit cell proliferation via actin cytoskeleton remodulation & microtubule network; Focal adhesion formation & maturation	Soenen et al., 2011
		Human macrophages	↓ Cell viability	Pawelczyk et al., 2008
		Normal human epidermal keratinocytes (HEK); Murine epidermal cells (JB6 P+)	↓ Cell viability, IL-8 & IL-6; Oxidative stress	Murray et al., 2013
		PC12 cells	↓ Cell viability; Cell death	Pisanic et al., 2007
	In vivo	Rats	Bioaccumulation in liver & spleen; Less cytotoxicity	Yu et al., 2018
		Rats	Cell lysis; Inflammation; Alter blood coagulation	Zhu et al., 2008
		Wistar rats	Weak pulmonary fibrosis; Moderate cytotoxicity	Szalay et al., 2012

Continued on following page

Table 3. Continued

Base Metal for Metal/ Metal Oxide NPs	Study Type	Study Models	Reported NPs Toxicity	References
Zn	*In vitro*	HepG2 cells	DNA damage via oxidative stress; ROS-triggered mitochondria-mediated apoptosis	Sharma et al., 2012
		Cultured primary astrocytes	↓ Cell viability; Induce apoptosis; ↑ Intracellular ROS	Wang et al., 2014
		RAW-264.7 mouse cells; BEAS-2B human cells; Primary bone marrow-derived dendritic mouse culture cells	↑ Release of ROS; ↑ Expression of CD80 & CD86; ↑ Pro-inflammatory cytokines (IL-6, TNF α); ↓ Cell growth	Heng et al., 2011
		SHSY5Y human neuronal cells	Cytotoxicity; Genotoxicity; Cell cycle alteration; Apoptosis; Micronuclei production; H2AX phosphorylation; DNA damage	Valdiglesias et al., 2013
		Rat retinal ganglion cells	Cytotoxicity; Pro-inflammatory cytokines induction	Guo et al., 2013
		Human epidermal cells		Sharma et al., 2009
		Human nasal mucosa cells		Hackenberg, 2011
		Murine macrophages		Roy et al., 2011
		Human bronchial epithelial cells		Heng et al., 2010
		MRC5 human lung fibroblasts	Genotoxicity; Cytotoxicity; Clastogenic effect; Actin depolymerisation; ↓ Lung cell viability (↑ LDH); ↑ ROS (↑DDIT3, ↑ERN1); ROS-mediated oxidative stress; Inflammation; Liver, lungs & kidney damage; Agglomeration & internalization into cells; Cell membrane damage; Complete cell death in culture at 49 mg/ml	Ng et al., 2017
		Murine macrophages		Pati et al., 2016
		3T3 fibroblasts; HeLa cells		Abdelmonem et al., 2015
		Human lung epithelial cells		Huang et al., 2010
		Human mesothelioma cells; Rodent fibroblast cells		Brunner et al., 2006
		Human hepatocytes & embryonic kidney cells	Alter cell morphology; DNA damage; Mitochondrial dysfunction	Guan et al., 2012
		Human dermal fibroblast cells	↓ Cell viability at 20 nm	Meyer et al., 2011
		HEp-2 cell line	Genotoxicity via DNA damage	Osman et al., 2010
		Human colon carcinoma cells	↑ Oxidative stress; ↓ Cell viability; ↑ Inflammatory biomarkers	De Berardis et al., 2010
		Chlorella vulgaris	ROS; Cytotoxicity; ↓ Cell viability; ↓ SOD; ↓ GSH; ↑ LDH	Suman et al., 2015
		Murine cell lines	Metabolic disbalance; Ionic homeostasis & transport defects	Namvar et al., 2015
		Wistar rats	Dose-dependent toxicity; ↑ ROS; ↑ Liver enzymes	Abbasalipourkabir et al., 2015
		MRC5 lung cells	Genotoxicity; DNA damage; ↑ Extracellular LDH	Ng et al., 2017
		Fruit fly (*Drosophila melanogaster*)	Liver Inflammation; Chromatin condensation; Apoptosis; Bioaccumulation in liver, intestine & faeces	Ryu et al., 2014
		Alveolar type II epithelial cells	Oxidative stress; Decrease viability	Xie et al., 2012
		Caco-2 intestinal epithelial cells	DNA damage; Cytotoxicity	Gerloff et al., 2009
	In vivo	Sprague-Dawley rats	Sustained renal periglomerular & interstitial inflammation via lymphocytic infiltration	Chien et al., 2017
		Zebrafish (*Danio rerio*)	Mini-chromosome maintenance failure; DNA replication disorder in different cell cycle periods	Hou et al., 2019
		Broncho-alveolar lavage (BAL) cells; WBC isolated from ZnO-NP-exposed rats	Disrupt ions homeostasis; Mitochondrial dysfunction; Caspase activation & apoptosis	Kao et al., 2011
		Fruit fly (*Drosophila melanogaster*)	↑ ROS & reduce Nrf2 production	Ng et al., 2017
		Human volunteers	No remarkable results, may be relatively safe though limited sample size	Mohammed et al., 2019

Continued on following page

Table 3. Continued

Base Metal for Metal/ Metal Oxide NPs	Study Type	Study Models	Reported NPs Toxicity	References
As	*in vitro*	Breast cancer cells (MDA-MB-231, MCF-7 & HEK-293)	Cytotoxicity	Subastri et al., 2018
		NB4 cells	Apoptosis; Cell inhibition	Wang et al., 2018
		NB4 cells	Cytotoxicity; Apoptosis	Dong et al., 2015
		Isolated rat hepatocytes	Apoptosis; Organelle impairment	Jahangirnejad et al., 2020
	In vivo	Rats	Hepatotoxicity; Histopathological changes; Intracytoplasmic/intranuclear inclusions; Apoptosis	Verma et al., 2021
		Labeo rohita Juveniles	Protein imbalance; Oxidative stress; DNA damage; Necrosis	Raza et al., 2021
Cd	*In vitro*	HFF PC-3 cell line; MCF-7 cell line	Cytotoxicity	Kominkova et al., 2017
		HCT-116 cells	↓ Cell viability & cell stability	Mirnajafizadeh et al., 2019
		Escherichia coli	Morphological changes; Cell surface damage; ↓ Cell division	Hossain & Mukherjee, 2013
		HeLa cells	Morphological changes with fragmented nuclei; ↑ Oxidative stress	
		MDA-MD-231 cells	Dose-dependent apoptosis & necrosis via ROS	Rodriguez- Fragoso et al., 2012
		Chick embryos	Developmental anomalies	
	In vivo	Zebrafish embryos	Developmental abnormalities during embryogenesis	Hallare et al., 2005
		Zebrafish	Intracellular oxidative stress; Cytotoxicity	Li et al., 2009
		Zebrafish embryo & larvae	↓ Hatching; Degenerate morphology; Cytotoxicity	Zhang et al., 2012
		Mice	Hepatotoxicity; Induction of ROS	Meng et al., 2019
		Rat	↑ Cd Metallothionein (Cd-MT); Lipid peroxidation & H_2O_2; Renal proximal tubular damage; ↓ Alkaline phosphatase; Altered mitochondria, nucleus & endoplasmic reticulum	Rana et al., 2018
		Mice	Bioaccumulation; Acute toxicity; Spermatozoa & organ damage	Liu et al., 2014
		Rats	Neurotoxicity	Papp et al., 2012
		Cerebellar cells	↓ Cell viability; ROS-mediated change; ↓ Total antioxidant	Varmazyari et al., 2020
		Cultures rat brain	↑ Glial fibrillary acidic protein (GFAP) & 8-hydroxy-2'-deoxyguanosine (8-OHdG)	

MECHANISM OF HEAVY METAL-BASED NANOPARTICLES' TOXICITY

As stated earlier, heavy metals frequently react with biological systems to generate metal cations, which are attracted to the nucleophilic sites of essential macromolecules (Balali-Mood et al., 2021) by losing one, or more electrons. The exact mechanism of action of specific heavy metal toxicity remains unclear; however, abnormal changes in genome as well as gene expression are proposed as an underlying process (Balali-Mood et al., 2021). Heavy metal exposure alters many cellular events, including proliferation, growth, differentiation, damage repair mechanisms and apoptosis, resulting in several negative effects, as summarized in **Table 3**. The overall mechanism of action to induce NPs toxicity are genomic instability by defects in DNA repair/synthesis followed by disbalance of redox homeostasis as the failing of normal physiological redox-regulated functions and selective binding to specific macromolecules (Fu et al., 2014; Attarilar et al., 2020; Egbuna et al., 2021). This, in turn, leads to the activation of inflammatory cytokines, unregulated cell signaling pathway and cytotoxicity, cell motility changes and apoptosis as manifested by initiation of cancer (Manuja et al., 2021). The initiation of oxidative stress concurrently activates the proinflammatory intermediaries via the standard signaling cascades namely, nuclear factor κβ (NF-κβ), MAPK and PI3K pathways (Azad et al., 2009). There is plethora of data as available in several websites emphasizing the toxic effects of these heavy metal-based NPs are mainly dependent on the original properties of the metal itself (Manuja et al., 2021).

However, the toxicity of the nanomaterials can be different on nanoscale level because of the changes of their magnetic properties, physical and chemical nature, and electrical conductivity as mentioned earlier such as, size, shape, surface/volume ratio, particle dissolution, particle surface, surface-containing groups, surface positive charges, metal ion releases from nanometals and their oxides, inflammation, aggregation and mode of interaction with host cells, UV light activation, and pH of the medium (Attarilar et al., 2020). Owing to long estimated half-life of about 700 days in human lungs, these NPs are consistent threat mainly to the respiratory system due to immunocompromised components of the matrix, which needs to be activated in order to clear these NPs from the body (Bahadar et al., 2016). It is reported in some animal and human studies that after inhalation and ingestion, the bio-distribution of NPs to liver, heart, lungs, GI tract, spleen and brain modulate their overall toxicity mechanism (Manuja et al., 2021; Raj and Kumar, 2020), the critical determinants being metal oxide-NPs-induced DNA fragmentation and formation of oxidation-induced DNA adducts. In this reference p53 is the key player (Raj and Kumar, 2020) though broad spectrum target-specific studies to explore the potential adverse outcomes of physico-chemical properties such as, shape, size and charge, etc. of the heavy metal-based NPs are still to be discovered. **Figure 2** briefly depicts the whole scenario.

Figure 2. Mechanism of heavy metal-based nanomaterials' toxicity on human health

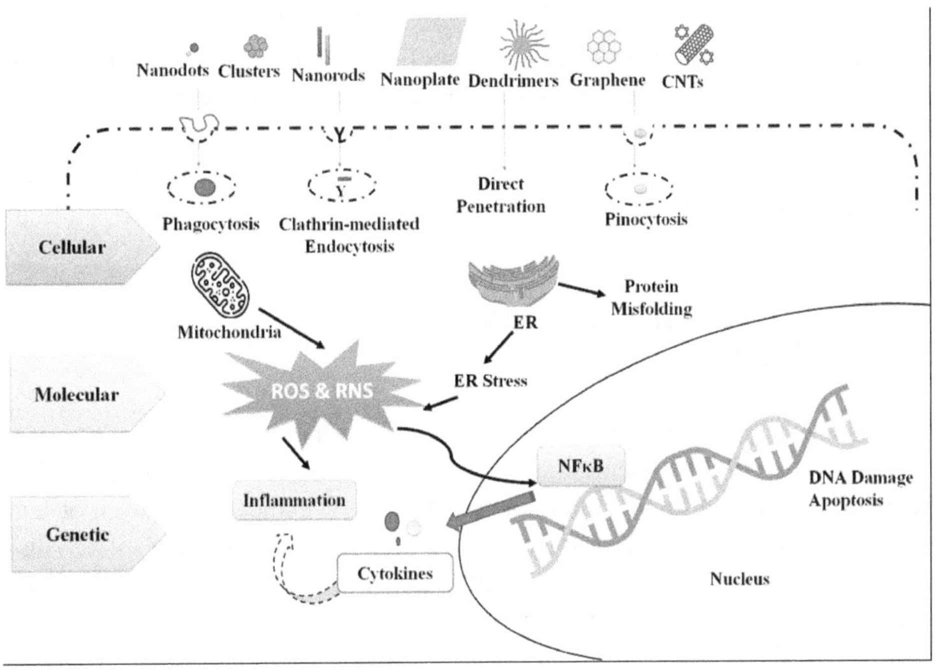

EARLY DETECTION BIOMARKERS OF HEAVY METAL-BASED NANOPARTICLES' TOXICITY: PRESENT STATUS

Extensive studies regarding the toxicity of heavy metal are documented. The mechanism of action of heavy metal on various systems is also well recognized, and its biological markers in blood and urine are identified on the basis of hindrance of metabolic pathways by these metals (Das et al., 2013). For example, delta-amino levulinic acid deydratase (δ-ALAD) is the potential biomarker of Pb-induced NPs toxicity. In most of the cases heavy metals exhibit their toxicity by binding to the sulfhydryl (-SH) groups; thus, inhibit enzyme activity, disrupt cellular transport, substitute essential ions from specific celullar sites and alter protein functions (Kakkar and Jaffery, 2005). Microarray analysis of gene expression has become an imperative tool for exploring the biological effects in this regard (Waring et al., 2002). Also, the use of proteomics and genomics with bioinformatics approaches have been expanded with technological advancement, measuring of pro- and anti- inflammatory cytokines, and monitoring the balance of TH1/TH2 cytokines, etc. may become an important evaluation tool for NP immunotoxicity (Elsabahy and Wooley, 2013). Although, screening a single reporter is inadequate, combinations of various cytokines as cytokine network system is much more effective in this regard and especially when comparing with unexposed controls in search of an early detection biomarker of heavy metal-based NPs toxicity.

However, measure of metal toxicity cannot be extrapolated to their miniature forms as the NPs have larger surface area, size, charge, etc. (Attarilar et al., 2020). There is a lacuna in understanding the behavior of heavy metals and their toxicity in NPs forms. Various *in vitro* studies on assessing the toxicity of metallic NPs is documented (Bushueva et al., 2019; Bulcke and Dringen, 2016; Sharma et al., 2012), though only a few *in vivo* studies documented experimental rodent and zebrafish models can be treated as an early biomarker for toxicity of heavy metal-based NPs, which is not sufficient for comment on human system (Li et al., 2016; Hou et al., 2019; Hallare et al., 2005). Therefore, more *in vivo* and epidemiological studies in order to extrapolate the preclinical rodent to human studies on targeted biological systems are warranted considering the toxicity of individual metal-NPs in terms of dissolution in the ionic forms and bioavailability. This kind of research may able to find out the early detection of biomarkers in conjunction with exposure analysis that ultimately ensures the safer use of heavy metal-based nanomaterials.

RISK FACTORS' EVALUATION AND SAFETY MEASURES: CURRENT SCENARIO

As discussed previously, there are various routes of exposures through which nanomaterials can enter into the environment and health systems, and exerts their effects, not only human but also plants and animals; hence, all biota are confronted risks implicated by nanotechnology at the levels of manufacturing, designing, usage and disposal (Jane Cypriyana et al., 2021). Globally, risks are increasing proportionately with the increase in the scope of nanotechnology, as due to their extremely smaller size (nanoscale level), nanomaterials as discussed here, have more easy and wide access to human and for spreading in the environment. It has been found by several researchers that both from the indirect exposure of nanomaterials present in environment and food, and the direct exposure through handling and uses of nanomaterials in cosmetics and other consumer products, use of novel nanodrugs, etc. have put a demanding need to explore the risk factors, especially heavy metal-based agents (Singh, 2017). Simultaneously, as the exposure causes serious health-related problems; therefore, attention on the safety measures is the urgent need of the hour considering the ultrafine sizes, chemical reactivity and extended atmospheric residence time of the heavy metal-based NPs as discussed here, which are substantially more dangerous than the coarser particles (Sonwani et al., 2021). Overall risk factors for manufacturing, processing and handling of several nanomaterials in the industries, associated hazards and common safety measures in different occupational sectors are summarized below: -

Risk factors	Safety measures
1. Workplace exposure including manufacturing, processing and handling of nanomaterials 2. Health hazards of workers including 'take home effect' on spouses, children and other family members	• Company health and safety policy • Medical screening and surveillance • Safety training programs • Awareness of stakeholders • Personal monitoring • Area monitoring • Personal protective equipment, gloves, air purifying • masks, etc. • Green technology • Proper ventilation and exhaust system • Engineering controls: Elimination of hazards

Thus, it is futile and impractical to rely on the same safety guidelines, which are designed by keeping in mind the harmful effects of particles in the core state, rather to formulate the specific guideline by the concerned authority. As defined by CDC, risk is the measure of the hazards multiplied by the exposure; hence, controlling the exposure ultimately leading to minimization of the potential risks in the industries. However, for underdeveloped and developing countries including India, this sector is neglected; thus, further strengthening the requirement of nation-wide large scale epidemiological studies involving different heavy metal-based nanomaterials and their toxicity in the environment and health.

INTERVENTION AND PREVENTION OF NANOMATERIALS' TOXICITY: GLOBAL SITUATION

The use of heavy metals in industry is well documented, and its toxicity to human and ecosystems are well established. Considering the detrimental effects and toxicity posed on human and environmental health, various interventions to increase public awareness is conducted by several government and non-government organizations, nationally as well as globally (Sharma et al., 2021). These include measures to reduce the exposure of these metals, finding sustainable nontoxic alternatives, educating general population regarding metal toxicity, health assessment of workers, etc. Agencies like WHO, USEPA carry out intervention programmers for the same and declare the permissible limits for all these hazardous agents in drinking water to reduce the disease burden. In contrast, uses of metallic NPs in industries are relatively new field, and there is paucity of data on nanotoxicity and health risk to humans and ecosystem, largely from the developing and underdeveloped countries. However, as the detrimental impacts of metal/metal oxides-NPs are attributed to size, morphology, concentration, aggregation mode, charges and surface properties, the variations of these properties of NPs may be on the right direction to mitigate their possible harmful effects from environment and human health. Some notable researches in this regard are tried to improve the structure and surface of NPs by using surface coating (Osmond-McLeod et al., 2013), core-shell (Davidson et al., 2016), doping-based methods (Wang et al., 2012), geometric control (Ji et al., 2012) and surface passivation methods (Cai et al., 2017), etc.

Overall, it can be said that proficient conceptualization and development of nano/micro-structures can pave the way to ameliorate the harmful impacts of metal/metal oxides-NPs to facilitate their role in therapeutics and diagnostics. Recently the concept of "Green Synthesis" has emerged, where NPs are produced *in vivo* by organism like *E. coli* to provide a safer alternative as compared to their chemical counterpart in terms of toxicity (Trabelsi et al., 2013; Sharma et al., 2015; Thiruvengadam et al., 2019). Other structural changes include delivery through metal-organic frameworks, surface alteration, polymers and artificial intelligence-based computational approaches (Manuja et al., 2021). Moreover, the toxicity of two different types of CdO-NPs produced by calcining $Cd(OH)_2$ with and without polymer coordination strongly suggests that carbon surface covering through polymerization of the organic unit might greatly lessen the harmful effects of CdO-NPs in zebrafish (Balmuri et al., 2017), but this requires validation in human systems. Valdiglesias et al (2016) also suggested that surface coating i.e., dimercapto succinic acid (DMSA) can suppress a potential cytotoxicity by inhibiting direct contact between human dermal fibroblasts and maghemite (Fe_2O_3)-NPs; however, further in-depth investigation to explore more sophisticated methods for commercial production of heavy metal-based NPs are the urgent medical need of the country from public health point of view in order to ensure Nation-wide precise and complete toxicity prevention program considering their day-to-day increasing demands in our lives.

CONCLUSION AND FUTURE RESEARCH NEEDS

Overall, there is a gradual increase in attention towards the nanotechnology and its applications in the commercial sector, which attracts a lot of business in developing a range of products that consecutively increases the number of workers in the occupational sector as the need of manpower will be in great demands. Moreover, this modern technology gains interest of a lot of consumers because of the fascinating opportunities and hope for the betterment of life. As per the age back, and still most conventional and agreed definition of toxicity, anything taken more than the normal, or the recommended amount and exerting a damaging effect has been referred to as toxicity (Pumera, 2010). However, large knowledge gaps still exist in nanotoxicology field because of the lack of study with respect to altered properties of specific, or targeted heavy metal-based NPs.

According to researchers by 2020, there are ~6 million people globally working in the field of nanotechnology; hence, it is becoming occupational and public health priority for understanding the impacts of NPs in exposed population as well as hazards to the environment (Dang et al., 2010; Boccuni et al., 2017; Iavicoli et al., 2017). Thus, from the scientific and sustainability point of view, it is the need of the hour for scientists, researchers and policy makers to provide utmost care and attention towards translational researches and remedial measures of nanomaterials' toxicity and related health hazards. However, it is an undeniable truth that every coin has two sides, similarly the current scenario only focuses on the growth of the nanotechnology in various branches though the elongated branches leave behind the deeper roots of health/environmental hazards and exposure either directly, or indirectly that affecting our surrounding as well as the living systems (Figure 3).

Figure 3. Pros and cons of nanotechnology where branches represent potential applications, and roots indicate possible threats to environment and health

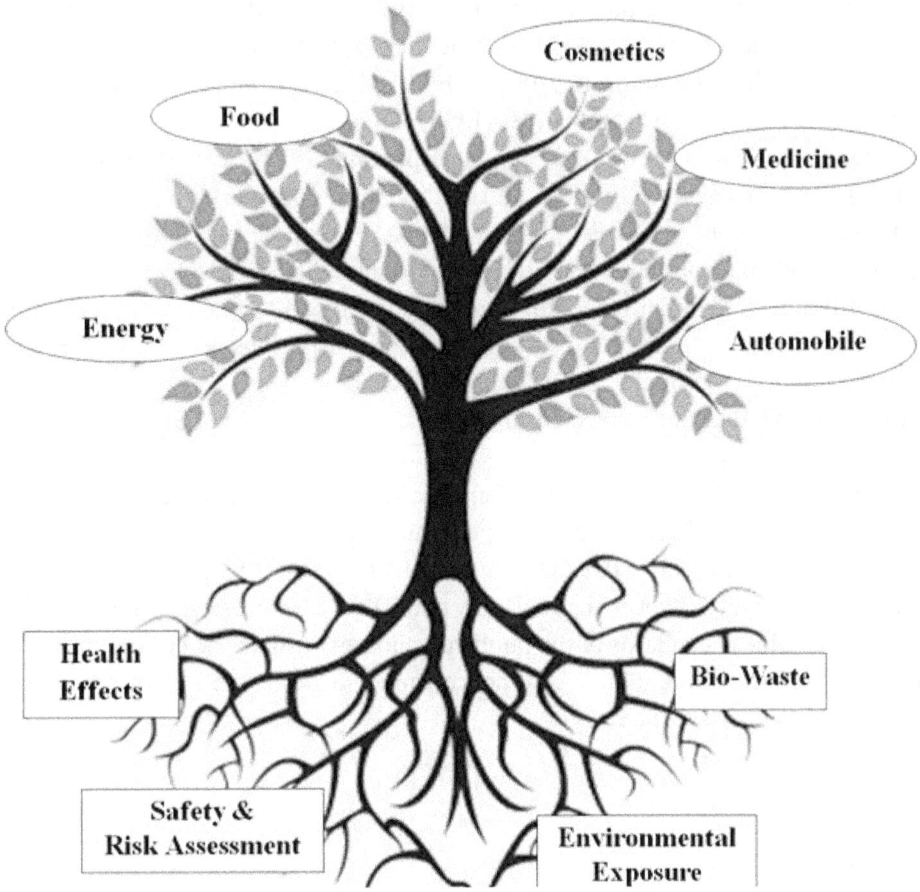

Thus, the branches of the tree (nanotechnology) has been spontaneously finding its way into the scope of the future, but as the shadow of these tree covers a large area that slowly affecting our lives at large putting sustainability at a risk; the study of the roots at a deeper level is necessary, which can provide better directions for the set-up of the workplace limiting the exposure to workers, change in the synthesis and manufacturing of nanomaterials by acquiring eco-friendly ways, beneficial for all the stakeholders as well as the environment. Very negligible to limited existing information regarding the toxicity of nanomaterials, further widen the scope of research for the development of newer, point-of-care assays, or early detection biomarkers for a particular material's nanotoxicity along with a safer and hopeful future.

Abbreviations: Aluminum oxide (Al_2O_3), Arsenic (As), Arsenic Oxide Nanoparticles (AsO-NPs), Arsenic trioxide (As_2O_3-NPs), Blood Brain Barrier (B-B-B), Cadmium (Cd), Cadmium hydroxide ($Cd(OH)_2$), Cadmium Sulfate Nanoparticles (Cds-NPs), Carbon nanotubes (CNTs), Cerium dioxide (CeO_2), Chromium (Cr), Cobalt (Co), Computed Tomography (CT), Copper (Cu), Copper oxide (CuO), Copper Oxide Nanoparticles (CuO- NPs), Cuprous oxide (Cu_2O), Delta-amino levulinic acid dehydratase (δ-ALAD), Dimercapto succinic acid (DMSA), Gastrointestinal tract (GI), Glutathione (GSH), Gold (Au), Graphene (Gr), Indium oxide (In_2O_3), Iron (Fe), Iron (III) oxide (Fe_2O_3), Iron oxide (Fe_2O_3), Iron Oxide nanoparticles (FeO-NPs), Lanthanum oxide (La_2O_3), Lead (Pb), Lead oxide (PbO), Maghemite (Fe_2O_3), Magnesium oxide (MgO), Magnetic Resonance Imaging (MRI), Mercury (Hg), Metal oxide nanoparticles (NPs), Mitogen-activated protein kinase (MAPK), Nanoparticles (NPs), Nanotubes (NT), Nickel (Ni), Nickel oxide (NiO), Nuclear factor κβ (NF κβ), Palladium (Pd), Phosphatidylinositol 3-kinase (PI3K), Phosphatidylinositol 3-kinase/protein kinase B (PTEN/AKT), Positron Emission Tomography (PET), Quantum dots (QDs), Reactive Oxygen Species (ROS), Silica (Si), Silicon dioxide (SiO_2), Silver (Ag), Surface-enhanced Raman Spectroscopy (SERS), Superparamagnetic iron oxide nanoparticles (SPION), Tetrahertz (THz), Time-of-flight (TOF),Titanium oxide (TiO_2), Toll-like receptors (TRP), United States Environment Protection Agency (USEPA), Ultra violet (UV), World Health Organization (WHO), Zinc oxide (ZnO), Zinc oxide nanoparticles (ZnO-NPs), Zirconium dioxide (ZrO_2).

Author's Contribution: NN, corresponding author; conceived, designed and edited the book chapter. **SNS, UK** equally contributed to literature survey and drew illustrations. **SNS, UK, NN** wrote the book chapter, read and approved the final version of the book chapter.

Fund Disclaimer: Grant not received from any funding agency in public, commercial, or not-for-profit sector.

Acknowledgment: Thanks to ICMR-NIOH for basic infrastructure support, and Ms. Deepsi Rathore, PhD student of Biochemistry Dept., for editing the bibliography.

Declaration: The authors declare that the manuscript, or portions thereof, is not under consideration by another book/journal, and has not been previously published. The research performed is in accordance with the statement of ethical standards of the publishing house. Further, the editors of the book have the right of reviewing and editing the submission; hence, authors transfer, assign, or otherwise conveys all copyright ownership to the publisher.

Conflict of Interest: None.

REFERENCES

Abbas, Q., Yousaf, B., Ali, M. U., Munir, M. A. M., El-Naggar, A., Rinklebe, J., & Naushad, M. (2020). Transformation pathways and fate of engineered nanoparticles (ENPs) in distinct interactive environmental compartments: A review. *Environment International, 138*, 105646. doi:10.1016/j.envint.2020.105646 PMID:32179325

Abbasalipourkabir, R., Moradi, H., Zarei, S., Asadi, S., Salehzadeh, A., Ghafourikhosroshahi, A., Mortazavi, M., & Ziamajidi, N. (2015). Toxicity of zinc oxide nanoparticles on adult male Wistar rats. *Food and Chemical Toxicology, 84*, 154–160. doi:10.1016/j.fct.2015.08.019 PMID:26316185

Abd Elkodous, M., El-Husseiny, H., El-Sayyad, G., Hashem, A., Doghish, A., Elfadil, D., Radwan, Y., El-Zeiny, H., Bedair, H., & Matsuda, A. (2021). Recent advances in waste-recycled nanomaterials for biomedical applications: Waste-to-wealth. *Nanotechnology Reviews, 10*(1), 1662–1739. doi:10.1515/ntrev-2021-0099

Abdelmonem, A. M., Pelaz, B., Kantner, K., Bigall, N. C., Del Pino, P., & Parak, W. J. (2015). Charge and agglomeration dependent in vitro uptake and cytotoxicity of zinc oxide nanoparticles. *Journal of Inorganic Biochemistry, 153*, 334–338. doi:10.1016/j.jinorgbio.2015.08.029 PMID:26387023

Acharya, K. P. (2009). *Photocurrent spectroscopy of CdS/plastic, CdS/glass, and ZnTe/GaAs hetero-pairs formed with pulsed-laser deposition* [Doctoral dissertation]. Bowling Green State University.

Ahamed, M., Siddiqui, M. A., Akhtar, M. J., Ahmad, I., Pant, A. B., & Alhadlaq, H. A. (2010). Genotoxic potential of copper oxide nanoparticles in human lung epithelial cells. *Biochemical and Biophysical Research Communications, 396*(2), 578–583. doi:10.1016/j.bbrc.2010.04.156 PMID:20447378

Akçan, R., Aydogan, H. C., Yildirim, M. Ş., Taştekin, B., & Sağlam, N. (2020). Nanotoxicity: A challenge for future medicine. *Turkish Journal of Medical Sciences, 50*(4), 1180–1196. doi:10.3906ag-1912-209 PMID:32283898

Akhtar, M. J., Kumar, S., Alhadlaq, H. A., Alrokayan, S. A., Abu-Salah, K. M., & Ahamed, M. (2016). Dose-dependent genotoxicity of copper oxide nanoparticles stimulated by reactive oxygen species in human lung epithelial cells. *Toxicology and Industrial Health, 32*(5), 809–821. doi:10.1177/0748233713511512 PMID:24311626

Akter, M., Sikder, M. T., Rahman, M. M., Ullah, A. K. M. A., Hossain, K. F. B., Banik, S., Hosokawa, T., Saito, T., & Kurasaki, M. (2018). A systematic review on silver nanoparticles-induced cytotoxicity: Physicochemical properties and perspectives. *Journal of Advanced Research, 9*, 1–16. doi:10.1016/j.jare.2017.10.008 PMID:30046482

Al Naggar, Y. (2016). Effects of metal and metal oxide nanoparticles on honey bees (Apis millefera L.). *Journal of Aquatic Pollution and Toxicology, 1.*

Al Naggar, Y., Dabour, K., Masry, S., Sadek, A., Naiem, E., & Giesy, J. P. (2020). Sublethal effects of chronic exposure to CdO or PbO nanoparticles or their binary mixture on the honey bee (Apis millefera L.). *Environmental Science and Pollution Research International, 27*(16), 19004–19015. doi:10.100711356-018-3314-2 PMID:30280346

Alarifi, S., Ali, D., Verma, A., Alakhtani, S., & Ali, B. A. (2013). Cytotoxicity and genotoxicity of copper oxide nanoparticles in human skin keratinocytes cells. *International Journal of Toxicology, 32*(4), 296–307. doi:10.1177/1091581813487563 PMID:23667135

Albukhaty, S., Naderi-Manesh, H., & Tiraihi, T. (2013). In vitro labeling of neural stem cells with poly-L-lysine coated super paramagnetic nanoparticles for green fluorescent protein transfection. *Iranian Biomedical Journal, 17*(2), 71–76. doi:10.6091/ibj.1114.2013 PMID:23567848

Aleksiichuk, V., Omelchuk, S., Sokurenko, L., Kaminsky, R., Kovalchuk, O., & Chaikovsky, Y. (2018). The influence of lead nanoparticles on the morpho-functional changes of rat liver during the postexposure period. *Microscopy Research and Technique, 81*(7), 781–788. doi:10.1002/jemt.23036 PMID:29652104

Al-Hakkani, M. F. (2020). Biogenic copper nanoparticles and their applications: A review. *SN Applied Sciences, 2*(3), 505. doi:10.100742452-020-2279-1

Amiri, A., Mohammadi, M., & Shabani, M. (2016). Synthesis and toxicity evaluation of lead oxide (PbO) nanoparticles in rats. *Electronic Journal of Biology, 12*(2), 110-114.

Amsath, J., Sugumaran, V., & Sugumaran, J. (2017). Effect of arsenic (As_2O_3) on hematological parameters of freshwater air breathing fish, Channa punctatus (bloch). *Current Trends in Biomedical Engineering & Biosciences*, *7*(1), 3–7. doi:10.19080/CTBEB.2017.07.555702

Apostolov, A. T., Apostolova, I. N., & Wesselinowa, J. M. (2014). Dielectric constant of multiferroic pure and doped CuO nanoparticles. *Solid State Communications*, *192*, 71–74. doi:10.1016/j.ssc.2014.05.014

Aruoja, V., Dubourguier, H. C., Kasemets, K., & Kahru, A. (2009). Toxicity of nanoparticles of CuO, ZnO and TiO2 to microalgae Pseudokirchneriella subcapitata. *The Science of the Total Environment*, *407*(4), 1461–1468. doi:10.1016/j.scitotenv.2008.10.053 PMID:19038417

Atha, D. H., Wang, H., Petersen, E. J., Cleveland, D., Holbrook, R. D., Jaruga, P., Dizdaroglu, M., Xing, B., & Nelson, B. C. (2012). Copper oxide nanoparticle mediated DNA damage in terrestrial plant models. *Environmental Science & Technology*, *46*(3), 1819–1827. doi:10.1021/es202660k PMID:22201446

Attarilar, S., Yang, J., Ebrahimi, M., Wang, Q., Liu, J., Tang, Y., & Yang, J. (2020). The Toxicity Phenomenon and the Related Occurrence in Metal and Metal Oxide Nanoparticles: A Brief Review from the Biomedical Perspective. *Frontiers in Bioengineering and Biotechnology*, *8*, 822. doi:10.3389/fbioe.2020.00822 PMID:32766232

Azad, M. B., Chen, Y., & Gibson, S. B. (2009). Regulation of autophagy by reactive oxygen species (ROS): Implications for cancer progression and treatment. *Antioxidants & Redox Signaling*, *11*(4), 777–790. doi:10.1089/ars.2008.2270 PMID:18828708

Babes, L., Denizot, B., Tanguy, G., Le Jeune, J. J., & Jallet, P. (1999). Synthesis of iron oxide nanoparticles used as MRI contrast agents: A parametric study. *Journal of Colloid and Interface Science*, *212*(2), 474–482. doi:10.1006/jcis.1998.6053 PMID:10092379

Bahadar, H., Maqbool, F., Niaz, K., & Abdollahi, M. (2016). Toxicity of Nanoparticles and an Overview of Current Experimental Models. *Iranian Biomedical Journal*, *20*(1), 1–11. doi:10.7508/ibj.2016.01.001 PMID:26286636

Baig, N., Kammakakam, I., & Falath, W. (2021). Nanomaterials: A review of synthesis methods, properties, recent progress, and challenges. *Materials Advances*, *2*(6), 1821–1871. doi:10.1039/D0MA00807A

Balali-Mood, M., Naseri, K., Tahergorabi, Z., Khazdair, M. R., & Sadeghi, M. (2021). Toxic mechanisms of five heavy metals: Mercury, lead, chromium, cadmium, and arsenic. *Frontiers in Pharmacology*, *12*, 643972. doi:10.3389/fphar.2021.643972 PMID:33927623

Balmuri, S. R., Selvaraj, U., Kumar, V. V., Anthony, S. P., Tsatsakis, A. M., Golokhvast, K. S., & Raman, T. (2017). Effect of surfactant in mitigating cadmium oxide nanoparticle toxicity: Implications for mitigating cadmium toxicity in environment. *Environmental Research*, *152*, 141–149. doi:10.1016/j.envres.2016.10.005 PMID:27771568

Bannunah, A. M., Vllasaliu, D., Lord, J., & Stolnik, S. (2014). Mechanisms of nanoparticle internalization and transport across an intestinal epithelial cell model: Effect of size and surface charge. *Molecular Pharmaceutics*, *11*(12), 4363–4373. doi:10.1021/mp500439c PMID:25327847

Barhoum, A., García-Betancourt, M. L., Jeevanandam, J., Hussien, E. A., Mekkawy, S. A., Mostafa, M., & Bechelany, M. (2022). Review on natural, incidental, bioinspired, and engineered nanomaterials: History, definitions, classifications, synthesis, properties, market, toxicities, risks, and regulations. *Nanomaterials (Basel, Switzerland)*, *12*(2), 177. doi:10.3390/nano12020177 PMID:35055196

Baruah, S., & Dutta, J. (2009). Nanotechnology applications in pollution sensing and degradation in agriculture: A review. *Environmental Chemistry Letters*, *7*(3), 191–204. doi:10.100710311-009-0228-8

Bergin, I. L., & Witzmann, F. A. (2013). Nanoparticle toxicity by the gastrointestinal route: Evidence and knowledge gaps. *International Journal of Biomedical Nanoscience and Nanotechnology*, *3*(1-2), 163. Advance online publication. doi:10.1504/IJBNN.2013.054515 PMID:24228068

Bhattacharjee, S., Macintyre, C. R., Bahl, P., Kumar, U., Wen, X., Aguey-Zinsou, K. F., & Joshi, R. (2020). Reduced graphene oxide and nanoparticles incorporated durable Electroconductive silk fabrics. *Advanced Materials Interfaces*, *7*(20), 2000814. doi:10.1002/admi.202000814

Boccuni, F., Gagliardi, D., Ferrante, R., Rondinone, B. M., & Iavicoli, S. (2017). Measurement techniques of exposure to nanomaterials in the workplace for low- and medium-income countries: A systematic review. *International Journal of Hygiene and Environmental Health*, *220*(7), 1089–1097. doi:10.1016/j.ijheh.2017.06.003 PMID:28711303

Borkow, G., & Gabbay, J. (2009). Copper, an ancient remedy returning to fight microbial, fungal and viral infections. *Current Chemical Biology*, *3*(3), 272–278. doi:10.2174/2212796810903030272

Brandelli, A. (2020). The interaction of nanostructured antimicrobials with biological systems: Cellular uptake, trafficking and potential toxicity. *Food Science and Human Wellness*, *9*(1), 8–20. doi:10.1016/j.fshw.2019.12.003

Bratovcic, A. (2020). *Synthesis, characterization, applications, and toxicity of lead oxide nanoparticles*. Lead Chemistry. doi:10.5772/intechopen.91362

Brock, D. A., Douglas, T. E., Queller, D. C., & Strassmann, J. E. (2011). Primitive agriculture in a social amoeba. *Nature*, *469*(7330), 393–396. doi:10.1038/nature09668 PMID:21248849

Brunner, T. J., Wick, P., Manser, P., Spohn, P., Grass, R. N., Limbach, L. K., Bruinink, A., & Stark, W. J. (2006). In vitro cytotoxicity of oxide nanoparticles: Comparison to asbestos, silica, and the effect of particle solubility. *Environmental Science & Technology*, *40*(14), 4374–4381. doi:10.1021/es052069i PMID:16903273

Bulcke, F., & Dringen, R. (2016). Handling of copper and copper oxide nanoparticles by astrocytes. *Neurochemical Research*, *41*(1), 33–43. doi:10.100711064-015-1688-9 PMID:26268441

Bulcke, F., Dringen, R., & Scheiber, I. F. (2017). Neurotoxicity of copper. *Advances in Neurobiology*, *18*, 313–343. doi:10.1007/978-3-319-60189-2_16 PMID:28889275

Bundschuh, M., Filser, J., Lüderwald, S., McKee, M. S., Metreveli, G., Schaumann, G. E., Schulz, R., & Wagner, S. (2018). Nanoparticles in the environment: Where do we come from, where do we go to? *Environmental Sciences Europe*, *30*(1), 6. doi:10.118612302-018-0132-6 PMID:29456907

Bushueva, T., Minigalieva, I., Panov, V., Kuznetsova, A., Naumova, A., Shur, V., Shishkina, E., Gurvich, V., Privalova, L., & Katsnelson, B. (2019). More data on in vitro assessment of comparative and combined toxicity of metal oxide nanoparticles. *Food and Chemical Toxicology*, *133*, 110753. doi:10.1016/j.fct.2019.110753 PMID:31400477

Buzea, C., Pacheco, I. I., & Robbie, K. (2007). Nanomaterials and nanoparticles: Sources and toxicity. *Biointerphases*, *2*(4), MR17–MR71. doi:10.1116/1.2815690 PMID:20419892

Cabral, H., Miyata, K., Osada, K., & Kataoka, K. (2018). Block Copolymer Micelles in Nanomedicine Applications. *Chemical Reviews*, *118*(14), 6844–6892. doi:10.1021/acs.chemrev.8b00199 PMID:29957926

Cai, X., Lee, A., Ji, Z., Huang, C., Chang, C. H., Wang, X., Liao, Y. P., Xia, T., & Li, R. (2017). Reduction of pulmonary toxicity of metal oxide nanoparticles by phosphonate-based surface passivation. *Particle and Fibre Toxicology*, *14*(1), 13. Advance online publication. doi:10.118612989-017-0193-5 PMID:28431555

Cao, Y., Chen, A., Radcliffe, J., Dietrich, K. N., Jones, R. L., Caldwell, K., & Rogan, W. J. (2009). Postnatal cadmium exposure, neurodevelopment, and blood pressure in children at 2, 5, and 7 years of age. *Environmental Health Perspectives*, *117*(10), 1580–1586. doi:10.1289/ehp.0900765 PMID:20019909

Chavali, M. S., & Nikolova, M. P. (2019). Metal oxide nanoparticles and their applications in nanotechnology. *SN Applied Sciences*, *1*(6), 607. doi:10.100742452-019-0592-3

Chawla, R., Sivakumar, S., & Kaur, H. (2021). Antimicrobial edible films in food packaging: Current scenario and recent nanotechnological advancements-a review. *Carbohydrate Polymer Technologies and Applications*, *2*, 100024. doi:10.1016/j.carpta.2020.100024

Chen, Y., Wubs, M., Mork, J., & Koenderink, A. F. (2011). Coherent single-photon absorption by single emitters coupled to one-dimensional nanophotonic waveguides. *New Journal of Physics*, *13*(10), 103010. doi:10.1088/1367-2630/13/10/103010

Chen, Z., Meng, H., Xing, G., Chen, C., Zhao, Y., Jia, G., Wang, T., Yuan, H., Ye, C., Zhao, F., Chai, Z., Zhu, C., Fang, X., Ma, B., & Wan, L. (2006). Acute toxicological effects of copper nanoparticles in vivo. *Toxicology Letters*, *163*(2), 109–120. doi:10.1016/j.toxlet.2005.10.003 PMID:16289865

Chertok, B., Moffat, B. A., David, A. E., Yu, F., Bergemann, C., Ross, B. D., & Yang, V. C. (2008). Iron oxide nanoparticles as a drug delivery vehicle for MRI monitored magnetic targeting of brain tumors. *Biomaterials*, *29*(4), 487–496. doi:10.1016/j.biomaterials.2007.08.050 PMID:17964647

Chien, C. C., Yan, Y. H., Juan, H. T., Cheng, T. J., Liao, J. B., Lee, H. P., & Wang, J. S. (2017). Sustained renal inflammation following 2 weeks of inhalation of occupationally relevant levels of zinc oxide nanoparticles in Sprague Dawley rats. *Journal of Toxicologic Pathology*, *30*(4), 307–314. doi:10.1293/tox.2017-0025 PMID:29097840

Cohen, D., Soroka, Y., Maor, Z., Oron, M., Portugal-Cohen, M., Berhanu, D., Valsami-Jones, E., Hai, N., & Milner, Y. (2013). Evaluation of topically applied copper(II) oxide nanoparticle cytotoxicity in human skin organ culture. *Toxicology In Vitro*, *27*(1), 292–298. doi:10.1016/j.tiv.2012.08.026 PMID:22954531

Colvin, V. L. (2003). The potential environmental impact of engineered nanomaterials. *Nature Biotechnology, 21*(10), 1166–1170. doi:10.1038/nbt875 PMID:14520401

Contado, C. (2015). Nanomaterials in consumer products: A challenging analytical problem. *Frontiers in Chemistry, 3*, 48. doi:10.3389/fchem.2015.00048 PMID:26301216

Dang, Y., Zhang, Y., Fan, L., Chen, H., & Roco, M. C. (2010). Trends in worldwide nanotechnology patent applications: 1991 to 2008. *Journal of Nanoparticle Research, 12*(3), 687–706. doi:10.100711051-009-9831-7 PMID:21170123

Das, & Jargar, Hattiwale, Yendigeri, Das, & Dhundasi. (2013). Serum vitamin E (α-tocopherol) estimation: A potential biomarker of antioxidant status evaluation on heavy metal toxicities. *Recent Patents on Biomarkers, 3*(1), 36–43. doi:10.2174/2210309011303010036

Das, S., Unni, B., Bhattacharjee, M., Wann, S. B., & Rao, P. G. (2012). Toxicological effects of arsenic exposure in a freshwater teleost fish, Channa punctatus. *African Journal of Biotechnology, 11*(19), 4447–4454. doi:10.5897/AJB11.2808

Davidson, D. C., Derk, R., He, X., Stueckle, T. A., Cohen, J., Pirela, S. V., Demokritou, P., Rojanasakul, Y., & Wang, L. (2016). Direct stimulation of human fibroblasts by nCeO2 in vitro is attenuated with an amorphous silica coating. *Particle and Fibre Toxicology, 13*(1), 23. doi:10.118612989-016-0134-8 PMID:27142434

De Berardis, B., Civitelli, G., Condello, M., Lista, P., Pozzi, R., Arancia, G., & Meschini, S. (2010). Exposure to ZnO nanoparticles induces oxidative stress and cytotoxicity in human colon carcinoma cells. *Toxicology and Applied Pharmacology, 246*(3), 116–127. doi:10.1016/j.taap.2010.04.012 PMID:20434478

Del Solar, V., & Contel, M. (2019). Metal-based antibody drug conjugates. Potential and challenges in their application as targeted therapies in cancer. *Journal of Inorganic Biochemistry, 199*, 110780. doi:10.1016/j.jinorgbio.2019.110780 PMID:31434020

Delcroix, G. J. R., Jacquart, M., Lemaire, L., Sindji, L., Franconi, F., Le Jeune, J. J., & Montero-Menei, C. N. (2009). Mesenchymal and neural stem cells labeled with HEDP-coated SPIO nanoparticles: In vitro characterization and migration potential in rat brain. *Brain Research, 1255*, 18–31. doi:10.1016/j.brainres.2008.12.013 PMID:19103182

Dolez, P. I. (2021). Energy Harvesting Materials and Structures for Smart Textile Applications: Recent Progress and Path Forward. *Sensors (Basel), 21*(18), 6297. doi:10.339021186297 PMID:34577509

Dong, X., Ma, N., Liu, M., & Liu, Z. (2015). Effects of As2O3 nanoparticles on cell growth and apoptosis of NB4 cells. *Experimental and Therapeutic Medicine, 10*(4), 1271–1276. doi:10.3892/etm.2015.2651 PMID:26622477

Drobne, D. (2007). Nanotoxicology for safe and sustainable nanotechnology. *Arhiv za Higijenu Rada i Toksikologiju, 58*(4), 471–478. doi:10.2478/v10004-007-0040-4 PMID:18063532

Duan, J., Yu, Y., Yu, Y., Li, Y., Wang, J., Geng, W., Jiang, L., Li, Q., Zhou, X., & Sun, Z. (2014). Silica nanoparticles induce autophagy and endothelial dysfunction via the PI3K/Akt/mTOR signaling pathway. *International Journal of Nanomedicine, 9*, 5131–5141. doi:10.2147/IJN.S71074 PMID:25395850

Duggal, D. (2011). Role of nanotechnology in new drug delivery systems. *International Journal of Drug Development and Research*, *3*(4), 4–8.

Dumkova, J., Smutna, T., Vrlikova, L., Docekal, B., Kristekova, D., Vecera, Z., Husakova, Z., Jakesova, V., Jedličková, A., Mikuška, P., Alexa, L., Coufalík, P., Tvrdoňová, M., Křůmal, K., Vaculovič, T., Kanický, V., Hampl, A., & Buchtová, M. (2020). A Clearance Period after Soluble Lead Nanoparticle Inhalation Did Not Ameliorate the Negative Effects on Target Tissues Due to Decreased Immune Response. *International Journal of Molecular Sciences*, *21*(22), 8738. doi:10.3390/ijms21228738 PMID:33228049

Dumkova, J., Smutna, T., Vrlíkova, L., Le Coustumer, P., Vecera, Z., Docekal, B., & Buchtova, M. (2017). Sub-chronic inhalation of lead oxide nanoparticles revealed their broad distribution and tissue-specific subcellular localization in target organs. *Particle and Fibre Toxicology*, *14*(1), 1–19. doi:10.118612989-017-0236-y PMID:29268755

Egbuna, C., Parmar, V. K., Jeevanandam, J., Ezzat, S. M., Patrick-Iwuanyanwu, K. C., Adetunji, C. O., Khan, J., Onyeike, E. N., Uche, C. Z., Akram, M., Ibrahim, M. S., El Mahdy, N. M., Awuchi, C. G., Saravanan, K., Tijjani, H., Odoh, U. E., Messaoudi, M., Ifemeje, J. C., Olisah, M. C., ... Ibeabuchi, C. G. (2021). Toxicity of Nanoparticles in Biomedical Application: Nanotoxicology. *Journal of Toxicology*, *9954443*, 1–21. Advance online publication. doi:10.1155/2021/9954443 PMID:34422042

Elias, A., & Tsourkas, A. (2009). Imaging circulating cells and lymphoid tissues with iron oxide nanoparticles. *Hematology*, *2009*(1), 720–726. doi:10.1182/asheducation-2009.1.720 PMID:20008258

El-Kemary, M., El-Shamy, H., & Mosaad, M. M. (2009). The role of capping agent on the interaction of cadmium sulphide nanoparticles with flufenamic acid drug. *Materials Chemistry and Physics*, *118*(1), 81–85. doi:10.1016/j.matchemphys.2009.07.007

Elsabahy, M., & Wooley, K. L. (2013). Cytokines as biomarkers of nanoparticle immunotoxicity. *Chemical Society Reviews*, *42*(12), 5552–5576. doi:10.1039/c3cs60064e PMID:23549679

El-Trass, A., ElShamy, H., El-Mehasseb, I., & El-Kemary, M. (2012). CuO nanoparticles: Synthesis, characterization, optical properties and interaction with amino acids. *Applied Surface Science*, *258*(7), 2997–3001. doi:10.1016/j.apsusc.2011.11.025

Ericson, B., Landrigan, P., Taylor, M. P., Frostad, J., Caravanos, J., Keith, J., & Fuller, R. (2016). The global burden of lead toxicity attributable to informal used lead-acid battery sites. *Annals of Global Health*, *82*(5), 686–699. doi:10.1016/j.aogh.2016.10.015 PMID:28283119

Fahmy, B., & Cormier, S. A. (2009). Copper oxide nanoparticles induce oxidative stress and cytotoxicity in airway epithelial cells. *Toxicology In Vitro*, *23*(7), 1365–1371. doi:10.1016/j.tiv.2009.08.005 PMID:19699289

Falsini, S., Bardi, U., Abou-Hassan, A., & Ristori, S. (2018). Sustainable strategies for large-scale nanotechnology manufacturing in the biomedical field. *Green Chemistry*, *20*(17), 3897–3907. doi:10.1039/C8GC01248B

Feng, W., Nie, W., Cheng, Y., Zhou, X., Chen, L., Qiu, K., Chen, Z., Zhu, M., & He, C. (2015). In vitro and in vivo toxicity studies of copper sulfide nanoplates for potential photothermal applications. *Nanomedicine; Nanotechnology, Biology, and Medicine, 11*(4), 901–912. doi:10.1016/j.nano.2014.12.015 PMID:25652899

Ferdous, Z., & Nemmar, A. (2020). Health Impact of Silver Nanoparticles: A Review of the Biodistribution and Toxicity Following Various Routes of Exposure. *International Journal of Molecular Sciences, 21*(7), 2375. doi:10.3390/ijms21072375 PMID:32235542

Fergusson, J. (1990). *The heavy elements: chemistry, environmental impact and health effects.* Pergamon Press.

Fu, P. P., Xia, Q., Hwang, H. M., Ray, P. C., & Yu, H. (2014). Mechanisms of nanotoxicity: Generation of reactive oxygen species. *Yao Wu Shi Pin Fen Xi, 22*(1), 64–75. doi:10.1016/j.jfda.2014.01.005 PMID:24673904

Ge, Y., Zhang, Y., He, S., Nie, F., Teng, G., & Gu, N. (2009). Fluorescence modified chitosan-coated magnetic nanoparticles for high-efficient cellular imaging. *Nanoscale Research Letters, 4*(4), 287–295. doi:10.100711671-008-9239-9 PMID:20596545

Gerloff, K., Albrecht, C., Boots, A. W., Forster, I., & Schins, R. P. (2009). Cytotoxicity and oxidative DNA damage by nanoparticles in human intestinal Caco-2 cells. *Nanotoxicology, 3*(4), 355–364. doi:10.3109/17435390903276933

Ghobadi, M. Z., Afsaneh, E., & Ghourchian, H. (2020). Nanotechnology: occupational health hazards of nanoparticles and legalization challenges. The ELSI handbook of nanotechnology: Risk, safety, ELSI and commercialization, 113-134. doi:10.1002/9781119592990.ch7

Gonzales-Weimuller, M., Zeisberger, M., & Krishnan, K. M. (2009). Size-dependant heating rates of iron oxide nanoparticles for magnetic fluid hyperthermia. *Journal of Magnetism and Magnetic Materials, 321*(13), 1947–1950. doi:10.1016/j.jmmm.2008.12.017 PMID:26405373

Gonzalez, H. O., Roling, J. A., Baldwin, W. S., & Bain, L. J. (2006). Physiological changes and differential gene expression in mummichogs (Fundulus heteroclitus) exposed to arsenic. *Aquatic Toxicology (Amsterdam, Netherlands), 77*(1), 43–52. doi:10.1016/j.aquatox.2005.10.014 PMID:16356559

Guan, R., Kang, T., Lu, F., Zhang, Z., Shen, H., & Liu, M. (2012). Cytotoxicity, oxidative stress, and genotoxicity in human hepatocyte and embryonic kidney cells exposed to ZnO nanoparticles. *Nanoscale Research Letters, 7*(1), 602. Advance online publication. doi:10.1186/1556-276X-7-602 PMID:23110934

Guo, D., Bi, H., Liu, B., Wu, Q., Wang, D., & Cui, Y. (2013). Reactive oxygen species-induced cytotoxic effects of zinc oxide nanoparticles in rat retinal ganglion cells. *Toxicology In Vitro, 27*(2), 731–738. doi:10.1016/j.tiv.2012.12.001 PMID:23232460

Gupta, R. (2019). Cadmium nanoparticles and its toxicity. *Journal of Critical Reviews, 2009*, 1–7. doi:10.22159/jcr.2019v6i5.34073

Gupta, R., & Xie, H. (2018). Nanoparticles in daily life: Applications, toxicity and regulations. *Journal of Environmental Pathology, Toxicology and Oncology*, *37*(3), 209–230. doi:10.1615/JEnvironPatholToxicolOncol.2018026009 PMID:30317972

Hackenberg, S., Scherzed, A., Technau, A., Kessler, M., Froelich, K., Ginzkey, C., Koehler, C., Burghartz, M., Hagen, R., & Kleinsasser, N. (2011). Cytotoxic, genotoxic and pro-inflammatory effects of zinc oxide nanoparticles in human nasal mucosa cells in vitro. *Toxicology In Vitro*, *25*(3), 657–663. doi:10.1016/j.tiv.2011.01.003 PMID:21232592

Hallare, A. V., Schirling, M., Luckenbach, T., Köhler, H. R., & Triebskorn, R. (2005). Combined effects of temperature and cadmium on developmental parameters and biomarker responses in zebrafish (Danio rerio) embryos. *Journal of Thermal Biology*, *30*(1), 7–17. doi:10.1016/j.jtherbio.2004.06.002

Hanagata, N., Zhuang, F., Connolly, S., Li, J., Ogawa, N., & Xu, M. (2011). Molecular responses of human lung epithelial cells to the toxicity of copper oxide nanoparticles inferred from whole genome expression analysis. *ACS Nano*, *5*(12), 9326–9338. doi:10.1021/nn202966t PMID:22077320

Hannah, W., & Thompson, P. B. (2008). Nanotechnology, risk and the environment: A review. *Journal of Environmental Monitoring*, *10*(3), 291–300. doi:10.1039/b718127m PMID:18392270

Hedberg, J., Blomberg, E., & Odnevall Wallinder, I. (2019). In the search for nanospecific effects of dissolution of metallic nanoparticles at freshwater-like conditions. *Critical Reviews in Environmental Science and Technology*, *53*(8), 4030–4044. doi:10.1021/acs.est.8b05012 PMID:30908015

Heng, B. C., Zhao, X., Tan, E. C., Khamis, N., Assodani, A., Xiong, S., Ruedl, C., Ng, K. W., & Loo, J. S. (2011). Evaluation of the cytotoxic and inflammatory potential of differentially shaped zinc oxide nanoparticles. *Archives of Toxicology*, *85*(12), 1517–1528. doi:10.100700204-011-0722-1 PMID:21656222

Hobson, D. W., Roberts, S. M., Shvedova, A. A., Warheit, D. B., Hinkley, G. K., & Guy, R. C. (2016). Applied nanotoxicology. *International Journal of Toxicology*, *35*(1), 5–16. doi:10.1177/1091581816628484 PMID:26957538

Hochella, M. F. Jr, Mogk, D. W., Ranville, J., Allen, I. C., Luther, G. W., Marr, L. C., McGrail, B. P., Murayama, M., Qafoku, N. P., Rosso, K. M., Sahai, N., Schroeder, P. A., Vikesland, P., Westerhoff, P., & Yang, Y. (2019). Natural, incidental, and engineered nanomaterials and their impacts on the Earth system. *Science*, *363*(6434), eaau8299. Advance online publication. doi:10.1126cience.aau8299 PMID:30923195

Hoffmann, M. R., Martin, S. T., Choi, W., & Bahnemann, D. W. (1995). Environmental Applications of Semiconductor Photocatalysis. *Chemical Reviews*, *95*(1), 69–96. doi:10.1021/cr00033a004

Honarvar, Z., Hadian, Z., & Mashayekh, M. (2016). Nanocomposites in food packaging applications and their risk assessment for health. *Electronic Physician*, *8*(6), 2531–2538. doi:10.19082/2531 PMID:27504168

Hossain, S. T., & Mukherjee, S. K. (2013). Toxicity of cadmium sulfide (CdS) nanoparticles against Escherichia coli and HeLa cells. *Journal of Hazardous Materials*, *260*, 1073–1082. doi:10.1016/j.jhazmat.2013.07.005 PMID:23892173

Hou, J., Liu, H., Zhang, S., Liu, X., Hayat, T., Alsaedi, A., & Wang, X. (2019). Mechanism of toxic effects of Nano-ZnO on cell cycle of zebrafish (Danio rerio). *Chemosphere*, *229*, 206–213. doi:10.1016/j.chemosphere.2019.04.217 PMID:31078877

Huang, C. C., Aronstam, R. S., Chen, D. R., & Huang, Y. W. (2010). Oxidative stress, calcium homeostasis, and altered gene expression in human lung epithelial cells exposed to ZnO nanoparticles. *Toxicology In Vitro*, *24*(1), 45–55. doi:10.1016/j.tiv.2009.09.007 PMID:19755143

Huang, G. G., Wang, C. T., Tang, H. T., Huang, Y. S., & Yang, J. (2006). ZnO nanoparticle-modified infrared internal reflection elements for selective detection of volatile organic compounds. *Analytical Chemistry*, *78*(7), 2397–2404. doi:10.1021/ac051930+ PMID:16579626

Hughes, G. A. (2005). Nanostructure-mediated drug delivery. *Nanomedicine; Nanotechnology, Biology, and Medicine*, *1*(1), 22–30. doi:10.1016/j.nano.2004.11.009 PMID:17292054

Humtsoe, N., Davoodi, R., Kulkarni, B. G., & Chavan, B. (2007). Effect of arsenic on the enzymes of the Rohu carp, Labeo rohita (Hamilton 1822). *Raffles Bulletin of Zoology, 14*, 17–19.

Iavicoli, I., Leso, V., Beezhold, D. H., & Shvedova, A. A. (2017). Nanotechnology in agriculture: Opportunities, toxicological implications, and occupational risks. *Toxicology and Applied Pharmacology*, *329*, 96–111. doi:10.1016/j.taap.2017.05.025 PMID:28554660

Izci, M., Maksoudian, C., Manshian, B. B., & Soenen, S. J. (2021). The use of alternative strategies for enhanced nanoparticle delivery to solid tumors. *Chemical Reviews*, *121*(3), 1746–1803. doi:10.1021/acs.chemrev.0c00779 PMID:33445874

Jahangirnejad, R., Goudarzi, M., Kalantari, H., Najafzadeh, H., & Rezaei, M. (2020). Subcellular organelle toxicity caused by arsenic nanoparticles in isolated rat hepatocytes. *The International Journal of Occupational and Environmental Medicine*, *11*(1), 41–52. doi:10.15171/ijoem.2020.1614 PMID:31905194

Jain, S., Jain, A., Kachhawah, P., & Devra, V. (2015). Synthesis and size control of copper nanoparticles and their catalytic application. *Transactions of Nonferrous Metals Society of China*, *25*(12), 3995–4000. doi:10.1016/S1003-6326(15)64048-1

Jamakala, O., & Rani, A. U. (2014). Mitigating role of zinc and iron against cadmium induced toxicity in liver and kidney of male albino rat: A study with reference to metallothionein quantification. *International Journal of Pharmacy and Pharmaceutical Sciences*, *6*(9), 411–417.

Jane Cypriyana, J. P. J., Saigeetha, S., Lavanya, A. A., Samrot, A. V., Ponniah, P., & Chakravarthi, S. (2021). Overview on toxicity of nanoparticles, it's mechanism, models used in toxicity studies and disposal methods–A review. *Biocatalysis and Agricultural Biotechnology*, *36*, 102117. doi:10.1016/j.bcab.2021.102117

Jeevanandam, J., Barhoum, A., Chan, Y. S., Dufresne, A., & Danquah, M. K. (2018). Review on nanoparticles and nanostructured materials: History, sources, toxicity and regulations. *Beilstein Journal of Nanotechnology*, *9*(1), 1050–1074. doi:10.3762/bjnano.9.98 PMID:29719757

Jeng, H. A., & Swanson, J. (2006). Toxicity of metal oxide nanoparticles in mammalian cells. *Journal of Environmental Science and Health*, *41*(12), 2699–2711. doi:10.1080/10934520600966177 PMID:17114101

Jeyaraj, M., Gurunathan, S., Qasim, M., Kang, M. H., & Kim, J. H. (2019). A Comprehensive Review on the Synthesis, Characterization, and Biomedical Application of Platinum Nanoparticles. *Nanomaterials (Basel, Switzerland)*, *9*(12), 1719. doi:10.3390/nano9121719 PMID:31810256

Ji, Z., Wang, X., Zhang, H., Lin, S., Meng, H., Sun, B., George, S., Xia, T., Nel, A. E., & Zink, J. I. (2012). Designed synthesis of CeO2 nanorods and nanowires for studying toxicological effects of high aspect ratio nanomaterials. *ACS Nano*, *6*(6), 5366–5380. doi:10.1021/nn3012114 PMID:22564147

Jing, X., Park, J. H., Peters, T. M., & Thorne, P. S. (2015). Toxicity of copper oxide nanoparticles in lung epithelial cells exposed at the air-liquid interface compared with in vivo assessment. *Toxicology In Vitro*, *29*(3), 502–511. doi:10.1016/j.tiv.2014.12.023 PMID:25575782

Joseph, P., Muchnok, T. K., Klishis, M. L., Roberts, J. R., Antonini, J. M., Whong, W. Z., & Ong, T. (2001). Cadmium-induced cell transformation and tumorigenesis are associated with transcriptional activation of c-fos, c-jun, and c-myc proto-oncogenes: Role of cellular calcium and reactive oxygen species. *Toxicological Sciences*, *61*(2), 295–303. doi:10.1093/toxsci/61.2.295 PMID:11353138

Joshi, A., Sharma, A., Bachheti, R.K., Husen, A., & Mishra, V.K. (2019). Plant-mediated synthesis of copper oxide nanoparticles and their biological applications. *Nanomaterials and Plant Potential*, 221–237. doi:10.1007/978-3-030-05569-1_8

Kakkar, P., & Jaffery, F. N. (2005). Biological markers for metal toxicity. *Environmental Toxicology and Pharmacology*, *19*(2), 335–349. doi:10.1016/j.etap.2004.09.003 PMID:21783494

Kao, Y. Y., Chen, Y. C., Cheng, T. J., Chiung, Y. M., & Liu, P. S. (2012). Zinc oxide nanoparticles interfere with zinc ion homeostasis to cause cytotoxicity. *Toxicological Sciences*, *125*(2), 462–472. doi:10.1093/toxsci/kfr319 PMID:22112499

Karami, H., Karimi, M. A., Haghdar, S., Sadeghi, A., Mir-Ghasemi, R., & Mahdi-Khani, S. (2008). Synthesis of lead oxide nanoparticles by sonochemical method and its application as cathode and anode of lead-acid batteries. *Materials Chemistry and Physics*, *108*(2-3), 337–344. doi:10.1016/j.matchemphys.2007.09.045

Karlsson, H. L., Cronholm, P., Gustafsson, J., & Moller, L. (2008). Copper oxide nanoparticles are highly toxic: A comparison between metal oxide nanoparticles and carbon nanotubes. *Chemical Research in Toxicology*, *21*(9), 1726–1732. doi:10.1021/tx800064j PMID:18710264

Khalil, A. T., Ovais, M., Ullah, I., Ali, M., Jan, S. A., Shinwari, Z. K., & Maaza, M. (2020). Bioinspired synthesis of pure massicot phase lead oxide nanoparticles and assessment of their biocompatibility, cytotoxicity and in-vitro biological properties. *Arabian Journal of Chemistry*, *13*(1), 916–931. doi:10.1016/j.arabjc.2017.08.009

Khan, I., Saeed, K., & Khan, I. (2019a). Nanoparticles: Properties, applications and toxicities. *Arabian Journal of Chemistry*, *12*(7), 908–931. doi:10.1016/j.arabjc.2017.05.011

Khan, M. R., Adam, V., Rizvi, T. F., Zhang, B., Ahamad, F., Josko, I., Zhu, Y., Yang, M., & Mao, C. (2019b). Nanoparticle-Plant Interactions: Two-Way Traffic. *Small (Germany)*, *15*(37), e1901794. doi:10.1002mll.201901794 PMID:31318142

Kim, S. D., Moon, C. K., Eun, S. Y., Ryu, P. D., & Jo, S. A. (2005). Identification of ASK1, MKK4, JNK, c-Jun, and caspase-3 as a signaling cascade involved in cadmium-induced neuronal cell apoptosis. *Biochemical and Biophysical Research Communications*, *328*(1), 326–334. doi:10.1016/j.bbrc.2004.11.173 PMID:15670787

Kolodziejczak-Radzimska, A., & Jesionowski, T. (2014, April 09). Zinc Oxide-From Synthesis to Application: A Review. *Materials (Basel)*, *7*(4), 2833–2881. doi:10.3390/ma7042833 PMID:28788596

Kominkova, M., Milosavljevic, V., Vitek, P., Polanska, H., Cihalova, K., Dostalova, S., Hynstova, V., Guran, R., Kopel, P., Richtera, L., Masarik, M., Brtnicky, M., Kynicky, J., Zitka, O., & Adam, V. (2017). Comparative study on toxicity of extracellularly biosynthesized and laboratory synthesized CdTe quantum dots. *Journal of Biotechnology*, *241*, 193–200. doi:10.1016/j.jbiotec.2016.10.024 PMID:27984119

Kooi, M. E., Cappendijk, V. C., Cleutjens, K. B., Kessels, A. G., Kitslaar, P. J., Borgers, M., Frederik, P. M., Daemen, M. J., & van Engelshoven, J. M. (2003). Accumulation of ultrasmall superparamagnetic particles of iron oxide in human atherosclerotic plaques can be detected by in vivo magnetic resonance imaging. *Circulation*, *107*(19), 2453–2458. doi:10.1161/01.CIR.0000068315.98705.CC PMID:12719280

Kozhevnikova, N. S., Vorokh, A. S., & Rempel, A. A. (2010). Preparation of stable colloidal solution of cadmium sulfide CdS using ethylenediaminetetraacetic acid. *Russian Journal of General Chemistry*, *80*(3), 391–394. doi:10.1134/S1070363210030035

Kumar, V., & Arora, K. (2020). Trends in nano-inspired biosensors for plants. *Materials Science for Energy Technologies*, *3*, 255–273. doi:10.1016/j.mset.2019.10.004

Kurwadkar, S., Pugh, K., Gupta, A., & Ingole, S. (2015a). Nanoparticles in the environment: Occurrence, distribution, and risks. *Journal of Hazardous, Toxic and Radioactive Waste*, *19*(3). Advance online publication. doi:10.1061/(ASCE)HZ.2153-5515.0000258

Kurwadkar, S., Zhang, X.J., Ramirez, D., Mitchell, F.L. (2015b). *Emerging micro-pollutants in the environment: Occurrence, fate, and distribution.* . doi:10.1021/bk-2015-1198.fw001

Lafront, A. M., Zhang, W., Ghali, E., & Houlachi, G. (2010). Electrochemical noise studies of the corrosion behaviour of lead anodes during zinc electrowinning maintenance. *Electrochimica Acta*, *55*(22), 6665–6675. doi:10.1016/j.electacta.2010.06.005

Laha, D., Pramanik, A., Maity, J., Mukherjee, A., Pramanik, P., Laskar, A., & Karmakar, P. (2014). Interplay between autophagy and apoptosis mediated by copper oxide nanoparticles in human breast cancer cells MCF7. *Biochimica et Biophysica Acta*, *1840*(1), 1–9. doi:10.1016/j.bbagen.2013.08.011 PMID:23962629

Lanone, S., Rogerieux, F., Geys, J., Dupont, A., Maillot-Marechal, E., Boczkowski, J., Lacroix, G., & Hoet, P. (2009). Comparative toxicity of 24 manufactured nanoparticles in human alveolar epithelial and macrophage cell lines. *Particle and Fibre Toxicology*, *6*(1), 14. doi:10.1186/1743-8977-6-14 PMID:19405955

Lee, I. C., Ko, J. W., Park, S. H., Shin, N. R., Shin, I. S., Moon, C., Kim, J. H., Kim, H. C., & Kim, J. C. (2016). Comparative toxicity and biodistribution assessments in rats following subchronic oral exposure to copper nanoparticles and microparticles. *Particle and Fibre Toxicology*, *13*(1), 56. doi:10.118612989-016-0169-x PMID:27788687

Lee, Y., Choi, J., Lee, K. J., Stott, N., & Kim, D. (2008). Large-scale synthesis of copper nanoparticles by chemically controlled reduction for applications of inkjet-printed electronics. *Nanotechnology*, *19*(1), 415604. doi:10.1088/0957-4484/19/41/415604 PMID:21832649

Lei, R., Wu, C., Yang, B., Ma, H., Shi, C., Wang, Q., & Liao, M. (2008). Integrated metabolomic analysis of the nano-sized copper particle-induced hepatotoxicity and nephrotoxicity in rats: A rapid in vivo screening method for nanotoxicity. *Toxicology and Applied Pharmacology*, *232*(2), 292–301. doi:10.1016/j.taap.2008.06.026 PMID:18706438

Lespes, G., Faucher, S., & Slaveykova, V. I. (2020). Natural nanoparticles, anthropogenic nanoparticles, where is the frontier? *Frontiers in Environmental Science*, *8*, 71. doi:10.3389/fenvs.2020.00071

Li, K. G., Chen, J. T., Bai, S. S., Wen, X., Song, S. Y., Yu, Q., Li, J., & Wang, Y. Q. (2009). Intracellular oxidative stress and cadmium ions release induce cytotoxicity of unmodified cadmium sulfide quantum dots. *Toxicology In Vitro*, *23*(6), 1007–1013. doi:10.1016/j.tiv.2009.06.020 PMID:19540911

Li, X., Jia, Y., Wei, J., Zhu, H., Wang, K., Wu, D., & Cao, A. (2010). Solar cells and light sensors based on nanoparticle-grafted carbon nanotube films. *ACS Nano*, *4*(4), 2142–2148. doi:10.1021/nn901563y PMID:20222675

Li, Y., Sun, Y., Zhang, G., He, Z., Wang, Y., & Cui, J. (2016). Effects of copper oxide nanoparticles on developing zebrafish embryos and larvae. *International Journal of Nanomedicine*, *11*, 905–918. doi:10.2147/IJN.S100350 PMID:27022258

Lin, C. F., Liang, E. Z., Shih, S. M., & Su, W. F. (2002). CdS nanoparticle light-emitting diode on Si. In Light-Emitting Diodes: Research, Manufacturing, and Applications. International Society for Optics and Photonics.

Liosis, C., Papadopoulou, A., Karvelas, E., Karakasidis, T. E., & Sarris, I. E. (2021). Heavy Metal adsorption using magnetic nanoparticles for water purification: A critical review. *Materials (Basel)*, *14*(24), 7500. doi:10.3390/ma14247500 PMID:34947096

Liu, G., Gao, J., Ai, H., & Chen, X. (2013). Applications and potential toxicity of magnetic iron oxide nanoparticles. *Small*, *9*(9-10), 1533–1545. doi:10.1002mll.201201531 PMID:23019129

Liu, L., Sun, M., Li, Q., Zhang, H., Alvarez, P. J., Liu, H., & Chen, W. (2014). Genotoxicity and Cytotoxicity of Cadmium Sulfide Nanomaterials to Mice: Comparison Between Nanorods and Nanodots. *Environmental Engineering Science*, *31*(7), 373–380. doi:10.1089/ees.2013.0417 PMID:25053877

Liu, S., Regulacio, M. D., Tee, S. Y., Khin, Y. W., Teng, C. P., Koh, L. D., Guan, G., & Han, M. Y. (2016). Preparation, functionality, and application of metal oxide-coated noble metal nanoparticles. *Chemical Record (New York, N.Y.)*, *6*(4), 1965–1990. doi:10.1002/tcr.201600028 PMID:27291595

Liu, Y., Hardie, J., Zhang, X., & Rotello, V. M. (2017). Effects of engineered nanoparticles on the innate immune system. *Seminars in Immunology*, *34*, 25–32. doi:10.1016/j.smim.2017.09.011 PMID:28985993

Liu, Y., Pang, N., Cai, Y., Yang, Y., Zeng, C., & Wang, Y. (2021). Application of Nanooptics in Photographic Imagery and Medical Imaging. *Journal of Chemistry*, *2021*, 1–15. Advance online publication. doi:10.1155/2021/2384322

Lopez, E., Figueroa, S., Oset-Gasque, M. J., & Gonzalez, M. P. (2003). Apoptosis and necrosis: Two distinct events induced by cadmium in cortical neurons in culture. *British Journal of Pharmacology*, *138*(5), 901–911. doi:10.1038j.bjp.0705111 PMID:12642392

Lu, X., Miousse, I. R., Pirela, S. V., Moore, J. K., Melnyk, S., Koturbash, I., & Demokritou, P. (2016). In vivo epigenetic effects induced by engineered nanomaterials: A case study of copper oxide and laser printer-emitted engineered nanoparticles. *Nanotoxicology*, *10*(5), 629–639. doi:10.3109/17435390.2015.1108473 PMID:26559097

Luh, M. D., Baker, R. A., & Henley, D. E. (1973). Arsenic analysis and toxicity—A review. *The Science of the Total Environment*, *2*(1), 1–12. doi:10.1016/0048-9697(73)90002-8 PMID:4804487

Ma, H., Williams, P. L., & Diamond, S. A. (2013). Ecotoxicity of manufactured ZnO nanoparticles—A review. *Environmental Pollution*, *172*, 76–85. doi:10.1016/j.envpol.2012.08.011 PMID:22995930

Ma, R. M., Dai, L., & Qin, G. G. (2007). Enhancement-mode metal-semiconductor field-effect transistors based on single n-Cd S nanowires. *Applied Physics Letters*, *90*(9), 093109. doi:10.1063/1.2710004

Ma, Y., Cai, F., Li, Y., Chen, J., Han, F., & Lin, W. (2020). A review of the application of nanoparticles in the diagnosis and treatment of chronic kidney disease. *Bioactive Materials*, *5*(3), 732–743. doi:10.1016/j.bioactmat.2020.05.002 PMID:32596555

Madhavi, V., Reddy, A. V. B., & Madhavi, G. (2022). Nanosorbents for the removal of heavy metal pollutants. In Nano-Bioremediation: Fundamentals and Applications (pp. 363-387). Elsevier. doi:10.1016/B978-0-12-823962-9.00004-0

Magdolenova, Z., Drlickova, M., Henjum, K., Runden-Pran, E., Tulinska, J., Bilanicova, D., Pojana, G., Kazimirova, A., Barancokova, M., Kuricova, M., Liskova, A., Staruchova, M., Ciampor, F., Vavra, I., Lorenzo, Y., Collins, A., Rinna, A., Fjellsbø, L., Volkovova, K., ... Dusinska, M. (2015). Coating-dependent induction of cytotoxicity and genotoxicity of iron oxide nanoparticles. *Nanotoxicology*, *9*(sup1), 44–56. doi:10.3109/17435390.2013.847505 PMID:24228750

Maharramov, A. M., Hasanova, U. A., Suleymanova, I. A., Osmanova, G. E., & Hajiyeva, N. E. (2019). The engineered nanoparticles in food chain: Potential toxicity and effects. *SN Applied Sciences*, *1*(11), 1362. doi:10.100742452-019-1412-5

Majhi, K. C., & Yadav, M. S. (2021). Synthesis of inorganic nanomaterials using carbohydrates. In Green Sustainable Process for Chemical and Environmental Engineering and Science (pp. 109-135). Elsevier. doi:10.1016/B978-0-12-821887-7.00003-3

Manuja, A., Kumar, B., Kumar, R., Chhabra, D., Ghosh, M., Manuja, M., Brar, B., Pal, Y., Tripathi, B. N., & Prasad, M. (2021). Metal/metal oxide nanoparticles: Toxicity concerns associated with their physical state and remediation for biomedical applications. *Toxicology Reports*, *8*, 1970–1978. doi:10.1016/j.toxrep.2021.11.020 PMID:34934635

Martos, M., Morales, J., Sanchez, L., Ayouchi, R., Leinen, D., Martin, F., & Barrado, J. R. (2001). Electrochemical properties of lead oxide films obtained by spray pyrolysis as negative electrodes for lithium secondary batteries. *Electrochimica Acta*, *46*(19), 2939–2948. doi:10.1016/S0013-4686(01)00512-6

Mathew, B. B., & Krishnamurthy, N. B. (2018). Evaluation of lead oxide nanoparticles synthesized by chemical and biological methods. *Journal of Nanomedicine Research*, *7*(3), 195–198. doi:10.15406/jnmr.2018.07.00187

Meng, C. Y., Han, Y. F., Liu, Y. L., Gao, H. X., Ren, Y. Y., Qian, Q. Z., Wang, Q., & Li, Q. Z. (2019). Resveratrol alleviate the injury of mice liver induced by cadmium sulfide nanoparticles. *The Kaohsiung Journal of Medical Sciences*, *35*(5), 297–302. doi:10.1002/kjm2.12056 PMID:30913377

Meng, H., Chen, Z., Xing, G., Yuan, H., Chen, C., Zhao, F., Zhang, C., & Zhao, Y. (2007). Ultrahigh reactivity provokes nanotoxicity: Explanation of oral toxicity of nano-copper particles. *Toxicology Letters*, *175*(1-3), 102–110. doi:10.1016/j.toxlet.2007.09.015 PMID:18024012

Meyer, K., Rajanahalli, P., Ahamed, M., Rowe, J. J., & Hong, Y. (2011). ZnO nanoparticles induce apoptosis in human dermal fibroblasts via p53 and p38 pathways. *Toxicology In Vitro*, *25*(8), 1721–1726. doi:10.1016/j.tiv.2011.08.011 PMID:21903158

Minigaliyeva, I. A., Sutunkova, M. P., Gurvich, V. B., Bushueva, T. V., Klinova, S. V., Solovyeva, S. N., Chernyshov, I. N., Valamina, I. E., Shur, V. Y., Shishkina, E. V., Makeyev, O. H., Panov, V. G., Privalova, L. I., & Katsnelson, B. A. (2020). An overview of experiments with lead-containing nanoparticles performed by the Ekaterinburg nanotoxicological research team. *Nanotoxicology*, *14*(6), 788–806. doi:10.1080/17435390.2020.1762132 PMID:32396411

Miri, A., Sarani, M., Hashemzadeh, A., Mardani, Z., & Darroudi, M. (2018). Biosynthesis and cytotoxic activity of lead oxide nanoparticles. *Green Chemistry Letters and Reviews*, *11*(4), 567–572. doi:10.1080/17518253.2018.1547926

Mirnajafizadeh, F., Ramsey, D., McAlpine, S., Wang, F., & Stride, J. A. (2019). Nanoparticles for Bioapplications: Study of the Cytotoxicity of Water Dispersible CdSe(S) and CdSe(S)/ZnO Quantum Dots. *Nanomaterials (Basel, Switzerland)*, *9*(3), 465. doi:10.3390/nano9030465 PMID:30897752

Mitchell, M. J., Billingsley, M. M., Haley, R. M., Wechsler, M. E., Peppas, N. A., & Langer, R. (2021). Engineering precision nanoparticles for drug delivery. *Nature Reviews. Drug Discovery*, *20*(2), 101–124. doi:10.103841573-020-0090-8 PMID:33277608

Mittal, D., Kaur, G., Singh, P., Yadav, K., & Ali, S. A. (2020). Nanoparticle-based sustainable agriculture and food science: Recent advances and future outlook. *Frontiers in Nanotechnology*, *2*, 10. doi:10.3389/fnano.2020.579954

Mody, C., & Choi, H. (2013). From materials science to nanotechnology: Institutions, communities, and disciplines at Cornell University, 1960-2000. *Historical Studies in the Natural Sciences*, *43*(2), 121–161. doi:10.1525/hsns.2013.43.2.121

Mody, V. V., Siwale, R., Singh, A., & Mody, H. R. (2010). Introduction to metallic nanoparticles. *Journal of Pharmacy & Bioallied Sciences*, *2*(4), 282–289. doi:10.4103/0975-7406.72127 PMID:21180459

Mohammed, Y. H., Holmes, A., Haridass, I. N., Sanchez, W. Y., Studier, H., Grice, J. E., Benson, H. A. E., & Roberts, M. S. (2019). Support for the Safe Use of Zinc Oxide Nanoparticle Sunscreens: Lack of Skin Penetration or Cellular Toxicity after Repeated Application in Volunteers. *The Journal of Investigative Dermatology*, *139*(2), 308–315. doi:10.1016/j.jid.2018.08.024 PMID:30448212

Monroe, R. K., & Halvorsen, S. W. (2006). Cadmium blocks receptor-mediated Jak/STAT signaling in neurons by oxidative stress. *Free Radical Biology & Medicine*, *41*(3), 493–502. doi:10.1016/j.freeradbiomed.2006.04.023 PMID:16843830

Moore, A., Marecos, E., Bogdanov Jr, A., & Weissleder, R. (2000). Tumoral distribution of long-circulating dextran-coated iron oxide nanoparticles in a rodent model. *Radiology, 214*(2), 568-574. doi:.r00fe19568 doi:10.1148/radiology.214.2

Morales, M. P., Bomati-Miguel, O., De Alejo, R. P., Ruiz-Cabello, J., Veintemillas-Verdaguer, S., & O'Grady, K. (2003). Contrast agents for MRI based on iron oxide nanoparticles prepared by laser pyrolysis. *Journal of Magnetism and Magnetic Materials*, *266*(1-2), 102–109. doi:10.1016/S0304-8853(03)00461-X

Murray, A. R., Kisin, E., Inman, A., Young, S. H., Muhammed, M., Burks, T., Uheida, A., Tkach, A., Waltz, M., Castranova, V., Fadeel, B., Kagan, V. E., Riviere, J. E., Monteiro-Riviere, N., & Shvedova, A. A. (2013). Oxidative stress and dermal toxicity of iron oxide nanoparticles in vitro. *Cell Biochemistry and Biophysics*, *67*(2), 461–476. doi:10.100712013-012-9367-9 PMID:22669739

Musial, J., Krakowiak, R., Mlynarczyk, D. T., Goslinski, T., & Stanisz, B. J. (2020). Titanium dioxide nanoparticles in food and personal care products -What do we know about their safety? *Nanomaterials (Basel, Switzerland)*, *10*(6), 1110. doi:10.3390/nano10061110 PMID:32512703

Nafees, M., Ikram, M., & Ali, S. (2017). Thermal stability of lead sulfide and lead oxide nano-crystalline materials. *Applied Nanoscience*, *7*(7), 399–406. doi:10.100713204-017-0578-7

Nair, R., Varghese, S., Nair, B., Maekawa, T., Yoshida, Y., & Kumar, D. (2010). Nanoparticulate Material Delivery To Plants. *Plant Science*, *179*(3), 154–163. doi:10.1016/j.plantsci.2010.04.012

Namvar, F., Rahman, H. S., Mohamad, R., Azizi, S., Tahir, P. M., Chartrand, M. S., & Yeap, S. K. (2015). Cytotoxic effects of biosynthesized zinc oxide nanoparticles on murine cell lines. *Evidence-Based Complementary and Alternative Medicine*, *593014*, 1–11. Advance online publication. doi:10.1155/2015/593014 PMID:25784947

Naqvi, S., Samim, M., Abdin, M. Z., Ahmed, F. J., Maitra, A. N., Prashant, C. K., & Dinda, A. K. (2010). Concentration-dependent toxicity of iron oxide nanoparticles mediated by increased oxidative stress. *International Journal of Nanomedicine*, *5*, 983. doi:10.2147/IJN.S13244 PMID:21187917

Navarro-Espinoza, S., Meza-Figueroa, D., Guzman, R., Duarte-Moller, A., Esparza-Ponce, H., Paz-Moreno, F., & Pedroza-Montero, M. (2021). Release of nanoparticles in the environment and catalytic converters ageing. *Nanomaterials (Basel, Switzerland), 11*(12), 3406. doi:10.3390/nano11123406 PMID:34947754

Neethirajan, S., & Jayas, D. S. (2011). Nanotechnology for the food and bioprocessing industries. *Food and Bioprocess Technology, 4*(1), 39–47. doi:10.100711947-010-0328-2 PMID:32215165

Ng, C. T., Yong, L. Q., Hande, M. P., Ong, C. N., Yu, L. E., Bay, B. H., & Baeg, G. H. (2017). Zinc oxide nanoparticles exhibit cytotoxicity and genotoxicity through oxidative stress responses in human lung fibroblasts and Drosophila melanogaster. *International Journal of Nanomedicine, 12*, 1621-1637. doi:.s124403 doi:10.2147/ijn

Ng, D. Q., Chu, Y., Tan, S. W., Wang, S. L., Lin, Y. P., Chu, C. H., Soo, Y.-L., Song, Y.-F., & Chen, P. J. (2019). In vivo evidence of intestinal lead dissolution from lead dioxide (PbO 2) nanoparticles and resulting bioaccumulation and toxicity in medaka fish. *Environmental Science. Nano, 6*(2), 580–591. doi:10.1039/C8EN00893K

Nihei, H., Matsuoka, F., & Okamoto, A. (2009). Targeted coherent control inside photonic band gaps for nanophotonics devices. In *2009 14th OptoElectronics and Communications Conference* (pp. 1-2). Institute of Electrical and Electronics Engineers. 10.1109/OECC.2009.5220789

Oberbek, P., Kozikowski, P., Czarnecka, K., Sobiech, P., Jakubiak, S., & Jankowski, T. (2019). Inhalation exposure to various nanoparticles in work environment—Contextual information and results of measurements. *Journal of Nanoparticle Research, 21*(11), 222. doi:10.100711051-019-4651-x

Omara, A. E. D., Elsakhawy, T., Alshaal, T., El-Ramady, H., Kovacs, Z., & Fari, M. (2019). Nanoparticles: A novel approach for sustainable agro-productivity. Environment. *Biodiversity and Soil Security, 3*(2019), 29–62. doi:10.21608/jenvbs.2019.7478.1050

Osman, I. F., Baumgartner, A., Cemeli, E., Fletcher, J. N., & Anderson, D. (2010). Genotoxicity and cytotoxicity of zinc oxide and titanium dioxide in HEp-2 cells. *Nanomedicine (London), 5*(8), 1193–1203. doi:10.2217/nnm.10.52 PMID:21039197

Osmond-McLeod, M. J., Osmond, R. I., Oytam, Y., McCall, M. J., Feltis, B., Mackay-Sim, A., Wood, S. A., & Cook, A. L. (2013). Surface coatings of ZnO nanoparticles mitigate differentially a host of transcriptional, protein and signalling responses in primary human olfactory cells. *Particle and Fibre Toxicology, 10*(1), 54. doi:10.1186/1743-8977-10-54 PMID:24144420

Oszlanczi, G., Papp, A., Szabo, A., Nagymajtenyi, L., Sapi, A., Konya, Z., Paulik, E., & Vezer, T. (2011). Nervous system effects in rats on subacute exposure by lead-containing nanoparticles via the airways. *Inhalation Toxicology, 23*(4), 173–181. doi:10.3109/08958378.2011.553248 PMID:21456951

Pandian, S. R. K., Deepak, V., Kalishwaralal, K., & Gurunathan, S. (2011). Biologically synthesized fluorescent CdS NPs encapsulated by PHB. *Enzyme and Microbial Technology, 48*(4-5), 319–325. doi:10.1016/j.enzmictec.2011.01.005 PMID:22112944

Papadaki, D., Kiriakidis, G., & Tsoutsos T. (2018). Applications of nanotechnology in construction industry. *Fundamentals of Nanoparticles*, 343-370. doi:10.1016/B978-0-323-51255-8.00011-2

Papp, A., Oszlanczi, G., Horvath, E., Paulik, E., Kozma, G., Sapi, A., Konya, Z., & Szabo, A. (2012). Consequences of subacute intratracheal exposure of rats to cadmium oxide nanoparticles: Electrophysiological and toxicological effects. *Toxicology and Industrial Health*, *28*(10), 933–941. doi:10.1177/0748233711430973 PMID:22246446

Passian, A., & Imam, N. (2019). Nanosystems, edge computing, and the next generation computing systems. *Sensors (Basel)*, *19*(18), 4048. doi:10.339019184048 PMID:31546907

Pati, R., Das, I., Mehta, R. K., Sahu, R., & Sonawane, A. (2016). Zinc-Oxide Nanoparticles Exhibit Genotoxic, Clastogenic, Cytotoxic and Actin Depolymerization Effects by Inducing Oxidative Stress Responses in Macrophages and Adult Mice. *Toxicological Sciences*, *150*(2), 454–472. doi:10.1093/toxsci/kfw010 PMID:26794139

Patra, J. K., & Baek, K. H. (2014). Green nanobiotechnology: Factors affecting synthesis and characterization techniques. *Journal of Nanomaterials*, *2014*, 1–12. doi:10.1155/2014/417305

Patra, J. K., & Gouda, S. (2013). Application of nanotechnology in textile engineering: An overview. *Journal of Engineering and Technology Research*, *5*(5), 104–111. doi:10.5897/JETR2013.0309

Patra, J. K., Das, G., Fraceto, L. F., Campos, E. V. R., Rodriguez-Torres, M. D. P., Acosta-Torres, L. S., & Shin, H. S. (2018). Nano based drug delivery systems: Recent developments and future prospects. *Journal of Nanobiotechnology*, *16*(1), 71. doi:10.118612951-018-0392-8 PMID:30231877

Pawelczyk, E., Arbab, A. S., Chaudhry, A., Balakumaran, A., Robey, P. G., & Frank, J. A. (2008). In vitro model of bromodeoxyuridine or iron oxide nanoparticle uptake by activated macrophages from labeled stem cells: Implications for cellular therapy. *Stem Cells (Dayton, Ohio)*, *26*(5), 1366–1375. doi:10.1634temcells.2007-0707 PMID:18276802

Pazhanisamy, K., & Indra, N. (2007). Toxic effects of arsenic on protein content in the fish, Labeo rohita (Hamilton). Nature. *Environment and Pollution Technology*, *6*(1), 113–116.

Peixe, T. S., de Souza Nascimento, E., Schofield, K. L., Arcuri, A. S., & Bulcao, R. P. (2015). Nanotoxicology and exposure in the occupational setting. *Occupational Diseases and Environmental Medicine*, *3*(03), 35–48. doi:10.4236/odem.2015.33005

Peng, X. H., Qian, X., Mao, H., Wang, A. Y., Nie, S., & Shin, D. M. (2008). Targeted magnetic iron oxide nanoparticles for tumor imaging and therapy. *International Journal of Nanomedicine*, *3*(3), 311–321. doi:10.2147/IJN.S2824 PMID:18990940

Perez-Hernandez, H., Perez-Moreno, A., Sarabia-Castillo, C. R., Garcia-Mayagoitia, S., Medina-Perez, G., Lopez-Valdez, F., Campos-Montiel, R. G., Jayanta-Kumar, P., & Fernandez-Luqueno, F. (2021). Ecological Drawbacks of Nanomaterials Produced on an Industrial Scale: Collateral Effect on Human and Environmental Health. *Water, Air, and Soil Pollution*, *232*(10), 435. doi:10.100711270-021-05370-2 PMID:34658457

Perlman, O., Weitz, I. S., & Azhari, H. (2015). Copper oxide nanoparticles as contrast agents for MRI and ultrasound dual-modality imaging. *Physics in Medicine and Biology*, *60*(15), 5767–5783. doi:10.1088/0031-9155/60/15/5767 PMID:26159685

Pesch, B., Haerting, J., Ranft, U., Klimpel, A., Oelschlägel, B., & Schill, W. (2000). Occupational risk factors for renal cell carcinoma: Agent-specific results from a case-control study in Germany. MURC Study Group. Multicenter urothelial and renal cancer study. *International Journal of Epidemiology*, *29*(6), 1014–1024. doi:10.1093/ije/29.6.1014 PMID:11101542

Pisanic, T. R. II, Blackwell, J. D., Shubayev, V. I., Finones, R. R., & Jin, S. (2007). Nanotoxicity of iron oxide nanoparticle internalization in growing neurons. *Biomaterials*, *28*(16), 2572–2581. doi:10.1016/j.biomaterials.2007.01.043 PMID:17320946

Pumera, M. (2010). Nanotoxicology: The molecular science point of view. *Chemistry, an Asian Journal*, *6*(2), 340–348. doi:10.1002/asia.201000398 PMID:20725923

Qiao, H., Liu, W., Gu, H., Wang, D., & Wang, Y. (2015). The transport and deposition of nanoparticles in respiratory system by inhalation. *Journal of Nanomaterials*, *2015*, 1–8. Advance online publication. doi:10.1155/2015/394507

Raj, S., & Kumar, D. (2020). Biochemical toxicology: Heavy metals and nanomaterials. *Biochemical Toxicology: Heavy Metals and Nanomaterials*, *139*. Advance online publication. doi:10.5772/intechopen.90928

Rajput, V., Minkina, T., Mazarji, M., Shende, S., Sushkova, S., Mandzhieva, S., Burachevskaya, M., Chaplygin, V., Singh, A., & Jatav, H. (2020). Accumulation of nanoparticles in the soil-plant systems and their effects on human health. *Annals of Agricultural Science*, *65*(2), 137–143. doi:10.1016/j.aoas.2020.08.001

Rana, K., Verma, Y., Rani, V., & Rana, S. V. S. (2018). Renal toxicity of nanoparticles of cadmium sulphide in rat. *Chemosphere*, *193*, 142–150. doi:10.1016/j.chemosphere.2017.11.011 PMID:29128560

Rana, S. V. S. (2020). A Comprehensive Assessment of Hepatotoxicity Induced by Engineered Nanoparticles-A Review. *Journal of Toxicology and Risk Assessment*, *6*(1). Advance online publication. doi:10.23937/2572-4061.1510035

Ranjha, M. M. A. N., Shafique, B., Rehman, A., Mehmood, A., Ali, A., Zahra, S. M., Roobab, U., Singh, A., Ibrahim, S. A., & Siddiqui, S. A. (2022). Biocompatible Nanomaterials in Food Science, Technology, and Nutrient Drug Delivery: Recent Developments and Applications. *Frontiers in Nutrition*, *8*, 778155. doi:10.3389/fnut.2021.778155 PMID:35127783

Raza, M. A., Kanwal, Z., Shahid, A., Fatima, S., Sajjad, A., Riaz, S., & Naseem, S. (2021). Toxicity Evaluation of Arsenic Nanoparticles on Growth, Biochemical, Hematological, and Physiological Parameters of Labeo rohita Juveniles. *Advances in Materials Science*, *5185425*, 1–10. Advance online publication. doi:10.1155/2021/5185425

Robinson, S., & Nakkeeran, R. (2013). *Photonic crystal ring resonator based optical filters*. Advances in Photonic Crystals. doi:10.5772/54533

Rodriguez-Fragoso, P., Reyes-Esparza, J., Leon-Buitimea, A., & Rodriguez-Fragoso, L. (2012). Synthesis, characterization and toxicological evaluation of maltodextrin capped cadmium sulfide nanoparticles in human cell lines and chicken embryos. *Journal of Nanobiotechnology*, *10*(1), 1–11. doi:10.1186/1477-3155-10-47 PMID:23270407

Rubilar, O., Rai, M., Tortella, G., Diez, M. C., Seabra, A. B., & Duran, N. (2013). Biogenic nanoparticles: Copper, copper oxides, copper sulphides, complex copper nanostructures and their applications. *Biotechnology Letters*, *35*(9), 1365–1375. doi:10.100710529-013-1239-x PMID:23690046

Ryu, H. W., Lee, D. H., Florens, L., Swanson, S. K., Washburn, M. P., & Kwon, S. H. (2014). Analysis of the heterochromatin protein 1 (HP1) interactome in Drosophila. *Journal of Proteomics*, *102*, 137–147. doi:10.1016/j.jprot.2014.03.016 PMID:24681131

Santos, A. C., Morais, F., Simoes, A., Pereira, I., Sequeira, J. A., Pereira-Silva, M., & Ribeiro, A. (2019). Nanotechnology for the development of new cosmetic formulations. *Expert Opinion on Drug Delivery*, *16*(4), 313–330. doi:10.1080/17425247.2019.1585426 PMID:30793641

Schirmer, K., & Auffan, M. (2015). Nanotoxicology in the environment. *Environmental Science. Nano*, *2*(6), 561–563. doi:10.1039/C5EN90023A

Schmalz, G., Hickel, R., van Landuyt, K. L., & Reichl, F. X. (2017). Nanoparticles in dentistry. *Dental Materials*, *33*(11), 1298–1314. doi:10.1016/j.dental.2017.08.193 PMID:28951037

Schulte, P. A., Kuempel, E. D., & Drew, N. M. (2018). Characterizing risk assessments for the development of occupational exposure limits for engineered nanomaterials. *Regulatory Toxicology and Pharmacology*, *95*, 207–219. doi:10.1016/j.yrtph.2018.03.018 PMID:29574195

Shah, M. A., Pirzada, B. M., Price, G., Shibiru, A. L., & Qurashi, A. (2022). Applications of nanotechnology in smart textile industry: A critical review. *Journal of Advanced Research*, *38*, 55–75. doi:10.1016/j.jare.2022.01.008 PMID:35572402

Shaikh, S., Shyama, S. K., & Desai, P. V. (2015). Absorption, LD50 and Effects of CoO, MgO and PbO Nanoparticles on Mice "Mus musculus". *IOSR Journal of Environmental Science, Toxicology and Food Technology*, *9*(2), 32-38. doi:10.9790/2402-09213238

Sharma, J. K., Akhtar, M. S., Ameen, S., Srivastava, P., & Singh, G. (2015). Green synthesis of CuO nanoparticles with leaf extract of Calotropis gigantea and its dye-sensitized solar cells applications. *Journal of Alloys and Compounds*, *632*, 321–325. doi:10.1016/j.jallcom.2015.01.172

Sharma, S., Mitra, P., Bhardwaj, P., & Sharma, P. (2021). Blood lead level in school going children of Jodhpur, Rajasthan, India. *Turkish Journal of Biochemistry*, *46*(4), 393–398. doi:10.1515/tjb-2020-0418

Sharma, V., Anderson, D., & Dhawan, A. (2012). Zinc oxide nanoparticles induce oxidative DNA damage and ROS-triggered mitochondria mediated apoptosis in human liver cells (HepG2). *Apoptosis*, *17*(8), 852–870. doi:10.100710495-012-0705-6 PMID:22395444

Sharma, V., Shukla, R. K., Saxena, N., Parmar, D., Das, M., & Dhawan, A. (2009). DNA damaging potential of zinc oxide nanoparticles in human epidermal cells. *Toxicology Letters*, *185*(3), 211–218. doi:10.1016/j.toxlet.2009.01.008 PMID:19382294

Sharma, V., Singh, P., Pandey, A. K., & Dhawan, A. (2012). Induction of oxidative stress, DNA damage and apoptosis in mouse liver after sub-acute oral exposure to zinc oxide nanoparticles. *Mutation Research*, *745*(1-2), 84–91. doi:10.1016/j.mrgentox.2011.12.009 PMID:22198329

Siddiqi, K. S., & Husen, A. (2020). Current status of plant metabolite-based fabrication of copper/copper oxide nanoparticles and their applications: A review. *Biomaterials Research, 24*(1), 1–15. doi:10.118640824-020-00188-1 PMID:32514371

Singh, N. A. (2017). Nanotechnology innovations, industrial applications and patents. *Environmental Chemistry Letters, 15*(2), 185–191. doi:10.100710311-017-0612-8

Smita, S., Gupta, S. K., Bartonova, A., Dusinska, M., Gutleb, A. C., & Rahman, Q. (2012). Nanoparticles in the environment: Assessment using the causal diagram approach. *Environmental Health, 11*(S1), S13. Advance online publication. doi:10.1186/1476-069X-11-S1-S13 PMID:22759495

Soenen, S. J., Himmelreich, U., Nuytten, N., & De Cuyper, M. (2011). Cytotoxic effects of iron oxide nanoparticles and implications for safety in cell labelling. *Biomaterials, 32*(1), 195–205. doi:10.1016/j.biomaterials.2010.08.075 PMID:20863560

Song, P., Hai, Y., Wang, X., Zhao, L., Chen, B., Cui, P., Xie, Q., Yu, L., Li, Y., Wu, Z., & Li, H. (2018). Realgar transforming solution suppresses angiogenesis and tumor growth by inhibiting VEGF receptor 2 signaling in vein endothelial cells. *Archives of Pharmacal Research, 41*(4), 467–480. doi:10.100712272-018-1014-6 PMID:29542005

Song, X., Yao, W., Zhang, B., & Wu, Y. (2011). Application of Pt/CdS for the photocatalytic flue gas desulfurization. *International Journal of Photoenergy, 2012*, 1–5. doi:10.1155/2012/684735

Sonwani, S., Madaan, S., Arora, J., Suryanarayan, S., Rangra, D., Mongia, N., Vats, T., & Saxena, P. (2021). Inhalation exposure to atmospheric nanoparticles and its associated impacts on human health: A review. *Frontiers in Sustainable Cities, 3*, 690444. doi:10.3389/frsc.2021.690444

Srikanth, K., Pereira, E., Duarte, A. C., & Rao, J. V. (2016). Evaluation of cytotoxicity, morphological alterations and oxidative stress in Chinook salmon cells exposed to copper oxide nanoparticles. *Protoplasma, 253*(3), 873–884. doi:10.100700709-015-0849-7 PMID:26115719

Srivastava, V., Gusain, D., & Sharma, Y. (2015). Critical review on the toxicity of some widely used engineered nanoparticles. *Industrial & Engineering Chemistry Research, 54*(24), 6209–6233. doi:10.1021/acs.iecr.5b01610

Strambeanu, N., Demetrovici, L., Dragos, D., & Lungu, M. (2015). Nanoparticles: Definition, classification and general physical properties. In Nanoparticles' promises and risks (pp. 3-8). Springer. doi:10.1007/978-3-319-11728-7_1

Studer, A. M., Limbach, L. K., Van Duc, L., Krumeich, F., Athanassiou, E. K., Gerber, L. C., Moch, H., & Stark, W. J. (2010). Nanoparticle cytotoxicity depends on intracellular solubility: Comparison of stabilized copper metal and degradable copper oxide nanoparticles. *Toxicology Letters, 197*(3), 169–174. doi:10.1016/j.toxlet.2010.05.012 PMID:20621582

Subastri, A., Arun, V., Sharma, P., Suyavaran, A., Nithyananthan, S., Alshammari, G. M., ... Thirunavukkarasu, C. (2018). Synthesis and characterisation of arsenic nanoparticles and its interaction with DNA and cytotoxic potential on breast cancer cells. *Chemico-Biological Interactions, 295*, 73–83. doi:10.1016/j.cbi.2017.12.025 PMID:29277637

Subhan, M. A., Choudhury, K. P., & Neogi, N. (2021). Advances with molecular nanomaterials in industrial manufacturing applications. *Nanomanufacturing*, *1*(2), 75–97. doi:10.3390/nanomanufacturing1020008

Suman, T. Y., Radhika Rajasree, S. R., & Kirubagaran, R. (2015). Evaluation of zinc oxide nanoparticles toxicity on marine algae chlorella vulgaris through flow cytometric, cytotoxicity and oxidative stress analysis. *Ecotoxicology and Environmental Safety*, *113*, 23–30. doi:10.1016/j.ecoenv.2014.11.015 PMID:25483368

Sun, J., Wang, S., Zhao, D., Hun, F. H., Weng, L., & Liu, H. (2011). Cytotoxicity, permeability, and inflammation of metal oxide nanoparticles in human cardiac microvascular endothelial cells: Cytotoxicity, permeability, and inflammation of metal oxide nanoparticles. *Cell Biology and Toxicology*, *27*(5), 333–342. doi:10.100710565-011-9191-9 PMID:21681618

Sutunkova, M. P., Solovyeva, S. N., Chernyshov, I. N., Klinova, S. V., Gurvich, V. B., Shur, V. Y., Shishkina, E. V., Zubarev, I. V., Privalova, L. I., & Katsnelson, B. A. (2020). Manifestation of Systemic Toxicity in Rats after a Short-Time Inhalation of Lead Oxide Nanoparticles. *International Journal of Molecular Sciences*, *21*(3), 690. doi:10.3390/ijms21030690 PMID:31973040

Szalay, B., Tatrai, E., Nyiro, G., Vezer, T., & Dura, G. (2012). Potential toxic effects of iron oxide nanoparticles in in vivo and in vitro experiments. *Journal of Applied Toxicology*, *32*(6), 446–453. doi:10.1002/jat.1779 PMID:22161551

Taghavi, S. M., Momenpour, M., Azarian, M., Ahmadian, M., Souri, F., Taghavi, S. A., Sadeghain, M., & Karchani, M. (2013). Effects of Nanoparticles on the Environment and Outdoor Workplaces. *Electronic Physician*, *5*(4), 706–712. doi:10.14661/2013.706-712 PMID:26120406

Tania, I. S., & Ali, M. (2021). Effect of the coating of zinc oxide (ZnO) nanoparticles with binder on the functional and mechanical properties of cotton fabric. *Materials Today: Proceedings*, *38*, 2607–2611. doi:10.1016/j.matpr.2020.08.171

Thakur, M. P., Phillips, H. R. P., Brose, U., De Vries, F. T., Lavelle, P., Loreau, M., Mathieu, J., Mulder, C., Van der Putten, W. H., Rillig, M. C., Wardle, D. A., Bach, E. M., Bartz, M. L. C., Bennett, J. M., Briones, M. J. I., Brown, G., Decaëns, T., Eisenhauer, N., Ferlian, O., ... Cameron, E. K. (2020). Towards an integrative understanding of soil biodiversity. *Biological Reviews of the Cambridge Philosophical Society*, *95*(2), 350–364. doi:10.1111/brv.12567 PMID:31729831

Thiruvengadam, M., Chung, I. M., Gomathi, T., Ansari, M. A., Gopiesh Khanna, V., Babu, V., & Rajakumar, G. (2019). Synthesis, characterization and pharmacological potential of green synthesized copper nanoparticles. *Bioprocess and Biosystems Engineering*, *42*(11), 1769–1777. doi:10.100700449-019-02173-y PMID:31372759

Tiwari, M., Jain, P., Hariharapura, R. C., Narayanan, K., Bhat, U., Udupa, N., & Rao, J. V. (2016). Biosynthesis of copper nanoparticles using copper-resistant Bacillus cereus, a soil isolate. *Process Biochemistry*, *51*(10), 1348–1356. doi:10.1016/j.procbio.2016.08.008

Trabelsi, H., Azzouz, I., Sakly, M., & Abdelmelek, H. (2013). Subacute toxicity of cadmium on hepatocytes and nephrocytes in the rat could be considered as a green biosynthesis of nanoparticles. *International Journal of Nanomedicine*, *8*, 1121–1128. doi:10.2147/IJN.S39426 PMID:23515701

Tudu, B. K., Sinhamahapatra, A., & Kumar, A. (2020). Surface modification of cotton fabric using TiO2 nanoparticles for self-cleaning, oil–water separation, antistain, anti-water absorption, and antibacterial properties. *ACS Omega*, *5*(14), 7850–7860. doi:10.1021/acsomega.9b04067 PMID:32309694

Tulinska, J., Krivosikova, Z., Liskova, A., Mikusova, M. L., Masanova, V., Rollerova, E., ... Docekal, B. (2022). Six-week inhalation of lead oxide nanoparticles in mice affects antioxidant defense, immune response, kidneys, intestine and bones. *Environmental Science. Nano*, *9*(2), 751–766. doi:10.1039/D1EN00957E

Ude, V. C., Brown, D. M., Viale, L., Kanase, N., Stone, V., & Johnston, H. J. (2017). Impact of copper oxide nanomaterials on differentiated and undifferentiated Caco-2 intestinal epithelial cells; assessment of cytotoxicity, barrier integrity, cytokine production and nanomaterial penetration. *Particle and Fibre Toxicology*, *14*(1), 31. Advance online publication. doi:10.118612989-017-0211-7 PMID:28835236

Umer, A., Naveed, S., Ramzan, N., & Rafique, M. S. (2012). Selection of a suitable method for the synthesis of copper nanoparticles. *Nano*, *7*(05), 1230005. doi:10.1142/S1793292012300058

USEPA. (2001). National primary drinking water regulations; arsenic and clarifications to compliance and new source contaminants monitoring" from the environmental protection agency; final rule. *Federal Register*, *66*, 6976–7066.

Valdiglesias, V., Costa, C., Kilic, G., Costa, S., Pasaro, E., Laffon, B., & Teixeira, J. P. (2013). Neuronal cytotoxicity and genotoxicity induced by zinc oxide nanoparticles. *Environment International*, *55*, 92–100. doi:10.1016/j.envint.2013.02.013 PMID:23535050

Valdiglesias, V., Fernandez-Bertolez, N., Kilic, G., Costa, C., Costa, S., Fraga, S., Bessa, M. J., Pasaro, E., Teixeira, J. P., & Laffon, B. (2016). Are iron oxide nanoparticles safe? Current knowledge and future perspectives. *Journal of Trace Elements in Medicine and Biology*, *38*, 53–63. doi:10.1016/j.jtemb.2016.03.017 PMID:27056797

Varmazyari, A., Taghizadehghalehjoughi, A., Sevim, C., Baris, O., Eser, G., Yildirim, S., Hacimuftuoglu, A., Buha, A., Wallace, D. R., Tsatsakis, A., Aschner, M., & Mezhuev, Y. (2020). Cadmium sulfide-induced toxicity in the cortex and cerebellum: In vitro and in vivo studies. *Toxicology Reports*, *7*, 637–648. doi:10.1016/j.toxrep.2020.04.011 PMID:32489905

Vasile, C. (2018). Polymeric nanocomposites and nanocoatings for food packaging: A review. *Materials (Basel)*, *11*(10), 1834. doi:10.3390/ma11101834 PMID:30261658

Velmurugan, P., Shim, J., Bang, K. S., & Oh, B. T. (2016). Gold nanoparticles mediated coloring of fabrics and leather for antibacterial activity. *Journal of Photochemistry and Photobiology. B, Biology*, *160*, 102–109. doi:10.1016/j.jphotobiol.2016.03.051 PMID:27104665

Verma, Y., Rana, K., Rani, V., & Rana, S. V. (2021). *Hepatotoxic Manifestations of Arsenic Trioxide Loaded Poly (Lactide-co-Glycolide)*. Nanoparticles in Wistar Rat., doi:10.21203/rs.3.rs-533921/v2

Vigneshwaran, N., Kathe, A. A., Varadarajan, P. V., Nachane, R. P., & Balasubramanya, R. H. (2007). Functional finishing of cotton fabrics using silver nanoparticles. *Journal of Nanoscience and Nanotechnology*, *7*(6), 1893–1897. doi:10.1166/jnn.2007.737 PMID:17654961

Vimercati, L., Cavone, D., Caputi, A., De Maria, L., Tria, M., Prato, E., & Ferri, G. M. (2020). Nanoparticles: An experimental study of zinc nanoparticles toxicity on marine crustaceans. General overview on the health implications in humans. *Frontiers in Public Health*, *8*, 192. doi:10.3389/fpubh.2020.00192 PMID:32509719

Viscardi, M., Arena, M., Guadagno, L., Vertuccio, L., & Barra, G. (2018). Multi-functional nanotechnology integration for aeronautical structures performance enhancement. *International Journal of Structural Integrity*.

Waalkes, M. P. (2000). Cadmium carcinogenesis in review. *Journal of Inorganic Biochemistry*, *79*(1-4), 241–244. doi:10.1016/S0162-0134(00)00009-X PMID:10830873

Wang, B., & Du, Y. (2013). Cadmium and its neurotoxic effects. *Oxidative Medicine and Cellular Longevity*, *898034*, 1–12. Advance online publication. doi:10.1155/2013/898034 PMID:23997854

Wang, F., Li, H., Yuan, Z., Sun, Y., Chang, F., Deng, H., Xie, L., & Li, H. (2016). A highly sensitive gas sensor based on CuO nanoparticles synthetized via a sol–gel method. *RSC Advances*, *6*(83), 79343–79349. doi:10.1039/C6RA13876D

Wang, J., Deng, X., Zhang, F., Chen, D., & Ding, W. (2014). ZnO nanoparticle-induced oxidative stress triggers apoptosis by activating JNK signaling pathway in cultured primary astrocytes. *Nanoscale Research Letters*, *9*(1), 117. doi:10.1186/1556-276X-9-117 PMID:24624962

Wang, K., Wang, X., Januszewski, B., Liu, Y., Li, D., Fu, R., Elimelech, M., & Huang, X. (2022). Tailored design of nanofiltration membranes for water treatment based on synthesis–property–performance relationships. *Chemical Society Reviews*, *51*(2), 672–719. doi:10.1039/D0CS01599G PMID:34932047

Wang, X., Xia, T., Duch, M. C., Ji, Z., Zhang, H., Li, R., Sun, B., Lin, S., Meng, H., Liao, Y. P., Wang, M., Song, T. B., Yang, Y., Hersam, M. C., & Nel, A. E. (2012). Pluronic F108 coating decreases the lung fibrosis potential of multiwall carbon nanotubes by reducing lysosomal injury. *Nano Letters*, *12*(6), 3050–3061. doi:10.1021/nl300895y PMID:22546002

Wang, Y., Dong, Y., Liu, Y., & Liu, Z. (2018). Arsenic trioxide nanoparticles inhibit acute promyelocytic leukemia cell proliferation and induce apoptosis via PTEN/AKT signalling pathway. *Biomedical Research*, *29*(7), 1450–1454. doi:10.4066/biomedicalresearch.29-17-3983

Wanunu, M., Dadosh, T., Ray, V., Jin, J., McReynolds, L., & Drndic, M. (2010). Rapid electronic detection of probe-specific microRNAs using thin nanopore sensors. *Nature Nanotechnology*, *5*(11), 807–814. doi:10.1038/nnano.2010.202 PMID:20972437

Waring, J. F., Gum, R., Morfitt, D., Jolly, R. A., Ciurlionis, R., Heindel, M., Gallenberg, L., Buratto, B., & Ulrich, R. G. (2002). Identifying toxic mechanisms using DNA microarrays: Evidence that an experimental inhibitor of cell adhesion molecule expression signals through the aryl hydrocarbon nuclear receptor. *Toxicology*, *181-182*, 537–550. doi:10.1016/S0300-483X(02)00477-8 PMID:12505364

Waser, O., Hess, M., Guntner, A., Novak, P., & Pratsinis, S. E. (2013). Size controlled CuO nanoparticles for Li-ion batteries. *Journal of Power Sources*, *241*, 415–422. doi:10.1016/j.jpowsour.2013.04.147

Wei, W., Xu, C., & Wu, H. (2006). Magnetic iron oxide nanoparticles mediated gene therapy for breast cancer-an in vitro study. *Journal of Huazhong University of Science and Technology*, *26*(6), 728–730. doi:10.100711596-006-0628-y PMID:17357502

Wongrakpanich, A., Mudunkotuwa, I. A., Geary, S. M., Morris, A. S., Mapuskar, K. A., Spitz, D. R., Grassian, V. H., & Salem, A. K. (2016). Size-dependent cytotoxicity of copper oxide nanoparticles in lung epithelial cells. *Environmental Science. Nano*, *3*(2), 365–374. doi:10.1039/C5EN00271K PMID:27347420

World Health Organization. (2011). *Guidelines for drinking-water quality*. WHO.

Wu, Y. N., Wu, P. C., Yang, L. X., Ratinac, K. R., Thordarson, P., Jahn, K. A., Chen, D. H., Shieh, D. B., & Braet, F. (2013). The anticancer properties of iron core-gold shell nanoparticles in colorectal cancer cells. *International Journal of Nanomedicine*, *8*, 3321–3331. doi:10.2147/IJN.S47742 PMID:24039416

Xi, G., Peng, Y., Xu, L., Zhang, M., Yu, W., & Qian, Y. (2004). Selected-control synthesis of PbO2 submicrometer-sized hollow spheres and Pb3O4 microtubes. *Inorganic Chemistry Communications*, *7*(5), 607–610. doi:10.1016/j.inoche.2004.03.001

Xie, Y., Williams, N. G., Tolic, A., Chrisler, W. B., Teeguarden, J. G., Maddux, B. L., Pounds, J. G., Laskin, A., & Orr, G. (2012). Aerosolized ZnO nanoparticles induce toxicity in alveolar type II epithelial cells at the air-liquid interface. *Toxicological Sciences*, *125*(2), 450–461. doi:10.1093/toxsci/kfr251 PMID:21964423

Yaqoob, S. B., Adnan, R., Rameez Khan, R. M., & Rashid, M. (2020). Gold, silver, and palladium nanoparticles: A chemical tool for biomedical applications. *Frontiers in Chemistry*, *8*, 376. doi:10.3389/fchem.2020.00376 PMID:32582621

Yigit, M. V., Mazumdar, D., & Lu, Y. (2008). MRI detection of thrombin with aptamer functionalized superparamagnetic iron oxide nanoparticles. *Bioconjugate Chemistry*, *19*(2), 412–417. doi:10.1021/bc7003928 PMID:18173225

Yildirimer, L., Thanh, N. T., Loizidou, M., & Seifalian, A. M. (2011). Toxicology and clinical potential of nanoparticles. *Nano Today*, *6*(6), 585–607. doi:10.1016/j.nantod.2011.10.001 PMID:23293661

Yu, Q., Xiong, X. Q., Zhao, L., Xu, T. T., Bi, H., Fu, R., & Wang, Q. H. (2018). Biodistribution and toxicity assessment of superparamagnetic iron oxide nanoparticles in vitro and in vivo. *Current Medical Science*, *38*(6), 1096–1102. doi:10.100711596-018-1989-8 PMID:30536075

Zahra, Z., Habib, Z., Hyun, S., & Sajid, M. (2022). Nanowaste: Another Future Waste, Its Sources, Release Mechanism, and Removal Strategies in the Environment. *Sustainability*, *14*(4), 2041. doi:10.3390u14042041

Zhang, L., Chu, M., Ji, C., Tan, J., & Yuan, Q. (2022). Preparation, applications, and challenges of functional DNA nanomaterials. *Nano Research*, 1–18. doi:10.100712274-022-4793-5 PMID:36065175

Zhang, W., Lin, K., Sun, X., Dong, Q., Huang, C., Wang, H., Guo, M., & Cui, X. (2012). Toxicological effect of MPA–CdSe QDs exposure on zebrafish embryo and larvae. *Chemosphere*, *89*(1), 52–59. doi:10.1016/j.chemosphere.2012.04.012 PMID:22595531

Zheng, X. G., Xu, C. N., Tomokiyo, Y., Tanaka, E., Yamada, H., & Soejima, Y. (2000). Observation of charge stripes in cupric oxide. *Physical Review Letters*, *85*(24), 5170–5173. doi:10.1103/PhysRevLett.85.5170 PMID:11102213

Zhou, H., Mayorga-Martinez, C. C., Pané, S., Zhang, L., & Pumera, M. (2021). Magnetically Driven Micro and Nanorobots. *Chemical Reviews*, *121*(8), 4999–5041. doi:10.1021/acs.chemrev.0c01234 PMID:33787235

Zhou, X., Zhao, L., Luo, J., Tang, H., Xu, M., Wang, Y., Yang, X., Chen, H., Li, Y., Ye, G., Shi, F., Lv, C., & Jing, B. (2019). The Toxic Effects and Mechanisms of Nano-Cu on the Spleen of Rats. *International Journal of Molecular Sciences*, *20*(6), 1469. doi:10.3390/ijms20061469 PMID:30909528

Zhu, H., Jiang, R., Xiao, L., Chang, Y., Guan, Y., Li, X., & Zeng, G. (2009). Photocatalytic decolorization and degradation of Congo Red on innovative crosslinked chitosan/nano-CdS composite catalyst under visible light irradiation. *Journal of Hazardous Materials*, *169*(1-3), 933–940. doi:10.1016/j.jhazmat.2009.04.037 PMID:19477069

Zhu, M. T., Feng, W. Y., Wang, B., Wang, T. C., Gu, Y. Q., Wang, M., Wang, Y., Ouyang, H., Zhao, Y. L., & Chai, Z. F. (2008). Comparative study of pulmonary responses to nano- and submicron-sized ferric oxide in rats. *Toxicology*, *247*(2-3), 102–111. doi:10.1016/j.tox.2008.02.011 PMID:18394769

Zrazhevskiy, P., Sena, M., & Gao, X. (2010). Designing multifunctional quantum dots for bioimaging, detection, and drug delivery. *Chemical Society Reviews*, *39*(11), 4326–4354. doi:10.1039/b915139g PMID:20697629

Compilation of References

Abbasalipourkabir, R., Moradi, H., Zarei, S., Asadi, S., Salehzadeh, A., Ghafourikhosroshahi, A., Mortazavi, M., & Ziamajidi, N. (2015). Toxicity of zinc oxide nanoparticles on adult male Wistar rats. *Food and Chemical Toxicology*, *84*, 154–160. doi:10.1016/j.fct.2015.08.019 PMID:26316185

Abbas, Q., Yousaf, B., Ali, M. U., Munir, M. A. M., El-Naggar, A., Rinklebe, J., & Naushad, M. (2020). Transformation pathways and fate of engineered nanoparticles (ENPs) in distinct interactive environmental compartments: A review. *Environment International*, *138*, 105646. doi:10.1016/j.envint.2020.105646 PMID:32179325

Abd Elkodous, M., El-Husseiny, H., El-Sayyad, G., Hashem, A., Doghish, A., Elfadil, D., Radwan, Y., El-Zeiny, H., Bedair, H., & Matsuda, A. (2021). Recent advances in waste-recycled nanomaterials for biomedical applications: Waste-to-wealth. *Nanotechnology Reviews*, *10*(1), 1662–1739. doi:10.1515/ntrev-2021-0099

Abdelmonem, A. M., Pelaz, B., Kantner, K., Bigall, N. C., Del Pino, P., & Parak, W. J. (2015). Charge and agglomeration dependent in vitro uptake and cytotoxicity of zinc oxide nanoparticles. *Journal of Inorganic Biochemistry*, *153*, 334–338. doi:10.1016/j.jinorgbio.2015.08.029 PMID:26387023

Abdel-Moniem, S. M., El-Liethy, M. A., Ibrahim, H. S., & Ali, M. E. M. (2021). Innovative Green/Non-Toxic Bi_2S_3@g-C_3N_4 Nanosheet for Dark Antimicrobial Activity and Photocatalytic Depollution: Turnover Assessment. *Ecotoxicology and Environmental Safety*, *226*, 112808. doi:10.1016/j.ecoenv.2021.112808 PMID:34600290

Abu Elella, M. H., Goda, E. S., Abdallah, H. M., Shalan, A. E., Gamal, H., & Yoon, K. R. (2021a). Innovative bactericidal adsorbents containing modified xanthan gum/montmorillonite nanocomposites for wastewater treatment. *International Journal of Biological Macromolecules*, *167*, 1113–1125. doi:10.1016/j.ijbiomac.2020.11.065 PMID:33197479

Abu Elella, M. H., Goda, E. S., Gamal, H., El-Bahy, S. M., Nour, M. A., & Yoon, K. R. (2021b). Green antimicrobial adsorbent containing grafted xanthan gum/SiO $_2$ nanocomposites for malachite green dye. *International Journal of Biological Macromolecules*, *191*, 385–395. doi:10.1016/j.ijbiomac.2021.09.040 PMID:34537301

Acharya, K. P. (2009). *Photocurrent spectroscopy of CdS/plastic, CdS/glass, and ZnTe/GaAs hetero-pairs formed with pulsed-laser deposition* [Doctoral dissertation]. Bowling Green State University.

Acharya, P., Jayaprakasha, G. K., Crosby, K. M., Jifon, J. L., & Patil, B. S. (2020). Nanoparticle-Mediated Seed Priming Improves Germination, Growth, Yield, and Quality of Watermelons (*Citrullus lanatus*) at multi-locations in Texas. *Scientific Reports*, *10*(1), 5037. doi:10.103841598-020-61696-7 PMID:32193449

Adetunji, C. O. (2021). *Bionanomaterials for green bionanotechnology. In Bionanomaterials*. IOP Publishing.

Adhikary, R. K. (2021). Risk of Human Illness from Recreational Exposure to Microbial Pathogens in Freshwater Bodies: A Systematic Review. *Exposure and Health*, 1–19.

Agnihotri, S., & Dhiman, N. K. (2017). Development of nano-antimicrobial biomaterials for biomedical applications. *Advances in Biomaterials for Biomedical Applications*, *66*, 479–545. doi:10.1007/978-981-10-3328-5_12

Aguilar-Moreno, G. S., Navarro-Cerón, E., Velázquez-Hernández, A., Hernández-Eugenio, G., Aguilar-Méndez, M. Á., & Espinosa-Solares, T. (2020). Enhancing methane yield of chicken litter in anaerobic digestion using magnetite nanoparticles. *Renewable Energy*, *147*, 204–213. doi:10.1016/j.renene.2019.08.111

Ahamed, M., Siddiqui, M. A., Akhtar, M. J., Ahmad, I., Pant, A. B., & Alhadlaq, H. A. (2010). Genotoxic potential of copper oxide nanoparticles in human lung epithelial cells. *Biochemical and Biophysical Research Communications*, *396*(2), 578–583. doi:10.1016/j.bbrc.2010.04.156 PMID:20447378

Ahghari, M. R., Soltaninejad, V., & Maleki, A. (2020). Synthesis of nickel nanoparticles by a green and convenient method as a magnetic mirror with antibacterial activities. *Scientific Reports*, *10*(1), 12627. doi:10.103841598-020-69679-4 PMID:32724123

Ahmad, S. A., Das, S. S., Khatoon, A., Ansari, M. T., Afzal, M., Hasnain, M. S., & Nayak, A. K. (2020). Bactericidal activity of silver nanoparticles: A mechanistic review. *Materials Science for Energy Technologies*, *3*, 756–769. doi:10.1016/j.mset.2020.09.002

Ahmad, S., Munir, S., Zeb, N., Ullah, A., Khan, B., Ali, J., Bilal, M., Omer, M., Alamzeb, M., Salman, S. M., & Ali, S. (2019). Green nanotechnology: A review on green synthesis of silver nanoparticles-an ecofriendly approach. *International Journal of Nanomedicine*, *14*, 5087–5107. doi:10.2147/IJN.S200254 PMID:31371949

Ahmed, M. M., Fatima, F., Anwer, M. K., Alshahrani, S. M., Alalaiwe, A., & Katakam, P. (2019). Biosynthesis, characterization and anti-microbial activity of silver nanoparticle based gel hand wash. *Green Processing and Synthesis*, *8*(1), 577–583. doi:10.1515/gps-2019-0027

Ahmed, S., Ahmad, M., Swami, B. L., & Ikram, S. (2016). A review on plants extract mediated synthesis of silver nanoparticles for antimicrobial applications: A green expertise. *Journal of Advanced Research*, *7*(1), 17–28. doi:10.1016/j.jare.2015.02.007 PMID:26843966

Ajay, C. M., Mohan, S., Dinesha, P., & Rosen, M. A. (2020). Review of impact of nanoparticle additives on anaerobic digestion and methane generation. *Fuel*, *277*, 118234. doi:10.1016/j.fuel.2020.118234

Ajayi-Banji, A. A., & Rahman, S. (2021). Efficacy of magnetite (Fe3O4) nanoparticles for enhancing solid-state anaerobic co-digestion: Focus on reactor performance and retention time. *Bioresource Technology*, *324*, 124670. doi:10.1016/j.biortech.2021.124670 PMID:33453521

Akçan, R., Aydogan, H. C., Yildirim, M. Ş., Taştekin, B., & Sağlam, N. (2020). Nanotoxicity: A challenge for future medicine. *Turkish Journal of Medical Sciences*, *50*(4), 1180–1196. doi:10.3906ag-1912-209 PMID:32283898

Akgönüllü, S., Bakhshpour, M., & Denizli, A. (2021). Molecularly imprinted bionanomaterials and their biomedical applications. Bionanomaterials Fundamentals and Biomedical Applications, 9-27.

Akhtar, M. J., Kumar, S., Alhadlaq, H. A., Alrokayan, S. A., Abu-Salah, K. M., & Ahamed, M. (2016). Dose-dependent genotoxicity of copper oxide nanoparticles stimulated by reactive oxygen species in human lung epithelial cells. *Toxicology and Industrial Health*, *32*(5), 809–821. doi:10.1177/0748233713511512 PMID:24311626

Akter, M., Sikder, M. T., Rahman, M. M., Ullah, A. K. M. A., Hossain, K. F. B., Banik, S., Hosokawa, T., Saito, T., & Kurasaki, M. (2017). A systematic review on silver nanoparticles-induced cytotoxicity: Physicochemical properties and perspectives. *Journal of Advanced Research*, *9*, 1–16. doi:10.1016/j.jare.2017.10.008 PMID:30046482

Al Naggar, Y. (2016). Effects of metal and metal oxide nanoparticles on honey bees (Apis millefera L.). *Journal of Aquatic Pollution and Toxicology, 1.*

Al Naggar, Y., Dabour, K., Masry, S., Sadek, A., Naiem, E., & Giesy, J. P. (2020). Sublethal effects of chronic exposure to CdO or PbO nanoparticles or their binary mixture on the honey bee (Apis millefera L.). *Environmental Science and Pollution Research International*, *27*(16), 19004–19015. doi:10.100711356-018-3314-2 PMID:30280346

Alabdallah, N. M., & Hasan, M. (2021). Plant based green synthesis of silver nanoparticles and its effective role in abiotic stress tolerance in crop plants. *Saudi Journal of Biological Sciences*, *28*(10), 5631–5639. doi:10.1016/j.sjbs.2021.05.081 PMID:34588874

Alaraby, M., Romero, S., Hernandez, A., & Marcos, R. (2019). Toxic and genotoxic effects of silver nanoparticles in Drosophila. *Environmental and Molecular Mutagenesis*, *60*(3), 277–285. doi:10.1002/em.22262 PMID:30353950

Alarçin, E., Guan, X., Kashaf, S. S., Elbaradie, K., Yang, H., Jang, H. L., & Khademhosseini, A. (2016). Recreating composition, structure, functionalities of tissues at nanoscale for regenerative medicine. *Regenerative Medicine*, *11*(8), 849–858. doi:10.2217/rme-2016-0120 PMID:27885900

Alarifi, S., Ali, D., Verma, A., Alakhtani, S., & Ali, B. A. (2013). Cytotoxicity and genotoxicity of copper oxide nanoparticles in human skin keratinocytes cells. *International Journal of Toxicology*, *32*(4), 296–307. doi:10.1177/1091581813487563 PMID:23667135

Al-Asheh, S., Banat, F. R., Al-Omari, Z., & Duvnjak, Z. (2000). Predictions of binary sorption isotherms for the sorption of heavy metals by pine bark using single isotherm data. *Chemosphere*, *41*(5), 659–665. doi:10.1016/S0045-6535(99)00497-X PMID:10834365

Albukhaty, S., Naderi-Manesh, H., & Tiraihi, T. (2013). In vitro labeling of neural stem cells with poly-L-lysine coated super paramagnetic nanoparticles for green fluorescent protein transfection. *Iranian Biomedical Journal*, *17*(2), 71–76. doi:10.6091/ibj.1114.2013 PMID:23567848

Albuquerque, D.C. (2021). Quan. *TTEES 5 Sponsors,* 34.

Al-Dousari, A., Al-Nassar, W., & Ahmed, M. (2020). Photovoltaic and Wind Energy: Challenges and Solutions in Desert Regions. *E3S Web of Conferences, 166,* 1–6. doi:10.1051/e3sconf/202016604003

Aleksiichuk, V., Omelchuk, S., Sokurenko, L., Kaminsky, R., Kovalchuk, O., & Chaikovsky, Y. (2018). The influence of lead nanoparticles on the morpho-functional changes of rat liver during the postexposure period. *Microscopy Research and Technique*, *81*(7), 781–788. doi:10.1002/jemt.23036 PMID:29652104

Al-Ezzi, A. S., & Ansari, M. N. M. (2022). Photovoltaic Solar Cells: A Review. *Applied System Innovation*, *5*(67), 1–17. doi:10.3390/asi5040067

Al-Hakkani, M. F. (2020). Biogenic copper nanoparticles and their applications: A review. *SN Applied Sciences*, *2*(3), 505. doi:10.100742452-020-2279-1

Alharbi, F. A., & Alarfaj, A. A. (2020). Green synthesis of silver nanoparticles from *Neuradaprocumbens* and its antibacterial activity against multi-drug resistant microbial pathogens. *Journal of King Saud University. Science*, *32*(2), 1346–1352. doi:10.1016/j.jksus.2019.11.026

Alhassid, Y. (2000). The statistical theory of quantum dots. *Reviews of Modern Physics*, *72*(4), 895–968. doi:10.1103/RevModPhys.72.895

Alic, J. A., & Sarewitz, D. (2016). Rethinking Innovation for Decarbonizing Energy Systems. *Energy Research & Social Science*, *21*, 212–221. doi:10.1016/j.erss.2016.08.005

Ali, S. A., Mittal, D., & Kaur, G. (2021). In-situ monitoring of xenobiotics using genetically engineered whole-cell-based microbial biosensors: Recent advances and outlook. *World Journal of Microbiology & Biotechnology*, *37*(5), 1–24. doi:10.100711274-021-03024-3 PMID:33843020

Allafchian, A., Mousavi, Z. S., & Hosseini, S. S. (2019). Application of cress seed musilage magnetic nanocomposites for removal of methylene blue dye from water. *International Journal of Biological Macromolecules*, *136*, 199–208. doi:10.1016/j.ijbiomac.2019.06.083 PMID:31201917

Almaary, K. S., Sayed, S. R. M., Abd-Elkader, O., Dawoud, T. M., El-Orabi, N. F., & Elgorban, A. M. (2020). Complete green synthesis of silver-nanoparticles applying seed-borne *Penicillium duclauxii*. *Saudi Journal of Biological Sciences*, *27*(5), 1333–1339. doi:10.1016/j.sjbs.2019.12.022 PMID:32346343

Alsbaiee, A. (2016). Rapid removal of organic micropollutants from water by a porous β-cyclodextrin polymer. *Nature*, *529*(7585), 190-194.

Aman, R. A., Mahas, A., & Mahfouz, M. (2020). Nucleic acid detection using CRISPR/Cas biosensing technologies. *ACS Synthetic Biology*, *9*(6), 1226–1233. doi:10.1021/acssynbio.9b00507 PMID:32159950

Amiri, A., Mohammadi, M., & Shabani, M. (2016). Synthesis and toxicity evaluation of lead oxide (PbO) nanoparticles in rats. *Electronic Journal of Biology*, *12*(2), 110-114.

Amsath, J., Sugumaran, V., & Sugumaran, J. (2017). Effect of arsenic (As2O3) on hematological parameters of fresh-water air breathing fish, Channa punctatus (bloch). *Current Trends in Biomedical Engineering & Biosciences*, *7*(1), 3–7. doi:10.19080/CTBEB.2017.07.555702

Anand, M. A. V., Vinayagam, R., Vijayakumar, S., Balupillai, A., Herbert, F. J., Kumar, S., Ghidan, A. Y., Al-Antary, T., & David, E. (2019). Green synthesis, characterization and antibacterial activity of silver nanoparticles by *Malus domestica* and its cytotoxic effect on (MCF-7) cell line. *Microbial Pathogenesis*, *135*, 103609. doi:10.1016/j.micpath.2019.103609 PMID:31247255

Andrady, A. L. (2015). Persistence of plastic litter in the oceans. In M. Bergmann, L. Gutow, & M. Klages (Eds.), *Marine anthropogenic litter* (pp. 57–72). Springer. doi:10.1007/978-3-319-16510-3_3

Anwar, A., Abdalla, S. A. O., Aslam, Z., Shah, M. R., Siddiqui, R., & Khan, N. A. (2019). Oleic acid-conjugated silver nanoparticles as efficient antiamoebic agent against*Acanthamoeba castellanii*. *Parasitology Research*, *118*(7), 2295–2304. doi:10.100700436-019-06329-3 PMID:31093751

Apostolov, A. T., Apostolova, I. N., & Wesselinowa, J. M. (2014). Dielectric constant of multiferroic pure and doped CuO nanoparticles. *Solid State Communications*, *192*, 71–74. doi:10.1016/j.ssc.2014.05.014

Araujo, P., Sayer, C., Poco, J., & Giudici, R. (2002). Techniques for reducing residual monomer content in polymers: A review. *Polymer Engineering and Science*, *42*(7), 1442–1468. doi:10.1002/pen.11043

Aruoja, V., Dubourguier, H. C., Kasemets, K., & Kahru, A. (2009). Toxicity of nanoparticles of CuO, ZnO and TiO2 to microalgae Pseudokirchneriella subcapitata. *The Science of the Total Environment*, *407*(4), 1461–1468. doi:10.1016/j.scitotenv.2008.10.053 PMID:19038417

Ashkavand, P., Zarafshar, M., Tabari, M., Mirzaie, J., Nikpour, A., Bordbar, S. K., Struve, D., & Striker, G. G. (2018). Application of SiO2 nanoparticles as pretreatment alleviates the impact of drought on the physiological performance of *Prunus mahaleb* (Rosaceae). *Boletín de la Sociedad Argentina de Botánica*, *53*(2), 207–219. doi:10.31055/1851.2372.v53.n2.20578

Ashraf, M. A., Akbar, A., Askari, S. H., Iqbal, M., Rasheed, R., & Hussain, I. (2018). Recent advances in abiotic stress tolerance of plants through chemical priming: An overview. In A. Rakshit & H. Singh (Eds.), *Advances in Seed Priming* (pp. 51–79). Springer. doi:10.1007/978-981-13-0032-5_4

Atha, D. H., Wang, H., Petersen, E. J., Cleveland, D., Holbrook, R. D., Jaruga, P., Dizdaroglu, M., Xing, B., & Nelson, B. C. (2012). Copper oxide nanoparticle mediated DNA damage in terrestrial plant models. *Environmental Science & Technology*, *46*(3), 1819–1827. doi:10.1021/es202660k PMID:22201446

Attarilar, S., Yang, J., Ebrahimi, M., Wang, Q., Liu, J., Tang, Y., & Yang, J. (2020). The Toxicity Phenomenon and the Related Occurrence in Metal and Metal Oxide Nanoparticles: A Brief Review from the Biomedical Perspective. *Frontiers in Bioengineering and Biotechnology*, *8*, 822. doi:10.3389/fbioe.2020.00822 PMID:32766232

Awaad, A., Nakamura, M., & Ishimura, K. (2012). *Imaging of size-dependent uptake and identification of novel pathways in mouse Peyer's patches using fluorescent organosilica particles*. Academic Press.

Awaad, A., Nakamura, M., & Ishimura, K. (2012, July). Imaging of size-dependent uptake and identification of novel pathways in mouse Peyer's patches using fluorescent organosilica particles. *Nanomedicine; Nanotechnology, Biology, and Medicine*, *8*(5), 627–636. doi:10.1016/j.nano.2011.08.009

Azad, M. B., Chen, Y., & Gibson, S. B. (2009). Regulation of autophagy by reactive oxygen species (ROS): Implications for cancer progression and treatment. *Antioxidants & Redox Signaling*, *11*(4), 777–790. doi:10.1089/ars.2008.2270 PMID:18828708

Babaei, K., Sharifi, R. S., Pirzad, A., & Khalilzadeh, R. (2017). Effects of bio fertilizer and nano Zn-Fe oxide on physiological traits, antioxidant enzymes activity and yield of wheat (*Triticum aestivum* L.) under salinity stress. *Journal of Plant Interactions*, *12*(1), 381–389. doi:10.1080/17429145.2017.1371798

Babes, L., Denizot, B., Tanguy, G., Le Jeune, J. J., & Jallet, P. (1999). Synthesis of iron oxide nanoparticles used as MRI contrast agents: A parametric study. *Journal of Colloid and Interface Science*, *212*(2), 474–482. doi:10.1006/jcis.1998.6053 PMID:10092379

Baby, T. T., & Sundara, R. (2013). Synthesis of silver nanoparticle decorated multiwalled carbon nanotubes-graphene mixture and its heat transfer studies in nanofluid. *AIP Advances*, *3*(1), 012111. doi:10.1063/1.4789404

Bahadar, H., Maqbool, F., Niaz, K., & Abdollahi, M. (2016). Toxicity of Nanoparticles and an Overview of Current Experimental Models. *Iranian Biomedical Journal*, *20*(1), 1–11. doi:10.7508/ibj.2016.01.001 PMID:26286636

Baig, N., Kammakakam, I., & Falath, W. (2021). Nanomaterials: A review of synthesis methods, properties, recent progress, and challenges. *Materials Advances*, *2*(6), 1821–1871. doi:10.1039/D0MA00807A

Bailey, R. E., Smith, A. M., & Nie, S. (2004). Quantum dots in biology and medicine. *Physica E, Low-Dimensional Systems and Nanostructures*, *25*(1), 1–12. doi:10.1016/j.physe.2004.07.013

Bai, S., Guo, W., Sun, J., Li, J., Tian, Y., Chen, A., Luo, R., & Li, D. (2016). Synthesis of SnO2–CuO heterojunction using electrospinning and application in detecting of CO. *Sensors and Actuators. B, Chemical*, *226*, 96–103. doi:10.1016/j.snb.2015.11.028

Bais, H. P., Weir, T. L., Perry, L. G., Gilroy, S., & Vivanco, J. M. (2006). The role of root exudates in rhizosphere interactions with plants and other organisms. *Annual Review of Plant Biology*, *57*(1), 233–266. doi:10.1146/annurev.arplant.57.032905.105159 PMID:16669762

Balali-Mood, M., Naseri, K., Tahergorabi, Z., Khazdair, M. R., & Sadeghi, M. (2021). Toxic mechanisms of five heavy metals: Mercury, lead, chromium, cadmium, and arsenic. *Frontiers in Pharmacology*, *12*, 643972. doi:10.3389/fphar.2021.643972 PMID:33927623

Bala, N., Dey, A., Das, S., Basu, R., & Nandy, P. (2014). Effect of hydroxyapatite nanorod on chickpea (*Cicer arietinum*) plant growth and its possible use as nano-fertilizer. *Indian Journal of Plant Physiology / Official Publication of the Indian Society for Plant Physiology*, *4*(3), 1061–1069.

Ballarin, B., Mignani, A., Mogavero, F., Gabbanini, S., & Morigi, M. (2015). Hybrid material based on ZnAl hydrolcite and silver nanoparticles for deodorant formulation. *Applied Clay Science*, *114*, 303–308. doi:10.1016/j.clay.2015.06.014

Balmuri, S. R., Selvaraj, U., Kumar, V. V., Anthony, S. P., Tsatsakis, A. M., Golokhvast, K. S., & Raman, T. (2017). Effect of surfactant in mitigating cadmium oxide nanoparticle toxicity: Implications for mitigating cadmium toxicity in environment. *Environmental Research*, *152*, 141–149. doi:10.1016/j.envres.2016.10.005 PMID:27771568

Banik, S., & Pérez-de-Luque, A. (2017). In vitro effects of copper nanoparticles on plant pathogens, beneficial microbes and crop plants. *Spanish Journal of Agricultural Research*, *15*(2), e1005. doi:10.5424jar/2017152-10305

Bannunah, A. M., Vllasaliu, D., Lord, J., & Stolnik, S. (2014). Mechanisms of nanoparticle internalization and transport across an intestinal epithelial cell model: Effect of size and surface charge. *Molecular Pharmaceutics*, *11*(12), 4363–4373. doi:10.1021/mp500439c PMID:25327847

Banu, H., Renuka, N., Faheem, S. M., Ismail, R., Singh, V., Saadatmand, Z., Khan, S. S., Narayanan, K., Raheem, A., Premkumar, K., & Vasanthakumar, G. (2018). Gold and silver NPs biomimetically synthesized using date palm pollen extract-induce apoptosis and regulate p53 and Bcl-2 expression in human breast adenocarcinoma cells. *Biological Trace Element Research*, *186*(1), 122–134. doi:10.100712011-018-1287-0 PMID:29552710

Barhoum, A., García-Betancourt, M. L., Jeevanandam, J., Hussien, E. A., Mekkawy, S. A., Mostafa, M., & Bechelany, M. (2022). Review on natural, incidental, bioinspired, and engineered nanomaterials: History, definitions, classifications, synthesis, properties, market, toxicities, risks, and regulations. *Nanomaterials (Basel, Switzerland)*, *12*(2), 177. doi:10.3390/nano12020177 PMID:35055196

Barik, S. K., & Singh, B. N. (2019). Nanoemulsion-loaded hydrogel coatings for inhibition of bacterial virulence and biofilm formation on solid surfaces. *Scientific Reports*, *9*(1), 1–13. PMID:30626917

Barot, T., Rawtani, D., & Kulkarni, P. (2020). Physicochemical and biological assessment of silver nanoparticles immobilized Halloysite nanotubes-based resin composite for dental applications. *Heliyon*, *6*(3, e03601), 1–10. doi:10.1016/j.heliyon.2020.e03601 PMID:32195405

Barrena, R., Vargas-García, M. del C., Capell, G., Barańska, M., Puntes, V., Moral-Vico, J., Sánchez, A., & Font, X. (2021). Sustained effect of zero-valent iron nanoparticles under semi-continuous anaerobic digestion of sewage sludge: Evolution of nanoparticles and microbial community dynamics. *The Science of the Total Environment*, *777*, 145969. Advance online publication. doi:10.1016/j.scitotenv.2021.145969 PMID:33676214

Baruah, S., & Dutta, J. (2009). Nanotechnology applications in pollution sensing and degradation in agriculture: A review. *Environmental Chemistry Letters*, *7*(3), 191–204. doi:10.100710311-009-0228-8

Baruwati, B., Kumar, D. K., & Manorama, S. V. (2006). Hydrothermal synthesis of highly crystalline ZnO nanoparticles: A competitive sensor for LPG and EtOH. *Sensors and Actuators. B, Chemical*, *119*(2), 676–682. doi:10.1016/j.snb.2006.01.028

Baudrimont, M., Andrei, J., Mornet, S., Gonzalez, P., Mesmer-Dudons, N., Gourves, P.-Y., Jaffal, A., Dedourge-Geffard, O., Geffard, A., Geffard, O., Garric, J., & Feurtet-Mazel, A. (2018). Trophic transfer and effects of gold nanoparticles (AuNPs) in Gammarus fossarum from contaminated periphytic biofilm. *Environmental Science and Pollution Research International*, *25*(12), 11181–11191. doi:10.100711356-017-8400-3 PMID:28091991

Bayat, M., Javanbakht, V., & Esmaili, J. (2018). Synthesis of zeolite/nickel ferrite/sodium alginate bionanocomposite via a co-precipitation technique for efficient removal of water-soluble methylene blue dye. *International Journal of Biological Macromolecules*, *116*, 607–619. doi:10.1016/j.ijbiomac.2018.05.012 PMID:29738864

Bayda, S., Adeel, M., Tuccinardi, T., Cordani, M., & Rizzolio, F. (2019). The History of Nanoscience and Nanotechnology: From Chemical-Physical Applications to Nanomedicine. *Molecules (Basel, Switzerland)*, *25*(1), 112. doi:10.3390/molecules25010112 PMID:31892180

Bazak, R., Houri, M., Achy, S. E., Hussein, W., & Refaat, T. (2014). Passive targeting of nanoparticles to cancer: A comprehensive review of the literature. *Molecular and Clinical Oncology*, *2*(6), 904–908. doi:10.3892/mco.2014.356 PMID:25279172

Bazina, L., Bitounis, D., Cao, X., DeLoid, G. M., Parviz, D., Strano, M. S., Greg Lin, H.-Y., Bell, D. C., Thrall, B. D., & Demkritou, P. (2021). Biotransformations and Cytotoxicity of Graphene and Inorganic Two-Dimensional Nanomaterials using Simulated Digestions coupled with a Triculture in vitro Model of the Human Gastrointestinal Epithelium. *Environmental Science. Nano*, *8*(11), 3233–3249. doi:10.1039/D1EN00594D

Bergin, I. L., & Witzmann, F. A. (2013). Nanoparticle toxicity by the gastrointestinal route: Evidence and knowledge gaps. *International Journal of Biomedical Nanoscience and Nanotechnology*, *3*(1-2), 163. Advance online publication. doi:10.1504/IJBNN.2013.054515 PMID:24228068

Bergin, M. H., Ghoroi, C., Dixit, D., Schauer, J. J., & Shindell, D. T. (2017). Large Reductions in Solar Energy Production Due to Dust and Particulate Air Pollution. *Environmental Science & Technology Letters*, *4*(8), 339–344. doi:10.1021/acs.estlett.7b00197

Besha, A. T., Liu, Y., Bekele, D. N., Dong, Z., Naidu, R., & Gebremariam, G. N. (2020). Sustainability and environmental ethics for the application of engineered nanoparticles. *Environmental Science & Policy*, *103*, 85–98. doi:10.1016/j.envsci.2019.10.013

Bhatia, S. (2016). Nanoparticles types, classification, characterization, fabrication methods and drug delivery applications. In Natural polymer drug delivery systems. Springer.

Bhattacharjee, S., Macintyre, C. R., Bahl, P., Kumar, U., Wen, X., Aguey-Zinsou, K. F., & Joshi, R. (2020). Reduced graphene oxide and nanoparticles incorporated durable Electroconductive silk fabrics. *Advanced Materials Interfaces*, *7*(20), 2000814. doi:10.1002/admi.202000814

Bhattacharyya, K. G., & Gupta, S. S. (2008). Adsorption of a few heavy metals on natural and modified kaolinite and montmorillonite: A review. *Advances in Colloid and Interface Science*, *140*(2), 114–131. doi:10.1016/j.cis.2007.12.008 PMID:18319190

Bianchera, A., Buttini, F., & Bettini, R. (2020). Micro/nanosystems and biomaterials for controlled delivery of antimicrobial and anti-biofilm agents. *Expert Opinion on Therapeutic Patents*, *30*(12), 983–1000. doi:10.1080/13543776.2020.1839415 PMID:33078643

bin Hussein, M. Z., Zainal, Z., Yahaya, A. H., & Foo, D. W. (2002). Controlled release of a plant growth regulator, alpha-naphthaleneacetate from the lamella of Zn-Al-layered double hydroxide nanocomposite. *Journal of Controlled Release, 82*(2-3), 417–427. doi:10.1016/S0168-3659(02)00172-4

Bindraban, P. S., Dimkpa, C. O., & Pandey, R. (2020). Exploring phosphorus fertilizers and fertilization strategies for improved human and environmental health. *Biology and Fertility of Soils, 56*(3), 299–317. doi:10.100700374-019-01430-2

Boccuni, F., Gagliardi, D., Ferrante, R., Rondinone, B. M., & Iavicoli, S. (2017). Measurement techniques of exposure to nanomaterials in the workplace for low- and medium-income countries: A systematic review. *International Journal of Hygiene and Environmental Health, 220*(7), 1089–1097. doi:10.1016/j.ijheh.2017.06.003 PMID:28711303

Bonefeld-Jørgensen, E. C., Long, M., Hofmeister, M. V., & Vinggaard, A. M. (2007). Endocrine disrupting potential of bisphenol A, bisphenol A dimethacrylate, 4-n-nonylphenol, and 4-n-octylphenol in vitro: New data and a brief review. *Environmental Health Perspectives, 115*(Suppl 1), 69–76. doi:10.1289/ehp.9368 PMID:18174953

Borkow, G., & Gabbay, J. (2009). Copper, an ancient remedy returning to fight microbial, fungal and viral infections. *Current Chemical Biology, 3*(3), 272–278. doi:10.2174/2212796810903030272

Boscaro, M. E., Marin, D. F. C., da Silva, D. C., & Maintinguer, S. I. (2022). Biomass and Bioenergy Effect of Fe 3 O 4 nanoparticles on microbial diversity and biogas production in anaerobic digestion of crude glycerol. *Biomass and Bioenergy, 160*, 106439. doi:10.1016/j.biombioe.2022.106439

Boykov, I. N., Shuford, E., & Zhang, B. (2019). Nanoparticle titanium dioxide affects the growth and microRNA expression of switchgrass (*Panicum virgatum*). *Genomics, 111*(3), 450–456. doi:10.1016/j.ygeno.2018.03.002 PMID:29524463

Brady, R. R. (1999). The Nature and Properties of Soils. Prentice Hall.

Brandelli, A. (2020). The interaction of nanostructured antimicrobials with biological systems: Cellular uptake, trafficking and potential toxicity. *Food Science and Human Wellness, 9*(1), 8–20. doi:10.1016/j.fshw.2019.12.003

Branton, D., Deamer, D. W., Marziali, A., Bayley, H., Benner, S. A., Butler, T., Ventra, M. D., Garaj, S., Hibbs, A., Huang, X., Jovanovich, S. B., Krstic, P. S., Lindsay, S., Ling, X. S., Mastrangelo, C. H., Meller, A., Oliver, J. S., Pershin, Y. V., Ramsey, J. M., ... Schloss, J. A. (2008). The potential and challenges of nanopore sequencing. *Nature Biotechnology, 10*(10), 1146–1153. doi:10.1038/nbt.1495 PMID:18846088

Bratovcic, A. (2020). *Synthesis, characterization, applications, and toxicity of lead oxide nanoparticles*. Lead Chemistry. doi:10.5772/intechopen.91362

Bredhult, C., Bäcklin, B. M., & Olovsson, M. (2007). Effects of some endocrine disruptors on the proliferation and viability of human endometrial endothelial cells in vitro. *Reproductive Toxicology (Elmsford, N.Y.), 23*(4), 550–559. doi:10.1016/j.reprotox.2007.03.006 PMID:17493787

Brennan, S. A., Fhoghlu, C. N., Devitt, B. M., Mahony, F. J. O., Brabazon, D., & Walsh, A. (2019). Review synthesis and application of silver nanoparticles (Ag NPs) for the prevention of infection in healthcare workers. *International Journal of Molecular Sciences, 20*(15), 3620. doi:10.3390/ijms20153620 PMID:31344881

Broadley, M. R., White, P. J., Hammond, J. P., Zelko, I., & Lux, A. (2007). Zinc in plants. *The New Phytologist, 173*(4), 677–702. doi:10.1111/j.1469-8137.2007.01996.x PMID:17286818

Brock, D. A., Douglas, T. E., Queller, D. C., & Strassmann, J. E. (2011). Primitive agriculture in a social amoeba. *Nature, 469*(7330), 393–396. doi:10.1038/nature09668 PMID:21248849

Brouwer, A. S., van Den Broek, M., Seebregts, A., & Faaij, A. (2014). Impacts of Large-Scale Intermittent Renewable Energy Sources on Electricity Systems, and How These Can Be Modeled. *Renewable & Sustainable Energy Reviews, 33*, 443–466. doi:10.1016/j.rser.2014.01.076

Browne, M. A., Crump, P., Niven, S., Teuten, E., Tonkin, A., Galloway, T. S., & Thompson, R. (2011). Accumulation of microplastic on shorelines woldwide: Sources and sinks. *Environmental Science & Technology*, *45*(21), 9175–9179. doi:10.1021/es201811s PMID:21894925

Brunner, T. J., Wick, P., Manser, P., Spohn, P., Grass, R. N., Limbach, L. K., Bruinink, A., & Stark, W. J. (2006). In vitro cytotoxicity of oxide nanoparticles: Comparison to asbestos, silica, and the effect of particle solubility. *Environmental Science & Technology*, *40*(14), 4374–4381. doi:10.1021/es052069i PMID:16903273

Buchta, C., Bittner, C., Heinzl, H., Höcker, P., Mocher, M., & Mayerhofer, M. (2005). Transfusion-related exposure to the plasticizer di(2-ethylhexyl)phthalate in patients receiving plateletpheresis concentrates. *Transfusion*, *45*(5), 798–802. doi:10.1111/j.1537-2995.2005.04380.x PMID:15847671

Bulcke, F., & Dringen, R. (2016). Handling of copper and copper oxide nanoparticles by astrocytes. *Neurochemical Research*, *41*(1), 33–43. doi:10.100711064-015-1688-9 PMID:26268441

Bulcke, F., Dringen, R., & Scheiber, I. F. (2017). Neurotoxicity of copper. *Advances in Neurobiology*, *18*, 313–343. doi:10.1007/978-3-319-60189-2_16 PMID:28889275

Bundschuh, M., Filser, J., Lüderwald, S., McKee, M. S., Metreveli, G., Schaumann, G. E., Schulz, R., & Wagner, S. (2018). Nanoparticles in the environment: Where do we come from, where do we go to? *Environmental Sciences Europe*, *30*(1), 6. doi:10.118612302-018-0132-6 PMID:29456907

Burdusel, A. C., Gherasim, O., Grumezescu, A. M., Mogoanta, L., Ficai, A., & Andronescu, E. (2018). Biomedical applications of silver nanoparticles: An Up-to-Date overview. *Nanomaterials (Basel, Switzerland)*, *8*(9), 681. doi:10.3390/nano8090681 PMID:30200373

Burello, E., & Worth, A. P. (2011). A theoretical framework for predicting the oxidative stress potential of oxide nanoparticles. *Nanotoxicology*, *5*(2), 228–235. doi:10.3109/17435390.2010.502980 PMID:21609138

Bushueva, T., Minigalieva, I., Panov, V., Kuznetsova, A., Naumova, A., Shur, V., Shishkina, E., Gurvich, V., Privalova, L., & Katsnelson, B. (2019). More data on in vitro assessment of comparative and combined toxicity of metal oxide nanoparticles. *Food and Chemical Toxicology*, *133*, 110753. doi:10.1016/j.fct.2019.110753 PMID:31400477

Buzea, C., Pacheco, I. I., & Robbie, K. (2007). Nanomaterials and nanoparticles: Sources and toxicity. *Biointerphases*, *2*(4), MR17–MR71. doi:10.1116/1.2815690 PMID:20419892

Cabral, H., Miyata, K., Osada, K., & Kataoka, K. (2018). Block Copolymer Micelles in Nanomedicine Applications. *Chemical Reviews*, *118*(14), 6844–6892. doi:10.1021/acs.chemrev.8b00199 PMID:29957926

Cai, X., Lee, A., Ji, Z., Huang, C., Chang, C. H., Wang, X., Liao, Y. P., Xia, T., & Li, R. (2017). Reduction of pulmonary toxicity of metal oxide nanoparticles by phosphonate-based surface passivation. *Particle and Fibre Toxicology*, *14*(1), 13. Advance online publication. doi:10.118612989-017-0193-5 PMID:28431555

Calafat, A. M., Ye, X., Wong, L.-Y., Reidy, J. A., & Needham, L. L. (2008). Exposure of the US population to bisphenol A and 4-tertiary-octylphenol: 2003–2004. *Environmental Health Perspectives*, *116*(1), 39–45. doi:10.1289/ehp.10753 PMID:18197297

Campoccia, D., Montanaro, L., & Arciola, C. R. (2013). A review of the biomaterials technologies for infection-resistant surfaces. *Biomaterials*, *34*(34), 8533–8554. doi:10.1016/j.biomaterials.2013.07.089 PMID:23953781

Cao, D., Gong, S., Shu, X., Zhu, D., & Liang, L. (2019). Preparation of ZnO nanoparticles with high dispersibility based on Oriented Attachment (OA) Process. *Nanoscale Research Letters*, *14*(1), 210. doi:10.118611671-019-3038-3 PMID:31222635

Cao, Y., Chen, A., Radcliffe, J., Dietrich, K. N., Jones, R. L., Caldwell, K., & Rogan, W. J. (2009). Postnatal cadmium exposure, neurodevelopment, and blood pressure in children at 2, 5, and 7 years of age. *Environmental Health Perspectives*, *117*(10), 1580–1586. doi:10.1289/ehp.0900765 PMID:20019909

Cao, Y., Xiao, W., Li, S., & Qiu, D. (2021). A Comparative Study of Graphdiyne and Graphene Oxide to Human Umbilical Vein Endothelial Cells. *Journal of Applied Toxicology*, *41*(12), 2021–2030. doi:10.1002/jat.4182 PMID:33973267

Carbone, M., Donia, D. T., Sabbatella, G., & Antiochia, R. (2016). Silver nanoparticles in polymeric matrices for fresh food packaging. *Journal of King Saud University. Science*, *28*(4), 273–279. doi:10.1016/j.jksus.2016.05.004

Carpio, I. E. M., Santos, C. M., Wei, X., & Rodrigues, D. F. (2012). Toxicity of a Polymer-Graphene Oxide Composite against Bacterial Planktonic Cells, Biofilms, and Mammalian Cells. *Nanoscale*, *4*(15), 4746–4756. doi:10.1039/c2nr30774j PMID:22751735

Carpita, N. C., & Gibeaut, D. M. (1993). Structural models of primary cell walls in flowering plants: Consistency of molecular structure with the physical properties of the walls during growth. *The Plant Journal*, *3*(1), 1–30. doi:10.1111/j.1365-313X.1993.tb00007.x PMID:8401598

Carvalho, S. M., Leonel, A. G., Mansur, A. A. P., Carvalho, I. C., Krambrock, K., & Mansur, H. S. (2019). Bifunctional magnetopolymersomes of iron oxide nanoparticles and carboxymethylcellulose conjugated with doxorubicin for hyperthermochemotherapy of brain cancer cells. *Biomaterials Science*, *7*(5), 2102–2122. doi:10.1039/C8BM01528G PMID:30869664

Castillo-Henriquez, L., Alfaro-Aguilar, K., Ugalde-Alvarez, J., Vega-Fernandez, L., de Oca-Vasquez, G. M., & Vega-Baudrit, J. R. (2020). Green synthesis of gold and silver nanoparticles from plant extracts and their possible applications as antimicrobial agents in the agricultural area. *Nanomaterials (Basel, Switzerland)*, *10*(9), 1763. doi:10.3390/nano10091763 PMID:32906575

Chakravarthy, P. V., & Balaji, N. S. (2010). Applications of nanotechnology in veterinary medicine. *Veterinary World*, *3*, 477–480.

Chandhru, M., Logesh, R., Rani, S. K., Ahmed, N., & Vasimalai, N. (2019). One-pot green route synthesis of silver nanoparticles from jack fruit seeds and their antibacterial activities with *Escherichia coli* and *Salmonella* bacteria. *Biocatalysis and Agricultural Biotechnology*, *20*, 101241. doi:10.1016/j.bcab.2019.101241

Chand, K., Ishaque Abro, M., Aftab, U., Shah, A. H., Lakhan, M. N., Cao, D., Mehdi, G., & Mohamed, A. M. A. (2019). Green synthesis characterization and antimicrobial activity against *Staphylococcus aureus* of silver nanoparticles using extracts of neem, onion and tomato. *RSC Advances*, *30*(30), 17002–17015. doi:10.1039/C9RA01407A PMID:35519862

Chandrasekhar, N., & Vinay, S. P. (2017). Yellow colored blooms of *Argemone mexicana* and *Turnera ulmifolia* mediated synthesis of silver nanoparticles and study of their antibacterial and antioxidant activity. *Applied Nanoscience*, *7*(8), 851–861. doi:10.100713204-017-0624-5

Charradi, K., Ahmed, Z., & BenMoussa, M.A. (2022) A facile approach for the synthesis of spinel zinc ferrite/cellulose as an effective photocatalyst for the degradation of methylene blue in aqueous solution. *Cellulose, 29*(9), 1-12. doi:10.1007/s10570-021-04334-3

Chaud, M., Souto, E. B., Zielinska, A., Severino, P., Batain, F., Oliveira-Junior, J., & Alves, T. (2021). Nanopesticides in Agriculture: Benefits and Challenge in Agricultural Productivity, Toxicological Risks to Human Health and Environment. *Toxics*, *9*(6), 131. Advance online publication. doi:10.3390/toxics9060131 PMID:34199739

Chauhan, A., Sillu, D., & Agnihotri, S. (2019). Removal of pharmaceutical contaminants in wastewater using nanomaterials: A comprehensive review. *Current Drug Metabolism, 20*(6), 483–505. doi:10.2174/1389200220666181127104812 PMID:30479212

Chavali, M. S., & Nikolova, M. P. (2019). Metal oxide nanoparticles and their applications in nanotechnology. *SN Applied Sciences, 1*(6), 607. doi:10.100742452-019-0592-3

Chawla, R., Sivakumar, S., & Kaur, H. (2021). Antimicrobial edible films in food packaging: Current scenario and recent nanotechnological advancements-a review. *Carbohydrate Polymer Technologies and Applications, 2*, 100024. doi:10.1016/j.carpta.2020.100024

Chen, G., Seo, J., Yang, C., & Prasad, P. N. (2013). Nanochemistry and nanomaterials for photovoltaics. *Chemical Society Reviews, 42*(21), 8304–8338. doi:10.1039/c3cs60054h PMID:23868557

Chen, H. (2018). Metal based nanoparticles in agricultural system: Behavior, transport, and interaction with plants. *Chemical Speciation and Bioavailability, 30*(1), 123–134. doi:10.1080/09542299.2018.1520050

Chen, Y., Wubs, M., Mork, J., & Koenderink, A. F. (2011). Coherent single-photon absorption by single emitters coupled to one-dimensional nanophotonic waveguides. *New Journal of Physics, 13*(10), 103010. doi:10.1088/1367-2630/13/10/103010

Chen, Z., Meng, H., Xing, G., Chen, C., Zhao, Y., Jia, G., Wang, T., Yuan, H., Ye, C., Zhao, F., Chai, Z., Zhu, C., Fang, X., Ma, B., & Wan, L. (2006). Acute toxicological effects of copper nanoparticles in vivo. *Toxicology Letters, 163*(2), 109–120. doi:10.1016/j.toxlet.2005.10.003 PMID:16289865

Chertok, B., Moffat, B. A., David, A. E., Yu, F., Bergemann, C., Ross, B. D., & Yang, V. C. (2008). Iron oxide nanoparticles as a drug delivery vehicle for MRI monitored magnetic targeting of brain tumors. *Biomaterials, 29*(4), 487–496. doi:10.1016/j.biomaterials.2007.08.050 PMID:17964647

Chhipa, H. (2017). Nanofertilizers and nanopesticides for agriculture. *Environmental Chemistry Letters, 15*(1), 15–22. doi:10.100710311-016-0600-4

Chia, H. L., Latiff, N. M., Gusmão, R., Sofer, Z., & Pumera, M. (2019). Cytotoxicity of Shear Exfoliated Pnictogen (As, Sb, Bi) Nanosheets. *Chemistry (Weinheim an der Bergstrasse, Germany), 25*(9), 2242–2249. doi:10.1002/chem.201804336 PMID:30637810

Chia, H. L., Latiff, N. M., Sofer, Z., & Pumera, M. (2018). Cytotoxicity of Group 5 Transition Metal Ditellurides (MTe_2; M = V, Nb, Ta). *Chemistry (Weinheim an der Bergstrasse, Germany), 24*(1), 206–211. doi:10.1002/chem.201704316 PMID:29027716

Chien, C. C., Yan, Y. H., Juan, H. T., Cheng, T. J., Liao, J. B., Lee, H. P., & Wang, J. S. (2017). Sustained renal inflammation following 2 weeks of inhalation of occupationally relevant levels of zinc oxide nanoparticles in Sprague Dawley rats. *Journal of Toxicologic Pathology, 30*(4), 307–314. doi:10.1293/tox.2017-0025 PMID:29097840

Chittora, D., Meena, M., Barupal, T., Swapnil, P., & Sharma, K. (2020). Cyanobacteria as a source of biofertilizers for sustainable agriculture. *Biochemistry and Biophysics Reports, 22*, 100737. doi:10.1016/j.bbrep.2020.100737 PMID:32083191

Chiu, T. C., & Huang, C. C. (2009). Aptamer-functionalized nano-biosensors. *Sensors (Basel), 9*(12), 10356–10388. doi:10.339091210356 PMID:22303178

Choi, W. S., Koo, H. Y., Park, J.-H., & Kim, D.-Y. (2005). Synthesis of two types of nanoparticles in polyelectrolyte capsule nanoreactors and their dual functionality. *Journal of the American Chemical Society, 127*(46), 16136–16142. doi:10.1021/ja053981u PMID:16287301

Chouhan, R. S., Vinayaka, A. C., & Thakur, M. S. (2010). Thiol-stabilized luminescent CdTe quantum dot as biological fluorescent probe for sensitive detection of methyl parathion by a fluoro-immune chromatographic technique. *Analytical and Bioanalytical Chemistry, 397*(4), 1467–1475. doi:10.100700216-009-3433-1 PMID:20101498

Chowdhury, S. M., Dasgupta, S., McElroy, A. E., & Sitharaman, B. (2014). Structural Disruption increases Toxicity of Graphene Nanoribbons. *Journal of Applied Toxicology, 34*(11), 1235–1246. doi:10.1002/jat.3066 PMID:25224919

Chozhavendhan, S., Karthigadevi, G., Bharathiraja, B., Kumar, R. P., Abo, L. D., Prabhu, S. V., Balachandar, R., & Jayakumar, M. (2023). Current and prognostic overview on the strategic exploitation of anaerobic digestion and digestate: A review. *Environmental Research, 216*, 114526. doi:10.1016/j.envres.2022.114526 PMID:36252837

Christopher, F. C., Kumar, P. S., Christopher, F. J., Joshiba, G. J., & Madhesh, P. (2020). Recent advancements in rapid analysis of pesticides using nano biosensors: A present and future perspective. *Journal of Cleaner Production, 269*, 122356. doi:10.1016/j.jclepro.2020.122356

Chung, S., Revia, R. A., & Zhang, M. (2021). Graphene quantum dots and their applications in bioimaging, biosensing, and therapy. *Advanced Materials, 33*(22), 1904362. doi:10.1002/adma.201904362 PMID:31833101

Cipelli, R., Harries, L., Okuda, K., Yoshihara, S., Melzer, D., & Galloway, T. S. (2013). Bisphenol A modulates the expression of Estrogen-Related Receptor-α in T-Cells. *Reproduction (Cambridge, England), 147*(4), 419–426. doi:10.1530/REP-13-0423 PMID:24231368

Claudio, L. (2012). Food packaging and public health. *Environmental Health Perspectives, 120*(6), A233–A237. doi:10.1289/ehp.120-a232 PMID:22659036

Cohen, D., Soroka, Y., Maor, Z., Oron, M., Portugal-Cohen, M., Berhanu, D., Valsami-Jones, E., Hai, N., & Milner, Y. (2013). Evaluation of topically applied copper(II) oxide nanoparticle cytotoxicity in human skin organ culture. *Toxicology In Vitro, 27*(1), 292–298. doi:10.1016/j.tiv.2012.08.026 PMID:22954531

Cole, M., Lindeque, P., Goodhead, R., Moger, J., Halsband-Lenk, C., & Galloway, T. S. (2013). Microplastic ingestion by zooplankton. *Environmental Science & Technology, 47*(12), 6646–6655. doi:10.1021/es400663f PMID:23692270

Cole, M., Lindeque, P., Halsband-Lenk, C., & Galloway, T. S. (2011). Microplastic as a contaminant in the marine environment: A review. *Marine Pollution Bulletin, 62*(12), 2588–2597. doi:10.1016/j.marpolbul.2011.09.025 PMID:22001295

Colvin, V. L. (2003). The potential environmental impact of engineered nanomaterials. *Nature Biotechnology, 21*(10), 1166–1170. doi:10.1038/nbt875 PMID:14520401

Contado, C. (2015). Nanomaterials in consumer products: A challenging analytical problem. *Frontiers in Chemistry, 3*, 48. doi:10.3389/fchem.2015.00048 PMID:26301216

Coropceanu, I., & Bawendi, M. G. (2014). Core/Shell Quantum Dot Based Luminescent Solar Concentrators with Reduced Reabsorption and Enhanced Efficiency. *Nano Letters, 14*(7), 4097–4101. doi:10.1021/nl501627e PMID:24902615

Corradini, E., Moura, M. R., & Mattoso, L. H. C. (2010). A preliminary study of the incorporation of NPK fertilizer into chitosan nanoparticles. *Polym. Lett., 4*(8), 509–515. doi:10.3144/expresspolymlett.2010.64

Corredor, E., Testillano, P. S., Coronado, M. J., González-Melendi, P., Fernández-Pacheco, R., Marquina, C., Ibarra, M. R., de la Fuente, J. M., Rubiales, D., Pérez-de-Luque, A., & Risueño, M. C. (2009). Nanoparticle penetration and transport in living pumpkin plants: In situ subcellular identification. *BMC Plant Biology, 9*(1), 45. doi:10.1186/1471-2229-9-45 PMID:19389253

Courtois, P., de Vaufleury, A., Grosser, A., Lors, C., & Vandenbulcke, F. (2021). Transfer of sulfidized silver from silver nanoparticles, in sewage sludge, to plants and primary consumers in agricultural soil environment. *The Science of the Total Environment, 777*, 145900. doi:10.1016/j.scitotenv.2021.145900 PMID:33676211

Cranford, S. W., & Buehler, M. J. (2011). Mechanical properties of graphyne. *Carbon, 49*(13), 4111–4121. doi:10.1016/j.carbon.2011.05.024

Creutz, S. E., Fainblat, R., Kim, Y., De Siena, M. C., & Gamelin, D. R. (2017). A Selective Cation Exchange Strategy for the Synthesis of Colloidal Yb3+-Doped Chalcogenide Nanocrystals with Strong Broadband Visible Absorption and Long-Lived Near-Infrared Emission. *Journal of the American Chemical Society, 139*(34), 11814–11824. doi:10.1021/jacs.7b04938 PMID:28750510

Crompton, T. (2007). *Additive migration from plastics into foods. A guide for analytical chemistry.* iSmithersRapra Publishing.

Daifa, M., Shmoeli, E., & Domb, A. J. (2019). Enhanced flocculation activity of polyacrylamide-based flocculant for purification of industrial wastewater. *Polymers for Advanced Technologies, 30*(10), 2636–2646. doi:10.1002/pat.4730

Dai, H. (2002). Carbon nanotubes: Opportunities and challenges. *Surface Science, 500*(1-3), 218–241. doi:10.1016/S0039-6028(01)01558-8

Dai, Y., & Liu, C. C. (2019). Recent advances on electrochemical biosensing strategies toward universal point-of-care systems. *Angewandte Chemie, 131*(36), 12483–12496. doi:10.1002/ange.201901879 PMID:30990933

Dakal, T. C., Kumar, A., Majumdar, R. S., & Yadav, V. (2016). Mechanistic basis of antimicrobial actions of silver nanoparticles. *Frontiers in Microbiology, 7*, 1–17. doi:10.3389/fmicb.2016.01831 PMID:27899918

Dang, Y., Zhang, Y., Fan, L., Chen, H., & Roco, M. C. (2010). Trends in worldwide nanotechnology patent applications: 1991 to 2008. *Journal of Nanoparticle Research, 12*(3), 687–706. doi:10.100711051-009-9831-7 PMID:21170123

Daorta, S. F., Proto, A., Fusco, R., Andreani, L. C., & Liscidini, M. (2014). Cascade Luminescent Solar Concentrators. *Applied Physics Letters, 104*(15), 10–14. doi:10.1063/1.4871481

Dasary, S. S. R., Rai, U. S., Yu, H. T., Anjaneyulu, Y., Dubey, M., & Ray, P. C. (2008). Gold nanoparticle-based surface enhanced fluorescence for detection of organophosphorus agents. *Chemical Physics Letters, 460*(1-3), 187–190. doi:10.1016/j.cplett.2008.05.082 PMID:24031096

Dashtbani-Roozbehani, A., & Brown, M. H. (2021). Efflux pump mediated antimicrobial resistance by staphylococci in health-related environments: Challenges and the quest for inhibition. *Antibiotics (Basel, Switzerland), 10*(12), 1502. doi:10.3390/antibiotics10121502 PMID:34943714

Das, & Jargar, Hattiwale, Yendigeri, Das, & Dhundasi. (2013). Serum vitamin E (α-tocopherol) estimation: A potential biomarker of antioxidant status evaluation on heavy metal toxicities. *Recent Patents on Biomarkers, 3*(1), 36–43. doi:10.2174/2210309011303010036

Das, S., Unni, B., Bhattacharjee, M., Wann, S. B., & Rao, P. G. (2012). Toxicological effects of arsenic exposure in a freshwater teleost fish, Channa punctatus. *African Journal of Biotechnology, 11*(19), 4447–4454. doi:10.5897/AJB11.2808

Davarpanah, S., Tehranifar, A., Davarynejad, G., Abadía, J., & Khorasani, R. (2016). Effects of foliar applications of zinc and boron nano-fertilizers on pomegranate (*Punica granatum* cv. *Ardestani*) fruit yield and quality. *Scientia Horticulturae, 210*, 57–64. doi:10.1016/j.scienta.2016.07.003

Davidson, D. C., Derk, R., He, X., Stueckle, T. A., Cohen, J., Pirela, S. V., Demokritou, P., Rojanasakul, Y., & Wang, L. (2016). Direct stimulation of human fibroblasts by nCeO2 in vitro is attenuated with an amorphous silica coating. *Particle and Fibre Toxicology, 13*(1), 23. doi:10.118612989-016-0134-8 PMID:27142434

de Abreu, D. A. P., Cruz, J. M., Angulo, L., & Losada, P. P. (2010). Mass transport studies of different additives in polyamide and exfoliated nanocomposite polyamide films for food industry. *Packaging Technology & Science, 23*(2), 59–68. doi:10.1002/pts.879

De Azeredo, H. M. C. (2013). Antimicrobial nanostructures in food packaging. *Trends in Food Science & Technology, 30*(1), 56–69. doi:10.1016/j.tifs.2012.11.006

De Berardis, B., Civitelli, G., Condello, M., Lista, P., Pozzi, R., Arancia, G., & Meschini, S. (2010). Exposure to ZnO nanoparticles induces oxidative stress and cytotoxicity in human colon carcinoma cells. *Toxicology and Applied Pharmacology, 246*(3), 116–127. doi:10.1016/j.taap.2010.04.012 PMID:20434478

Deekala, V., Babu, B. K., & Rudraraju, R. (2019). Pharmacological studies of zinc oxide nanoparticles. *Indian Journal of Biochemistry & Biophysics, 56*, 500–505.

Del Solar, V., & Contel, M. (2019). Metal-based antibody drug conjugates. Potential and challenges in their application as targeted therapies in cancer. *Journal of Inorganic Biochemistry, 199*, 110780. doi:10.1016/j.jinorgbio.2019.110780 PMID:31434020

Delcroix, G. J. R., Jacquart, M., Lemaire, L., Sindji, L., Franconi, F., Le Jeune, J. J., & Montero-Menei, C. N. (2009). Mesenchymal and neural stem cells labeled with HEDP-coated SPIO nanoparticles: In vitro characterization and migration potential in rat brain. *Brain Research, 1255*, 18–31. doi:10.1016/j.brainres.2008.12.013 PMID:19103182

Delfani, M., Firouzabadi, M. B., Farrokhi, N., & Makarian, H. (2014). Some physiological responses of black-eyed pea to iron and magnesium nanofertilizers. *Communications in Soil Science and Plant Analysis, 45*(4), 530–540. doi:10.1080/00103624.2013.863911

De, M., Ghosh, P. S., & Rotello, V. M. (2008). Applications of nanoparticles in biology. *Advanced Materials, 20*(22), 4225–4241. doi:10.1002/adma.200703183

Denizli, A., Kesenci, K., Arica, Y., & Piskin, E. (2000). Dithiocarbamate-incorporated monosize polystyrene microspheres for selective removal of mercury ions. *Reactive & Functional Polymers, 44*(3), 235–243. doi:10.1016/S1381-5148(99)00099-1

DeRosa, M. C., Monreal, C., Schnitzer, M., Walsh, R., & Sultan, Y. (2010). Nanotechnology in fertilizers. *Nature Nanotechnology, 5*(2), 91. doi:10.1038/nnano.2010.2 PMID:20130583

Deshmukh, S. P., Patil, S. M., Mullani, S. B., & Delekar, S. D. (2019). Silver nanoparticles as an effective disinfectant: A review. *Materials Science and Engineering C, 97*, 954–965. doi:10.1016/j.msec.2018.12.102 PMID:30678983

Devarajan, P. V., Jindal, A. B., Patil, R. R., Mulla, F., Gaikwad, R. V., & Samad, A. (2010). Particle shape: A new design parameter for passive targeting in splenotropic drug delivery. *Journal of Pharmaceutical Sciences, 99*(6), 2576–2581. doi:10.1002/jps.22052 PMID:20091830

Dhuper, S., Panda, D., & Nayak, P. L. (2012). Green synthesis and characterization of zero valent iron nanoparticles from the leaf extract of *Mangifera indica. Nano Trends: A Journal of Nanotechnology and Its Applications, 13*(2), 16–22.

Díaz, I., Fdz-Polanco, F., Mutsvene, B., & Fdz-Polanco, M. (2020). Effect of operating pressure on direct biomethane production from carbon dioxide and exogenous hydrogen in the anaerobic digestion of sewage sludge. *Applied Energy, 280*, 115915. doi:10.1016/j.apenergy.2020.115915

Diederich, F., & Gómez-López, M. (1999). Supramolecular fullerene chemistry. *Chemical Society Reviews, 28*(5), 263–277. doi:10.1039/a804248i

Dikshit, A., Shukla, S. K., & Mishra, R. K. (2013). *Exploring Nanomaterials with PGPR in Current Agricultural Scenario*. Lap Lambert Academic Publishing.

Dinu, M. V., & Dragan, E. S. (2010). Evaluation of Cu2+, Co2+, and Ni2+ ions removal from aqueous solution using a novel chitosan/clinoptilolite composites: Kinetics and isotherms. *Chemical Engineering Journal, 160*(1), 157–163. doi:10.1016/j.cej.2010.03.029

Dodds, E. C., & Lawson, W. (1936). Synthetic estrogenic agents without the phenanthrene nucleus. *Nature, 137*(3476), 996. doi:10.1038/137996a0

Dolez, P. I. (2021). Energy Harvesting Materials and Structures for Smart Textile Applications: Recent Progress and Path Forward. *Sensors (Basel), 21*(18), 6297. doi:10.339021186297 PMID:34577509

Dong, J., Zhao, H., Qiao, F., Liu, P., Wang, X. D., & Ai, S. Y. (2013). Quantum dot immobilized acetylcholinesterase for the determination of organophosphate pesticides using graphene-chitosan nanocomposite modified electrode. *Analytical Methods, 5*(11), 2866–2872. doi:10.1039/c3ay26599d

Dong, X., Ma, N., Liu, M., & Liu, Z. (2015). Effects of As2O3 nanoparticles on cell growth and apoptosis of NB4 cells. *Experimental and Therapeutic Medicine, 10*(4), 1271–1276. doi:10.3892/etm.2015.2651 PMID:26622477

Dreaden, E. C., Alkilany, A. M., Huang, X., Murphy, C. J., & El-Sayed, M. A. (2012). The golden age: Gold nanoparticles for biomedicine. *Chemical Society Reviews, 41*(7), 2740–2779. doi:10.1039/C1CS15237H PMID:22109657

Drobne, D. (2007). Nanotoxicology for safe and sustainable nanotechnology. *Arhiv za Higijenu Rada i Toksikologiju, 58*(4), 471–478. doi:10.2478/v10004-007-0040-4 PMID:18063532

Duan, J., Yu, Y., Yu, Y., Li, Y., Wang, J., Geng, W., Jiang, L., Li, Q., Zhou, X., & Sun, Z. (2014). Silica nanoparticles induce autophagy and endothelial dysfunction via the PI3K/Akt/mTOR signaling pathway. *International Journal of Nanomedicine, 9*, 5131–5141. doi:10.2147/IJN.S71074 PMID:25395850

Dubaish, F., & Liebezeit, G. (2013). Suspended microplastics and black carbon particles in the Jade system, southern North Sea. *Water, Air, and Soil Pollution, 224*(2), 1–8. doi:10.100711270-012-1352-9

Duggal, D. (2011). Role of nanotechnology in new drug delivery systems. *International Journal of Drug Development and Research, 3*(4), 4–8.

Duhan, J. S., Kumar, R., Kumar, N., Kaur, P., Nehra, K., & Duhan, S. (2017). Nanotechnology: The new perspective in precision agriculture. *Biotechnology Reports (Amsterdam, Netherlands), 15*, 11–23. doi:10.1016/j.btre.2017.03.002 PMID:28603692

Duhan, J. S., Kumar, R., Kumar, N., Kaur, P., Nehra, K., & Duhan, S. (2017). *Nanotechnology: The new perspective in precision agriculture. In Biotechnology Reports*. Elsevier.

Du, J., Hu, Z., Yu, Z., Li, H., Pan, J., Zhao, D., & Bai, Y. (2019). Antibacterial activity of a novel *Forsythia suspensa* fruit mediated green silver nanoparticles against food-borne pathogens and mechanisms investigation. *Materials Science and Engineering C, 102*, 247–253. doi:10.1016/j.msec.2019.04.031 PMID:31146997

Duman, O., Tunç, S., Polat, T. G., & Bozoğlan, B. K. (2016). Synthesis of magnetic oxidized multiwalled carbon nanotube-κ-carrageenan-Fe_3O_4 nanocomposite adsorbent and its application in cationic Methylene Blue dye adsorption. *Carbohydrate Polymers, 147*, 79–88. doi:10.1016/j.carbpol.2016.03.099 PMID:27178911

Dumkova, J., Smutna, T., Vrlikova, L., Docekal, B., Kristekova, D., Vecera, Z., Husakova, Z., Jakesova, V., Jedličková, A., Mikuška, P., Alexa, L., Coufalík, P., Tvrdoňová, M., Křůmal, K., Vaculovič, T., Kanický, V., Hampl, A., & Buchtová, M. (2020). A Clearance Period after Soluble Lead Nanoparticle Inhalation Did Not Ameliorate the Negative Effects on Target Tissues Due to Decreased Immune Response. *International Journal of Molecular Sciences*, *21*(22), 8738. doi:10.3390/ijms21228738 PMID:33228049

Dumkova, J., Smutna, T., Vrlíkova, L., Le Coustumer, P., Vecera, Z., Docekal, B., & Buchtova, M. (2017). Sub-chronic inhalation of lead oxide nanoparticles revealed their broad distribution and tissue-specific subcellular localization in target organs. *Particle and Fibre Toxicology*, *14*(1), 1–19. doi:10.118612989-017-0236-y PMID:29268755

Eastman. (2010). *Eastman material data sheet*. http://www.eastman.com, http://www. eastman.com/lierature.center/T/TRS252.pdf

Ebbesen, T. W. (1994). Carbon nanotubes. *Annual Review of Materials Science*, *24*(1), 235–264. doi:10.1146/annurev.ms.24.080194.001315

Ebel, J. P. (1990). A method for quantifying particle absorption from the small intenstine. *Pharmaceutical Research*, *7*(8), 848–851. doi:10.1023/A:1015964916486 PMID:2235881

Effiong, D., Uwah, T., Jumbo, E., & Akpabio, A. (2020). Nanotechnology in cosmetics: Basics, Current Trends and Safety Concerns- A Review. *Advances in Nanoparticles*, *9*, 1–22.

EFSA. (2006). Opinion of the scientific panel on food additives, flavourings, processing aids and materials in contact with food on a request from the commission related to 2,2-bis(4-hydroxyphenyl)propane (bisphenol A). *EFSA Journal*, *428*, 13–75.

EFSA. (2011). Scientific opinion: Guidance on the risk assessment of the application of nanoscience and nanotechnologies in the food and feed chain. *EFSA Journal*, *9*, 2140.

EFSA. (2015). Scientific Opinion on the risks to public health related to the presence of bisphenol A (BPA) in foodstuffs. *EFSA Journal*, *13*(1), 3978. doi:10.2903/j.efsa.2015.3978

Egbuna, C., Parmar, V. K., Jeevanandam, J., Ezzat, S. M., Patrick-Iwuanyanwu, K. C., Adetunji, C. O., Khan, J., Onyeike, E. N., Uche, C. Z., Akram, M., Ibrahim, M. S., El Mahdy, N. M., Awuchi, C. G., Saravanan, K., Tijjani, H., Odoh, U. E., Messaoudi, M., Ifemeje, J. C., Olisah, M. C., ... Ibeabuchi, C. G. (2021). Toxicity of Nanoparticles in Biomedical Application: Nanotoxicology. *Journal of Toxicology*, *9954443*, 1–21. Advance online publication. doi:10.1155/2021/9954443 PMID:34422042

Ehrlich, S., Calafat, A. M., Hunblet, O., Smith, T., & Hauser, R. (2014). Handling of thermal receipts as a source of exposure to bisphenol A. *Journal of the American Medical Association*, *311*(8), 859–860. doi:10.1001/jama.2013.283735 PMID:24570250

Eke, J., Mills, P. A., Page, J. R., Wrigth, G. P., Tsyusko, O. V., & Escobar, I. C. (2020). Nanohybrid Membrane Synthesis with Phosphorene Nanoparticles: A Study of the Addition, Stability and Toxicity. *Polymers*, *12*(7), 1555. doi:10.3390/polym12071555 PMID:32674304

Ekinci, M., Dursun, A., Yildirim, E., & Parlakova, F. (2014). Effects of nanotechnology liquid fertilizers on the plant growth and yield of cucumber (*Cucumis sativus* L.). *Acta Scientiarum Polonorum. Hortorum Cultus*, *13*(3), 135–141.

El Bakari, K., & Kling, W. L. (2010). Virtual Power Plants: An Answer to Increasing Distributed Generation. *IEEE PES Innovative Smart Grid Technologies Conference Europe*, 1–6. doi:10.1109/ISGTEUROPE.2010.5638984

Elahi, N., Kamali, M., Baghersad, M. H., & Amini, B. (2019). A fluorescence Nano-biosensors immobilization on Iron (MNPs) and gold (AuNPs) nanoparticles for detection of Shigella spp. *Materials Science and Engineering C, 105*, 110113. doi:10.1016/j.msec.2019.110113 PMID:31546438

Elbeshehy, E. K. F., Elazzazy, A. M., & Aggelis, G. (2015). [REMOVED HYPERLINK FIELD]Silver nanoparticles synthesis mediated by new isolates of *Bacillus* spp., nanoparticle characterization and their activity against Bean Yellow Mosaic Virus and human pathogens. *Frontiers in Microbiology*, *6*, 453. doi:10.3389/fmicb.2015.00453 PMID:26029190

Eleraky, N. E., Allam, A., Hassan, S. B., & Omar, M. M. (2020). Nanomedicine fight against antibacterial resistance: An overview of the recent pharmaceutical innovations. *Pharmaceutics*, *12*(2), 142. doi:10.3390/pharmaceutics12020142 PMID:32046289

Eletskii, A. V. (1997). Carbon nanotubes. *Physics Uspekhi*, *40*(9), 899–924. doi:10.1070/PU1997v040n09ABEH000282

Elias, A., & Tsourkas, A. (2009). Imaging circulating cells and lymphoid tissues with iron oxide nanoparticles. *Hematology*, *2009*(1), 720–726. doi:10.1182/asheducation-2009.1.720 PMID:20008258

El-Kemary, M., El-Shamy, H., & Mosaad, M. M. (2009). The role of capping agent on the interaction of cadmium sulphide nanoparticles with flufenamic acid drug. *Materials Chemistry and Physics*, *118*(1), 81–85. doi:10.1016/j.matchemphys.2009.07.007

Elmer, W., & White, J. C. (2008). The future of nanotechnology in plant pathology. *Annual Review of Phytopathology*, *56*(1), 111–133. doi:10.1146/annurev-phyto-080417-050108 PMID:30149792

El-Saadony, M. T., Lmoshadak, A., Shafi, A. S., Albaqami, M. E., Saad, N. M., El-Tahan, A. M., Desoky, A. M., Elnahal, E. M., Almakas, A., Abd, A., El-Mageed, T. A., Taha, A. E., Elrys, A. S., & Helmy, A. M. (2021). Vital roles of sustainable nano-fertilizers in improving plant quality and quantity-an updated review. *Saudi Journal of Biological Sciences*, *28*(12), 7349–7359. doi:10.1016/j.sjbs.2021.08.032 PMID:34867037

Elsabahy, M., & Wooley, K. L. (2013). Cytokines as biomarkers of nanoparticle immunotoxicity. *Chemical Society Reviews*, *42*(12), 5552–5576. doi:10.1039/c3cs60064e PMID:23549679

El-Trass, A., ElShamy, H., El-Mehasseb, I., & El-Kemary, M. (2012). CuO nanoparticles: Synthesis, characterization, optical properties and interaction with amino acids. *Applied Surface Science*, *258*(7), 2997–3001. doi:10.1016/j.apsusc.2011.11.025

Elzoghby, A. O., Samy, W. M., & Elgindy, N. A. (2012). Albumin-based nanoparticles as potential controlled release drug delivery systems. *Journal of Controlled Release*, *157*(2), 168–182. doi:10.1016/j.jconrel.2011.07.031 PMID:21839127

Emadian, S.E. (2017). *Physiological Responses of Loblolly Pine (Finustaeda L.) to Silicon and Water Stress*. Academic Press.

Endo, M., Iijima, S., & Dresselhaus, M.S. (1996). *Carbon nanotubes.* Elsevier.

Engler, R. E. (2012). The complex interaction between marine debris and toxic chemicals in the ocean. *Environmental Science & Technology*, *46*(22), 12302–12315. doi:10.1021/es3027105 PMID:23088563

Ericson, B., Landrigan, P., Taylor, M. P., Frostad, J., Caravanos, J., Keith, J., & Fuller, R. (2016). The global burden of lead toxicity attributable to informal used lead-acid battery sites. *Annals of Global Health*, *82*(5), 686–699. doi:10.1016/j.aogh.2016.10.015 PMID:28283119

Esim, O., & Hascicek, C. (2021). Albumin-based nanoparticles as promising drug delivery systems for cancer treatment. *Current Pharmaceutical Analysis*, *17*(3), 346–359.

European Commission. (2014). *EC guidance document. Guidance to hazard analysis and critical control point (HACCP) application and to regulation 852/2004 and 853/2004*. https://europa.eu/comm/food/food/biosafety/hygienelegislation/index_en.htm

Fahmi, T., Branch, L. D., Nima, Z. A., Jang, D. S., Savenka, A. V., Biris, A. S., & Basnakian, A. G. (2017). Mechanism of Graphene-induced Cytotoxicity: Role of Endonucleases. *Journal of Applied Toxicology*, *37*(11), 1325–1332. doi:10.1002/jat.3462 PMID:28543094

Fahmy, B., & Cormier, S. A. (2009). Copper oxide nanoparticles induce oxidative stress and cytotoxicity in airway epithelial cells. *Toxicology In Vitro*, *23*(7), 1365–1371. doi:10.1016/j.tiv.2009.08.005 PMID:19699289

Falsini, S., Bardi, U., Abou-Hassan, A., & Ristori, S. (2018). Sustainable strategies for large-scale nanotechnology manufacturing in the biomedical field. *Green Chemistry*, *20*(17), 3897–3907. doi:10.1039/C8GC01248B

Farahani, M., Rezaei-Tavirani, M., Zali, H., Hatamie, S., & Ghasemi, N. (2020). Systems Toxicology Assessment Revealed the Impact of Graphene-based Materials on Cell Cycle Regulators. *Journal of Biomedical Materials Research. Part A*, *108*(7), 1520–1533. doi:10.1002/jbm.a.36923 PMID:32170903

Fasiku, V. O., Omolo, C. A., Devnarain, N., Ibrahim, U. H., Rambharose, S., Faya, M., Mocktar, C., Singh, S. D., & Govender, T. (2021). Chitosan-based hydrogel for the dual delivery of antimicrobial agents against bacterial methicillin-resistant Staphylococcus aureus biofilm-infected wounds. *ACS Omega*, *6*(34), 21994–22010. doi:10.1021/acsomega.1c02547 PMID:34497894

Fatima, F., Bajpai, P., Pathak, N., Singh, S., Priya, S., & Verma, S. R. (2015). Antimicrobial and immunomodulatory efficacy of extracellularly synthesized silver and gold nanoparticles by a novel phosphate solubilizing fungus *Bipolaris tetramera*. *BMC Microbiology*, *15*(1), 52. doi:10.118612866-015-0391-y PMID:25881309

Fatima, F., Hashim, A., & Anees, S. (2021). Efficacy of nanoparticles as nanofertilizer production: A review. *Environmental Science and Pollution Research International*, *28*(2), 1292–1303. doi:10.100711356-020-11218-9 PMID:33070292

Fatima, F., Pathak, N., Verma, S. R., & Bajpai, P. (2018). Toxicity and immunomodulatory efficacy of biosynthesized silver myconanosomes on pathogenic microbes and macrophage cells. *Artificial Cells, Nanomedicine, and Biotechnology*, *46*(8), 1637–1645. doi:10.1080/21691401.2017.1388247 PMID:29022370

Fatima, F., Verma, S. R., Pathak, N., & Bajpai, P. (2016). Extracellular mycosynthesis of silver nanoparticles and their microbicidal activity. *Journal of Global Antimicrobial Resistance*, *7*, 88–92. doi:10.1016/j.jgar.2016.07.013 PMID:27689341

Fazzino, F., Folino, A., Mauriello, F., Pedullà, A., & Calabrò, P. S. (2021). Biofuel production from fruit and vegetable market waste and mature landfill leachate by an active filter-anaerobic digestion integrated system. *Energy Conversion and Management: X*, *12*, 100130. doi:10.1016/j.ecmx.2021.100130

Fehaid, A., Fujii, R., Sato, T., & Taniguchi, A. (2020). Silver nanoparticles affect the inflammatory response in a lung epithelial cell line. *The Open Biotechnology Journal*, *14*(1), 113–123. doi:10.2174/1874070702014010113

Feizi, H., Rezvani Moghaddam, P., Shahtahmassebi, N., & Fotovat, A. (2012). Impact of bulk and nanosized titanium dioxide (TiO_2) on wheat seed germination and seedling growth. *Biological Trace Element Research*, *146*(1), 101–106. doi:10.100712011-011-9222-7 PMID:21979242

Feng, D., Aldrich, C., & Tan, H. (2000). Treatment of acid mine water by use of heavy metal precipitation and ion exchange. *Minerals Engineering*, *13*(6), 623–642. doi:10.1016/S0892-6875(00)00045-5

Feng, W., Nie, W., Cheng, Y., Zhou, X., Chen, L., Qiu, K., Chen, Z., Zhu, M., & He, C. (2015). In vitro and in vivo toxicity studies of copper sulfide nanoplates for potential photothermal applications. *Nanomedicine; Nanotechnology, Biology, and Medicine, 11*(4), 901–912. doi:10.1016/j.nano.2014.12.015 PMID:25652899

Ferdous, Z., & Nemmar, A. (2020). Health Impact of Silver Nanoparticles: A Review of the Biodistribution and Toxicity Following Various Routes of Exposure. *International Journal of Molecular Sciences, 21*(7), 2375. doi:10.3390/ijms21072375 PMID:32235542

Feregrino-Pérez, A. A., Magaña-López, E., Guzmán, C., & Esquivel, K. (2018). A general overview of the benefits and possible negative effects of the nanotechnology in horticulture. *Scientia Horticulturae, 238*, 126–137. doi:10.1016/j.scienta.2018.03.060

Fergusson, J. (1990). *The heavy elements: chemistry, environmental impact and health effects.* Pergamon Press.

Fernandez, V., & Eichert, T. (2009). Uptake of hydrophilic solutes through plant leaves: Current state of knowledge and perspectives of foliar fertilization. *Critical Reviews in Plant Sciences, 28*(1-2), 36–68. doi:10.1080/07352680902743069

Fischer, D., Li, Y., Ahlemeyer, B., Krieglstein, J., & Kissel, T. (2003). In vitro cytotoxicity testing of polycations: Influence of polymer structure on cell viability and hemolysis. *Biomaterials, 24*(7), 1121–1131. doi:10.1016/S0142-9612(02)00445-3 PMID:12527253

Fischer, M., & Vögtle, F. (1999). Dendrimers: From design to application—a progress report. *Angewandte Chemie International Edition, 38*(7), 884–905. doi:10.1002/(SICI)1521-3773(19990401)38:7<884::AID-ANIE884>3.0.CO;2-K PMID:29711851

Fleitas-Salazar, N., Silva-Campa, E., Pedroso-Santana, S., Tanori, J., Pedroza-Montero, M. R., & Riera, R. (2017). Effect of temperature on the synthesis of silver nanoparticles with polyethylene glycol: New insights into the reduction mechanism. *Journal of Nanoparticle Research, 19*(3), 113. doi:10.100711051-017-3780-3

Florence, A. T., & Hussain, N. (2001). Transcytosis of nanoparticle and dendrimer delivery systems, evolving vistas. *Advanced Drug Delivery Reviews, 50*, S69–S89. doi:10.1016/S0169-409X(01)00184-3 PMID:11576696

Florence, A. T., Sakthivel, T., & Toth, I. (2000). Oral uptake and translocation of a polylysinedendrimer with a lipid surface. *Journal of Controlled Release, 65*(1-2), 253–259. doi:10.1016/S0168-3659(99)00237-0 PMID:10699285

Fojtů, M., Teo, W. Z., & Pumera, M. (2017). Environmental Impact and Potential Health Risks of 2D Nanomaterials. *Environmental Science. Nano, 4*(8), 1617–1633. doi:10.1039/C7EN00401J

Forero López, A. D., Fabiani, M., Lassalle, V. L., Spetter, C. V., & Fernandez Severini, M. D. (2022). Critical review of the characteristics, interactions, and toxicity of micro/nanomaterials pollutants in aquatic environments. *Marine Pollution Bulletin, 174*, 113276. doi:10.1016/j.marpolbul.2021.113276 PMID:35090270

Fosso-Kankeu, E., Mittal, H., Waanders, F., Ntwampe, I. O., & Ray, S. S. (2016). Preparation and characterization of gum karaya hydrogel nanocomposite flocculant for metal ions removal from mine effluents. *International Journal of Environmental Science and Technology, 13*(2), 711–724. doi:10.100713762-015-0915-x

Fraceto, L. F., Grillo, R., de Medeiros, G. A., Scognamiglio, V., Rea, G., & Bartolucci, C. (2016). Nanotechnology in agriculture: Which innovation potential does it have? *Frontiers in Environmental Science, 4*, 20. doi:10.3389/fenvs.2016.00020

Freixa, A., Acuña, V., Sanchís, J., Farré, M., Barceló, D., & Sabater, S. (2018). Ecotoxicological effects of carbon based nanomaterials in aquatic organisms. *The Science of the Total Environment, 619–620*, 328–337. doi:10.1016/j.scitotenv.2017.11.095 PMID:29154051

Fruijtier-Pölloth, C. (2012). The toxicological mode of action and the safety of synthetic amorphoussilica- a nanostructured material. *Toxicology*, *294*(2-3), 61–79. doi:10.1016/j.tox.2012.02.001 PMID:22349641

Fu, P. P., Xia, Q., Hwang, H. M., Ray, P. C., & Yu, H. (2014). Mechanisms of nanotoxicity: Generation of reactive oxygen species. *Yao Wu Shi Pin Fen Xi*, *22*(1), 64–75. doi:10.1016/j.jfda.2014.01.005 PMID:24673904

Gacem, M. A., Telli, A., & Khelil, A. O. E. H. (2021). Nanomaterials for detection, degradation, and adsorption of pesticides from water and wastewater. Aquananotechnology, 325-346.

Galloway, T., Cipelli, R., Guralnik, J., Ferrucci, L., Bandinelli, S., Corsi, A. M., Money, C., McCormack, P., & Melzer, D. (2010, November). Daily bisphenolA excretion and associations with sex hormone concentrations: Results from the InCHIANTI adult population study. *Environmental Health Perspectives*, *118*(11), 1603–1608. doi:10.1289/ehp.1002367

Gangadoo, S., Dinev, I., Willson, N.-L., Moore, R. J., Chapman, J., & Stanley, D. (2020). Nanoparticles of selenium as high bioavailable and non-toxic supplement alternatives for broiler chickens. *Environmental Science and Pollution Research International*, *27*(14), 16159–16166. doi:10.100711356-020-07962-7 PMID:32107689

Gangapuram, B. R., Bandi, R., Dadigala, R., Kotu, D. M., & Guttena, V. (2017). Facile green synthesis of gold nanoparticles with carboxymethyl gum karaya, selective and sensitive colorimetric detection of copper (II) ions. *Journal of Cluster Science*, *28*(5), 2873–2890. doi:10.100710876-017-1264-3

Gao, Z., Wang, L., Su, R., Huang, R., Qi, W., & He, Z. (2015). A carbon dot-based "off–on" fluorescent probe for highly selective and sensitive detection of phytic acid. *Biosensors & Bioelectronics*, *70*, 232–238. doi:10.1016/j.bios.2015.03.043 PMID:25829220

Garrett, N. L., Lalatsa, A., Uchegbu, I., Schätzlein, A., & Moger, J. (2012). Exploring uptake mechanisms of oral nanomedicines using multimodal nonlinear optical microscopy. *Journal of Biophotonics*, *5*(5-6), 458–468. doi:10.1002/jbio.201200006 PMID:22389316

Gayrard, V., Lacroix, M. Z., Collet, S. H., Viguié, C., Bousquet-Melou, A., Toutain, P. L., & Picard-Hagen, N. (2013). High bioavailability of bisphenol A from sublingual exposure. *Environmental Health Perspectives*, *121*(8), 951–956. doi:10.1289/ehp.1206339 PMID:23761051

Geim, A.K. (2009). Graphene: Status and prospects. *Science, 324(*5934), 1530-1534.

George, J. M., Antony, A., & Mathew, B. (2018). Metal oxide nanoparticles in electrochemical sensing and biosensing: A review. *Mikrochimica Acta*, *185*(7), 1–26. doi:10.100700604-018-2894-3 PMID:29974265

Gerloff, K., Albrecht, C., Boots, A. W., Forster, I., & Schins, R. P. (2009). Cytotoxicity and oxidative DNA damage by nanoparticles in human intestinal Caco-2 cells. *Nanotoxicology*, *3*(4), 355–364. doi:10.3109/17435390903276933

Ge, S., Lu, J., Ge, L., Yan, N., & Yu, J. (2011). Development of a novel deltamethrin sensor based on molecularly imprinted silica nanospheres embedded CdTe quantum dots. *Spectrochimica Acta. Part A: Molecular and Biomolecular Spectroscopy*, *79*(5), 1704–1709. doi:10.1016/j.saa.2011.05.040 PMID:21684806

Ge, Y., Zhang, Y., He, S., Nie, F., Teng, G., & Gu, N. (2009). Fluorescence modified chitosan-coated magnetic nanoparticles for high-efficient cellular imaging. *Nanoscale Research Letters*, *4*(4), 287–295. doi:10.100711671-008-9239-9 PMID:20596545

Geys, J., Nemmar, A., Verbeken, E., Smolders, E., Ratoi, M., Hoylaerts, M. F., Nemery, B., & Hoet, P. H. M. (2008). Acute toxicity and prothrombotic effects of quantum dots: Impact of surface charge. *Environmental Health Perspectives*, *116*(12), 1607–1613. doi:10.1289/ehp.11566 PMID:19079709

Ghafariyan, M. H., Malakouti, M. J., Dadpour, M. R., Stroeve, P., & Mahmoudi, M. (2013). Effects of magnetite nanoparticles on soybean chlorophyll. *Environmental Science & Technology, 47*(18), 10645–10652. doi:10.1021/es402249b PMID:23951999

Ghobadi, M. Z. Afsaneh, et al. Nanotechnology. In C. M. Hussain (Ed.), *The ELSI Handbook of nanotechnology* (pp. 113–134). doi:10.1002/9781119592990.ch7

Ghorai, S., Sarkar, A. K., Panda, A. B., & Pal, S. (2013). Effective removal of Congo red dye from aqueous solution using modified xanthan gum/silica hybrid nanocomposite as adsorbent. *Bioresource Technology, 144*, 485–491. doi:10.1016/j.biortech.2013.06.108 PMID:23896441

Ghorai, S., Sarkar, A., Raoufi, M., Panda, A. B., Schönherr, H., & Pal, S. (2014). Enhanced removal of methylene blue and methyl violet dyes from aqueous solution using a nanocomposite of hydrolyzed polyacrylamide grafted xanthan gum and incorporated nanosilica. *ACS Applied Materials & Interfaces, 6*(7), 4766–4777. doi:10.1021/am4055657 PMID:24579659

Giljohann, D.A., & Seferos, D.S. (2020). Gold nanoparticles for biology and medicine. *Spherical Nucleic Acids,* 55-90.

Glick, B. R. (2012). Plant growth-promoting bacteria: Mechanisms and applications. *Scientifica, 963401*, 1–15. Advance online publication. doi:10.6064/2012/963401 PMID:24278762

Glinski, A., de Souza, T. L., da Luz, J. Z., Junior, A. G. B., de Oliveira, C. C., Ribeiro, C. A. O., & Neto, F. F. (2021). Toxicological effects of silver nanoparticles and cadmium chloride in macrophage cell line (RAW 264.7): An in vitro approach. *Journal of Trace Elements in Medicine and Biology, 68*, 126854. doi:10.1016/j.jtemb.2021.126854 PMID:34488184

Gogotsi, Y., & Presser, V. (2006). *Carbon nanomaterials.* CRC Press. doi:10.1201/9781420009378

Gomathi, A. C., Xavier Rajarathinam, S. R., Mohammed Sadiq, A., & Rajeshkumar, S. (2020). Anticancer activity of silver nanoparticles synthesized using aqueous fruit shell extract of *Tamarindus indica* on MCF-7 human breast cancer cell line. *Journal of Drug Delivery Science and Technology, 55*, 101376. doi:10.1016/j.jddst.2019.101376

Gondil, V. K., Kalaiyarasan, T., Bharti, V. K., & Chhibber, S. (2019). Antibiofilm potential of Seabuckthorn silver nanoparticles (SBT@AgNPs) against *Pseudomonas aeruginosa. 3Biotech, 9*, 402.

Gonzales, P., Felix, O., Alexander, C., Lutz, E., Ela, W., & Eduardo Sáez, A. (2014). Laboratory dust generation and size-dependent characterization of metal and metalloid-contaminated mine tailings deposits. *Journal of Hazardous Materials, 280*, 619–626. doi:10.1016/j.jhazmat.2014.09.002 PMID:25222928

Gonzales-Weimuller, M., Zeisberger, M., & Krishnan, K. M. (2009). Size-dependant heating rates of iron oxide nanoparticles for magnetic fluid hyperthermia. *Journal of Magnetism and Magnetic Materials, 321*(13), 1947–1950. doi:10.1016/j.jmmm.2008.12.017 PMID:26405373

Gonzalez, H. O., Roling, J. A., Baldwin, W. S., & Bain, L. J. (2006). Physiological changes and differential gene expression in mummichogs (Fundulus heteroclitus) exposed to arsenic. *Aquatic Toxicology (Amsterdam, Netherlands), 77*(1), 43–52. doi:10.1016/j.aquatox.2005.10.014 PMID:16356559

Gould-Fogerite, S., Mannino, R. J., & Margolis, D. (2003). Cochleate delivery vehicles: Applications to gene therapy. *Drug Delivery Technology, 3*, 40–47.

Grob, K., Biedermann, M., Scherbaum, E., Roth, M., & Rieger, K. (2006). Food contamination with organic materials in perspective: Packaging materials as the largest and least controlled source? A view focusing on the European situation. *Critical Reviews in Food Science and Nutrition, 46*(7), 529–536. doi:10.1080/10408390500295490 PMID:16954061

Groso, A., Petri-Fink, A., Magrez, A., Riediker, M., & Meyer, T. (2010). Management of nanomaterials safety in research environment. *Particle and Fibre Toxicology, 7*(1), 40. doi:10.1186/1743-8977-7-40 PMID:21143952

Grosser, A., Grobelak, A., Rorat, A., Courtois, P., Vandenbulcke, F., Lemière, S., Guyoneaud, R., Attard, E., & Celary, P. (2021). Effects of silver nanoparticles on performance of anaerobic digestion of sewage sludge and associated microbial communities. *Renewable Energy*, *171*, 1014–1025. doi:10.1016/j.renene.2021.02.127

Guan, R., Kang, T., Lu, F., Zhang, Z., Shen, H., & Liu, M. (2012). Cytotoxicity, oxidative stress, and genotoxicity in human hepatocyte and embryonic kidney cells exposed to ZnO nanoparticles. *Nanoscale Research Letters*, *7*(1), 602. Advance online publication. doi:10.1186/1556-276X-7-602 PMID:23110934

Guart, A., Wagner, M., Mezquida, A., Lacorte, S., Oehlmann, J., & Borrell, A. (2013). Migration of plasticisers from TritanTM and polycarbonate bottles and toxicological evaluation. *Food Chemistry*, *141*(1), 373–380. doi:10.1016/j.foodchem.2013.02.129 PMID:23768370

Guerra, J. D., Sandoval, G., Avalos-Borja, M., Pestryakov, A., Garibo, D., Susarrey-Arce, A., & Bogdanchikova, N. (2020). Selective antifungal activity of silver nanoparticles: A comparative study between *Candida tropicalis* and *Saccharomyces boulardii. Colloid and Interface Science Communications*, *37*, 100280. doi:10.1016/j.colcom.2020.100280

Gu, F. X., & Karmik, R. (2007). Targeted nanoparticles for cancer therapy. *Nano Today*, *2*(3), 14–21. doi:10.1016/S1748-0132(07)70083-X

Guilger-Casagrande, M., Germano-Costa, T., Pasquoto-Stigliani, T., Fraceto, L., & Lima, R. (2019). Biosynthesis of silver nanoparticles employing *Trichoderma harzianum* with enzymatic stimulation for the control of *Sclerotinia sclerotiorum. Scientific Reports*, *9*(1), 14351. doi:10.103841598-019-50871-0 PMID:31586116

Guilger-Casagrande, M., & Lima, R. (2019). Synthesis of Silver Nanoparticles Mediated by Fungi: A Review. *Frontiers in Bioengineering and Biotechnology*, *7*, 287. doi:10.3389/fbioe.2019.00287 PMID:31696113

Gul, M., Kotak, Y., & Muneer, T. (2016). Review on Recent Trend of Solar Photovoltaic Technology. *Energy Exploration and Exploitation*, *34*(4), 485–526. doi:10.1177/0144598716650552

Gu, M., Dai, Z., Yan, X., Yan, X., Ma, J., Niu, Y., Lan, W., Wang, X., & Xu, Q. (2021). Comparison of Toxicity of Ti_3C_2 and Nb_2C Mxene Quantum dots (QDs) to Human Umbilical Vein Endothelial Cells. *Journal of Applied Toxicology*, *41*(5), 745–754. doi:10.1002/jat.4085 PMID:33048420

Gumustas, M., & Sengel-Turk, C. T. (2017). Effect of polymer-based nanoparticles on the assay of antimicrobial drug delivery systems. In Multifunctional systems for combined delivery, biosensing and diagnostics. Elsevier.

Guo, D., Bi, H., Liu, B., Wu, Q., Wang, D., & Cui, Y. (2013). Reactive oxygen species-induced cytotoxic effects of zinc oxide nanoparticles in rat retinal ganglion cells. *Toxicology In Vitro*, *27*(2), 731–738. doi:10.1016/j.tiv.2012.12.001 PMID:23232460

Gupta, B. S., & Ako, J. E. (2005). Application of guar gum as a flocculant aid in food processing and potable water treatment. *European Food Research and Technology*, *221*(6), 746–751. doi:10.100700217-005-0056-4

Gupta, R. (2019). Cadmium nanoparticles and its toxicity. *Journal of Critical Reviews*, *2009*, 1–7. doi:10.22159/jcr.2019v6i5.34073

Gupta, R., & Xie, H. (2018). Nanoparticles in daily life: Applications, toxicity and regulations. *Journal of Environmental Pathology, Toxicology and Oncology*, *37*(3), 209–230. doi:10.1615/JEnvironPatholToxicolOncol.2018026009 PMID:30317972

Gupta, S., Murthy, C., & Prabha, C. R. (2018). Recent advances in carbon nanotube based electrochemical biosensors. *International Journal of Biological Macromolecules*, *108*, 687–703. doi:10.1016/j.ijbiomac.2017.12.038 PMID:29223757

Gurunathan, S., Qasim, M., Park, C., Yoo, H., Kim, J. H., & Hong, K. (2018). Cytotoxic potential and molecular pathway analysis of silver nanoparticles in human colon cancer cells HCT116. *International Journal of Molecular Sciences*, *19*(8), 2269. doi:10.3390/ijms19082269 PMID:30072642

Guru, T., Veronica, N., Thatikunta, R., & Reddy, S. N. (2015). Crop nutrition management with nano fertilizers. *International Journal of Environmental Science and Technology*, *1*(1), 4–6.

Hackenberg, S., Scherzed, A., Technau, A., Kessler, M., Froelich, K., Ginzkey, C., Koehler, C., Burghartz, M., Hagen, R., & Kleinsasser, N. (2011). Cytotoxic, genotoxic and pro-inflammatory effects of zinc oxide nanoparticles in human nasal mucosa cells in vitro. *Toxicology In Vitro*, *25*(3), 657–663. doi:10.1016/j.tiv.2011.01.003 PMID:21232592

Hafeez, B., Khanif, M., & Saleem, M. (2013). Role of zinc in plant nutrition: A review. *American Journal of Experimental Agriculture*, *3*(2), 374–391. doi:10.9734/AJEA/2013/2746

Hajam, Y. A., Rai, S., Kumar, R., Bashir, M., & Malik, J. A. (2020). Phenolic Compounds from Medicinal Herbs: Their Role in Animal Health and Diseases – A New Approach for Sustainable Welfare and Development. In R. Lone, R. Shuab, & A. Kamili (Eds.), *Plant Phenolics in Sustainable Agriculture*. Springer. doi:10.1007/978-981-15-4890-1_10

Hallare, A. V., Schirling, M., Luckenbach, T., Köhler, H. R., & Triebskorn, R. (2005). Combined effects of temperature and cadmium on developmental parameters and biomarker responses in zebrafish (Danio rerio) embryos. *Journal of Thermal Biology*, *30*(1), 7–17. doi:10.1016/j.jtherbio.2004.06.002

Hammond, J. L., Formisano, N., Estrela, P., Carrara, S., & Tkac, J. (2016). Electrochemical biosensors and nanobiosensors. *Essays in Biochemistry*, *60*(1), 69–80. doi:10.1042/EBC20150008 PMID:27365037

Hanagata, N., Zhuang, F., Connolly, S., Li, J., Ogawa, N., & Xu, M. (2011). Molecular responses of human lung epithelial cells to the toxicity of copper oxide nanoparticles inferred from whole genome expression analysis. *ACS Nano*, *5*(12), 9326–9338. doi:10.1021/nn202966t PMID:22077320

Han, H.-V., Lin, C.-C., Tsai, Y.-L., Chen, H.-C., Chen, K.-J., Yeh, Y.-L., Lin, W.-Y., Kuo, H.-C., & Yu, P. (2014). A Highly Efficient Hybrid GaAs Solar Cell Based on Colloidal-Quantum-Dot- Sensitization. *Scientific Reports*, *4*(1), 1–9. doi:10.1038rep05734 PMID:25034623

Han, M., Zhu, L., Mo, J., Wei, W., Yuan, B., Zhao, J., & Cao, C. (2020). Protein Corona and Immune Responses of Borophene: A Comparison of Nanosheet-Plasma Interface with Graphene and Phosphorene. *ACS Applied Bio Materials*, *3*(7), 4220–4229. doi:10.1021/acsabm.0c00306 PMID:35025423

Hannah, W., & Thompson, P. B. (2008). Nanotechnology risk and the environment: A review. *Journal of Environmental Monitoring*, *10*(3), 291–300. doi:10.1039/b718127m PMID:18392270

Hanning, I. B., Nutt, J. D., & Ricke, S. C. (2009). Salmonellosis outbreaks in the United States due to fresh produce: Sources and potential intervention measures. *Foodborne Pathogens and Disease*, *6*(6), 635–648. doi:10.1089/fpd.2008.0232 PMID:19580447

Harra, J., Makitalo, J., Siikanen, R., Virkki, M., Genty, G., Kobayashi, T., Kobayashi, T., Kauranen, M., & Makela, J. M. (2012). Size-controlled aerosol synthesis of silver nanoparticles for plasmonic materials. *Journal of Nanoparticle Research*, *14*(6), 870. doi:10.100711051-012-0870-0 PMID:22844206

Harush-Frenkel, O., Bivas-Benita, M., Nassar, T., Springer, C., Sherman, Y., Avital, A., Altschuler, Y., Borlak, J., & Benita, S. (2010). A safety and tolerability study of differently-charged nanoparticles for local pulmonary drug delivery. *Toxicology and Applied Pharmacology*, *246*(1-2), 83–90. doi:10.1016/j.taap.2010.04.011 PMID:20417650

Hasaneen, M. N. A., Abdel-Aziz, H. M. M., El-Bialy, D. M. A., & Omer, A. M. (2014). Preparation of chitosan nanoparticles for loading with NPK fertilizer. *African Journal of Biotechnology*, *13*(31), 3158–3164. doi:10.5897/AJB2014.13699

Hasan, S. (2015). A review on nanoparticles: Their synthesis and types. *Research Journal of Recent Sciences*, *2277*, 2502.

Hasanuzzaman, M., Nahar, K., & Fujita, M. (2013). Plant response to salt stress and role of exogenous protectants to mitigate salt induced damages. In P. Ahmad, M. Azooz, & M. Prasad (Eds.), *Ecophysiology and Responses of Plants under Salt Stress* (pp. 25–87). Springer. doi:10.1007/978-1-4614-4747-4_2

Hatami, M. (2017). Toxicity assessment of multi-walled carbon nanotubes on *Cucurbita pepo* L. under well-watered and water-stressed conditions. *Ecotoxicology and Environmental Safety*, *142*, 274–283. doi:10.1016/j.ecoenv.2017.04.018 PMID:28433592

Hedberg, J., Blomberg, E., & Odnevall Wallinder, I. (2019). In the search for nanospecific effects of dissolution of metallic nanoparticles at freshwater-like conditions. *Critical Reviews in Environmental Science and Technology*, *53*(8), 4030–4044. doi:10.1021/acs.est.8b05012 PMID:30908015

Heidari, M., Salmanpour, I., Ghorbani, H., & Asghari, H. R. (2018). Iron chelate and rhizobacteria changed growth, grain yield, and physiological characteristics in maize. *Scientia Agriculturae Bohemica*, *49*(4), 245–254. doi:10.2478ab-2018-0031

Helmroth, E., Rijk, R., Dekker, M., & Jongen, W. (2002). Predictive modelling of migration from packaging materials into food products for regulatory purposes. *Trends in Food Science & Technology*, *13*(3), 102–109. doi:10.1016/S0924-2244(02)00031-6

Heng, B. C., Zhao, X., Tan, E. C., Khamis, N., Assodani, A., Xiong, S., Ruedl, C., Ng, K. W., & Loo, J. S. (2011). Evaluation of the cytotoxic and inflammatory potential of differentially shaped zinc oxide nanoparticles. *Archives of Toxicology*, *85*(12), 1517–1528. doi:10.100700204-011-0722-1 PMID:21656222

Hengstler, J. G., Foth, H., Gebel, T., Kramer, P. J., Lilienblum, W., & Schweinfurth, H. (2011). Critical evaluation of key evidence on the human health hazards of exposure to bisphenol A. *Critical Reviews in Toxicology, 41*, 263–291.

Hidalgo-Ruz, V., Gutow, L., Thompson, R. C., & Thiel, M. (2012). Microplastics in the marine environment: A review of the methods used for identification and quantification. *Environmental Science & Technology*, *46*(6), 3060–3075. doi:10.1021/es2031505 PMID:22321064

Hlavkova, D., Caloudova, H., Palikova, P., Kopel, P., Plhalova, L., Beklova, M., & Havelkova, B. (2020). Effect of Gold Nanoparticles and Ions Exposure on the Aquatic Organisms. *Bulletin of Environmental Contamination and Toxicology*, *105*(4), 530–537. doi:10.100700128-020-02988-6 PMID:32940716

Hobson, D. W., Roberts, S. M., Shvedova, A. A., Warheit, D. B., Hinkley, G. K., & Guy, R. C. (2016). Applied nanotoxicology. *International Journal of Toxicology*, *35*(1), 5–16. doi:10.1177/1091581816628484 PMID:26957538

Hochella, M. F. Jr, Mogk, D. W., Ranville, J., Allen, I. C., Luther, G. W., Marr, L. C., McGrail, B. P., Murayama, M., Qafoku, N. P., Rosso, K. M., Sahai, N., Schroeder, P. A., Vikesland, P., Westerhoff, P., & Yang, Y. (2019). Natural, incidental, and engineered nanomaterials and their impacts on the Earth system. *Science*, *363*(6434), eaau8299. Advance online publication. doi:10.1126cience.aau8299 PMID:30923195

Hoffmann, M. R., Martin, S. T., Choi, W., & Bahnemann, D. W. (1995). Environmental Applications of Semiconductor Photocatalysis. *Chemical Reviews*, *95*(1), 69–96. doi:10.1021/cr00033a004

Holt, K. B., & Bard, A. J. (2005). Interaction of silver(I) ions with the respiratory chain of Escherichia coli: An electrochemical and scanning electrochemical microscopy study of the antimicrobial mechanism of micromolar Ag^{+}. *Biochemistry*, *44*(39), 13214–13223. doi:10.1021/bi0508542 PMID:16185089

Honarvar, Z., Hadian, Z., & Mashayekh, M. (2016). Nanocomposites in food packaging applications and their risk assessment for health. *Electronic Physician*, *8*(6), 2531–2538. doi:10.19082/2531 PMID:27504168

Hong, F., Yang, F., Liu, C., Gao, Q., Wan, Z., Gu, F., Wu, C., Ma, Z., Zhou, J., & Yang, P. (2005). Influences of nano-TiO_2 on the chloroplast aging of spinach under light. *Biological Trace Element Research*, *104*(3), 249–260. doi:10.1385/BTER:104:3:249 PMID:15930594

Hossain, S. T., & Mukherjee, S. K. (2013). Toxicity of cadmium sulfide (CdS) nanoparticles against Escherichia coli and HeLa cells. *Journal of Hazardous Materials*, *260*, 1073–1082. doi:10.1016/j.jhazmat.2013.07.005 PMID:23892173

Hossain, Z., Mustafa, G., & Komatsu, S. (2015). Plant Responses to Nanoparticle Stress. *International Journal of Molecular Sciences*, *16*(11), 26644–26653. doi:10.3390/ijms161125980 PMID:26561803

Hosseinzadeh, H., & Ramin, S. (2018). Fabrication of starch-graft-poly(acrylamide)/graphene oxide/hydroxyapatite nanocomposite hydrogel adsorbent for removal of malachite green dye from aqueous solution. *International Journal of Biological Macromolecules*, *106*, 101–115. doi:10.1016/j.ijbiomac.2017.07.182 PMID:28778526

Hosseinzadeh, H., Zoroufi, S., & Mahdavinia, G. R. (2015). Study on adsorption of cationic dye on novel kappa-carrageenan/poly(vinyl alcohol)/montmorillonite nanocomposite hydrogels. *Polymer Bulletin*, *72*(6), 1339–1363. doi:10.100700289-015-1340-5

Hou, J., Liu, H., Zhang, S., Liu, X., Hayat, T., Alsaedi, A., & Wang, X. (2019). Mechanism of toxic effects of Nano-ZnO on cell cycle of zebrafish (Danio rerio). *Chemosphere*, *229*, 206–213. doi:10.1016/j.chemosphere.2019.04.217 PMID:31078877

Huang, C. C., Aronstam, R. S., Chen, D. R., & Huang, Y. W. (2010). Oxidative stress, calcium homeostasis, and altered gene expression in human lung epithelial cells exposed to ZnO nanoparticles. *Toxicology In Vitro*, *24*(1), 45–55. doi:10.1016/j.tiv.2009.09.007 PMID:19755143

Huang, G. G., Wang, C. T., Tang, H. T., Huang, Y. S., & Yang, J. (2006). ZnO nanoparticle-modified infrared internal reflection elements for selective detection of volatile organic compounds. *Analytical Chemistry*, *78*(7), 2397–2404. doi:10.1021/ac051930+ PMID:16579626

Huang, H., Jiang, R., Feng, Y., Ouyang, H., Zhou, N., Zhang, X., & Wei, Y. (2020). Recent Development and Prospects of Surface Modification and Biomedical Applications of MXenes. *Nanoscale*, *12*(3), 1325–1338. doi:10.1039/C9NR07616F PMID:31872839

Huang, W., Zhou, J., Hu, Q., Qiu, B., Huang, M., Murugadoss, V., & Guo, Z. (2022). Improved methanogenesis in anaerobic wastewater treatment by magnetite @ polyaniline (Fe 3 O 4 @ PANI) composites. *Chemosphere*, *296*, 133953. doi:10.1016/j.chemosphere.2022.133953 PMID:35157884

Huang, X., Zhu, Y., & Kianfar, E. (2021). Nano biosensors: Properties, applications and electrochemical techniques. *Journal of Materials Research and Technology*, *12*, 1649–1672. doi:10.1016/j.jmrt.2021.03.048

Hua, Y., Heal, K. V., & Friesl-Hanl, W. (2017). The use of red mud as an immobiliser for metal/metalloid-contaminated soil: A review. *Journal of Hazardous Materials*, *325*, 17–30. doi:10.1016/j.jhazmat.2016.11.073 PMID:27914288

Huff, C., Long, J. M., & Abdel-Fattah, T. M. (2020). Beta-cyclodextrin-assisted synthesis of silver nanoparticle network and its application in a hydrogen generation reaction. *Catalysts*, *10*(9), 1014. doi:10.3390/catal10091014

Hughes, G. A. (2005). Nanostructure-mediated drug delivery. *Nanomedicine; Nanotechnology, Biology, and Medicine*, *1*(1), 22–30. doi:10.1016/j.nano.2004.11.009 PMID:17292054

Hulikere, M., Chandrashekhar, M., & Joshi, G. (2019). Characterization, antioxidant and antimicrobial activity of silver nanoparticles synthesized using marine endophytic fungus-*Cladosporium cladosporioides. Process Biochemistry*, *82*, 199–204. doi:10.1016/j.procbio.2019.04.011

Humtsoe, N., Davoodi, R., Kulkarni, B. G., & Chavan, B. (2007). Effect of arsenic on the enzymes of the Rohu carp, Labeo rohita (Hamilton 1822). *Raffles Bulletin of Zoology, 14*, 17–19.

Hussain, M. S., Musharraf, S. G., Bhanger, M. I., & Malik, M. I. (2020). Salicylaldehyde derivative of nano-chitosan as an efficient adsorbent for lead (II), copper (II), and cadmium (II) ions. *International Journal of Biological Macromolecules*, *147*, 643–652. doi:10.1016/j.ijbiomac.2020.01.091 PMID:31931059

Hussein, H. A., & Abdullah, M. A. (2020). Biosynthesis, mechanisms, and biomedical applications of silver nanoparticles. In D. Thangadurai, J. Sangeetha, & R. Prasad (Eds.), *Functional Bionanomaterials. Nanotechnology in the Life Sciences*. Springer. doi:10.1007/978-3-030-41464-1_14

Hussein, M. M., & Abou-Baker, N. H. (2018). The contribution of nano-zinc to alleviate salinity stress on cotton plants. *Royal Society Open Science*, *5*(8), 171809. doi:10.1098/rsos.171809 PMID:30224982

Hutasoit, S., Suada, J.K., & Susrama, I.G.K. (2013). Antifungal activity test extracts some type of marine life link to Aspergillus flavus and Penicillium sp. *E J. Trop.*, 27–38.

Iavicoli, I., Leso, V., Beezhold, D. H., & Shvedova, A. A. (2017). Nanotechnology in agriculture: Opportunities, toxicological implications, and occupational risks. *Toxicology and Applied Pharmacology*, *329*, 96–111. doi:10.1016/j.taap.2017.05.025 PMID:28554660

Ihoeghian, N. A., Amenaghawon, A. N., Ajieh, M. U., Oshoma, C. E., Ogofure, A., Erhunmwunse, N. O., Edosa, V. I. O., Tongo, I., Obuekwe, I. S., Isagba, E. S., Emokaro, C., Ezemonye, L. I. N., Lag-Brotons, A. J., Semple, K. T., & Martin, A. D. (2022). Anaerobic co-digestion of cattle rumen content and food waste for biogas production: Establishment of co-digestion ratios and kinetic studies. *Bioresource Technology Reports*, *18*, 101033. doi:10.1016/j.biteb.2022.101033

Imhof, H. K., Ivleva, N. P., Schmid, J., Niessner, R., & Laforsch, C. (2013). Contamination of beach sediments of a subalpine lake with microplastic particles. *Current Biology*, *23*(19), R867–R868. doi:10.1016/j.cub.2013.09.001 PMID:24112978

Inada, T., Kasai, Y., Wakabayashi, T., Shibata, H., & Isobe, A. (2007). Ammonium magnesium phosphate containing nanocomposite and its manufacture. Academic Press.

Ingle, A. P., Duran, N., & Rai, M. (2014). Bioactivity, mechanism of action, and cytotoxicity of copper-based nanoparticles: A review. *Applied Microbiology and Biotechnology*, *98*(3), 1001–1009. doi:10.100700253-013-5422-8 PMID:24305741

Iravani, S., Korbekandi, H., Mirmohammadi, S. V., & Zolfaghari, B. (2014). Synthesis of silver nanoparticles: Chemical, physical and biological methods. *Research in Pharmaceutical Sciences*, *9*(6), 385–406. PMID:26339255

Islam, M. S., Rahaman, M. S., & Yeum, J. H. (2015). Electrospun novel super-absorbent based on polysaccharide–polyvinyl alcohol–montmorillonite clay nanocomposites. *Carbohydrate Polymers*, *115*, 69–77. doi:10.1016/j.carbpol.2014.08.086 PMID:25439870

Israelachvili, J. (1992). *Intermolecular and surface forces* (2nd ed.). Academic Press.

Izci, M., Maksoudian, C., Manshian, B. B., & Soenen, S. J. (2021). The use of alternative strategies for enhanced nanoparticle delivery to solid tumors. *Chemical Reviews*, *121*(3), 1746–1803. doi:10.1021/acs.chemrev.0c00779 PMID:33445874

Jadhav, P., Khalid, Z. B., Zularisam, A. W., Krishnan, S., & Nasrullah, M. (2022). The role of iron-based nanoparticles (Fe-NPs) on methanogenesis in anaerobic digestion (AD) performance. *Environmental Research, 204*(PB), 112043. doi:10.1016/j.envres.2021.112043

Jafari, S. M., & McClements, D. J. (2018). *Nanoemulsions: formulation, applications, and characterization*. Academic Press.

Jahangirnejad, R., Goudarzi, M., Kalantari, H., Najafzadeh, H., & Rezaei, M. (2020). Subcellular organelle toxicity caused by arsenic nanoparticles in isolated rat hepatocytes. *The International Journal of Occupational and Environmental Medicine*, *11*(1), 41–52. doi:10.15171/ijoem.2020.1614 PMID:31905194

Jain, S., Jain, A., Kachhawah, P., & Devra, V. (2015). Synthesis and size control of copper nanoparticles and their catalytic application. *Transactions of Nonferrous Metals Society of China*, *25*(12), 3995–4000. doi:10.1016/S1003-6326(15)64048-1

Jalali, H. B., Sadeghi, S., Baylam, I., & Han, M. (2021). Exciton Recycling via InP Quantum Dot Funnels for Luminescent Solar Concentrators. *Nano Research*, *14*(5), 1488–1494. doi:10.100712274-020-3207-9

Jamakala, O., & Rani, A. U. (2014). Mitigating role of zinc and iron against cadmium induced toxicity in liver and kidney of male albino rat: A study with reference to metallothionein quantification. *International Journal of Pharmacy and Pharmaceutical Sciences*, *6*(9), 411–417.

Jamal, M., Ahmad, W., Andleeb, S., Jalil, F., Imran, M., Nawaz, M. A., & Kamil, M. A. (2018). Bacterial biofilm and associated infections. *Journal of the Chinese Medical Association*, *81*(1), 7–11. doi:10.1016/j.jcma.2017.07.012 PMID:29042186

Jambeck, J. R., Geyer, R., Wilcox, C., Siegler, T. R., Perryman, M., Andrady, A., Narayan, R., & Law, K. L. (2015). Plastic waste inputs from land into the ocean. *Science*, *347*(6223), 768–771. doi:10.1126cience.1260352 PMID:25678662

Jamdagni, P., Rana, J. S., & Khatri, P. (2018). Comparative study of antifungal effect of green and chemically synthesised silver nanoparticles in combination with carbendazim, mancozeb, and thiram. *IET Nanobiotechnology*, *12*(8), 1102–1107. doi:10.1049/iet-nbt.2018.5087 PMID:30964021

Jampilek, J., & Kralova, K. (2022). Advances in Nanostructures for Antimicrobial Therapy. *Materials (Basel)*, *15*(7), 2388. doi:10.3390/ma15072388 PMID:35407720

Janaki, V., Vijayaraghavan, K., Oh, B., Lee, K., Muthuchelian, K., Ramasamy, A. K., & Kamal-Kannan, S. (2012). Starch/polyaniline nanocomposite for enhanced removal of reactive dyes from synthetic effluent. *Carbohydrate Polymers*, *90*(4), 1437–1444. doi:10.1016/j.carbpol.2012.07.012 PMID:22944400

Jane Cypriyana, J. P. J., Saigeetha, S., Lavanya, A. A., Samrot, A. V., Ponniah, P., & Chakravarthi, S. (2021). Overview on toxicity of nanoparticles, it's mechanism, models used in toxicity studies and disposal methods–A review. *Biocatalysis and Agricultural Biotechnology*, *36*, 102117. doi:10.1016/j.bcab.2021.102117

Jani, P. U., Florence, A. T., & McCarthy, D E. (1992). Further histological evidence of gastrointestinal absorption of polystyrene nanospheres in the rat. *International Journal of Pharmaceutics*, *84*, 245–252.

Jani, P. U., Nomura, T., Yamashita, F., Takakura, A., Florence, A. T., & Hashida, M. (1996). Biliary excretion of polystyrene microspheres with covalently linked FITC fluorescence afteroral and parenteral administration to male wistar rats. *Journal of Drug Targeting*, *4*(2), 87–93. doi:10.3109/10611869609046266 PMID:8894968

Janmohammadi, M., Navid, A., Ebadi Segherloo, A., & Sabaghnia, N. (2016). Impact of nano-chelated micronutrients and biological fertilizers on growth performance and grain yield of maize under deficit irrigation condition. *Biologija (Vilnius, Lithuania)*, *62*(2), 134–147. doi:10.6001/biologija.v62i2.3339

Jeevanandam, J., Barhoum, A., Chan, Y. S., Dufresne, A., & Danquah, M. K. (2018). Review on nanoparticles and nanostructured materials: History, sources, toxicity and regulations. *Beilstein Journal of Nanotechnology*, *9*(1), 1050–1074. doi:10.3762/bjnano.9.98 PMID:29719757

Jeng, H. A., & Swanson, J. (2006). Toxicity of metal oxide nanoparticles in mammalian cells. *Journal of Environmental Science and Health*, *41*(12), 2699–2711. doi:10.1080/10934520600966177 PMID:17114101

Jennings, T., & Strouse, G. (2007). Past, present, and future of gold nanoparticles. *Bio-Applications of Nanoparticles*, 34-47.

Jeyaraj, M., Gurunathan, S., Qasim, M., Kang, M. H., & Kim, J. H. (2019). A Comprehensive Review on the Synthesis, Characterization, and Biomedical Application of Platinum Nanoparticles. *Nanomaterials (Basel, Switzerland)*, *9*(12), 1719. doi:10.3390/nano9121719 PMID:31810256

Jeyasubramanian, K., Thoppey, U. U. G., Hikku, G. S., Selvakumar, N., Subramania, A., & Krishnamoorthy, K. (2016). Enhancement in growth rate and productivity of spinach grown in hydroponics with iron oxide nanoparticles. *RSC Advances*, *6*(19), 15451–15459. doi:10.1039/C5RA23425E

Jha, M. N., & Prasad, A. N. (2006). Efficacy of new inexpensive cyanobacterial biofertilizer including its shelf-life. *World Journal of Microbiology & Biotechnology*, *22*(1), 73–79. doi:10.100711274-005-7024-9

Jing, X., Park, J. H., Peters, T. M., & Thorne, P. S. (2015). Toxicity of copper oxide nanoparticles in lung epithelial cells exposed at the air-liquid interface compared with in vivo assessment. *Toxicology In Vitro*, *29*(3), 502–511. doi:10.1016/j.tiv.2014.12.023 PMID:25575782

Ji, Z., Wang, X., Zhang, H., Lin, S., Meng, H., Sun, B., George, S., Xia, T., Nel, A. E., & Zink, J. I. (2012). Designed synthesis of CeO2 nanorods and nanowires for studying toxicological effects of high aspect ratio nanomaterials. *ACS Nano*, *6*(6), 5366–5380. doi:10.1021/nn3012114 PMID:22564147

Johnson, A. (2005). *Agriculture and nanotechnology.* The University of Wisconsin-Madison.

Jones, P. B. C. (2006). A Nanotech Revolution in Agriculture and the Food Industry. Information Systems for Biotechnology.

Jongeijk, E., & Verheesen, P. (2011). Specific delivery of agrochemicals. Academic Press.

Josef, J., & Katarína, K. (2015). Application of nanotechnology in agriculture and food industry, its prospects and risks. *Ecological Chemistry and Engineering*, *22*(3), 321–361. doi:10.1515/eces-2015-0018

Joseph, T., & Morrison, M. (2006). *Nanotechnology in Agriculture and Food.* A Nano forum report, Institute of Nanotechnology.

Joseph, P., Muchnok, T. K., Klishis, M. L., Roberts, J. R., Antonini, J. M., Whong, W. Z., & Ong, T. (2001). Cadmium-induced cell transformation and tumorigenesis are associated with transcriptional activation of c-fos, c-jun, and c-myc proto-oncogenes: Role of cellular calcium and reactive oxygen species. *Toxicological Sciences*, *61*(2), 295–303. doi:10.1093/toxsci/61.2.295 PMID:11353138

Joshi, A., Sharma, A., Bachheti, R.K., Husen, A., & Mishra, V.K. (2019). Plant-mediated synthesis of copper oxide nanoparticles and their biological applications. *Nanomaterials and Plant Potential*, 221–237. doi:10.1007/978-3-030-05569-1_8

Juntupally, S., Begum, S., Arelli, V., Mamindlapelli, N. K., Srinivasan, S., & Anupoju, G. R. (2022). Evaluating the impact of Iron Oxide nanoparticles (IO-NPs) and IO-NPs doped granular activated carbon on the anaerobic digestion of food waste at mesophilic and thermophilic temperature. *Journal of Environmental Chemical Engineering*, *10*(3), 107388. Advance online publication. doi:10.1016/j.jece.2022.107388

Kah, M. (2015). Nanopesticides and nanofertilizers: Emerging contaminants or opportunities for risk mitigation? *Frontiers in Chemistry*, *3*, 64. doi:10.3389/fchem.2015.00064 PMID:26636068

Kah, M., Kookana, R. S., Gogos, A., & Bucheli, T. D. (2018). A critical evaluation of nanopesticides and nanofertilizers against their conventional analogues. *Nature Nanotechnology, 13*(8), 677–684. doi:10.103841565-018-0131-1 PMID:29736032

Kakkar, P., & Jaffery, F. N. (2005). Biological markers for metal toxicity. *Environmental Toxicology and Pharmacology, 19*(2), 335–349. doi:10.1016/j.etap.2004.09.003 PMID:21783494

Kalishwaralal, K., Deepak, V., Ram Kumar Pandian, S., Kottaisamy, M., BarathManiKanth, S., Kartikeyan, B., & Gurunathan, S. (2010). Biosynthesis of silver and gold nanoparticles using *Brevibacterium casei. Colloids and Surfaces. B, Biointerfaces, 77*(2), 257–262. doi:10.1016/j.colsurfb.2010.02.007 PMID:20197229

Kamikoriyama, Y., Imamura, H., Muramatsu, A., & Kanie, K. (2019). Ambient aqueous-phase synthesis of copper nanoparticles and nanopastes with low-temperature sintering and ultra-high bonding abilities. *Scientific Reports, 9*(1), 899. doi:10.103841598-018-38422-5 PMID:30692589

Kao, Y. Y., Chen, Y. C., Cheng, T. J., Chiung, Y. M., & Liu, P. S. (2012). Zinc oxide nanoparticles interfere with zinc ion homeostasis to cause cytotoxicity. *Toxicological Sciences, 125*(2), 462–472. doi:10.1093/toxsci/kfr319 PMID:22112499

Karami, H., Karimi, M. A., Haghdar, S., Sadeghi, A., Mir-Ghasemi, R., & Mahdi-Khani, S. (2008). Synthesis of lead oxide nanoparticles by sonochemical method and its application as cathode and anode of lead-acid batteries. *Materials Chemistry and Physics, 108*(2-3), 337–344. doi:10.1016/j.matchemphys.2007.09.045

Karlsson, H. L., Cronholm, P., Gustafsson, J., & Moller, L. (2008). Copper oxide nanoparticles are highly toxic: A comparison between metal oxide nanoparticles and carbon nanotubes. *Chemical Research in Toxicology, 21*(9), 1726–1732. doi:10.1021/tx800064j PMID:18710264

Karunakaran, G., Suriyaprabha, R., Rajendran, V., & Kannan, N. (2016). Influence of ZrO_2, SiO_2, Al_2O_3 and TiO_2 nanoparticles on maize seed germination under different growth conditions. *IET Nanobiotechnology / IET, 10*(4), 171–177. doi:10.1049/iet-nbt.2015.0007 PMID:27463785

Karuppaiah, A., Rajan, R., Hariharan, S., Balasubramaniam, D. K., Gregory, M., & Sankar, V. (2020). Synthesis and characterization of folic acid conjugated gemcitabine tethered silver nanoparticles (FA-GEM-AgNPs) for targeted delivery. *Current Pharmaceutical Design, 26*(26), 3141–3146. doi:10.2174/1381612826666200316143239 PMID:32175835

Kasana, R. C., Panwar, N. R., Kaul, R. K., & Kumar, P. (2017). Biosynthesis and effects of copper nanoparticles on plants. *Environmental Chemistry Letters, 15*(2), 233–240. doi:10.100710311-017-0615-5

Kasgo, H., Durmus¸, A., & Kasgoz, A. (2008). Enhanced swelling and adsorption properties of AAm-AMPSNa/clay hydrogel nanocomposites for heavy metal ion removal. *Polymers for Advanced Technologies, 19*(3), 213–220. doi:10.1002/pat.999

Kassem, Y., Zoubi, R. A., & Hüseyin, G. (2019). The Possibility of Generating Electricity Using Small-Scale Wind Turbines and Solar Photovoltaic Systems for Households in Northern Cyprus: A Comparative Study. *Environments – MDPI, 6*(4), 1–22. doi:10.3390/environments6040047

Kaur, P., Thakur, R., Duhan, J. S., & Chaudhury, A. (2018). Management of wilt disease of chickpea in vivo by silver nanoparticles biosynthesized by rhizospheric microflora of chickpea (*Cicer arietinum*). *Journal of Chemical Technology and Biotechnology (Oxford, Oxfordshire), 93*(11), 3233–3243. doi:10.1002/jctb.5680

Kayan, G. Ö., & Kayan, A. (2021). Composite of natural polymers and their adsorbent properties on the dyes and heavy metal ions. *Journal of Polymers and the Environment, 29*(11), 3477–3496. doi:10.100710924-021-02154-x

Kazemi, J., & Javanbakht, V. (2020). Alginate beads impregnated with magnetic Chitosan@Zeolite nanocomposite for cationic methylene blue dye removal from aqueous solution. *International Journal of Biological Macromolecules*, *154*, 1426–1437. doi:10.1016/j.ijbiomac.2019.11.024 PMID:31733254

Kedziora, A., Speruda, M., Krzyzewska, K., Rybka, J., Lukowiak, A., & Bugla-Ploskonska, G. (2018). Similarities and differences between silver ions and silver in nanoforms as antibacterial agents. *International Journal of Molecular Sciences*, *19*(2), 444. doi:10.3390/ijms19020444 PMID:29393866

Kenry, & Lim, C.T. (2016). Biocompatibility and Nanotoxicity of Layered Two-Dimensional Nanomaterials. *ChemNanoMat*, *3*(1), 5-16. doi:10.1002/cnma.201600290

Khalil, A. T., Ovais, M., Ullah, I., Ali, M., Jan, S. A., Shinwari, Z. K., & Maaza, M. (2020). Bioinspired synthesis of pure massicot phase lead oxide nanoparticles and assessment of their biocompatibility, cytotoxicity and in-vitro biological properties. *Arabian Journal of Chemistry*, *13*(1), 916–931. doi:10.1016/j.arabjc.2017.08.009

Khan, I., Saeed, K., & Khan, I. (2019a). Nanoparticles: Properties, applications and toxicities. *Arabian Journal of Chemistry*, *12*(7), 908–931. doi:10.1016/j.arabjc.2017.05.011

Khan, K., Tareen, A. K., Iqbal, M., Wang, L., Ma, C., Shi, Z., Ye, Z., Ahmad, W., Sagar, R. U. R., Shams, S. S., Sophia, P. J., Ullah, Z., Xie, Z., & Zhang, H. (2021). Navigating Recent Advances in Monoelemental Materials (Xenes) - Fundamental to Biomedical Applications. *Progress in Solid State Chemistry*, *63*, 100326. doi:10.1016/j.progsolidstchem.2021.100326

Khan, M. R., Adam, V., Rizvi, T. F., Zhang, B., Ahamad, F., Josko, I., Zhu, Y., Yang, M., & Mao, C. (2019b). Nanoparticle-Plant Interactions: Two-Way Traffic. *Small (Germany)*, *15*(37), e1901794. doi:10.1002mll.201901794 PMID:31318142

Khan, N. T., Jameel, M., & Jameel, J. (2017). Silver nanoparticles biosynthesis by *Fusarium oxysporum* and determination of its antimicrobial potency. *Journal of Nanomedicine & Biotherapeutic Discovery*, *7*(01), 1. doi:10.4172/2155-983X.1000145

Khanzada, N. K., Farid, M. U., Kharraz, J. A., Choi, J., Tang, C. Y., Nghiem, L. D., Jang, A., & An, A. K. (2020). Removal of organic micropollutants using advanced membrane-based water and wastewater treatment: A review. *Journal of Membrane Science*, *598*, 117672. doi:10.1016/j.memsci.2019.117672

Khaydarov, R. R., & Khaydarov, R. A. (2009). Silver nanoparticles. In Nanomaterials: Risks and benefits. Springer.

Kholoud, M. M. (2010). Synthesis and applications of silver nanoparticles. *Arabian Journal of Chemistry*, *3*(3), 135–140. doi:10.1016/j.arabjc.2010.04.008

Khorrami, S., Zarrabi, A., Khaleghi, M., Danaei, M., & Mozafari, M. R. (2018). Selective cytotoxicity of green synthesized silver nanoparticles against the MCF-7 tumor cell line and their enhanced antioxidant and antimicrobial properties. *International Journal of Nanomedicine*, *2018*(13), 8013–8024. doi:10.2147/IJN.S189295 PMID:30568442

Kim, B., Rutka, J., & Chan, W. (2010). Nanomedicine. *The New England Journal of Medicine*, *363*(25), 2434–2443. doi:10.1056/NEJMra0912273 PMID:21158659

Kim, S. D., Moon, C. K., Eun, S. Y., Ryu, P. D., & Jo, S. A. (2005). Identification of ASK1, MKK4, JNK, c-Jun, and caspase-3 as a signaling cascade involved in cadmium-induced neuronal cell apoptosis. *Biochemical and Biophysical Research Communications*, *328*(1), 326–334. doi:10.1016/j.bbrc.2004.11.173 PMID:15670787

Klaine, S. J., Alvarez, P. J., Batley, G. E., Fernandes, T. F., Handy, R. D., Lyon, D. Y., Mahendra, S., McLaughlin, M. J., & Lead, J. R. (2008). Nanomaterials in the environment: Behavior, fate, bioavailability, and effects. *Environmental Toxicology and Chemistry*, *27*(9), 1825–1851. doi:10.1897/08-090.1 PMID:19086204

Koch, H. M., & Calafat, A. M. (2009). Human body burdens of chemicals used in plastics manufacture. *Philosophical Transactions of the Royal Society of London. Series B, Biological Sciences*, *364*(1526), 2063–2078. doi:10.1098/rstb.2008.0208 PMID:19528056

Koelmans, A. A. (2015). Modeling the role of microplastics in bioaccumulation of organic chemicals to marine aquatic organisms. Critical review. In M. Bergmann, L. Gutow, & M. Klages (Eds.), *Marine anthropogenic litter* (pp. 313–328). Springer. doi:10.1007/978-3-319-16510-3_11

Koelmans, A. A., Besseling, E., & Foekema, E. M. (2014). Leaching of plastic additives tomarine organisms. *Environmental Pollution*, *187*, 49–54. doi:10.1016/j.envpol.2013.12.013 PMID:24440692

Koelmans, A. A., Besseling, E., & Shim, W. J. (2015). Nanoplastics in the aquatic environment. In M. Bergmann, L. Gutow, & M. Klages (Eds.), *Marine anthropogenic litter* (pp. 329–344). Springer. doi:10.1007/978-3-319-16510-3_12

Kolodziejczak-Radzimska, A., & Jesionowski, T. (2014, April 09). Zinc Oxide-From Synthesis to Application: A Review. *Materials (Basel)*, *7*(4), 2833–2881. doi:10.3390/ma7042833 PMID:28788596

Kominkova, M., Milosavljevic, V., Vitek, P., Polanska, H., Cihalova, K., Dostalova, S., Hynstova, V., Guran, R., Kopel, P., Richtera, L., Masarik, M., Brtnicky, M., Kynicky, J., Zitka, O., & Adam, V. (2017). Comparative study on toxicity of extracellularly biosynthesized and laboratory synthesized CdTe quantum dots. *Journal of Biotechnology*, *241*, 193–200. doi:10.1016/j.jbiotec.2016.10.024 PMID:27984119

Konappa, N., Krishnamurthy, S., Arakere, U. C., Chowdappa, S., Akbarbasha, R., & Ramachandrappab, N. S. (2021). Nanofertilizers and nanopesticides: Recent trends, future prospects in agriculture. Woodhead Publishing.

Konate, A., He, X., Zhang, Z., Ma, Y., Zhang, P., Alugongo, G. M., & Rui, Y. (2017). Magnetic (Fe_3O_4) nanoparticles reduce heavy metals uptake and mitigate their toxicity in wheat seedling. *Sustainability*, *9*(5), 790. doi:10.3390u9050790

Kong, X., Squire, K., Li, E., LeDuff, P., Rorrer, G. L., Tang, S., Chen, B., McKay, C. P., Navarro-Gonzalez, R., & Wang, A. X. (2016). Chemical and biological sensing using diatom photonic crystal biosilica with in-situ growth plasmonic nanoparticles. *IEEE Transactions on Nanobioscience*, *15*(8), 828–834. doi:10.1109/TNB.2016.2636869 PMID:27959817

Kooi, M. E., Cappendijk, V. C., Cleutjens, K. B., Kessels, A. G., Kitslaar, P. J., Borgers, M., Frederik, P. M., Daemen, M. J., & van Engelshoven, J. M. (2003). Accumulation of ultrasmall superparamagnetic particles of iron oxide in human atherosclerotic plaques can be detected by in vivo magnetic resonance imaging. *Circulation*, *107*(19), 2453–2458. doi:10.1161/01.CIR.0000068315.98705.CC PMID:12719280

Korus, I., & Loska, K. (2009). Removal of Cr(III) and Cr(VI) ions from aqueous solutions by means of polyelectrolyte-enhanced ultrafiltration. *Desalination*, *247*(1-3), 390–395. doi:10.1016/j.desal.2008.12.036

Kottegoda, N., Munaweera, I., Madusanka, N., & Karunaratne, V. (2011). A green slow-release fertilizer composition based on urea-modified hydroxyapatite nanoparticles encapsulated wood. *Current Science*, *101*(1), 73–78.

Kottegoda, N., Sandaruwan, C., Priyadarshana, G., Siriwardhana, A., Rathnayake, U. A., Berugoda Arachchige, D. M., Kumarasinghe, A. R., Dahanayake, D., Karunaratne, V., & Amaratunga, G. A. (2017). Urea-hydroxyapatite nanohybrids for slow release of nitrogen. *ACS Nano*, *11*(2), 1214–1221. doi:10.1021/acsnano.6b07781 PMID:28121129

Kozhevnikova, N. S., Vorokh, A. S., & Rempel, A. A. (2010). Preparation of stable colloidal solution of cadmium sulfide CdS using ethylenediaminetetraacetic acid. *Russian Journal of General Chemistry*, *80*(3), 391–394. doi:10.1134/S1070363210030035

Koziara, J. M., Lockman, P. R., Allen, D. D., & Mumper, R. J. (2003). In situ blood–brain barrier transport of nanoparticles. *Pharmaceutical Research*, *20*(11), 1772–1778. doi:10.1023/B:PHAM.0000003374.58641.62 PMID:14661921

Krause, B. C., Kriegel, F. L., Rosenkranz, D., Dreiack, N., Tentschert, J., Jungnickel, H., Jalili, P., Fessard, V., Laux, P., & Luch, A. (2020). Aluminum and aluminum oxide nanomaterials uptake after oral exposure - a comparative study. *Scientific Reports, 10*(1), 2698. doi:10.103841598-020-59710-z PMID:32060369

Kreuter, J. (1991). Peroral administration of nanoparticles. *Advanced Drug Delivery Reviews, 7*(1), 71–86. doi:10.1016/0169-409X(91)90048-H

Krutyakov, Y. A., Kudrinskiy, A. A., Olenin, A. Y., & Lisichkin, G. V. (2008). Synthesis and properties of silver nanoparticles: Advances and prospects. *Russian Chemical Reviews, 77*(3), 233–257. doi:10.1070/RC2008v077n03ABEH003751

Kumar, A., Srivastava, A., Galaev, I. Y., & Mattiasson, B. (2007). Smart polymers: Physical forms and bioengineering applications. *Progress in Polymer Science, 2007*(32), 1205–1237. doi:10.1016/j.progpolymsci.2007.05.003

Kumari, R. M., Kumar, V., Kumar, M., Pareek, N., & Nimesh, S. (2020). Assessment of antibacterial and anticancer capability of silver nanoparticles extracellularly biosynthesized using *Aspergillus terreus. Nano Express, 1*(3), 030011. doi:10.1088/2632-959X/abc2e4

Kumari, V., & Sangal, A. (2020). Antimicrobial study of *Arjuna Terminalia* loaded PLGA nanoparticle. *Indian Journal of Biochemistry & Biophysics, 58*, 253–260.

Kumar, S. S., Ghosh, P., Kataria, N., Kumar, D., Thakur, S., Pathania, D., Kumar, V., Nasrullah, M., & Singh, L. (2021). The role of conductive nanoparticles in anaerobic digestion: Mechanism, current status and future perspectives. *Chemosphere, 280*, 130601. doi:10.1016/j.chemosphere.2021.130601 PMID:33945900

Kumar, S., Prasad, S., Yadav, K. K., Shrivastava, M., Gupta, N., Nagar, S., Bach, Q.-V., Kamyab, H., Khan, S. A., Yadav, S., & Malav, L. C. (2019). Hazardous heavy metals contamination of vegetables and food chain: Role of sustainable remediation approaches - A review. *Environmental Research, 179*, 108792. doi:10.1016/j.envres.2019.108792 PMID:31610391

Kumar, V., & Arora, K. (2020). Trends in nano-inspired biosensors for plants. *Materials Science for Energy Technologies, 3*, 255–273. doi:10.1016/j.mset.2019.10.004

Kurepa, J., Paunesku, T., Vogt, S., Arora, H., Rabatic, B. M., Lu, J., Wanzer, M. B., Woloschak, G. E., & Smalle, J. A. (2010). Uptake and distribution of ultrasmall anatase TiO_2 Alizarin red S nanoconjugates in *Arabidopsis thaliana. Nano Letters, 10*(7), 2296–2302. doi:10.1021/nl903518f PMID:20218662

Kurwadkar, S., Zhang, X.J., Ramirez, D., Mitchell, F.L. (2015b). *Emerging micro-pollutants in the environment: Occurrence, fate, and distribution.* . doi:10.1021/bk-2015-1198.fw001

Kurwadkar, S., Pugh, K., Gupta, A., & Ingole, S. (2015a). Nanoparticles in the environment: Occurrence, distribution, and risks. *Journal of Hazardous, Toxic and Radioactive Waste, 19*(3). Advance online publication. doi:10.1061/(ASCE)HZ.2153-5515.0000258

La Mantia, F. (2002). Handbook of plastics recycling. iSmithersRapra Publishing.

Lafront, A. M., Zhang, W., Ghali, E., & Houlachi, G. (2010). Electrochemical noise studies of the corrosion behaviour of lead anodes during zinc electrowinning maintenance. *Electrochimica Acta, 55*(22), 6665–6675. doi:10.1016/j.electacta.2010.06.005

Lagaron, J. M., & Lopez-Rubio, A. (2011). Nanotechnology for bioplastics: Opportunities, challenges and strategies. *Trends in Food Science & Technology, 22*(11), 611–617. doi:10.1016/j.tifs.2011.01.007

Laha, D., Pramanik, A., Maity, J., Mukherjee, A., Pramanik, P., Laskar, A., & Karmakar, P. (2014). Interplay between autophagy and apoptosis mediated by copper oxide nanoparticles in human breast cancer cells MCF7. *Biochimica et Biophysica Acta, 1840*(1), 1–9. doi:10.1016/j.bbagen.2013.08.011 PMID:23962629

Lakouraj, M. M., Mojerlou, F., & Zare, H. (2014). Nanogel and superparamagnetic nanocomposite based on sodium alginate for sorption of heavy metal ions. *Carbohydrate Polymers, 106*, 34–41. doi:10.1016/j.carbpol.2014.01.092 PMID:24721048

Landsiedel, R., Fabian, E., Ma-Hock, L., Wohlleben, W., Wiench, K., Oesch, F., & van Ravenzwaay, B. (2012). Toxico-/ biokinetics of nanomaterials. *Archives of Toxicology, 86*(7), 1021–1060. doi:10.100700204-012-0858-7 PMID:22576463

Lang, I. A., Galloway, T. S., Scarlett, A., Henley, W. E., Depledge, M., Wallace, R. B., & … . (2008). Association of urinary bisphenolA concentration with medical disorders and laboratory abnormalities in adults. *Journal of the American Medical Association, 300*(11), 1353–1355. doi:10.1001/jama.300.11.1303 PMID:18799442

Lanone, S., Rogerieux, F., Geys, J., Dupont, A., Maillot-Marechal, E., Boczkowski, J., Lacroix, G., & Hoet, P. (2009). Comparative toxicity of 24 manufactured nanoparticles in human alveolar epithelial and macrophage cell lines. *Particle and Fibre Toxicology, 6*(1), 14. doi:10.1186/1743-8977-6-14 PMID:19405955

Latiff, N. M., Sofer, Z., Fisher, A. C., & Pumera, M. (2017). Cytotoxicity of Exfoliated Layered Vanadium Dichalcogenides. *Chemistry (Weinheim an der Bergstrasse, Germany), 23*(3), 684–690. doi:10.1002/chem.201604430 PMID:27781318

Latiff, N. M., Teo, W. Z., Sofer, Z., Fisher, A. C., & Pumera, M. (2015). The Cytotoxicity of Layered Black Phosphorus. *Chemistry (Weinheim an der Bergstrasse, Germany), 21*(40), 13991–13995. doi:10.1002/chem.201502006 PMID:26291565

Le Pera, A., Sellaro, M., Bencivenni, E., & D'Amico, F. (2022). Environmental sustainability of an integrate anaerobic digestion-composting treatment of food waste: Analysis of an Italian plant in the circular bioeconomy strategy. *Waste Management, 139*(July), 341–351. doi:10.1016/j.wasman.2021.12.042

Lee, H. J., Chattopadhyay, S., Gong, E. Y., Ahn, R. S., & Lee, K. (2003). Antiandrogeniceffects of bisphenol A and nonylphenol on the function of androgen receptor. *Toxicological Sciences, 75*(1), 40–47. doi:10.1093/toxsci/kfg150 PMID:12805653

Lee, I. C., Ko, J. W., Park, S. H., Shin, N. R., Shin, I. S., Moon, C., Kim, J. H., Kim, H. C., & Kim, J. C. (2016). Comparative toxicity and biodistribution assessments in rats following subchronic oral exposure to copper nanoparticles and microparticles. *Particle and Fibre Toxicology, 13*(1), 56. doi:10.118612989-016-0169-x PMID:27788687

Lee, W., Goh, P. S., Lau, W. J., & Ismail, A. F. (2020). Removal of pharmaceutical contaminants from aqueous medium: A state-of-the-art review based on paracetamol. *Arabian Journal for Science and Engineering, 45*(9), 7109–7135. doi:10.100713369-020-04446-1

Lee, Y., Choi, J., Lee, K. J., Stott, N., & Kim, D. (2008). Large-scale synthesis of copper nanoparticles by chemically controlled reduction for applications of inkjet-printed electronics. *Nanotechnology, 19*(1), 415604. doi:10.1088/0957-4484/19/41/415604 PMID:21832649

Lei, R., Wu, C., Yang, B., Ma, H., Shi, C., Wang, Q., & Liao, M. (2008). Integrated metabolomic analysis of the nano-sized copper particle-induced hepatotoxicity and nephrotoxicity in rats: A rapid in vivo screening method for nanotoxicity. *Toxicology and Applied Pharmacology, 232*(2), 292–301. doi:10.1016/j.taap.2008.06.026 PMID:18706438

Leitão, D., Torres, J. P. N., & Fernandes, J. F. P. (2020). Spectral Irradiance Influence on Solar Cells Efficiency. *Energies, 13*(19), 1–18. doi:10.3390/en13195017

Lemos, S. C. S., Rezende, T. K. L., Assis, M., Romeiro, F. C., Peixoto, D. A., Gomes, E. O., Jacobsen, G. M., Teodoro, M. D., Gracia, L., Ferrari, J. L., Longo, E., Andrés, J., & de Lima, R. C. (2022). Efficient Ni and Fe doping process in ZnO with enhanced photocatalytic activity: A theoretical and experimental investigation. *Materials Research Bulletin, 111849*. Advance online publication. doi:10.1016/j.materresbull.2022.111849

Lemos, S. C. S., Romeiro, F. C., de Paula, L. F., Goncalves, R. F., de Moura, A. P., Ferrer, M. M., Longo, E., Patrocinio, A. O. T., & Lima, R. C. (2017). Effect of Er3+ ions on the phase formation and properties of In2O3 nanostructures crystallized upon microwave heating. *Journal of Solid State Chemistry*, *249*, 58–63. doi:10.1016/j.jssc.2017.02.011

Lespes, G., Faucher, S., & Slaveykova, V. I. (2020). Natural nanoparticles, anthropogenic nanoparticles, where is the frontier? *Frontiers in Environmental Science*, *8*, 71. doi:10.3389/fenvs.2020.00071

Liao, S., Zhang, Y., Pan, X., Zhu, F., Jiang, C., Liu, Q., Cheng, Z., Dai, G., Wu, G., Wang, L., & Chen, L. (2019). Antibacterial activity and mechanism of silver nanoparticles against multidrug-resistant *Pseudomonas aeruginosa*. *International Journal of Nanomedicine*, *2019*, 1469–1487. doi:10.2147/IJN.S191340 PMID:30880959

Lichtenberg, J. Y., Ling, Y., & Kim, S. (2019). Non-specific adsorption reduction methods in biosensing. *Sensors (Basel)*, *19*(11), 2488. doi:10.339019112488 PMID:31159167

Li, H., Cui, Z., & Han, C. (2009). Glutathione-stabilized silver nanoparticles as colorimetric sensor for Ni2+ ion. *Sensors and Actuators. B, Chemical*, *143*(1), 87–92. doi:10.1016/j.snb.2009.09.013

Li, K. G., Chen, J. T., Bai, S. S., Wen, X., Song, S. Y., Yu, Q., Li, J., & Wang, Y. Q. (2009). Intracellular oxidative stress and cadmium ions release induce cytotoxicity of unmodified cadmium sulfide quantum dots. *Toxicology In Vitro*, *23*(6), 1007–1013. doi:10.1016/j.tiv.2009.06.020 PMID:19540911

Li, K., Ding, D., Huo, D., Pu, K.-Y., Thao, N. N. P., Hu, Y., Li, Z., & Liu, B. (2012). Conjugated polymer based nanoparticles as dual-modal probes for targeted in vivo fluorescence and magnetic resonance imaging. *Advanced Functional Materials*, *22*(15), 3107–3115. doi:10.1002/adfm.201102234

Li, L., Liu, H., Chen, Y., Yang, D., Cai, C., Yuan, S., & Dai, X. (2022). Effect of Magnet-Fe3O4 composite structure on methane production during anaerobic sludge digestion: Establishment of direct interspecies electron transfer. *Renewable Energy*, *188*, 52–60. doi:10.1016/j.renene.2022.01.101

Lim, S. A., Lim, T. H., & Ahmad, A. N. (2020). The Applications of Biosensing and Artificial Intelligence Technologies for Rapid Detection and Diagnosis of COVID-19 in Remote Setting. In Diagnostic Strategies for COVID-19 and other Coronaviruses. Springer.

Lim, G. P., Soon, C. F., Ma, N. L., Morsin, M., Nayan, N., Ahmad, M. K., & Tee, K. S. (2021). Cytotoxicity of MXene-based Nanomaterials for Biomedical Applications: A Mini Review. *Environmental Research*, *201*, 111592. doi:10.1016/j.envres.2021.111592 PMID:34175291

Lim, S. Y., Shen, W., & Gao, Z. (2015). Carbon quantum dots and their applications. *Chemical Society Reviews*, *44*(1), 362–381. doi:10.1039/C4CS00269E PMID:25316556

Lin, C. F., Liang, E. Z., Shih, S. M., & Su, W. F. (2002). CdS nanoparticle light-emitting diode on Si. In Light-Emitting Diodes: Research, Manufacturing, and Applications. International Society for Optics and Photonics.

Liosis, C., Papadopoulou, A., Karvelas, E., Karakasidis, T. E., & Sarris, I. E. (2021). Heavy Metal adsorption using magnetic nanoparticles for water purification: A critical review. *Materials (Basel)*, *14*(24), 7500. doi:10.3390/ma14247500 PMID:34947096

Lithner, D., Larsson, Å., & Dave, G. (2011). Environmental and health hazard ranking andassessment of plastic polymers based on chemical composition. *The Science of the Total Environment*, *409*(18), 3309–3324. doi:10.1016/j.scitotenv.2011.04.038 PMID:21663944

Liu, X., & Zhang, M. (2005). *Characteristics of nano-subnanocomposites and response of soil and plant nutrition to them* [Ph.D. Dissertation]. Chinese Academy of Agricultural Sciences.

Liu, C., Shin, J., Son, S., Choe, Y., Farokhzad, N., Tang, Z., Xiao, Y., Kong, N., Kim, J. S., & Tao, W. (2021). Pnictogens in Medical Chemistry: Evolution from Erstwhile Drugs to Emerging Layered Photonic Nanomedicine. *Chemical Society Reviews*, *50*(4), 2260–2279. doi:10.1039/D0CS01175D PMID:33367452

Liu, G., Gao, J., Ai, H., & Chen, X. (2013). Applications and potential toxicity of magnetic iron oxide nanoparticles. *Small*, *9*(9-10), 1533–1545. doi:10.1002mll.201201531 PMID:23019129

Liu, H., Li, S., Chen, W., Wang, D., Li, C., Wu, C., Hao, J., Zhou, Z., Wang, X., & Wang, K. (2018). Scattering Enhanced Quantum Dots Based Luminescent Solar Concentrators by Silica Microparticles. *Solar Energy Materials and Solar Cells*, *179*, 380–385. doi:10.1016/j.solmat.2018.01.029

Liu, H., Wang, X., Wu, Y., Hou, J., Zhang, S., Zhou, N., & Wang, X. (2019). (Toxicity responses of dierent organs of zebrafish (*Daniorerio*) to silver nanoparticles with different particle sizes and surface coatings. *Environmental Pollution*, *246*, 414–422. doi:10.1016/j.envpol.2018.12.034 PMID:30579210

Liu, L., Sun, M., Li, Q., Zhang, H., Alvarez, P. J., Liu, H., & Chen, W. (2014). Genotoxicity and Cytotoxicity of Cadmium Sulfide Nanomaterials to Mice: Comparison Between Nanorods and Nanodots. *Environmental Engineering Science*, *31*(7), 373–380. doi:10.1089/ees.2013.0417 PMID:25053877

Liu, M., Feng, S., Ma, Y., Xie, C., He, X., Ding, Y., Zhang, J., Luo, W., Zheng, L., Chen, D., Yang, F., Chai, Z., Zhao, Y., & Zhang, Z. (2019). Influence of surface charge on the phytotoxicity, transformation, and translocation of CeO_2 nanoparticles in cucumber plants. *ACS Applied Materials & Interfaces*, *11*(18), 16905–16913. doi:10.1021/acsami.9b01627 PMID:30993970

Liu, R., & Lal, R. (2015). Potentials of engineered nanoparticles as fertilizers for increasing agronomic productions. *The Science of the Total Environment*, *514*, 131–139. doi:10.1016/j.scitotenv.2015.01.104 PMID:25659311

Liu, S., Regulacio, M. D., Tee, S. Y., Khin, Y. W., Teng, C. P., Koh, L. D., Guan, G., & Han, M. Y. (2016). Preparation, functionality, and application of metal oxide-coated noble metal nanoparticles. *Chemical Record (New York, N.Y.)*, *6*(4), 1965–1990. doi:10.1002/tcr.201600028 PMID:27291595

Liu, X., Zhao-bin, F., Fu-dao, Z., & Shu-qing, Z., & Xu-sheng, H. E. (2005). Responses of peanut to nanocalcium carbonate. *Plant Nutrition and Fertitizer Science*, *11*(3), 385–389.

Liu, Y., Ding, J., Wang, Q.-Q., Wen, M.-L., Tang, T.-T., Liu, Y., Yuan, R., Li, Y.-F., & An, M.-W. (2021). Research Progress on the Biomedical Uses of Graphene and Its Derivatives. *New Carbon Materials*, *36*(4), 779–793. doi:10.1016/S1872-5805(21)60073-2

Liu, Y., Hardie, J., Zhang, X., & Rotello, V. M. (2017). Effects of engineered nanoparticles on the innate immune system. *Seminars in Immunology*, *34*, 25–32. doi:10.1016/j.smim.2017.09.011 PMID:28985993

Liu, Y., Pang, N., Cai, Y., Yang, Y., Zeng, C., & Wang, Y. (2021). Application of Nanooptics in Photographic Imagery and Medical Imaging. *Journal of Chemistry*, *2021*, 1–15. Advance online publication. doi:10.1155/2021/2384322

Liu, Y., Xiao, Q., Ye, X., Wang, C., Jia, Z., Du, J., Kong, X., & Xi, Y. (2021). Effect of different charged Fe 3 O 4 nanoparticles on methane production for anaerobic digestion of wheat straw. *Journal of Cleaner Production*, *328*, 129655. doi:10.1016/j.jclepro.2021.129655

Li, X., Chen, D., & Xie, S. (2021). Current progress and prospects of organic nanoparticles against bacterial biofilm. *Advances in Colloid and Interface Science*, *294*, 102475. doi:10.1016/j.cis.2021.102475 PMID:34280601

Li, X., Jia, Y., Wei, J., Zhu, H., Wang, K., Wu, D., & Cao, A. (2010). Solar cells and light sensors based on nanoparticle-grafted carbon nanotube films. *ACS Nano*, *4*(4), 2142–2148. doi:10.1021/nn901563y PMID:20222675

Li, Y., Feng, L., Shi, X., Wang, X., Yang, Y., Yang, K., Liu, T., Yang, G., & Liu, Z. (2014). Surface Coating-dependent Cytotoxicity and Degradation of Graphene Derivatives: Towards the Design of Non-Toxic, Degradable Nano-Graphene. *Small, 10*(8), 1544–1554. doi:10.1002mll.201303234 PMID:24376215

Li, Y., Hong-Cu, Y. T., & Luo, D. (2005). Multiplexed detection of pathogen DNA with DNA-based fluorescence nano barcodes. *Nature Biotechnology, 23*(7), 885–889. doi:10.1038/nbt1106 PMID:15951805

Li, Y., Sun, Y., Zhang, G., He, Z., Wang, Y., & Cui, J. (2016). Effects of copper oxide nanoparticles on developing zebrafish embryos and larvae. *International Journal of Nanomedicine, 11*, 905–918. doi:10.2147/IJN.S100350 PMID:27022258

Li, Y., Zhu, N., Liang, X., Bai, X., Zheng, L., Zhao, J., Li, Y.-F., Zhang, Z., & Gao, Y. (2020). Silica nanoparticles alleviate mercury toxicity via immobilization and inactivation of Hg(ii) in soybean (*Glycine max*). *Environmental Science. Nano, 7*(6), 1807–1817. doi:10.1039/D0EN00091D

Lohani, A., Verma, A., Joshi, H., Yadav, N., & Karki, N. (2014). Nanotechnology-based cosmeceuticals. *ISRN Dermatology, 2014*, 1–14. doi:10.1155/2014/843687 PMID:24963412

Loiseau, A., Asila, V., Boitel-Aullen, G., Lam, M., Salmainm, M., & Boujday, S. (2019). Silver-based plasmonic nanoparticles for and their use in biosensing. *Biosensors (Basel), 9*(2), 78. doi:10.3390/bios9020078 PMID:31185689

Lopez, E., Figueroa, S., Oset-Gasque, M. J., & Gonzalez, M. P. (2003). Apoptosis and necrosis: Two distinct events induced by cadmium in cortical neurons in culture. *British Journal of Pharmacology, 138*(5), 901–911. doi:10.1038j.bjp.0705111 PMID:12642392

López-Moreno, M. L., Cassé, C., & Correa-Torres, S. N. (2018). Engineered nanomaterials interactions with living plants: Benefits, hazards and regulatory policies. *Current Opinion in Environmental Science & Health, 6*, 36–41. doi:10.1016/j.coesh.2018.07.013

López-Vargas, E. R., Ortega-Ortíz, H., Cadenas-Pliego, G., De Alba Romenus, K., Cabrera de la Fuente, M., Benavides-Mendoza, A., & Juárez-Maldonado, A. (2018). Foliar application of copper nanoparticles increases the fruit quality and the content of bioactive compounds in tomatoes. *Applied Sciences (Basel, Switzerland), 8*(7), 1020. doi:10.3390/app8071020

Loss, D., & DiVincenzo, D. P. (1998). Quantum computation with quantum dots. *Physical Review A, 57*(1), 120–126. doi:10.1103/PhysRevA.57.120

Lowry, G. V., Gregory, K. B., Apte, S. C., & Lead, J. R. (2012). Transformations of nanomaterials in the environment. *Environmental Science & Technology, 46*(13), 6893–6899. doi:10.1021/es300839e PMID:22582927

Lu, J. S., Chang, J. S., & Lee, D. J. (2020). Adding carbon-based materials on anaerobic digestion performance: A mini-review. *Bioresource Technology, 300*(November), 122696. doi:10.1016/j.biortech.2019.122696

Luan, B., Huynh, T., Zhao, L., & Zhou, R. (2015). Potential Toxicity of Graphene to Cell Functions via Disrupting Protein-Protein Interactions. *ACS Nano, 9*(1), 663–669. doi:10.1021/nn506011j PMID:25494677

Lu, C., Li, X., Guo, X., Fu, S., Zhang, W., Yuan, H., & Fang, J. (2023). Efficient inverted CsPbI3 perovskite solar cells fabricated in common air. *Chemical Engineering Journal, 452*, 139495. doi:10.1016/j.cej.2022.139495

Luh, M. D., Baker, R. A., & Henley, D. E. (1973). Arsenic analysis and toxicity—A review. *The Science of the Total Environment, 2*(1), 1–12. doi:10.1016/0048-9697(73)90002-8 PMID:4804487

Lundqvist, M., Stigler, J., Elia, G., Lynch, I., Cedervall, T., & Dawson, K. A. (2008). Nanoparticle size and surface properties determine the protein corona with possible implications for biological impacts. *Proceedings of the National Academy of Sciences of the United States of America, 105*(38), 14265–14270. doi:10.1073/pnas.0805135105 PMID:18809927

Luo, Y., Guo, W., Ngo, H. H., Nghiem, L. D., Hai, F. I., Zhang, J., Liang, S., & Wang, X. C. (2014). A review on the occurrence of micropollutants in the aquatic environment and their fate and removal during wastewater treatment. *The Science of the Total Environment, 473*, 619–641. doi:10.1016/j.scitotenv.2013.12.065 PMID:24394371

Luo, Y.-H., Chang, L. W., & Lin, P. (2015). Metal-based nanoparticles and the immune system: Activation, inflammation, and potential applications. *BioMed Research International, 2015*, 143720. doi:10.1155/2015/143720 PMID:26125021

Luo, Z., Wang, Z., Yan, Y., Li, J., Yan, C., & Xing, B. (2018). Titanium dioxide nanoparticles enhance inorganic arsenic bioavailability and methylation in two freshwater algae species. *Environmental Pollution, 238*, 631–637. doi:10.1016/j.envpol.2018.03.070 PMID:29614472

Lusher, A. (2015). Microplastics in the marine environment: distribution, interactions and effects. In M. Bergmann, L. Gutow, & M. Klages (Eds.), *Marine anthropogenic litter* (pp. 245–308). Springer. doi:10.1007/978-3-319-16510-3_10

Lu, X., Miousse, I. R., Pirela, S. V., Moore, J. K., Melnyk, S., Koturbash, I., & Demokritou, P. (2016). In vivo epigenetic effects induced by engineered nanomaterials: A case study of copper oxide and laser printer-emitted engineered nanoparticles. *Nanotoxicology, 10*(5), 629–639. doi:10.3109/17435390.2015.1108473 PMID:26559097

Lu, X., Sun, J., & Sun, X. (2020). Recent advances in biosensors for the detection of estrogens in the environment and food. *Trends in Analytical Chemistry, 2020*(127), 115882. doi:10.1016/j.trac.2020.115882

Lyon, D. Y., Fortner, J. D., Sayes, C. M., Colvin, V. L., & Hughe, J. B. (2005). Bacterial cell association and antimicrobial activity of a C60 water suspension. *Environmental Toxicology and Chemistry, 24*(11), 2757–2762. doi:10.1897/04-649R.1 PMID:16398110

Machmudah, S., Sato, T., Wahyudiono, Sasaki, M., & Goto, M. (2012). Silver nanoparticles generated by pulsed laser ablation in supercritical CO_2 medium. *High Pressure Research, 32*(1), 60–66. doi:10.1080/08957959.2011.649277

Madhavi, V., Reddy, A. V. B., & Madhavi, G. (2022). Nanosorbents for the removal of heavy metal pollutants. In Nano-Bioremediation: Fundamentals and Applications (pp. 363-387). Elsevier. doi:10.1016/B978-0-12-823962-9.00004-0

Maduraiveeran, G. (2020). Bionanomaterial-based electrochemical biosensing platforms for biomedical applications. *Analytical Methods, 12*(13), 1688–1701. doi:10.1039/D0AY00171F

Magar, H. S., Hassan, R. Y., & Mulchandani, A. (2021). Electrochemical impedance spectroscopy (EIS): Principles, construction, and biosensing applications. *Sensors (Basel), 21*(19), 6578. doi:10.339021196578 PMID:34640898

Magdolenova, Z., Drlickova, M., Henjum, K., Runden-Pran, E., Tulinska, J., Bilanicova, D., Pojana, G., Kazimirova, A., Barancokova, M., Kuricova, M., Liskova, A., Staruchova, M., Ciampor, F., Vavra, I., Lorenzo, Y., Collins, A., Rinna, A., Fjellsbø, L., Volkovova, K., ... Dusinska, M. (2015). Coating-dependent induction of cytotoxicity and genotoxicity of iron oxide nanoparticles. *Nanotoxicology, 9*(sup1), 44–56. doi:10.3109/17435390.2013.847505 PMID:24228750

Ma, H., Williams, P. L., & Diamond, S. A. (2013). Ecotoxicity of manufactured ZnO nanoparticles—A review. *Environmental Pollution, 172*, 76–85. doi:10.1016/j.envpol.2012.08.011 PMID:22995930

Maharramov, A. M., Hasanova, U. A., Suleymanova, I. A., Osmanova, G. E., & Hajiyeva, N. E. (2019). The engineered nanoparticles in food chain: Potential toxicity and effects. *SN Applied Sciences, 1*(11), 1362. doi:10.100742452-019-1412-5

Mahdavinia, G. R., & Asgari, A. (2013). Synthesis of kappa-carrageenan-g-poly(acrylamide)/sepiolite nanocomposite hydrogels and adsorption of cationic dye. *Polymer Bulletin, 70*(8), 2451–2470. doi:10.100700289-013-0966-4

Mahjoubian, M., Naeemi, A. S., & Sheykhan, M. (2021). Toxicological effects of Ag_2O and Ag_2CO_3 doped TiO_2 nanoparticles and pure TiO_2 particles on zebrafish (*Danio rerio*). *Chemosphere, 263*, 128182. doi:10.1016/j.chemosphere.2020.128182 PMID:33297149

Mahmoodi, M., & Javanbakht, V. (2021). Fabrication of Zn-based magnetic zeolitic imidazolate framework bionano-composite using basil seed mucilage for removal of azo cationic and anionic dyes from aqueous solution. *International Journal of Biological Macromolecules*, *167*, 1076–1090. doi:10.1016/j.ijbiomac.2020.11.062 PMID:33186651

Mahmoodzadeh, H., Nabavi, M., & Kashefi, H. (2015). Effect of nanoscale titanium dioxide particles on the germination and growth of canola (*Brassica napus*). *Journal of Ornamental Plants*, *3*(1), 25–32.

Ma-Hock, L., Landsiedel, R., Wiench, K., Geiger, D., Strauss, V., Gröters, S., Ravenzwaay, B., Gerst, M., Wohlleben, W., & Scherer, G. (2012). Shortterm rat inhalation study with aerosols of acrylic ester-based polymer dispersions containinga fraction of nanoparticles. *International Journal of Toxicology*, *31*(1), 46–57. doi:10.1177/1091581811424778 PMID:22267870

Maity, P., Bepari, M., Pradhan, A., Baral, R., Roy, S., & Choudhury, S. M. (2018). Synthesis and characterization of biogenic metal NPs and its cytotoxicity and anti-neoplasticity through the induction of oxidative stress, mitochondrial dysfunction and apoptosis. *Colloids and Surfaces. B, Biointerfaces*, *161*, 111–120. doi:10.1016/j.colsurfb.2017.10.040 PMID:29055863

Majhi, K. C., & Yadav, M. S. (2021). Synthesis of inorganic nanomaterials using carbohydrates. In Green Sustainable Process for Chemical and Environmental Engineering and Science (pp. 109-135). Elsevier. doi:10.1016/B978-0-12-821887-7.00003-3

Makhado, E., Pandey, S., Nomngongo, P. N., & Ramontja, J. (2018b). Preparation and characterization of xanthan gum-cl-poly(acrylic acid)/o-MWCNTs hydrogel nanocomposite as highly effective re-usable adsorbent for removal of methylene blue from aqueous solutions. *Journal of Colloid and Interface Science*, *513*, 700–714. doi:10.1016/j.jcis.2017.11.060 PMID:29216578

Makhado, E., Pandey, S., & Ramontja, J. (2018a). Microwave assisted synthesis of xanthan gum-cl-poly (acrylic acid) based-reduced graphene oxide hydrogel composite for adsorption of methylene blue and methyl violet from aqueous solution. *International Journal of Biological Macromolecules*, *119*, 255–269. doi:10.1016/j.ijbiomac.2018.07.104 PMID:30031820

Malaekeh-Nikouei, B., Bazzaz, B. S. F., Mirhadi, E., Tajani, A. S., & Khameneh, B. (2020). The role of nanotechnology in combating biofilm-based antibiotic resistance. *Journal of Drug Delivery Science and Technology*, *60*, 101880. doi:10.1016/j.jddst.2020.101880

Malassis, L., Dreyfus, R., Murphy, R. J., Hough, L. A., Donnio, B., & Murray, C. B. (2016). One-step green synthesis of gold and silver nanoparticles with ascorbic acid and their versatile surface post-functionalization. *RSC Advances*, *6*(39), 33092–33100. doi:10.1039/C6RA00194G

Malatji, N., Makhado, E., Ramohlola, K. E., Modibane, K. D., Maponya, T. C., Monama, G. R., & Hato, M. J. (2020). Synthesis and characterization of magnetic clay-based carboxymethyl cellulose-acrylic acid hydrogel nanocomposite for methylene blue dye removal from aqueous solution. *Environmental Science and Pollution Research International*, *27*(35), 44089–44105. doi:10.100711356-020-10166-8 PMID:32761344

Malik, J. A., Bhadauria, M., Lone, R., & Hajam, Y. A. (2020). Exploitation of Plant Phenolics in Animal Farming. In R. Lone, R. Shuab, & A. Kamili (Eds.), *Plant Phenolics in Sustainable Agriculture*. Springer. doi:10.1007/978-981-15-4890-1_3

Malik, J. A., & Lone, R. (2021). Heat shock proteins with an emphasis on HSP 60. *Molecular Biology Reports*, *48*(10), 6959–6969. doi:10.100711033-021-06676-4 PMID:34498161

Malik, J. A., Mishra, G., Hajam, Y. A., Lone, R., & Quazi, S. (2022). Metabolome Analyses in Response to Diverse Abiotic Stress. In A. Roychoudhury, T. Aftab, & K. Acharya (Eds.), *Omics Approach to Manage Abiotic Stress in Cereals*. Springer. doi:10.1007/978-981-19-0140-9_6

Malinauskaite, J., Jouhara, H., Czajczyńska, D., Stanchev, P., Katsou, E., Rostkowski, P., Thorne, R. J., Colón, J., Ponsá, S., Al-Mansour, F., Anguilano, L., Krzyżyńska, R., López, I. C., Vlasopoulos, A., & Spencer, N. (2017). Municipal Solid Waste Management and Waste-to-Energy in the Context of a Circular Economy and Energy Recycling in Europe. *Energy*, *141*, 2013–2044. doi:10.1016/j.energy.2017.11.128

Mandal, D., & Lalrinchhani. (2021). Nanofertilizer and its application in horticulture. *Journal of Applied Horticulture*, *23*(1), 70–77. doi:10.37855/jah.2021.v23i01.14

Manikandan, A., & Subramanian, K. S. (2014). Fabrication and characterisation of nanoporous zeolite based N fertilizer. *African Journal of Agricultural Research*, *9*(2), 276–284. doi:10.5897/AJAR2013.8236

Manikandan, A., & Subramanian, K. S. (2016). Evaluation of zeolite-based nitrogen nano-fertilizers on maize growth, yield and quality on inceptisols and alfisols. *International Journal of Plant and Soil Science*, *9*(4), 1–9. doi:10.9734/IJPSS/2016/22103

Manuja, A., Kumar, B., Kumar, R., Chhabra, D., Ghosh, M., Manuja, M., Brar, B., Pal, Y., Tripathi, B. N., & Prasad, M. (2021). Metal/metal oxide nanoparticles: Toxicity concerns associated with their physical state and remediation for biomedical applications. *Toxicology Reports*, *8*, 1970–1978. doi:10.1016/j.toxrep.2021.11.020 PMID:34934635

Ma, R. M., Dai, L., & Qin, G. G. (2007). Enhancement-mode metal-semiconductor field-effect transistors based on single n-Cd S nanowires. *Applied Physics Letters*, *90*(9), 093109. doi:10.1063/1.2710004

Marimuthu, S., Antonisamy, A. J., Malayandi, S., Rajendran, K., Pugazhendhi, A., & Ponnusamy, V. K. (2020). Silver nanoparticles in dye effluent treatment: A review on synthesis, treatment methods, mechanisms, photocatalytic degradation, toxic effects and mitigation of toxicity. *Journal of Photochemistry and Photobiology. B, Biology*, *205*, 111823. doi:10.1016/j.jphotobiol.2020.111823 PMID:32120184

Martos, M., Morales, J., Sanchez, L., Ayouchi, R., Leinen, D., Martin, F., & Barrado, J. R. (2001). Electrochemical properties of lead oxide films obtained by spray pyrolysis as negative electrodes for lithium secondary batteries. *Electrochimica Acta*, *46*(19), 2939–2948. doi:10.1016/S0013-4686(01)00512-6

Marzouk, N. M., Abd-Alrahman, H. A., EL-Tanahy, A. M. M., & Mahmoud, S. H. (2019). Impact of foliar spraying of nano micronutrient fertilizers on the growth, yield, physical quality, and nutritional value of two snap bean cultivars in sandy soils. *Bulletin of the National Research Center*, *43*(1), 84. doi:10.118642269-019-0127-5

Masilompane, T.M., Chaukura, N., Mishra, S.B., & Mishra, A.K. (2018). Chitosan-lignin-titania nanocomposites for the removal of brilliant black dye from aqueous solution. *International Journal of Biological Macromolecules, 120*(Pt B), 1659-1666. . doi:10.1016/j.ijbiomac.2018.09.129

Mastronardi, E., Foster, A., Zhang, X., & DeRosa, M. C. (2014). Smart materials based on DNA aptamers: Taking apta sensing to the next level. *Sensors (Basel)*, *14*(2), 3156–3171. doi:10.3390140203156 PMID:24553083

Mastronardi, E., Tsae, P., Zhang, X., Monreal, C., & DeRosa, M. C. (2015). Strategic Role of Nanotechnology in Fertilizers: Potential and Limitations. In M. Rai, C. Ribeiro, L. Mattoso, & N. Duran (Eds.), *Nanotechnologies in Food and Agriculture* (pp. 25–67). Springer International Publishing. doi:10.1007/978-3-319-14024-7_2

Mathew, B. B., & Krishnamurthy, N. B. (2018). Evaluation of lead oxide nanoparticles synthesized by chemical and biological methods. *Journal of Nanomedicine Research*, *7*(3), 195–198. doi:10.15406/jnmr.2018.07.00187

Matteis, V. D., Rizzello, L., Ingrosso, C., Liatsi-Douvitsa, E., Giorgi, M. L. D., Matteis, G. D., & Rinaldi, R. (2019). Cultivar-dependent anticancer and antibacterial properties of silver nanoparticles synthesized using leaves of different *Olea Europaea* trees. *Nanomaterials (Basel, Switzerland), 2019*(9), 1544. doi:10.3390/nano9111544 PMID:31671618

Mattsson, K., Johnson, E. V., Malmendal, A., Linse, S., Hansson, L.-A., & Cedervall, T. (2017). Brain damage and behavioural disorders in fish induced by plastic nanoparticles delivered through the food chain. *Scientific Reports, 7*(1), 11452. doi:10.103841598-017-10813-0 PMID:28904346

Mauricio-Inglesias, M., Peyron, S., Guillard, V., & Gontard, N. (2010). Wheat gluten nanocomposite films as food contact materials: Migration tests and impact of a novel food stabilizing technology (high pressure). *Journal of Applied Polymer Science, 116*, 2526–2535.

Ma, Y., Cai, F., Li, Y., Chen, J., Han, F., & Lin, W. (2020). A review of the application of nanoparticles in the diagnosis and treatment of chronic kidney disease. *Bioactive Materials, 5*(3), 732–743. doi:10.1016/j.bioactmat.2020.05.002 PMID:32596555

Ma, Y., He, X., Zhang, P., Zhang, Z., Ding, Y., Zhang, J., Wang, G., Xie, C., Luo, W., Zhang, J., Zheng, L., Chai, Z., & Yang, K. (2017). Xylem and phloem based transport of CeO_2 nanoparticles in hydroponic cucumber plants. *Environmental Science & Technology, 51*(9), 5215–5221. doi:10.1021/acs.est.6b05998 PMID:28383248

McCall, M. J. (2011). Nanoparticles in the real world. *Nature Nanotechnology, 6*(10), 613–614. doi:10.1038/nnano.2011.169 PMID:21979234

McGrath, S. P., & Zhao, F. J. (2003). Phytoextraction of metals and metalloids from contaminated soils. *Current Opinion in Biotechnology, 14*(3), 277–282. doi:10.1016/S0958-1669(03)00060-0 PMID:12849780

Meade, R. H. (1995). Contaminants in the Mississippi River. *Geological Survey Circular*, 1133.

Mehrangiz, J. M., Sirous, B., & Ebrahim, A. (2014). Study the effect of foliar application of nano chelate molybdenum fertilizer on the yield and yield components of peanut. *Egyptian Academic Journal of Biological Sciences, 5*(1), 67–71. doi:10.21608/eajbsh.2014.16829

Mehwish, H. M., Rajoka, M. S. R., Xiong, Y., Cai, H., Aadil, R. M., Mahmood, Q., He, Z., & Zhu, Q. (2021). Green synthesis of a silver nanoparticle using *Moringa oleifera* seed and its applications for antimicrobial and sun-light mediated photocatalytic water detoxification. *Journal of Environmental Chemical Engineering, 9*(4), 105290. doi:10.1016/j.jece.2021.105290

Mekkdad, A. A. A. (2017). Response of peanut nitrogen fertilizer levels and foliar zinc spraying rates in newly reclaimed sandy soils. *Journal of Plant Production, 8*(2), 153–159. doi:10.21608/jpp.2017.39240

Melzer, D., & Galloway, T. S. (2010, Oct.). Burden of proof. New Scientist, 26–27.

Melzer, D., Osborne, N. J., Henley, W. E., Cipelli, R., Young, A., Money, C., McCormack, P., Luben, R., Khaw, K.-T., Wareham, N. J., & Galloway, T. S. (2012). Urinary bisphenol A: Concentration and risk of future coronary artery disease in apparently healthy men and women. *Circulation, 125*(12), 1482–1490. doi:10.1161/CIRCULATIONAHA.111.069153 PMID:22354940

Melzer, D., Rice, N. E., Lewis, C., Henley, W. E., & Galloway, T. S. (2010). Association of urinary bisphenol. a concentration with heart disease: Evidence from NHANES 2003/06. *PLoS One, 5*(1), e8673. doi:10.1371/journal.pone.0008673 PMID:20084273

Meng, C. Y., Han, Y. F., Liu, Y. L., Gao, H. X., Ren, Y. Y., Qian, Q. Z., Wang, Q., & Li, Q. Z. (2019). Resveratrol alleviate the injury of mice liver induced by cadmium sulfide nanoparticles. *The Kaohsiung Journal of Medical Sciences*, *35*(5), 297–302. doi:10.1002/kjm2.12056 PMID:30913377

Meng, H., Chen, Z., Xing, G., Yuan, H., Chen, C., Zhao, F., Zhang, C., & Zhao, Y. (2007). Ultrahigh reactivity provokes nanotoxicity: Explanation of oral toxicity of nano-copper particles. *Toxicology Letters*, *175*(1-3), 102–110. doi:10.1016/j.toxlet.2007.09.015 PMID:18024012

Meychik, N., Nikolaeva, J., & Yermakov, I. (2005). Ion exchange properties of the root cell walls isolated from the halophyte plants (*Suaeda altissima* L.) grown under conditions of different salinity. *Plant and Soil*, *277*(1-2), 163–174. doi:10.100711104-005-6806-z

Meyer, K., Rajanahalli, P., Ahamed, M., Rowe, J. J., & Hong, Y. (2011). ZnO nanoparticles induce apoptosis in human dermal fibroblasts via p53 and p38 pathways. *Toxicology In Vitro*, *25*(8), 1721–1726. doi:10.1016/j.tiv.2011.08.011 PMID:21903158

Miculescu, F., Maidaniuc, A., Voicu, S. I., Thakur, V. K., Stan, G. E., & Ciocan, L. T. (2017). Progress in hydroxyapatite–starch based sustainable biomaterials for biomedical bone substitution applications. *ACS Sustainable Chemistry & Engineering*, *5*(10), 8491–8512. doi:10.1021/acssuschemeng.7b02314

Mielke, H., & Gundert-Remy, U. (2009). Bisphenol A levels in blood depend on age and exposure. *Toxicology Letters*, *190*(1), 32–40. doi:10.1016/j.toxlet.2009.06.861 PMID:19560527

Mikhailova, E. O. (2020). Silver nanoparticles: Mechanism of action and probable bio-application. *Journal of Functional Biomaterials*, *11*(4), 84. doi:10.3390/jfb11040084 PMID:33255874

Minghetti, M., & Schirmer, K. (2019). Interference of silver nanoparticles with essential metal homeostasis in a novel enterohepatic fish in vitro system. *Environmental Science. Nano*, *6*(6), 1777–1790. doi:10.1039/C9EN00310J

Minigaliyeva, I. A., Sutunkova, M. P., Gurvich, V. B., Bushueva, T. V., Klinova, S. V., Solovyeva, S. N., Chernyshov, I. N., Valamina, I. E., Shur, V. Y., Shishkina, E. V., Makeyev, O. H., Panov, V. G., Privalova, L. I., & Katsnelson, B. A. (2020). An overview of experiments with lead-containing nanoparticles performed by the Ekaterinburg nanotoxicological research team. *Nanotoxicology*, *14*(6), 788–806. doi:10.1080/17435390.2020.1762132 PMID:32396411

Miri, A., Sarani, M., Hashemzadeh, A., Mardani, Z., & Darroudi, M. (2018). Biosynthesis and cytotoxic activity of lead oxide nanoparticles. *Green Chemistry Letters and Reviews*, *11*(4), 567–572. doi:10.1080/17518253.2018.1547926

Mirnajafizadeh, F., Ramsey, D., McAlpine, S., Wang, F., & Stride, J. A. (2019). Nanoparticles for Bioapplications: Study of the Cytotoxicity of Water Dispersible CdSe(S) and CdSe(S)/ZnO Quantum Dots. *Nanomaterials (Basel, Switzerland)*, *9*(3), 465. doi:10.3390/nano9030465 PMID:30897752

Mirzajani, F., Askari, H., Hamzelou, S., Schober, Y., Römpp, A., Ghassempour, A., & Spengler, B. (2014). Proteomics study of silver nanoparticles toxicity on *Bacillus thuringiensis*. *Ecotoxicology and Environmental Safety*, *100*, 122–130. doi:10.1016/j.ecoenv.2013.10.009 PMID:24290895

Mitchell, M. J., Billingsley, M. M., Haley, R. M., Wechsler, M. E., Peppas, N. A., & Langer, R. (2021). Engineering precision nanoparticles for drug delivery. *Nature Reviews. Drug Discovery*, *20*(2), 101–124. doi:10.103841573-020-0090-8 PMID:33277608

Mittal, D., Kaur, G., Singh, P., Yadav, K., & Ali, S. A. (2020). Nanoparticle-based sustainable agriculture and food science: Recent advances and future outlook. *Frontiers in Nanotechnology*, *2*, 10. doi:10.3389/fnano.2020.579954

Mittal, H., & Ray, S. S. (2016). A study on the adsorption of methylene blue onto gum ghatti/TiO2 nanoparticles-based hydrogel nanocomposite. *International Journal of Biological Macromolecules*, *88*, 66–80. doi:10.1016/j.ijbiomac.2016.03.032 PMID:26997239

Mody, C., & Choi, H. (2013). From materials science to nanotechnology: Institutions, communities, and disciplines at Cornell University, 1960-2000. *Historical Studies in the Natural Sciences*, *43*(2), 121–161. doi:10.1525/hsns.2013.43.2.121

Mody, V. V., Siwale, R., Singh, A., & Mody, H. R. (2010). Introduction to metallic nanoparticles. *Journal of Pharmacy & Bioallied Sciences*, *2*(4), 282–289. doi:10.4103/0975-7406.72127 PMID:21180459

Moghaddam, A.Z., Ghiamati, E., Pourashuri, A., & Allahresani, A. (2018). Modified nickel ferrite nanocomposite/functionalized chitosan as a novel adsorbent for the removal of acidic dyes. *International Journal of Biological Macromolecules*, *120*(Pt B), 1714-1725. . doi:10.1016/j.ijbiomac.2018.09.198

Mogren, C. L., & Trumble, J. T. (2010). The impacts of metals and metalloids on insect behavior. *Entomologia Experimentalis et Applicata*, *135*(1), 1–17. doi:10.1111/j.1570-7458.2010.00967.x

Mohammed, Y. H., Holmes, A., Haridass, I. N., Sanchez, W. Y., Studier, H., Grice, J. E., Benson, H. A. E., & Roberts, M. S. (2019). Support for the Safe Use of Zinc Oxide Nanoparticle Sunscreens: Lack of Skin Penetration or Cellular Toxicity after Repeated Application in Volunteers. *The Journal of Investigative Dermatology*, *139*(2), 308–315. doi:10.1016/j.jid.2018.08.024 PMID:30448212

Mokarram-Kashtiban, S., Hosseini, S. M., Tabari Kouchaksaraei, M., & Younesi, H. (2019). The impact of nanoparticles zero-valent iron (nZVI) and rhizosphere microorganisms on the phytoremediation ability of white willow and its response. *Environmental Science and Pollution Research International*, *26*(11), 10776–10789. doi:10.100711356-019-04411-y PMID:30778927

Molavi, H., Neshastehgar, M., Shojaei, A., & Ghashghaeinejad, H. (2020). Ultrafast and simultaneous removal of anionic and cationic dyes by nanodiamond/UiO-66 hybrid nanocomposite. *Chemosphere*, *247*, 125882. doi:10.1016/j.chemosphere.2020.125882 PMID:32069713

Monroe, R. K., & Halvorsen, S. W. (2006). Cadmium blocks receptor-mediated Jak/STAT signaling in neurons by oxidative stress. *Free Radical Biology & Medicine*, *41*(3), 493–502. doi:10.1016/j.freeradbiomed.2006.04.023 PMID:16843830

Moon, J., Kwak, J. I., & An, Y. J. (2019). The effects of silver nanomaterial shape and size on toxicity to *Caenorhabditis elegans* in soil media. *Chemosphere*, *215*, 50–56. doi:10.1016/j.chemosphere.2018.09.177 PMID:30312916

Moore, A., Marecos, E., Bogdanov Jr, A., & Weissleder, R. (2000). Tumoral distribution of long-circulating dextran-coated iron oxide nanoparticles in a rodent model. *Radiology, 214*(2), 568-574. doi:.r00fe19568 doi:10.1148/radiology.214.2

Morales, M. P., Bomati-Miguel, O., De Alejo, R. P., Ruiz-Cabello, J., Veintemillas-Verdaguer, S., & O'Grady, K. (2003). Contrast agents for MRI based on iron oxide nanoparticles prepared by laser pyrolysis. *Journal of Magnetism and Magnetic Materials*, *266*(1-2), 102–109. doi:10.1016/S0304-8853(03)00461-X

Morales-Narváez, E., & Merkoçi, A. (2019). Graphene oxide as an optical biosensing platform: A progress report. *Advanced Materials*, *31*(6), 1805043. PMID:30549101

Moraru, C. I., Panchapakesan, C. P., Huang, Q., Takhistov, P., Liu, S. J., & Kokini, L. (2003). Nanotechnology: A new frontier in food science. *Food Technology*, *57*, 24–29.

Moriyama, K., Tagami, T., Akamizu, T., Usui, T., Saijo, M., Kanamoto, N., Hataya, Y., Shimatsu, A., Kuzuya, H., & Nakao, K. (2002). Thyroidhormone action is disrupted by bisphenol A as an antagonist. *The Journal of Clinical Endocrinology and Metabolism*, *87*(11), 5185–5190. doi:10.1210/jc.2002-020209 PMID:12414890

Morteza, E., Moaveni, P., Farahani, H. A., & Kiyani, M. (2013). Study of photosynthetic pigments changes of maize (*Zea mays* L.) under nano TiO_2 spraying at various growth stages. *SpringerPlus*, *2*(1), 247. doi:10.1186/2193-1801-2-247 PMID:23847752

Moura, M. R. D., Aouda, F. A., & Mattoso, L. H. C. (2008). Preparation of chitosan nanoparticles using methacrylic acid. *Journal of Colloid and Interface Science*, *321*(2), 477–483. doi:10.1016/j.jcis.2008.02.006 PMID:18295778

Mousavi, S.R., & Rezaei, M. (2011). Nanotechnology in Agriculture and Food Production. *Journal of Applied Environmental and Biological Sciences,* 415.

Mozafari, A. A., Ghadakchi Asl, A., & Ghaderi, N. (2018). Grape response to salinity stress and role of iron nanoparticle and potassium silicate to mitigate salt induced damage under in vitro conditions. *Physiology and Molecular Biology of Plants*, *24*(1), 25–35. doi:10.100712298-017-0488-x PMID:29398836

Mukherjee, R., Kumar, R., Sinha, A., Lama, Y., & Saha, A. K. (2016). A review on synthesis, characterization, and applications of nano zero valent iron (nZVI) for environmental remediation. *Critical Reviews in Environmental Science and Technology*, *46*(5), 443–466. doi:10.1080/10643389.2015.1103832

Muncke, J. (2011). Endocrine disrupting chemicals and other substances of concern in food contact materials: An updated review of exposure, effect and risk assessment. *The Journal of Steroid Biochemistry and Molecular Biology, 127*, 118–127.

Mura, S., Seddaiu, G., Bacchini, F., Roggero, P.P., & Greppi, G.F. (2013). Advances of nanotechnology in agro-environmental studies. *Italian Journal of Agronomy, 8,* 131-134, 135-136.

Murray, A. R., Kisin, E., Inman, A., Young, S. H., Muhammed, M., Burks, T., Uheida, A., Tkach, A., Waltz, M., Castranova, V., Fadeel, B., Kagan, V. E., Riviere, J. E., Monteiro-Riviere, N., & Shvedova, A. A. (2013). Oxidative stress and dermal toxicity of iron oxide nanoparticles in vitro. *Cell Biochemistry and Biophysics*, *67*(2), 461–476. doi:10.100712013-012-9367-9 PMID:22669739

Musial, J., Krakowiak, R., Mlynarczyk, D. T., Goslinski, T., & Stanisz, B. J. (2020). Titanium dioxide nanoparticles in food and personal care products -What do we know about their safety? *Nanomaterials (Basel, Switzerland)*, *10*(6), 1110. doi:10.3390/nano10061110 PMID:32512703

Muthu, K., & Priya, S. (2017). Green synthesis, characterization and catalytic activity of silver nanoparticles using *Cassia auriculata* flower extract separated fraction. *Spectrochimica Acta A. Molecular & Biomolecular*, *15*(179), 66–72. doi:10.1016/j.saa.2017.02.024

Mu, X., Wang, J.-Y., Bai, X., Xu, F., Liu, H., Yang, J., Jing, Y., Liu, L., Xue, X., Dai, H., Liu, Q., Sun, Y.-M., Liu, C., & Zhang, X.-D. (2017). Black Phosphorus Quantum Dot induced Oxidative Stress and Toxicity in Living Cells and Mice. *ACS Applied Materials & Interfaces*, *9*(24), 20399–20409. doi:10.1021/acsami.7b02900 PMID:28553710

Nafees, M., Ikram, M., & Ali, S. (2017). Thermal stability of lead sulfide and lead oxide nano-crystalline materials. *Applied Nanoscience*, *7*(7), 399–406. doi:10.100713204-017-0578-7

Nagarajan, R. (2008). *Nanoparticles: Building blocks for nanotechnology*. ACS Publications.

Naikoo, G. A., Arshad, F., Almas, M., Hassan, I. U., Pedram, M. Z., Aljabali, A. A. A., Mishra, V., Serrano-Aroca, A., Birkett, M., Charbe, N. B., Goyal, R., Negi, P., El-Tanani, M., & Tambuwala, M. M. (2022). 2D Materials, Synthesis, Characterization and Toxicity: A Critical Review. *Chemico-Biological Interactions*, *365*, 110081. doi:10.1016/j.cbi.2022.110081 PMID:35948135

Nair, P. M. G. (2018). Toxicological impact of carbon nanomaterials on plants. In K. M. Gothandam, S. Ranjan, N. Dasgupta, C. Ramalingam, & E. Lichtfouse (Eds.), *Nanotechnology, Food Security and Water Treatment* (pp. 163–183). Springer International Publishing. doi:10.1007/978-3-319-70166-0_5

Nair, P., & Alam, M. (2006). Performance limits of nanobiosensors. *Applied Physics Letters, 88*(23), 233120. doi:10.1063/1.2211310

Nair, P., & Chung, I. M. (2017). Regulation of morphological, molecular and nutrient status in Arabidopsis thaliana seedlings in response to ZnO nanoparticles and Zn ion exposure. *The Science of the Total Environment, 575*, 187–198. doi:10.1016/j.scitotenv.2016.10.017 PMID:27741454

Nair, R., Varghese, S. H., Nair, B. G., Maekawa, T., Yoshida, Y., & Kumar, D. S. (2010). Nanoparticulate material delivery to plants. *Plant Science, 179*(3), 154–163. doi:10.1016/j.plantsci.2010.04.012

Najahi-Missaoui, W., Arnold, R. D., & Cummings, B. S. (2021). Safe Nanoparticles: Are We There Yet? *International Journal of Material Science, 22*(1), 385. doi:10.3390/ijms22010385 PMID:33396561

Namvar, F., Rahman, H. S., Mohamad, R., Azizi, S., Tahir, P. M., Chartrand, M. S., & Yeap, S. K. (2015). Cytotoxic effects of biosynthesized zinc oxide nanoparticles on murine cell lines. *Evidence-Based Complementary and Alternative Medicine, 593014*, 1–11. Advance online publication. doi:10.1155/2015/593014 PMID:25784947

Naqvi, S., Samim, M., Abdin, M. Z., Ahmed, F. J., Maitra, A. N., Prashant, C. K., & Dinda, A. K. (2010). Concentration-dependent toxicity of iron oxide nanoparticles mediated by increased oxidative stress. *International Journal of Nanomedicine, 5*, 983. doi:10.2147/IJN.S13244 PMID:21187917

Naraghi, L., Negahban, M., Heydari, A., Razavi, M., & Afshari-Azad, H. (2018). Growth inhibition of *Fusarium oxysporum* f. sp. *lycopercisi*, the causal agent of Tomato Fusarium Wilt disease by nano formulations containing *Talaromyces flavus. Ekoloji, 27*(106), 103–112.

Navarro Gallon, S. M., Alpaslan, E., Wang, M., Larese-Casanova, P., Londono, M. E., Atehortua, L., Pavon, J. J., & Webster, T. J. (2019). Characterization and study of the antibacterial mechanisms of silver nanoparticles prepared with microalgal exopolysaccharides. *Materials Science and Engineering C, 99*, 685–695. doi:10.1016/j.msec.2019.01.134 PMID:30889742

Navarro, E., Baun, A., Behra, R., Hartmann, N. B., Filser, J., Miao, A. J., Quigg, A., Santschi, P. H., & Sigg, L. (2008). Environmental behavior and ecotoxicity of engineered nanoparticles to algae, plants, and fungi. *Ecotoxicology (London, England), 17*(5), 372–386. doi:10.100710646-008-0214-0 PMID:18461442

Navarro-Espinoza, S., Meza-Figueroa, D., Guzman, R., Duarte-Moller, A., Esparza-Ponce, H., Paz-Moreno, F., & Pedroza-Montero, M. (2021). Release of nanoparticles in the environment and catalytic converters ageing. *Nanomaterials (Basel, Switzerland), 11*(12), 3406. doi:10.3390/nano11123406 PMID:34947754

Navarro-León, E., Albacete, A., de la Torre-González, A., Ruiz, J. M., & Blasco, B. (2016). Phytohormone profile in *Lactuca sativa* and *Brassica oleracea* plants grown under Zn deficiency. *Phytochemistry, 130*, 85–89. doi:10.1016/j.phytochem.2016.08.003 PMID:27543253

Navrotsky, A. (2004). Environmental nanoparticles. In S. C. I. Contescu (Ed.), *Dekker encyclopedia of nanoscience and nanotechnology* (Vol. 2, pp. 1147–1155). CRC Press.

Neethirajan, S., & Jayas, D. S. (2011). Nanotechnology for the food and bioprocessing industries. *Food and Bioprocess Technology, 4*(1), 39–47. doi:10.100711947-010-0328-2 PMID:32215165

Nel, A. E., Madler, L., Velegol, D., Xia, T., Hoek, E. M. V., Somasundaran, P., Klaessig, F., Castranova, V., & Thompson, M. (2009). Understanding bio physicochemical interactions at the nano- bio- interface. *Nature Materials*, *8*(7), 543–557. doi:10.1038/nmat2442 PMID:19525947

Nel, A. E., Xia, T., Mädler, L., & Li, N. (2006). Toxic potential of materials at the nanolevel. *Science*, *311*(5761), 622–627. doi:10.1126cience.1114397 PMID:16456071

Newbold, R. R., Padilla-Banks, E., Jefferson, W. N., & Heindel, J. J. (2008). Effects of endocrinedisruptors on obesity. *International Journal of Andrology*, *31*(2), 201–208. doi:10.1111/j.1365-2605.2007.00858.x PMID:18315718

Ng, C. T., Yong, L. Q., Hande, M. P., Ong, C. N., Yu, L. E., Bay, B. H., & Baeg, G. H. (2017). Zinc oxide nanoparticles exhibit cytotoxicity and genotoxicity through oxidative stress responses in human lung fibroblasts and Drosophila melanogaster. *International Journal of Nanomedicine, 12*, 1621-1637. doi:.s124403 doi:10.2147/ijn

Ng, D. Q., Chu, Y., Tan, S. W., Wang, S. L., Lin, Y. P., Chu, C. H., Soo, Y.-L., Song, Y.-F., & Chen, P. J. (2019). In vivo evidence of intestinal lead dissolution from lead dioxide (PbO 2) nanoparticles and resulting bioaccumulation and toxicity in medaka fish. *Environmental Science. Nano*, *6*(2), 580–591. doi:10.1039/C8EN00893K

Nguyen, K.T. (2011). *Targeted nanoparticles for cancer therapy: Promises and challenge*. Academic Press.

Nicholson, J. W. (1996). The chemistry of polymers. The Royal Society of Chemistry.

Nihei, H., Matsuoka, F., & Okamoto, A. (2009). Targeted coherent control inside photonic band gaps for nanophotonics devices. In *2009 14th OptoElectronics and Communications Conference* (pp. 1-2). Institute of Electrical and Electronics Engineers. 10.1109/OECC.2009.5220789

Noreen, S., Fatima, Z., Ahmad, S., Athar, H. R., & Ashraf, M. (2018). Foliar application of micronutrients in mitigating abiotic stress in crop plants. In M. Hasanuzzaman, M. Fujita, H. Oku, K. Nahar & B. Hawrylak-Nowak (Eds.), Plant nutrients and abiotic stress tolerance (pp. 95-117). Springer. doi:10.1007/978-981-10-9044-8_3

O'Hagan, D. T. (1996). The intestinal uptake of particles and the implications for drug and antigen delivery. *Journal of Anatomy*, *189*, 477–482. PMID:8982819

Oberbek, P., Kozikowski, P., Czarnecka, K., Sobiech, P., Jakubiak, S., & Jankowski, T. (2019). Inhalation exposure to various nanoparticles in work environment—Contextual information and results of measurements. *Journal of Nanoparticle Research*, *21*(11), 222. doi:10.100711051-019-4651-x

OCED. (2004). *Emission scenario document on plastic additives. Series on emission scenariodocuments No. 3*. Environmental Directorate. OECD Environmental Health and SafetyPublications.

Odzak, N., Kistler, D., Behra, R., & Sigg, L. (2015). Dissolution of metal and metal oxide nanoparticles under natural freshwater conditions. *Environmental Chemistry*, *12*(2), 138–148. doi:10.1071/EN14049

Ogolo, N., Olafuyi, O., & Onyekonwu, M. (2012). Enhanced oil recovery using nanoparticles. In *SPE Saudi Arabia section technical symposium and exhibition*. OnePetro.

Okada, H., Tokunaga, T., Liu, X., Takayanagi, S., Matsushima, A., & Shimohigashi, Y. (2008). Direct evidence revealing structural elements essential for the high binding ability of bisphenol A to human estrogen-related receptor. *Environmental Health Perspectives*, *116*(1), 32–38. doi:10.1289/ehp.10587 PMID:18197296

Oliva, J., & Pablo, J. (2011). Removal of cadmium, copper, nickel, cobalt and mercury from water by Apatite II (TM): Column experiments. *Journal of Hazardous Materials*, *194*, 312–323. doi:10.1016/j.jhazmat.2011.07.104 PMID:21871722

Omara, A. E. D., Elsakhawy, T., Alshaal, T., El-Ramady, H., Kovacs, Z., & Fari, M. (2019). Nanoparticles: A novel approach for sustainable agro-productivity. Environment. *Biodiversity and Soil Security*, *3*(2019), 29–62. doi:10.21608/jenvbs.2019.7478.1050

Ombódi, A., & Saigusa, M. (2000). Broadcast application versus band application of polyolefin- coated fertilizer on green peppers grown on andisol. *Journal of Plant Nutrition*, *23*(10), 1485–1493. doi:10.1080/01904160009382116

Ong, W. T. J., & Nyam, L. L. (2022). Evaluation of silver nanoparticles in cosmeceutical and potential biosafety complications. *Saudi Journal of Biological Sciences*, *29*(4), 2085–2094. doi:10.1016/j.sjbs.2022.01.035 PMID:35531241

Osman, I. F., Baumgartner, A., Cemeli, E., Fletcher, J. N., & Anderson, D. (2010). Genotoxicity and cytotoxicity of zinc oxide and titanium dioxide in HEp-2 cells. *Nanomedicine (London)*, *5*(8), 1193–1203. doi:10.2217/nnm.10.52 PMID:21039197

Osmond-McLeod, M. J., Osmond, R. I., Oytam, Y., McCall, M. J., Feltis, B., Mackay-Sim, A., Wood, S. A., & Cook, A. L. (2013). Surface coatings of ZnO nanoparticles mitigate differentially a host of transcriptional, protein and signalling responses in primary human olfactory cells. *Particle and Fibre Toxicology*, *10*(1), 54. doi:10.1186/1743-8977-10-54 PMID:24144420

Oszlanczi, G., Papp, A., Szabo, A., Nagymajtenyi, L., Sapi, A., Konya, Z., Paulik, E., & Vezer, T. (2011). Nervous system effects in rats on subacute exposure by lead-containing nanoparticles via the airways. *Inhalation Toxicology*, *23*(4), 173–181. doi:10.3109/08958378.2011.553248 PMID:21456951

Othman, A. M., Elsayed, M. A., Al-Balakocy, N. G., Hassan, M. M., & Elshafei, A. M. (2019). Biosynthesis and characterization of silver nanoparticles induced by fungal proteins and its application in different biological activities. *Journal of Genetic Engineering and Biotechnology*, *17*(1), 8. doi:10.118643141-019-0008-1 PMID:31673864

Otles, S., & Yalcin, B. (2010). Nano-biosensors as new tool for detection of food quality and safety. *Log Forum*, *6*, 67–70.

Ou, L., Song, B., Liang, H., Liu, J., Feng, X., Deng, B., Sun, T., & Shao, L. (2016). Toxicity of Graphene-Family Nanoparticles: A General Review of the Origins and Mechanisms. *Particle and Fibre Toxicology*, *13*(1), 57. doi:10.118612989-016-0168-y PMID:27799056

Oves, M., Aslam, M., Rauf, M. A., Qayyum, S., Qari, H. A., Khan, M. S., Alam, M. Z., Tabrez, S., Pugazhendhi, A., & Ismail, I. M. (2018). Antimicrobial and anticancer activities of silver NPs synthesized from the root hair extract of *Phoenix dactylifera*. *Materials Science and Engineering C*, *89*, 429–443. doi:10.1016/j.msec.2018.03.035 PMID:29752116

Pachiega, R., Rodrigues, M. F., Rodrigues, C. V., Sakamoto, I. K., de Oliveira, J. E., Varesche, M. B. A., & Maintinguer, S. I. (2019). Hydrogen bioproduction with anaerobic bacteria consortium from brewery wastewater. *International Journal of Hydrogen Energy*, *44*(1), 155–163. doi:10.1016/j.ijhydene.2018.02.107

Padinjarathil, H., Joseph, M. M., Unnikrishnan, B. S., Preethi, G. U., Shiji, R., Archana, M. G., Maya, S., Syama, H. P., & Sreelekha, T. T. (2018). Galactomannan endowed biogenic silver NPs exposed enhanced cancer cytotoxicity with excellent biocompatibility. *International Journal of Biological Macromolecules*, *118*, 1174–1182. doi:10.1016/j.ijbiomac.2018.06.194 PMID:30001604

Pai, S., Kini, M. S., & Selvaraj, R. (2021). A review on adsorptive removal of dyes from wastewater by hydroxyapatite nanocomposites. *Environmental Science and Pollution Research International*, *28*(10), 11835–11849. doi:10.100711356-019-07319-9 PMID:31867692

Paladini, F., & Pollini, M. (2019). Antimicrobial silver nanoparticles for wound healing application: Progress and future trends. *Materials (Basel)*, *12*(16), 2540. doi:10.3390/ma12162540 PMID:31404974

Palmqvist, N., Seisenbaeva, G. A., Svedlindh, P., & Kessler, V. G. (2017). Maghemite nanoparticles acts as nanozymes, improving growth and abiotic stress tolerance in *Brassica napus*. *Nanoscale Research Letters*, *12*(1), 631. doi:10.118611671-017-2404-2 PMID:29260423

Pandey, G. (2018). Challenges and future prospects of agri-nanotechnology for sustainable agriculture in India. *Environmental Technology & Innovation*, *11*, 299–307. doi:10.1016/j.eti.2018.06.012

Pandey, M. D. (2022). Perspective of nanomaterials for sustainable biofuel and bioenergy production. *Materials Letters*, *313*, 131686. doi:10.1016/j.matlet.2022.131686

Pandian, S. R. K., Deepak, V., Kalishwaralal, K., & Gurunathan, S. (2011). Biologically synthesized fluorescent CdS NPs encapsulated by PHB. *Enzyme and Microbial Technology*, *48*(4-5), 319–325. doi:10.1016/j.enzmictec.2011.01.005 PMID:22112944

Pandit, S., Karunakaran, S., Boda, S. K., Basu, B., & De, M. (2016). High Antibacterial Activity of Functionalized Chemically Exfoliated MoS_2. *ACS Applied Materials & Interfaces*, *8*(46), 31567–31573. doi:10.1021/acsami.6b10916 PMID:27933975

Pandurangi, R. S., Seehra, M. S., Razzaboni, B. L., & Bolsaitis, P. (1990). Surface and bulk infrared modes of crystalline and amorphorous silica particles: A study of the relation of surfacestructure to cytotoxicity of respirable silica. *Environmental Health Perspectives*, *86*, 327–336. doi:10.1289/ehp.9086327 PMID:2169410

Papadaki, D., Kiriakidis, G., & Tsoutsos T. (2018). Applications of nanotechnology in construction industry. *Fundamentals of Nanoparticles*, 343-370. doi:10.1016/B978-0-323-51255-8.00011-2

Papp, A., Oszlanczi, G., Horvath, E., Paulik, E., Kozma, G., Sapi, A., Konya, Z., & Szabo, A. (2012). Consequences of subacute intratracheal exposure of rats to cadmium oxide nanoparticles: Electrophysiological and toxicological effects. *Toxicology and Industrial Health*, *28*(10), 933–941. doi:10.1177/0748233711430973 PMID:22246446

Pappo, J., Ermak, T. H., & Steger, H. J. (1989). Monoclonal antibody directed targeting of fluorescent microspheres to Peyer's patches M cells. *Immunology*, *73*, 277–280. PMID:1879877

Papp, T., Schiffmann, D., Weiss, D., Castranova, V., Vallyathan, D. V., & Rahman, Q. (2008). Human health implications of nanomaterial exposure. *Nanotoxicology*, *2*(1), 9–27. doi:10.1080/17435390701847935

Parada, J., Rubilar, O., Sousa, D. Z., Martínez, M., Fernández-Baldo, M. A., & Tortella, G. R. (2019). Short term changes in the abundance of nitrifying microorganisms in a soil-plant system simultaneously exposed to copper nanoparticles and atrazine. *The Science of the Total Environment*, *670*, 1068–1074. doi:10.1016/j.scitotenv.2019.03.221 PMID:31018422

Park, M., Kim, C. Y., Lee, D. H., Choi, C. L., Choi, J., Lee, S. R., & Choy, J. H. (2004). Intercalation of magnesium urea complex into swelling clay. *Journal of Physics and Chemistry of Solids*, *65*(2-3), 409–412. doi:10.1016/j.jpcs.2003.09.011

Park, S., Croteau, P., Boering, K. A., Etheridge, D. M., Ferretti, D., Fraser, P. J., Kim, K. R., Krumme, P. B., Langenfelds, R. L., Ommen, T. D. V., Steele, L. P., & Trudinger, C. M. (2012). Trends and seasonal cycles in the isotopic composition of nitrous oxide since 1940. *Nature Geoscience*, *5*(4), 261–265. doi:10.1038/ngeo1421

Park, S., Sung, H. K., & Kim, Y. (2016). Green synthesis of metal nanoparticles using sprout plants: Pros and cons. *Journal of Nanoscience and Nanotechnology*, *6*(5), 4444–4449. doi:10.1166/jnn.2016.10970 PMID:27483771

Parolo, C., & Merkoçi, A. (2013). Paper- based nanobiosensors for diagnostics. *Chemical Society Reviews*, *42*(2), 450–457. doi:10.1039/C2CS35255A PMID:23032871

Parsons, J. G., Lopez, M. L., Gonzalez, C. M., Peralta-Videa, J. R., & Gardea-Torresdey, J. L. (2010). Toxicity and biotransformation of uncoated and coated nickel hydroxide nanoparticles on mesquite plants. *Environmental Toxicology and Chemistry*, *29*(5), 1146–1154. doi:10.1002/etc.146 PMID:20821551

Parveen, A., Yalagatti, M. S., Abbaraju, V., & Deshpande, R. (2018). Emphasized mechanistic antimicrobial study of biofunctionalized silver nanoparticles on model *Proteus mirabilis*. *Journal of Drug Delivery*, *2018*, 3850139. doi:10.1155/2018/3850139 PMID:29951316

Passian, A., & Imam, N. (2019). Nanosystems, edge computing, and the next generation computing systems. *Sensors (Basel)*, *19*(18), 4048. doi:10.339019184048 PMID:31546907

Pathak, R. K., Kolishetti, N., & Dhar, S. (2015). Targeted nanoparticles in mitochondrial medicine. *Wiley Interdisciplinary Reviews. Nanomedicine and Nanobiotechnology*, *7*(3), 315–329. doi:10.1002/wnan.1305 PMID:25348382

Patil, S. S., Kore, K., & Kumar, P. (2009). Nanotechnology and its applications in veterinary and animal science. *Veterinary World*, *2*(1), 475–477. doi:10.5455/vetworld.2009.475-477

Patil, S., & Muthusamy, O. (2020). A bio-inspired approach of formulation and evaluation of *Aegle marmelos* fruit extract mediated silver nanoparticle gel and comparison of its antibacterial activity with antiseptic cream. *European Journal of Integrative Medicine*, *33*, 101025. doi:10.1016/j.eujim.2019.101025

Pati, R., Das, I., Mehta, R. K., Sahu, R., & Sonawane, A. (2016). Zinc-Oxide Nanoparticles Exhibit Genotoxic, Clastogenic, Cytotoxic and Actin Depolymerization Effects by Inducing Oxidative Stress Responses in Macrophages and Adult Mice. *Toxicological Sciences*, *150*(2), 454–472. doi:10.1093/toxsci/kfw010 PMID:26794139

Patolsky, F., Zheng, G., & Lieber, C. M. (2006). *Nanowire-based biosensors*. ACS Publications. doi:10.1021/ac069419j

Patra, J. K., & Baek, K. H. (2014). Green nanobiotechnology: Factors affecting synthesis and characterization techniques. *Journal of Nanomaterials*, *2014*, 1–12. doi:10.1155/2014/417305

Patra, J. K., Das, G., Fraceto, L. F., Campos, E. V. R., Rodriguez-Torres, M. del P., Acosta-Torres, L. S., Diaz-Torres, L. A., Grillo, R., Swamy, M. K., Sharma, S., Habtemariam, S., & Shin, H.-S. (2018). Nano based drug delivery systems: Recent developments and future prospects. *Journal of Nanobiotechnology*, *16*(1), 71. doi:10.118612951-018-0392-8 PMID:30231877

Patra, J. K., & Gouda, S. (2013). Application of nanotechnology in textile engineering: An overview. *Journal of Engineering and Technology Research*, *5*(5), 104–111. doi:10.5897/JETR2013.0309

Patra, S., Mishra, P., Mahapatra, S. C., & Mithun, S. K. (2016). Modelling impacts of chemical fertilizer on agricultural production: A case study on Hooghly district, West Bengal, India. *Modeling Earth Systems and Environment*, *2*(4), 1–11. doi:10.100740808-016-0223-6

Pavlicek, A., Part, F., Rose, G., Praetorius, A., Miernicki, M., Gazsó, A., & Huber-Humer, M. (2021). A European nano-registry as a reliable database for quantitative risk assessment of nanomaterials? A comparison of national approaches. *NanoImpact*, *21*, 100276. doi:10.1016/j.impact.2020.100276 PMID:35559789

Pawelczyk, E., Arbab, A. S., Chaudhry, A., Balakumaran, A., Robey, P. G., & Frank, J. A. (2008). In vitro model of bromodeoxyuridine or iron oxide nanoparticle uptake by activated macrophages from labeled stem cells: Implications for cellular therapy. *Stem Cells (Dayton, Ohio)*, *26*(5), 1366–1375. doi:10.1634temcells.2007-0707 PMID:18276802

Pazhanisamy, K., & Indra, N. (2007). Toxic effects of arsenic on protein content in the fish, Labeo rohita (Hamilton). Nature. *Environment and Pollution Technology*, *6*(1), 113–116.

Peighambardoust, S. J., Aghamohammadi-Bavil, O., Foroutan, R., & Arsalani, N. (2020). Removal of malachite green using carboxymethyl cellulose-g-polyacrylamide/montmorillonite nanocomposite hydrogel. *International Journal of Biological Macromolecules*, *159*, 1122–1131. doi:10.1016/j.ijbiomac.2020.05.093 PMID:32422256

Peixe, T. S., de Souza Nascimento, E., Schofield, K. L., Arcuri, A. S., & Bulcao, R. P. (2015). Nanotoxicology and exposure in the occupational setting. *Occupational Diseases and Environmental Medicine*, *3*(03), 35–48. doi:10.4236/odem.2015.33005

Peng, X. H., Qian, X., Mao, H., Wang, A. Y., Nie, S., & Shin, D. M. (2008). Targeted magnetic iron oxide nanoparticles for tumor imaging and therapy. *International Journal of Nanomedicine*, *3*(3), 311–321. doi:10.2147/IJN.S2824 PMID:18990940

Peppas, N. A. (1991). Physiologically responsive hydrogels. *Journal of Bioactive and Compatible Polymers*, *6*(3), 241–246. doi:10.1177/088391159100600303

Peralta-Videa, J. R., Zhao, L., Lopez-Moreno, M. L., de la Rosa, G., Hong, J., & Gardea-Torresdey, J. L. (2011). Nanomaterials and the environment: A review for the biennium 2008–2010. *Journal of Hazardous Materials*, *186*(1), 1–15. doi:10.1016/j.jhazmat.2010.11.020 PMID:21134718

Pérez-de-Luque, A. (2017). Interaction of nanomaterials with plants: What do we need for real applications in agriculture? *Frontiers in Environmental Science*, *5*, 12. doi:10.3389/fenvs.2017.00012

Perez-Hernandez, H., Perez-Moreno, A., Sarabia-Castillo, C. R., Garcia-Mayagoitia, S., Medina-Perez, G., Lopez-Valdez, F., Campos-Montiel, R. G., Jayanta-Kumar, P., & Fernandez-Luqueno, F. (2021). Ecological Drawbacks of Nanomaterials Produced on an Industrial Scale: Collateral Effect on Human and Environmental Health. *Water, Air, and Soil Pollution*, *232*(10), 435. doi:10.100711270-021-05370-2 PMID:34658457

Perlman, O., Weitz, I. S., & Azhari, H. (2015). Copper oxide nanoparticles as contrast agents for MRI and ultrasound dual-modality imaging. *Physics in Medicine and Biology*, *60*(15), 5767–5783. doi:10.1088/0031-9155/60/15/5767 PMID:26159685

Pesch, B., Haerting, J., Ranft, U., Klimpel, A., Oelschlägel, B., & Schill, W. (2000). Occupational risk factors for renal cell carcinoma: Agent-specific results from a case-control study in Germany. MURC Study Group. Multicenter urothelial and renal cancer study. *International Journal of Epidemiology*, *29*(6), 1014–1024. doi:10.1093/ije/29.6.1014 PMID:11101542

Pham, V. T. H., Truong, V. K., Quinn, M. D. J., Notley, S. M., Guo, Y., Baulin, V. A., Al Kobaisi, M., Crawford, R. J., & Ivanova, E. P. (2015). Graphene induces Formation of Pores that Kill Spherical and Rod-shaped Bacteria. *ACS Nano*, *9*(8), 8458–8467. doi:10.1021/acsnano.5b03368 PMID:26166486

Pirzada, M., & Altintas, Z. (2019). Nanomaterials for healthcare biosensing applications. *Sensors (Basel)*, *19*(23), 5311. doi:10.339019235311 PMID:31810313

Pisanic, T. R. II, Blackwell, J. D., Shubayev, V. I., Finones, R. R., & Jin, S. (2007). Nanotoxicity of iron oxide nanoparticle internalization in growing neurons. *Biomaterials*, *28*(16), 2572–2581. doi:10.1016/j.biomaterials.2007.01.043 PMID:17320946

PlasticsEurope. (2013). *Plastics, the facts 2013*. www.plasticseurope.org

Poland, C. A., Duffin, R., Kinloch, I., Maynard, A., Wallace, W. A. H., Seaton, A., Stone, V., Brown, S., MacNee, W., & Donaldson, K. (2008). Carbon nanotubes introduced into the abdominal cavity of mice show asbestos-like pathogenicity in a pilot study. *Nature Nanotechnology*, *23*(7), 423–428. doi:10.1038/nnano.2008.111 PMID:18654567

Popov, V. N. (2004). Carbon nanotubes: Properties and application. *Materials Science and Engineering R Reports*, *43*(3), 61–102. doi:10.1016/j.mser.2003.10.001

Pozo, D. (2010). *Silver nanoparticles*. BoD–Books on Demand. doi:10.5772/186

Prabhu, S., & Poulose, E. K. (2012). Silver nanoparticles: Mechanism of antimicrobial action, synthesis, medical applications, and toxicity effects. *International Nano Letters*, *2*(1), 32. doi:10.1186/2228-5326-2-32

Pradhan, S., & Mailapalli, D. R. (2017). Interaction of engineered nanoparticles with the agri-environment. *Journal of Agricultural and Food Chemistry*, *65*(38), 8279–8294. doi:10.1021/acs.jafc.7b02528 PMID:28876911

Prakash, B., Kujur, A., Yadav, A., Kumar, A., Singh, P. P., & Dubey, N. K. (2018). Nanoencapsulation: An efficient technology to boost the antimicrobial potential of plant essential oils in food system. *Food Control*, *89*, 1–11. doi:10.1016/j.foodcont.2018.01.018

Prasad, R., Bhattacharyya, A., & Nguyen, Q. D. (2017). Nanotechnology in sustainable agriculture: Recent developments, challenges, and perspectives. *Frontiers in Microbiology*, *8*, 1014. doi:10.3389/fmicb.2017.01014 PMID:28676790

Prasad, R., Kumar, V., & Prasad, K. S. (2014). Nanotechnology in sustainable agriculture: Present concerns and future aspects. *African Journal of Biotechnology*, *13*, 709.

Prasad, T. N. V. K. V., Sudhakar, P., Sreenivasulu, Y., Latha, P., Munaswamy, V., Reddy, R. K. T. S., Sreeprasad, P. R., Sajanlal, P. T., & Pradeep, T. (2012). Effect of nanoscale zinc oxide particles on the germination, growth and yield of Peanut. *Journal of Plant Nutrition*, *35*(6), 905–927. doi:10.1080/01904167.2012.663443

Prasanna, B.M. (2007). *Nanotechnology in agriculture*. ICAR National Fellow, Division of Genetics, I.A.R.I.

Prazak, R., Święciło, A., Krzepiłko, A., Michałek, S., & Arczewska, M. (2020). Impact of Ag nanoparticles on seed germination and seedling growth of green beans in normal and chill temperatures. *Agriculture*, *10*(8), 312. doi:10.3390/agriculture10080312

Predicala, B. (2009). *Nanotechnology: Potential for agriculture*. Prairie Swine Centre Inc., University of Saskatchewan.

Price, R. R., & Wagner, A. (2008). A method for treating agricultural crops using materials associated with tubular carriers. Academic Press.

Pumera, M. (2010). Nanotoxicology: The molecular science point of view. *Chemistry, an Asian Journal*, *6*(2), 340–348. doi:10.1002/asia.201000398 PMID:20725923

Pumera, M., Sanchez, S., Ichinose, I., & Tang, J. (2007). Electrochemical nanobiosensors. *Sensors and Actuators. B, Chemical*, *123*(2), 1195–1205. doi:10.1016/j.snb.2006.11.016

Qafoku, N. P. (2010). Terrestrial nanoparticles and their controls on soil geo-processes and reactions. *Advances in Agronomy*, *107*, 33–91. doi:10.1016/S0065-2113(10)07002-1

Qiao, H., Liu, W., Gu, H., Wang, D., & Wang, Y. (2015). The transport and deposition of nanoparticles in respiratory system by inhalation. *Journal of Nanomaterials*, *2015*, 1–8. Advance online publication. doi:10.1155/2015/394507

Qing, Y., Cheng, L., Li, R., Liu, G., Zhang, Y., Tang, X., Wang, J., Liu, H., & Qin, Y. (2018). Potential antibacterial mechanism of silver nanoparticles and the optimization of orthopedic implants by advanced modification technologies. *International Journal of Nanomedicine*, *13*, 3311–3327. doi:10.2147/IJN.S165125 PMID:29892194

Quazi, S. (2021a). *Anti-cancer activity of human gastrointestinal bacteria*. Academic Press.

Quazi, S. (2021b). *Anti-cancer activity of human gastrointestinal bacteria*. Academic Press.

Quazi, S. (2021b). *TNFR2 antagonist and agonist: A potential therapeutics in Cancer Immunotherapy*. Academic Press.

Quazi, S. (2022b). *Application of Biosensors in Cancers, An Overview*. Academic Press.

Quazi, S. (2022e). *The potential implementation of biosensors for the diagnosis of biomarkers of various cancer*. Academic Press.

Quazi, S. (2022f). *Application of Biosensors in Cancers, An Overview*. Academic Press.

Quazi, S. (2022f). *The potential implementation of biosensors for the diagnosis of biomarkers of various cancer*. Academic Press.

Quazi, S. (2022g). Artificial intelligence and machine learning in medicinal chemistry and validation of emerging drug targets. *Advancements in Controlled Drug Delivery Systems,* 27-43.

Quazi, S. (2021a). TNFR2 antagonist and agonist: A potential therapeutics in Cancer. *Immunotherapy*.

Quazi, S. (2021c). In-silico pharmacophore and Molecular Docking based drug discovery against Marburg Virus's Viral Protein 35; A potent of MAVD. *bioRxiv*. Advance online publication. doi:10.1101/2021.07.01.450693

Quazi, S. (2021d). In-silico Structural and Molecular Docking-Based Drug Discovery Against Viral Protein (VP35) of Marburg Virus: A potent Agent of MAVD. *bioRxiv*. Advance online publication. doi:10.1101/2021.02.09.430405

Quazi, S. (2022a). Artificial intelligence and machine learning in precision and genomic medicine. *Medical Oncology, 39*(8), 1–18. doi:10.100712032-022-01711-1 PMID:35704152

Quazi, S. (2022b). Elucidation of CRISPR-Cas9 application in novel cellular immunotherapy. *Molecular Biology Reports, 49*(7), 1–9. doi:10.100711033-022-07147-0 PMID:35122203

Quazi, S. (2022c). An overview of CAR T cell mediated B cell Maturation Antigen therapy. *Clinical Lymphoma, Myeloma & Leukemia, 22*(6), e392–e404. doi:10.1016/j.clml.2021.12.003 PMID:34992008

Quazi, S. (2022d). Telomerase gene therapy: A remission toward cancer. *Medical Oncology (Northwood, London, England), 39*(6), 1–20. doi:10.100712032-022-01702-2 PMID:35429243

Quazi, S., & Malik, J. (2021). Discovery of potential drug-like compounds against Viral protein (VP40) of Marburg Virus using pharmacophoric based virtual screening from ZINC database. *bioRxiv*. Advance online publication. doi:10.1101/2021.05.13.444037

Quintero-Quiroz, C., Acevedo, N., Zapata-Giraldo, J., Botero, L. E., Quintero, J., Zarate-Trivino, D., Saldarriaga, J., & Perez, V. Z. (2019). Optimization of silver nanoparticle synthesis by chemical reduction and evaluation of its antimicrobial and toxic activity. *Biomaterials Research, 23*(1), 27. doi:10.118640824-019-0173-y PMID:31890269

Rad, F., Mohsenifar, A., Tabatabaei, M., Safarnejad, M. R., Shahryari, F., Safarpour, H., Foroutan, A., Mardi, M., Davoudi, D., & Fotokian, M. (2012). Detection of Candidatus Phytoplasma aurantifolia with a quantum dots fret-based biosensor. *Journal of Plant Pathology, 94*, 525–534.

Radzuan, N. A., Sulong, A. B., & Sahari, J. (2017). A review of electrical conductivity models for conductive polymer composite. *International Journal of Hydrogen Energy, 42*(14), 9262–9273. doi:10.1016/j.ijhydene.2016.03.045

Rafique, M., Sadaf, I., Rafique, M. S., & Tahir, M. B. (2017). A review on green synthesis of silver nanoparticles and their applications. *Artificial Cells, Nanomedicine, and Biotechnology, 45*(7), 1272–1291. doi:10.1080/21691401.2016.1241792 PMID:27825269

Rahimdokht, M., Pajootan, E., & Ranjbar Mohammadi, M. (2019). Titania/gum tragacanth nanohydrogel for methylene blue dye removal from textile wastewater using response surface methodology. *Polymer International*, *68*(1), 134–140. doi:10.1002/pi.5706

Rai, V.S., Acharya, C.C., & Dey, N. (2012). *Implications of nanobiosensors in agriculture*. Academic Press.

Rai, M., Yadav, A., & Gade, A. (2009). Silver nanoparticles as a new generation of antimicrobials. *Biotechnology Advances*, *27*(1), 76–83. doi:10.1016/j.biotechadv.2008.09.002 PMID:18854209

Rai, V., Acharya, S., & Dey, N. (2012). Implications of nano biosensors in agriculture. *J Biomatter Nano Biotechnol*, *3*, 315–324.

Rajonee, A. A., Zaman, S., & Imamul Huq, S. M. (2017). Preparation, characterization and evaluation of efficacy of phosphorus and potassium incorporated nano fertilizer. *Advances in Nanoparticles*, *6*(02), 62–74. doi:10.4236/anp.2017.62006

Rajora, N., Kaushik, S., Jyoti, A., & Kothari, S. L. (2016). Rapid synthesis of silver nanoparticles by *Pseudomonas stutzeri* isolated from textile soil under optimised conditions and evaluation of their antimicrobial and cytotoxicity properties. *IET Nanobiotechnology / IET*, *10*(6), 367–373. doi:10.1049/iet-nbt.2015.0107 PMID:27906136

Rajput, V. D., Minkina, T., Feizi, M., Kumari, A., Khan, M., Mandzhieva, S., Sushkova, S., El-Ramady, H., Verma, K. K., Singh, A., Hullebusch, E., Singh, R. K., Jatav, H. S., & Choudhary, R. (2021). Effects of silicon and silicon-based nanoparticles on rhizosphere microbiome. *Plant Stress and Growth. Biology (Basel)*, *10*(8), 791. doi:10.3390/biology10080791 PMID:34440021

Rajput, V., Minkina, T., Mazarji, M., Shende, S., Sushkova, S., Mandzhieva, S., Burachevskaya, M., Chaplygin, V., Singh, A., & Jatav, H. (2020). Accumulation of nanoparticles in the soil-plant systems and their effects on human health. *Annals of Agricultural Science*, *65*(2), 137–143. doi:10.1016/j.aoas.2020.08.001

Raj, S., & Kumar, D. (2020). Biochemical toxicology: Heavy metals and nanomaterials. *Biochemical Toxicology: Heavy Metals and Nanomaterials*, *139*. Advance online publication. doi:10.5772/intechopen.90928

Raliya, R., Nair, R., Chavalmane, S., Wang, W. N., & Biswas, P. (2015). Mechanistic evaluation of translocation and physiological impact of titanium dioxide and zinc oxide nanoparticles on the tomato (*Solanum lycopersicum* L.) plant. *Metallomics*, *7*(12), 1584–1594. doi:10.1039/C5MT00168D PMID:26463441

Ramirez-Garcia, G., Gutierrez-Granados, S., Gallegos-Corona, M. A., Palma-Tirado, L., Orlye, F., Varenne, A., Mignet, N., Richard, C., & Martinez-Alfaro, M. (2017). Long-term toxicological effects of persistent luminescence nanoparticles after intravenous injection in mice. *International Journal of Pharmaceutics*, *532*(2), 686–695. doi:10.1016/j.ijpharm.2017.07.015 PMID:28705622

Rana, K., Verma, Y., Rani, V., & Rana, S. V. S. (2018). Renal toxicity of nanoparticles of cadmium sulphide in rat. *Chemosphere*, *193*, 142–150. doi:10.1016/j.chemosphere.2017.11.011 PMID:29128560

Rana, S. V. S. (2020). A Comprehensive Assessment of Hepatotoxicity Induced by Engineered Nanoparticles-A Review. *Journal of Toxicology and Risk Assessment*, *6*(1). Advance online publication. doi:10.23937/2572-4061.1510035

Rana, V. S., & Sharma, N. (2020). A Review on polysaccharide based nanocomposite hydrogel systems fabrication using diverse reinforcing materials. *Journal of Polymer & Composites*, *8*(1), 6–17.

Ranjha, M. M. A. N., Shafique, B., Rehman, A., Mehmood, A., Ali, A., Zahra, S. M., Roobab, U., Singh, A., Ibrahim, S. A., & Siddiqui, S. A. (2022). Biocompatible Nanomaterials in Food Science, Technology, and Nutrient Drug Delivery: Recent Developments and Applications. *Frontiers in Nutrition*, *8*, 778155. doi:10.3389/fnut.2021.778155 PMID:35127783

Rao, C., & Cheetham, A. (2001). Science and technology of nanomaterials: Current status and future prospects. *Journal of Materials Chemistry*, *11*(12), 2887–2894. doi:10.1039/b105058n

Rashid, M. M., Forte Tavčer, P., & Tomšič, B. (2021). Influence of Titanium Dioxide Nanoparticles on Human Health and the Environment. *Nanomaterials (Basel, Switzerland)*, *11*(9), 2354. doi:10.3390/nano11092354 PMID:34578667

Rasool, K., Helal, M., Ali, A., Ren, C. E., Gogotsi, Y., & Mahmoud, K. A. (2016). Antibacterial Activity of $Ti_3C_2T_x$ Mxene. *ACS Nano*, *10*(3), 3674–3684. doi:10.1021/acsnano.6b00181 PMID:26909865

Rastogi, A., Tripathi, D. K., Yadav, S., Chauhan, D. K., Živčák, M., Ghorbanpour, M., El-Sheery, N. I., & Brestic, M. (2019). Application of silicon nanoparticles in agriculture. *Biotech*, *9*(3), 90. doi:10.1007/s13205-019-1626-7

Raval, N.P., Shah, P.U., & Shah, N. K. (2016b) Adsorptive amputation of hazardous azo dye Congo red from wastewater: a critical review. *Environ Sci Pollut Res Int, A, 23*(15), 14810-14853. . doi:10.1007/s11356-016-6970-0

Raval, N. P., Shah, P. U., & Shah, N. K. (2016a). Adsorptive removal of Ni^{2+} ions from aqueous environment: A review. *Journal of Environmental Management*, *179*, 1–20. doi:10.1016/j.jenvman.2016.04.045 PMID:27149285

Ray, P. C., Yu, H., & Fu, P. P. (2009). Toxicity and environmental risks of nanomaterials: Challenges and future needs. *Journal of Environmental Science and Health. Part C, Environmental Carcinogenesis & Ecotoxicology Reviews*, *27*(1), 1–35. doi:10.1080/10590500802708267 PMID:19204862

Ray, S. S., & Okamoto, M. (2003). Polymeric/layered silicate nanocomposites: A review frompreparation to processing. *Progress in Polymer Science*, *28*(11), 1539–1641. doi:10.1016/j.progpolymsci.2003.08.002

Raza, M. A., Kanwal, Z., Shahid, A., Fatima, S., Sajjad, A., Riaz, S., & Naseem, S. (2021). Toxicity Evaluation of Arsenic Nanoparticles on Growth, Biochemical, Hematological, and Physiological Parameters of Labeo rohita Juveniles. *Advances in Materials Science*, *5185425*, 1–10. Advance online publication. doi:10.1155/2021/5185425

Reda, F. M., El-Saadony, M. T., El-Rayes, T. K., Attia, A. I., El-Sayed, S. A., Ahmed, S. Y., Madkour, M., & Alagawany, M. (2021). Use of biological nano zinc as a feed additive in quail nutrition: Biosynthesis, antimicrobial activity and its effect on growth, feed utilisation, blood metabolites and intestinal microbiota. *Italian Journal of Animal Science*, *20*(1), 324–335. doi:10.1080/1828051X.2021.1886001

Reddy Pullagurala, V. L., Adisa, I. O., Rawat, S., Kalagara, S., Hernandez-Viezcas, J. A., Peralta-Videa, J. R., & Gardea-Torresdey, J. L. (2018). ZnO nanoparticles increase photosynthetic pigments and decrease lipid peroxidation in soil grown cilantro (*Coriandrum sativum*). *Plant Physiology and Biochemistry*, *132*, 120–127. doi:10.1016/j.plaphy.2018.08.037 PMID:30189415

Reed, M. A. (1993). Quantum dots. *Scientific American*, *268*(1), 118–123. doi:10.1038cientificamerican0193-118

Reimann, S. M., & Manninen, M. (2002). Electronic structure of quantum dots. *Reviews of Modern Physics*, *74*(4), 1283–1342. doi:10.1103/RevModPhys.74.1283

Rezaei, M., & Abbasi, H. (2014). Foliar application of nanochelate and non-nanochelate of zinc on plant resistance physiological processes in cotton (*Gossipium hirsutum* L.). *Indian Journal of Plant Physiology / Official Publication of the Indian Society for Plant Physiology*, *4*(4), 1137–1144.

Rezić, I., Majdak, M., Ljoljić Bilić, V., Pokrovac, I., Martinaga, L., Somogyi Škoc, M., & Kosalec, I. (2021). Development of Antibacterial Protective Coatings Active against MSSA and MRSA on Biodegradable Polymers. *Polymers*, *13*(4), 659. doi:10.3390/polym13040659 PMID:33672118

Rice, V. A. (1942). Breeding and Improvement of Farm Animals (3rd ed.). Academic Press.

Rico, C. M., Majumdar, S., Duarte-Gardea, M., Peralta-Videa, J. R., & Gardea-Torresdey, J. L. (2011). Interaction of nanoparticles with edible plants and their possible implications in the food chain. *Journal of Agricultural and Food Chemistry*, *59*(8), 3485–3498. doi:10.1021/jf104517j PMID:21405020

Riedel, K., Kunze, G., & Baronian, K. H. (2006). *Biosensors and bioassays based on microorganisms*. Research Signpost.

Rizwan, M., & Singh, M. (2014). Ecofriendly application of nanomaterials: Nanobioremediation. *Journal ofNanoparticles*.

Robinson, S., & Nakkeeran, R. (2013). *Photonic crystal ring resonator based optical filters*. Advances in Photonic Crystals. doi:10.5772/54533

Rocha-Meneses, L., Hari, A., Inayat, A., Shanableh, A., Abdallah, M., Ghenai, C., Shanmugam, S., & Kikas, T. (2022). Application of nanomaterials in anaerobic digestion processes: A new strategy towards sustainable methane production. *Biochemical Engineering Journal*, *188*, 108694. doi:10.1016/j.bej.2022.108694

Rochester, J. R. (2013). Bisphenol A and human health: A review of the literature. *ReproductiveToxicology*, *42C*, 132–155. doi:10.1016/j.reprotox.2013.08.008 PMID:23994667

Rodrigues, A. V., de Souza, D. A. R., Garcia, F. D. R., & Ribeiro, S. J. L. (2022). Renewable Energy for a Green Future: Electricity Produced from Efficient Luminescent Solar Concentrators. *Solar Energy Advances*, *2*, 1–13. doi:10.1016/j.seja.2022.100013

Rodrigues, C. V., Alcaraz, F. A. R., Nespeca, M. G., Rodrigues, A. V., Motteran, F., Adorno, M. A. T., Varesche, M. B. A., & Maintinguer, S. I. (2020). Biohydrogen production in an integrated biosystem using crude glycerol from waste cooking oils. *Renewable Energy*, *162*, 701–711. doi:10.1016/j.renene.2020.08.061

Rodrigues, C. V., Santana, K. O., Nespeca, M. G., Rodrigues, A. V., Pires, L. O., & Maintinguer, S. I. (2020). Energy valorization of crude glycerol and sanitary sewage in hydrogen generation by biological processes. *International Journal of Hydrogen Energy*, *45*(21), 11943–11953. doi:10.1016/j.ijhydene.2020.02.168

Rodriguez-Fragoso, P., Reyes-Esparza, J., Leon-Buitimea, A., & Rodriguez-Fragoso, L. (2012). Synthesis, characterization and toxicological evaluation of maltodextrin capped cadmium sulfide nanoparticles in human cell lines and chicken embryos. *Journal of Nanobiotechnology*, *10*(1), 1–11. doi:10.1186/1477-3155-10-47 PMID:23270407

Roduner, E. (2006). Size matters: Why nanomaterials are different. *Chemical Society Reviews*, *35*(7), 583–592. doi:10.1039/b502142c PMID:16791330

Rogers, K. R., Navratilova, J., Stefaniak, A., Bowers, L., Knepp, A. K., Al-Abed, S. R., Potter, P., Gitipour, A., Radwan, I., Nelson, C., & Bradham, K. D. (2018). Characterization of engineered nanoparticles in commercially available spray disinfectant products advertised to contain colloidal silver. *The Science of the Total Environment*, *619–620*, 1375–1384. doi:10.1016/j.scitotenv.2017.11.195 PMID:29723948

Roh, J. Y., Eom, H. J., & Choi, J. (2012). Involvement of *Caenohabditis elegans* MAPK signaling pathways in oxidative stress response induced by silver nanoparticles exposure. *Toxicological Research*, *28*(1), 19–24. doi:10.5487/TR.2012.28.1.019 PMID:24278585

Romeiro, F. C., & Rodrigues, A. V. (2022). Green Synthesis of Semiconductors and Environmental Applications: From Single to High Added Value Products. Handbook of Research on Green Synthesis and Applications of Nanomaterials, 27–50. doi:10.4018/978-1-7998-8936-6.ch002

Romeiro, F. C., Silva, B. C., Martins, A. S., Zanoni, M. V. B., & Orlandi, M. O. (2021). Superior performance of rGO-tin oxide nanocomposite for selective reduction of CO2 to methanol. *Journal of CO2 Utilization, 46*. doi:10.1016/j.jcou.2021.101460

Romeiro, F. C., Castro, N. S., Scolfaro, L., Borges, P. D., & Lima, R. C. (2022). Theoretical and experimental study of effects of Co2+ doping on structural and electronic properties of ZnO. *Journal of Physics and Chemistry of Solids*, *162*, 110501. doi:10.1016/j.jpcs.2021.110501

Ropero, A. B., Alonso Magdalena, P., García-García, E., Ripoll, C., Fuentes, E., & Nadal, A. (2008). Bisphenol A disruption of the endocrine pancreas and blood glucose homeostasis. *International Journal of Andrology*, *31*(2), 194–200. doi:10.1111/j.1365-2605.2007.00832.x PMID:17971160

Rubilar, O., Rai, M., Tortella, G., Diez, M. C., Seabra, A. B., & Duran, N. (2013). Biogenic nanoparticles: Copper, copper oxides, copper sulphides, complex copper nanostructures and their applications. *Biotechnology Letters*, *35*(9), 1365–1375. doi:10.100710529-013-1239-x PMID:23690046

Rui, M., Ma, C., Hao, Y., Guo, J., Rui, Y., Tang, X., Zhao, Q., Fan, X., Zhang, Z., Hou, T., & Zhu, S. (2016). Iron oxide nanoparticles as a potential iron fertilizer for peanut (*Arachis hypogaea*). *Frontiers in Plant Science*, *7*, 815. doi:10.3389/fpls.2016.00815 PMID:27375665

Ryu, H. W., Lee, D. H., Florens, L., Swanson, S. K., Washburn, M. P., & Kwon, S. H. (2014). Analysis of the heterochromatin protein 1 (HP1) interactome in Drosophila. *Journal of Proteomics*, *102*, 137–147. doi:10.1016/j.jprot.2014.03.016 PMID:24681131

S, R., Vasa, T. N., & Sabumon, P. C. (2022). A mini review on effect of nano particles of Fe in the anaerobic digestion of waste activated sludge. *Materials Today: Proceedings*, *51*, 1482–1488. doi:10.1016/j.matpr.2021.10.265

Sadiq, A. C., Olasupo, A., Ngah, W. S. W., Rahim, N. Y., & Mohd Suah, F. B. (2021). A decade development in the application of chitosan-based materials for dye adsorption: A short review. *International Journal of Biological Macromolecules*, *191*, 1151–1163. doi:10.1016/j.ijbiomac.2021.09.179 PMID:34600954

Safarpour, H., Safarnejad, M. R., Tabatabaei, M., Mohsenifar, A., Rad, F., Basra, M., Shahryari, F., & Hasanzadeh, F. (2012). Development of a quantum dots FRET-based biosensor for efficient detection of Polymyxa betae. *Canadian Journal of Plant Pathology*, *34*(4), 507–515. doi:10.1080/07060661.2012.709885

Safyan, N., Reza, M., Darbaghshahi, N., & Bahari, B. (2012). The effect of microelements spraying on growth, qualitative and quantitative grain corn in Iran. *International Research Journal of Applied and Basic Sciences*, *3*, 2780–2784.

Sahoo, M., Vishwakarma, S., Panigrahi, C., & Kumar, J. (2021). Nanotechnology: Current applications and future scope in food. *Food Frontiers*, *2*(1), 3–22. doi:10.1002/fft2.58

Saif, S., Tahir, A., & Chen, Y. (2016). Green synthesis of iron nanoparticles and their environmental applications and implications. *Nanomaterials (Basel, Switzerland)*, *6*(11), 209. doi:10.3390/nano6110209 PMID:28335338

Saifullah, M., Shishir, M. R. I., Ferdowsi, R., Rahman, M. R. T., & Van-Vuong, Q. (2019). Micro and nano encapsulation, retention and controlled release of flavor and aroma compounds: A critical review. *Trends in Food Science & Technology*, *86*, 230–251. doi:10.1016/j.tifs.2019.02.030

Sá, L. R. V., Cammarota, M. C., & Ferreira-Leitão, V. S. (2014). Produção de Hidrogênio via Fermentação anaeróbia - aspectos gerais e possibilidade de utilização de resíduos agroindustriais brasileiros. *Quimica Nova*, *37*(5), 857–867. doi:10.5935/0100-4042.20140138

Salama, H. M. H. (2012). Effects of silver nanoparticles in some crop plants, common bean (*Phaseolus vulgaris* L.) and corn (*Zea mays* L.). *International Research Journal of Biotechnology*, *3*, 190–197.

Sampaio, P. G. V., & González, M. O. A. (2017). Photovoltaic Solar Energy: Conceptual Framework. *Renewable & Sustainable Energy Reviews*, *74*, 590–601. doi:10.1016/j.rser.2017.02.081

Sánchez-López, E., Gomes, D., Esteruelas, G., Bonilla, L., Lopez-Machado, A. L., Galindo, R., Cano, A., Espina, M., Ettcheto, M., Camins, A., Silva, A. M., Durazzo, A., Santini, A., Garcia, M. L., & Souto, E. B. (2020). Metal-based nanoparticles as antimicrobial agents: An overview. *Nanomaterials (Basel, Switzerland)*, *10*(2), 292. doi:10.3390/nano10020292 PMID:32050443

Santos, A. C., Morais, F., Simoes, A., Pereira, I., Sequeira, J. A., Pereira-Silva, M., & Ribeiro, A. (2019). Nanotechnology for the development of new cosmetic formulations. *Expert Opinion on Drug Delivery*, *16*(4), 313–330. doi:10.1080/17425247.2019.1585426 PMID:30793641

Santos, A., Veiga, F., & Figueiras, A. (2019). Dendrimers as pharmaceutical excipients: Synthesis, properties, toxicity and biomedical applications. *Materials (Basel)*, *13*(1), 65. doi:10.3390/ma13010065 PMID:31877717

Santoso, D., Lefroy, R. D. B., & Blair, G. J. (1995). Sulphur and phosphorus dynamics in an acid soil/crop system. *Australian Journal of Soil Research*, *33*(1), 113–124. doi:10.1071/SR9950113

Saod, W.M., Ghafoor, D. A., Najeeb, L. M., Al-Taee, M. J. M. (2020). Study of analytical characterization and antibacterial activity of silver nanoparticles synthesized by using pineapple juice as a reducing agent. *Indian Journal of Forensic Medicine & Toxicology*, 14.

Saravanan, A., Kumar, P. S., Rangasamy, G., Hariharan, R., Hemavathy, R. V., Deepika, P. D., Anand, K., & Karthika, S. (2023). Strategies for enhancing the efficacy of anaerobic digestion of food industry wastewater: An insight into bioreactor types, challenges, and future scope. *Chemosphere*, *310*, 136856. doi:10.1016/j.chemosphere.2022.136856 PMID:36243094

Saravanan, M., Barik, S. K., MubarakAli, D., Prakash, P., & Pugazhendhi, A. (2018). Synthesis of silver nanoparticles from *Bacillus brevis* (NCIM 2533) and their antibacterial activity against pathogenic bacteria. *Microbial Pathogenesis*, *116*, 221–226. doi:10.1016/j.micpath.2018.01.038 PMID:29407231

Sardar, R., Funston, A. M., Mulvaney, P., & Murray, R. W. (2009). Gold nanoparticles: Past, present, and future. *Langmuir*, *25*(24), 13840–13851. doi:10.1021/la9019475 PMID:19572538

Say, R., Garipcan, B., Emir, S., Patir, S., & Denizli, A. (2002). Preparation of poly(hydroxyethyl methacrylate-co-methacrylamidohistidine) beads and its design as a affinity adsorbent for Cu(II) removal from aqueous solutions. *Colloids and Surfaces. A, Physicochemical and Engineering Aspects*, *196*(2-3), 199–207. doi:10.1016/S0927-7757(01)00891-3

Sayyid, N. H., & Zghair, Z. R. (2021). Biosynthesis of silver nanoparticles produced by *Klebsiella pneumonia*. *Materials Today: Proceedings*, *42*, 2045–2049. doi:10.1016/j.matpr.2020.12.257

Schirmer, K., & Auffan, M. (2015). Nanotoxicology in the environment. *Environmental Science. Nano*, *2*(6), 561–563. doi:10.1039/C5EN90023A

Schmalz, G., Hickel, R., van Landuyt, K. L., & Reichl, F. X. (2017). Nanoparticles in dentistry. *Dental Materials*, *33*(11), 1298–1314. doi:10.1016/j.dental.2017.08.193 PMID:28951037

Schmidt, B., Petersen, J. H., Bender Koch, C., Plackett, D., Johansen, N. R., Katiyar, V., & Larsen, E. H. (2009). Combining asymmetrical flow field-flow fractionation with light-scattering andinductively coupled plasma mass spectrometric detection for characterization of nanoclayused in biopolymer nanocomposites. *Food Additives and Contaminants A*, *26*(12), 1619–1627. doi:10.1080/02652030903225740 PMID:19927249

Schulte, P. A., Kuempel, E. D., & Drew, N. M. (2018). Characterizing risk assessments for the development of occupational exposure limits for engineered nanomaterials. *Regulatory Toxicology and Pharmacology*, *95*, 207–219. doi:10.1016/j.yrtph.2018.03.018 PMID:29574195

Schulze, A., Mitterer, F., Pombo, J. P., & Schild, S. (2021). Biofilms by bacterial human pathogens: Clinical relevance-development, composition and regulation-therapeutical strategies. *Microbial Cell*, *8*(2), 28–56. doi:10.15698/mic2021.02.741 PMID:33553418

Schwab, F., Zhai, G., Kern, M., Turner, A., Schnoor, J. L., & Wiesner, M. R. (2016). Barriers, pathways and processes for uptake, translocation and accumulation of nanomaterials in plants—Critical review. *Nanotoxicology*, *10*(3), 257–278. doi:10.3109/17435390.2015.1048326 PMID:26067571

Schwarzenbach, R. P., Escher, B. I., Fenner, K., Hofstetter, T. B., Johnson, C. A., von Gunten, U., & Wehrli, B. (2006). The challenge of micropollutants in aquatic systems. *Science*, *313*(5790), 1072–1077. doi:10.1126cience.1127291 PMID:16931750

Scott, N. R. (2005). Nanotechnology and animal health. Revue scient fique et technique. International Office of Epizootics. doi:10.20506/rst.24.1.1579

Seghatoleslami, M., Feizi, H., Mousavi, G., & Berahmand, A. (2015). Effect of magnetic field and silver nanoparticles on yield and water use efficiency of *Carum copticum* under water stress conditions. *Polish Journal of Chemical Technology*, *17*(1), 110–114. doi:10.1515/pjct-2015-0016

Sekhon, B. S. (2014). Nanotechnology in agri-food production: An overview. *Nanotechnology, Science and Applications*, *7*, 31–53. doi:10.2147/NSA.S39406 PMID:24966671

Seleiman, M. F., Almutairi, K. F., Alotaibi, M., Shami, A., Alhammad, B. A., & Battaglia, M. L. (2021). Nano-Fertilization as an Emerging Fertilization Technique: Why Can Modern Agriculture Benefit from Its Use? *Plants*, *10*(1), 2. doi:10.3390/plants10010002 PMID:33375026

Selva Preetha, P., & Balakrishnan, N. (2017). A Review of nano fertilizers and their use and functions in soil. *International Journal of Current Microbiology and Applied Sciences*, *6*(12), 3117–3133. doi:10.20546/ijcmas.2017.612.364

Semida, W. M., Abdelkhalik, A., Mohamed, G. F., Abd El-Mageed, T. A., Abd El-Mageed, S. A., Rady, M. M., & Ali, E. F. (2021). Foliar application of zinc oxide nanoparticles promotes drought stress tolerance in eggplant (*Solanum melongena* L.). *Plants*, *10*(2), 421. doi:10.3390/plants10020421 PMID:33672429

Sengul, A. B., & Asmatulu, E. (2020). Toxicity of metal and metal oxide nanoparticles: A review. *Environmental Chemistry Letters*, *18*(5), 1659–1683. doi:10.100710311-020-01033-6

Setälä, O., Fleming-Lehtinen, V., & Lehtiniemi, M. (2012). Ingestion and transfer of microplastics in the planktonic food web. *Environmental Pollution*, *185*, 77–83. doi:10.1016/j.envpol.2013.10.013 PMID:24220023

Sexton, K., Needham, L., & Pickles, J. (2004). Human biomonitoring of environmental chemicals. *American Scientist*, *92*(1), 38–45. doi:10.1511/2004.45.921

Shaaban, H. A., & Farouk, A. (2022). Encapsulation of Essential Oils and Their Use in Food Applications. In M. a. de Oliveira & E. e. de Aguiar Andrade (Eds.), *Essential Oils -Advances in Extractions and Biological Applications*. Intech Open. doi:10.5772/intechopen.103147

Shafique, M., & Luo, X. (2019). Nanotechnology in transportation vehicles: An overview of its applications, environmental, health and safety concerns. *Materials (Basel)*, *12*(15), 2493. doi:10.3390/ma12152493 PMID:31390752

Shah, M. A., Pirzada, B. M., Price, G., Shibiru, A. L., & Qurashi, A. (2022). Applications of nanotechnology in smart textile industry: A critical review. *Journal of Advanced Research*, *38*, 55–75. doi:10.1016/j.jare.2022.01.008 PMID:35572402

Shaikh, S., Shyama, S. K., & Desai, P. V. (2015). Absorption, LD50 and Effects of CoO, MgO and PbO Nanoparticles on Mice "Mus musculus". *IOSR Journal of Environmental Science, Toxicology and Food Technology, 9*(2), 32-38. doi:10.9790/2402-09213238

Shang, Y., Hasan, M. K., Ahammed, G. J., Li, M., Yin, H., & Zhou, J. (2019). Applications of nanotechnology in plant growth and crop protection: A review. *Molecules (Basel, Switzerland), 24*(14), 2558. doi:10.3390/molecules24142558 PMID:31337070

Sharaf, M. A., Arida, H. A. S., Sayed, A., Younis, A. A., & Farag, A. B. (2009). Separation and preconcentration of some heavy-metal ions using new chelating polymeric hydrogels. *Journal of Applied Polymer Science, 113*(2), 1335–1344. doi:10.1002/app.30137

Sharifi, R. S. (2016). Application of biofertilizers and zinc increases yield, nodulation and unsaturated fatty acids of soybean. *Zemdirbyste-Agriculture, 103*(3), 251–258. doi:10.13080/z-a.2016.103.032

Sharifi, R., Mohammadi, K., & Rokhzadi, A. (2016). Effect of seed priming and foliar application with micronutrients on quality of forage corn (*Zea mays*). *Environmental and Experimental Biology, 14*(4), 151–156. doi:10.22364/eeb.14.21

Sharifi, S., Behzadi, S., Laurent, S., Laird Forrest, M., Stroeve, P., & Mahmoudi, M. (2012). Toxicity of nanomaterials. *Chemical Society Reviews, 41*(6), 2323–2343. doi:10.1039/C1CS15188F PMID:22170510

Sharma, J. K., Akhtar, M. S., Ameen, S., Srivastava, P., & Singh, G. (2015). Green synthesis of CuO nanoparticles with leaf extract of Calotropis gigantea and its dye-sensitized solar cells applications. *Journal of Alloys and Compounds, 632*, 321–325. doi:10.1016/j.jallcom.2015.01.172

Sharma, M., Gungor, K., Yeltik, A., Olutas, M., Guzelturk, B., Kelestemur, Y., Erdem, T., Delikanli, S., McBride, J. R., & Demir, H. V. (2017). Near-Unity Emitting Copper-Doped Colloidal Semiconductor Quantum Wells for Luminescent Solar Concentrators. *Advanced Materials, 29*(30), 1–10. doi:10.1002/adma.201700821 PMID:28605062

Sharma, P., Sharma, A., Sharma, M., Bhalla, N., Estrela, P., Jain, A., Thakur, P., & Thakur, A. (2017). Nanomaterial fungicides: In vitro and in vivo antimycotic activity of cobalt and nickel nanoferrites on phytopathogenic fungi. *Global Challenges (Hoboken, NJ), 1*(9), 1700041. doi:10.1002/gch2.201700041 PMID:31565296

Sharma, S., Mitra, P., Bhardwaj, P., & Sharma, P. (2021). Blood lead level in school going children of Jodhpur, Rajasthan, India. *Turkish Journal of Biochemistry, 46*(4), 393–398. doi:10.1515/tjb-2020-0418

Sharma, V., Anderson, D., & Dhawan, A. (2012). Zinc oxide nanoparticles induce oxidative DNA damage and ROS-triggered mitochondria mediated apoptosis in human liver cells (HepG2). *Apoptosis, 17*(8), 852–870. doi:10.100710495-012-0705-6 PMID:22395444

Sharma, V., Shukla, R. K., Saxena, N., Parmar, D., Das, M., & Dhawan, A. (2009). DNA damaging potential of zinc oxide nanoparticles in human epidermal cells. *Toxicology Letters, 185*(3), 211–218. doi:10.1016/j.toxlet.2009.01.008 PMID:19382294

Sharma, V., Singh, P., Pandey, A. K., & Dhawan, A. (2012). Induction of oxidative stress, DNA damage and apoptosis in mouse liver after sub-acute oral exposure to zinc oxide nanoparticles. *Mutation Research, 745*(1-2), 84–91. doi:10.1016/j.mrgentox.2011.12.009 PMID:22198329

Sharmila, R. C. (2010). *Nutrient Release Pattern of Nano-Fertilizer Formulations*. Tamil Nadu Agricultural University.

Sharon, M., Choudhary, A. K., & Kumar, R. (2010). Nanotechnology in agricultural diseases and food safety. *Journal of Phytology, 2*, 83–92.

Sharonova, N. L., Yapparov, A. Kh., Khisamutdinov, N. Sh., Ezhkova, A. M., Yapparov, I. A., Ezhkov, V. O., Degtyareva, I. A., & Babynin, E. V. (2015). Nanostructured water-phosphorite suspension is a new promising fertilizer. *Nanotechnologies in Russia*, *10*(7-8), 651–661. doi:10.1134/S1995078015040187

Shaviv, A. (2000). Advances in controlled release of fertilizers. *Advances in Agronomy*, *71*, 1–49.

Shaw, B. J., Al-Bairuty, G., & Handy, R. D. (2012). Effects of waterborne copper nanoparticles and copper sulphate on rainbow trout, (Oncorhynchus mykiss): Physiology and accumulation. *Aquatic Toxicology (Amsterdam, Netherlands)*, *116–117*, 90–101. doi:10.1016/j.aquatox.2012.02.032 PMID:22480992

Shaw, Z. L., Kuriakose, S., Cheeseman, S., Dickey, M. D., Genzer, J., Christofferson, A. J., Crawford, R. J., McConville, C. F., Chapman, J., Truong, V. K., Elbourne, A., & Walia, S. (2021). Antipathogenic Properties and Applications of Low-Dimensional Materials. *Nature Communications*, *12*(1), 3897. doi:10.103841467-021-23278-7 PMID:34162835

Shejawal, K. P., Randive, D. S., Bhinge, S. D., Bhutkar, M. A., Todkar, S. S., Mulla, A. S., & Jadhav, N. R. (2021). Green synthesis of silver, iron and gold nanoparticles of lycopene extracted from tomato: Their characterization and cytotoxicity against COLO320DM, HT29 and Hella cell. *Journal of Materials Science. Materials in Medicine*, *32*(2), 1–12. doi:10.100710856-021-06489-8 PMID:33576907

Sheykhbaglou, R., Sedghi, M., Tajbakhsh-Shishevan, M., & Seyed-Sharifi, R. (2010). Effects of nano-iron oxide particles on agronomic traits of soybean. *Notulae Scientia Biologicae*, *2*(2), 112–113. doi:10.15835/nsb224667

Shukla, S. K., Kumar, R., Mishra, R. K., Pandey, A., Pathak, A., Zaidi, M., Srivastava, S. K., & Dikshit, A. (2015). Prediction and validation of gold nanoparticles (GNPs) on plant growth promoting rhizobacteria (PGPR): A step toward the development of nano-biofertilizers. *Nano Reviews*, *4*, 439–448.

Shukla, S., Kumar, R., Mishra, R., Pandey, A., Pathak, A., Zaidi, M., Srivastava, S., & Dikshit, A. (2015). Prediction and validation of gold nanoparticles (GNPs) on plant growth promoting rhizobacteria (PGPR): A step toward development of nano-biofertilizers. *Nanotechnology Reviews*, *4*(5), 439–448. doi:10.1515/ntrev-2015-0036

Siddiqi, K. S., & Husen, A. (2020). Current status of plant metabolite-based fabrication of copper/copper oxide nanoparticles and their applications: A review. *Biomaterials Research*, *24*(1), 1–15. doi:10.118640824-020-00188-1 PMID:32514371

Siddiqui, M. H., Al-Whaibi, M. H., Faisal, M., & Al Sahli, A. A. (2014). Nano-silicon dioxide mitigates the adverse effects of salt stress on *Cucurbita pepo* L. *Environmental Toxicology and Chemistry*, *33*(11), 2429–2437. doi:10.1002/etc.2697 PMID:25066835

Silva, A. C. N., Deda, D. K., da Róz, A. L., Prado, R. A., Carvalho, C. C., Viviani, V., & Leite, F. L. (2013). Nanobiosensors based on chemically modified AFM probes: A useful tool for metsulfuron-methyl detection. *Sensors (Basel)*, *13*(2), 1477–1489. doi:10.3390130201477 PMID:23348034

Silvan, J. M., Zorraquin-Pena, I., Gonzalez de Llano, D., Moreno-Arribas, M. V., & Martinez-Rodriguez, A. J. (2018). Antibacterial activity of glutathione-stabilized silver nanoparticles against *Campylobacter* multidrug-resistant strains. *Frontiers in Microbiology*, *9*, 458. doi:10.3389/fmicb.2018.00458 PMID:29615993

Silvestre, C., Duraccio, D., & Cimmino, S. (2011). Food packaging based on polymer nanomaterials. *Progress in Polymer Science*, *36*(12), 1766–1782. doi:10.1016/j.progpolymsci.2011.02.003

Sim, W., Barnard, R. T., Blaskovich, M. A. T, & Ziora, Z. M. (2018). Antimicrobial silver in medicinal and consumer applications: A patent review of the past decade (2007–2017). *Antibiotics (Basel)*, *7*(4), 93.

Simarmata, T., Hersanti, Turmuktini, T., Fitriatin, B. N., Setiawati, M. R., & Purwanto. (2018). Application of bioameliorant and biofertilizers to increase the soil health and rice productivity. *Hayati Journal of Biosciences*, *23*(4), 181–184. doi:10.1016/j.hjb.2017.01.001

Simbine, E. O., Rodrigues, L. C., Lapa-guimaraes, J., Kamimura, E. S., Corassin, C. H., & Oliveira, C. A. F. (2019). Application of silver nanoparticles in food packages: A review. *Food Science and Technology (Campinas)*, *39*(4), 793–802. doi:10.1590/fst.36318

Simonet, B. M., & Valcárcel, M. (2009). Monitoring nanoparticles in the environment. *Analytical and Bioanalytical Chemistry*, *393*(1), 17–21. doi:10.100700216-008-2484-z PMID:18974979

Šimon, P., Chaudrey, Q., & Bakoš, D. (2008). Migration of engineered nanoparticles from polymer packaging to food—A physicochemical view. *Journal of Food and Nutrition Research*, *47*, 105–113.

Singh, K.R., Nayak, V., & Singh, R.P. (2021). Introduction to bionanomaterials: An overview. In *Bionanomaterials: Fundamentals and biomedical applications*. IOP Science.

Singh, M. D. (2015). *Studies on the effect of time of application and concentration of nano zinc sulphide (nZS) on the growth and yield of sunflower (Helianthus annuus L.)* [M. Sc. (Agri.) Thesis]. University of Agricultural Sciences, Dharwad, Karnataka, India.

Singh, A., Gautam, P. K., Verma, A., Singh, V., Shivapriya, P. M., Shivalkar, S., Sahoo, A. K., & Samanta, S. K. (2020). Green synthesis of metallic nanoparticles as effective alternatives to treat antibiotics resistant bacterial infections: A review. *Biotechnology Reports (Amsterdam, Netherlands)*, *25*, e00427. doi:10.1016/j.btre.2020.e00427 PMID:32055457

Singh, B. N., Prateeksha, N., Upreti, D. K., Singh, B. R., Defoirdt, T., Gupta, V. K., & Vahabi, K. (2017). Bactericidal, quorum quenching and anti-biofilm nanofactories: A new niche for nanotechnologists. *Critical Reviews in Biotechnology*, *37*(4), 525–540. doi:10.1080/07388551.2016.1199010 PMID:27684212

Singh, N. A. (2017). Nanotechnology innovations, industrial applications and patents. *Environmental Chemistry Letters*, *15*(2), 185–191. doi:10.100710311-017-0612-8

Singh, R., Vora, I., Nadhe, S. B., Wadhwani, S. A., Shedbalkar, U. U., & Chopade, B. A. (2018). Antibacterial activities of bacteriagenic silver nanoparticles against nosocomial *Acinetobacter baumannii. Journal of Nanoscience and Nanotechnology*, *18*(6), 3806–3815. doi:10.1166/jnn.2018.15013 PMID:29442713

Singh, S. P., Mishra, A., Shyanti, R. K., Singh, R. P., & Acharya, A. (2021). Silver nanoparticles synthesized using *Carica papaya* leaf extract (AgNPs-PLE) causes cell cycle arrest and apoptosis in human prostate (DU145) cancer cells. *Biological Trace Element Research*, *199*(4), 1316–1331. doi:10.100712011-020-02255-z PMID:32557113

Smita, S., Gupta, S. K., Bartonova, A., Dusinska, M., Gutleb, A. C., & Rahman, Q. (2012). Nanoparticles in the environment: Assessment using the causal diagram approach. *Environmental Health*, *11*(S1), S13. Advance online publication. doi:10.1186/1476-069X-11-S1-S13 PMID:22759495

Sobhan, A., Oh, J.-H., Park, M.-K., Kim, S. W., Park, C., & Lee, J. (2018). Single walled carbon nanotube based biosensor for detection of peanut allergy-inducing protein ara h1. *Korean Journal of Chemical Engineering*, *35*(1), 172–178. doi:10.100711814-017-0259-y

Soenen, S. J., Himmelreich, U., Nuytten, N., & De Cuyper, M. (2011). Cytotoxic effects of iron oxide nanoparticles and implications for safety in cell labelling. *Biomaterials*, *32*(1), 195–205. doi:10.1016/j.biomaterials.2010.08.075 PMID:20863560

Sohair, E. E. D., Abdall, A. A., Amany, A. M., Faruque, H. M. D., & Houda, R. A. (2018). Evaluation of nitrogen, phosphorus and potassium nano-fertilizers on yield, yield components and fiber properties of egyptian cotton (*Gossyppium barbadense* L.). *Journal of Plant Sciences and Crop Protection*, *1*(3), 1–10.

Sohni, S., Hashim, R., Nidaullah, H., Lamaming, J., & Sulaiman, O. (2019). Chitosan/nano-lignin based composite as a new sorbent for enhanced removal of dye pollution from aqueous solutions. *International Journal of Biological Macromolecules*, *132*, 1304–1317. doi:10.1016/j.ijbiomac.2019.03.151 PMID:30922916

Solaimuthu, A., Nishitha, A., Murali, P., & Sai, K. P. (2020). Nano-biosensors and their relevance in tissue engineering. *Current Opinion in Biomedical Engineering*, *13*, 84–93. doi:10.1016/j.cobme.2019.12.005

Solanki, R. G., & Rajaram, P. (2017). Structural, optical and morphological properties of CdS nanoparticles synthesized using hydrazine hydrate as a complexing agent. *Nano-Structures & Nano-Objects*, *12*, 157–165. doi:10.1016/j.nanoso.2017.10.003

Song, P., Hai, Y., Wang, X., Zhao, L., Chen, B., Cui, P., Xie, Q., Yu, L., Li, Y., Wu, Z., & Li, H. (2018). Realgar transforming solution suppresses angiogenesis and tumor growth by inhibiting VEGF receptor 2 signaling in vein endothelial cells. *Archives of Pharmacal Research*, *41*(4), 467–480. doi:10.100712272-018-1014-6 PMID:29542005

Song, S.-J., Shin, Y. C., Lee, H. U., Kim, B., Han, D.-W., & Lim, D. (2018). Dose- and Time-Dependent Cytotoxicity of Layered Black Phosphorus in Fibroblastic Cells. *Nanomaterials (Basel, Switzerland)*, *8*(6), 408. doi:10.3390/nano8060408 PMID:29882820

Song, X., Yao, W., Zhang, B., & Wu, Y. (2011). Application of Pt/CdS for the photocatalytic flue gas desulfurization. *International Journal of Photoenergy*, *2012*, 1–5. doi:10.1155/2012/684735

Sonwani, S., Madaan, S., Arora, J., Suryanarayan, S., Rangra, D., Mongia, N., Vats, T., & Saxena, P. (2021). Inhalation exposure to atmospheric nanoparticles and its associated impacts on human health: A review. *Frontiers in Sustainable Cities*, *3*, 690444. doi:10.3389/frsc.2021.690444

Srikanth, K., Pereira, E., Duarte, A. C., & Rao, J. V. (2016). Evaluation of cytotoxicity, morphological alterations and oxidative stress in Chinook salmon cells exposed to copper oxide nanoparticles. *Protoplasma*, *253*(3), 873–884. doi:10.100700709-015-0849-7 PMID:26115719

Srivastava, A. K., & Malhotra, S. K. (2017). Nutrient use efficiency in perennial fruit crops – A review. *Journal of Plant Nutrition*, *40*(13), 1928–1953. doi:10.1080/01904167.2016.1249798

Srivastava, V., Gusain, D., & Sharma, Y. (2015). Critical review on the toxicity of some widely used engineered nanoparticles. *Industrial & Engineering Chemistry Research*, *54*(24), 6209–6233. doi:10.1021/acs.iecr.5b01610

Strambeanu, N., Demetrovici, L., Dragos, D., & Lungu, M. (2015). Nanoparticles: Definition, classification and general physical properties. In Nanoparticles' promises and risks (pp. 3-8). Springer. doi:10.1007/978-3-319-11728-7_1

Studer, A. M., Limbach, L. K., Van Duc, L., Krumeich, F., Athanassiou, E. K., Gerber, L. C., Moch, H., & Stark, W. J. (2010). Nanoparticle cytotoxicity depends on intracellular solubility: Comparison of stabilized copper metal and degradable copper oxide nanoparticles. *Toxicology Letters*, *197*(3), 169–174. doi:10.1016/j.toxlet.2010.05.012 PMID:20621582

Stuer-Lauridsen, F. (2005). Review of passive accumulation devices for monitoring organic micropollutants in the aquatic environment. *Environmental Pollution*, *136*(3), 503–524. doi:10.1016/j.envpol.2004.12.004 PMID:15862404

Subastri, A., Arun, V., Sharma, P., Suyavaran, A., Nithyananthan, S., Alshammari, G. M., ... Thirunavukkarasu, C. (2018). Synthesis and characterisation of arsenic nanoparticles and its interaction with DNA and cytotoxic potential on breast cancer cells. *Chemico-Biological Interactions*, *295*, 73–83. doi:10.1016/j.cbi.2017.12.025 PMID:29277637

Subhan, M. A., Choudhury, K. P., & Neogi, N. (2021). Advances with molecular nanomaterials in industrial manufacturing applications. *Nanomanufacturing*, *1*(2), 75–97. doi:10.3390/nanomanufacturing1020008

Subramanian, K. S., Manikandan, A., Thirunavukkarasu, M., & Rahale, C. S. (2015). Nano-fertilizers for Balanced Crop Nutrition. In M. Rai, C. Ribeiro, L. Mattoso, & N. Duran (Eds.), *Nanotechnologies in Food and Agriculture* (pp. 69–80). Springer International Publishing. doi:10.1007/978-3-319-14024-7_3

Subramanian, K. S., & Tarafdar, J. C. (2011). Prospects of nanotechnology in Indian farming. *International Journal of Agricultural Sciences*, *81*, 887–893.

Suchomel, P., Kvitek, L., Prucek, R., Panacek, A., Halder, A., Vajda, S., & Zboril, R. (2018). Simple size-controlled synthesis of Au nanoparticles and their size-dependent catalytic activity. *Scientific Reports*, *8*(1), 4589. doi:10.103841598-018-22976-5 PMID:29545580

Su, L. Z., Liu, Y., Li, Y. F., An, Y. L., & Shi, L. Q. (2021). Responsive Polymeric Nanoparticles for Biofilm-infection Control. *Chinese Journal of Polymer Science*, *39*(11), 1376–1391. doi:10.100710118-021-2610-3

Sulaiman, G. M., Hussien, H. T., & Saleem, M. M. N. M. (2015). Biosynthesis of silver nanoparticles synthesized by *Aspergillus flavus* and their antioxidant, antimicrobial and cytotoxicity properties. *Bulletin of Materials Science*, *38*(3), 639–644. doi:10.100712034-015-0905-0

Suman, P. H., Felix, A. A., Tuller, H. L., Varela, J. A., & Orlandi, M. O. (2015). Comparative gas sensor response of SnO2, SnO and Sn3O4 nanobelts to NO2 and potential interferents. *Sensors and Actuators. B, Chemical*, *208*, 122–127. doi:10.1016/j.snb.2014.10.119

Suman, T. Y., Radhika Rajasree, S. R., & Kirubagaran, R. (2015). Evaluation of zinc oxide nanoparticles toxicity on marine algae chlorella vulgaris through flow cytometric, cytotoxicity and oxidative stress analysis. *Ecotoxicology and Environmental Safety*, *113*, 23–30. doi:10.1016/j.ecoenv.2014.11.015 PMID:25483368

Sun, J., Wang, S., Zhao, D., Hun, F. H., Weng, L., & Liu, H. (2011). Cytotoxicity, permeability, and inflammation of metal oxide nanoparticles in human cardiac microvascular endothelial cells: Cytotoxicity, permeability, and inflammation of metal oxide nanoparticles. *Cell Biology and Toxicology*, *27*(5), 333–342. doi:10.100710565-011-9191-9 PMID:21681618

Sun, X., Zhai, C., & Wang, X. Y. (2013). A novel and highly sensitive acetyl-cholinesterase biosensor modified with hollow gold nanospheres. *Bioprocess and Biosystems Engineering*, *36*(3), 273–283. doi:10.100700449-012-0782-5 PMID:22829062

Suriyaprabha, R., Karunakaran, G., Yuvakkumar, R., Rajendran, V., & Kannan, N. (2012). Silica nanoparticles for increased silica availability in maize (*Zea mays* L) seeds under hydroponic conditions. *Current Nanoscience*, *8*(6), 902–908. doi:10.2174/157341312803989033

Su, S., Sun, Q., Gu, X., Xu, Y., Shen, J., Zhu, D., Chao, J., Fan, C., & Wang, L. (2019). Two-dimensional nanomaterials for biosensing applications. *Trends in Analytical Chemistry*, *119*, 115610. doi:10.1016/j.trac.2019.07.021

Sutunkova, M. P., Solovyeva, S. N., Chernyshov, I. N., Klinova, S. V., Gurvich, V. B., Shur, V. Y., Shishkina, E. V., Zubarev, I. V., Privalova, L. I., & Katsnelson, B. A. (2020). Manifestation of Systemic Toxicity in Rats after a Short-Time Inhalation of Lead Oxide Nanoparticles. *International Journal of Molecular Sciences*, *21*(3), 690. doi:10.3390/ijms21030690 PMID:31973040

Svenson, S., & Tomalia, D. A. (2012). Dendrimers in biomedical applications—Reflections on the field. *Advanced Drug Delivery Reviews*, *64*, 102–115. doi:10.1016/j.addr.2012.09.030 PMID:16305813

Syu, Y. C., Hsu, W. E., & Lin, C. T. (2018). Field-effect transistor biosensing: Devices and clinical applications. *ECS Journal of Solid State Science and Technology: JSS*, *7*(7), Q3196–Q3207. doi:10.1149/2.0291807jss

Szalay, B., Tatrai, E., Nyiro, G., Vezer, T., & Dura, G. (2012). Potential toxic effects of iron oxide nanoparticles in in vivo and in vitro experiments. *Journal of Applied Toxicology*, *32*(6), 446–453. doi:10.1002/jat.1779 PMID:22161551

Szuplewska, A., Kulpińska, D., Jakubczak, M., Dybko, A., Chudy, M., Olszyna, A., Brzózka, Z., & Hastrzębska, A. M. (2022). The 10th Anniversary of MXenes: Challenges and Prospects for Their Surface Modification toward Future Biotechnological Applications. *Advanced Drug Delivery Reviews*, *182*, 114099. doi:10.1016/j.addr.2021.114099 PMID:34990793

Taghavi, S. M., Momenpour, M., Azarian, M., Ahmadian, M., Souri, F., Taghavi, S. A., Sadeghain, M., & Karchani, M. (2013). Effects of Nanoparticles on the Environment and Outdoor Workplaces. *Electronic Physician*, *5*(4), 706–712. doi:10.14661/2013.706-712 PMID:26120406

Tahmasbi, D., Zarghami, R., Azghandi, A. V., & Chaichi, M. (2011). Effects of nanosilver and nitroxin biofertilizer on yield and yield components of potato minitubers. *International Journal of Agriculture and Biology*, *13*, 986–990.

Tailor, G., Yadav, B. L., Chaudhary, J., Joshi, M., & Suvalka, C. (2020). Green synthesis of silver nanoparticles using *Ocimum canum* and their anti-bacterial activity. *Biochemistry and Biophysics Reports*, *24*, 100848. doi:10.1016/j.bbrep.2020.100848 PMID:33305022

Talsness, C. E., Andrade, A. J. M., Kuriyama, S. N., Taylor, J. A., & vomSaal, F. S. (2009). Components of plastic: Experimental studies in animals and relevance for human health. *Philosophical Transactions of the Royal Society of London. Series B, Biological Sciences*, *364*(1526), 2079–2096. doi:10.1098/rstb.2008.0281 PMID:19528057

Tan, E., Li, B. L., Ariga, K., Lim, C.-T., Garaj, S., & Leong, D. T. (2019). Toxicity of Two-Dimensional Layered Materials and Their Heterostructures. *Bioconjugate Chemistry*, *30*(9), 2287–2299. doi:10.1021/acs.bioconjchem.9b00502 PMID:31381854

Tania, I. S., & Ali, M. (2021). Effect of the coating of zinc oxide (ZnO) nanoparticles with binder on the functional and mechanical properties of cotton fabric. *Materials Today: Proceedings*, *38*, 2607–2611. doi:10.1016/j.matpr.2020.08.171

Tan, P., Li, H. S., Wang, J., & Gopinath, S. C. B. (2020). Silver nanoparticle in biosensor and bioimaging: Clinical perspectives. *Biotechnology and Applied Biochemistry*, 1–7. doi:10.1002/bab.2045 PMID:33043496

Tantawy, A. S., Salama, Y. A. M., Abdel-Mawgoud, A. M. R., & Ghoname, A. A. (2014). Comparison of chelated calcium with nano calcium on alleviation of salinity negative effects on tomato plants. *Middle East Journal of Agriculture Research*, *3*(4), 912–916.

Tan, Y. L., & Ho, H. K. (2018). Navigating albumin-based nanoparticles through various drug delivery routes. *Drug Discovery Today*, *23*(5), 1108–1114. doi:10.1016/j.drudis.2018.01.051 PMID:29408437

Tarafdar, J. C., Raliya, R., Mahawar, H., & Rathore, I. (2014). Development of zinc nanofertilizer to enhance crop production in pearl millet (*Pennisetum americanum*). *Agricultural Research*, *3*(3), 257–262. doi:10.100740003-014-0113-y

Ten Hulscher, T. E., & Cornelissen, G. (2005). Effect of temperature on sorption equilibrium and sorption kinetics of organic micropollutants-a review. *Chemosphere*, *32*(4), 609–626. doi:10.1016/0045-6535(95)00345-2

Tenzer, S., Docter, D., Kuharev, J., Musyanovych, A., Fetz, V., Hecht, R., Schlenk, F., Fischer, D., Kiouptsi, K., Reinhardt, C., Landfester, K., Schild, H., Maskos, M., Knauer, S. K., & Stauber, R. H. (2013). Rapidformation of plasma protein corona critically affects nanoparticle pathophysiology. *Nature Nanotechnology*, *8*(10), 772–781. doi:10.1038/nnano.2013.181 PMID:24056901

Teo, W. Z., Chng, E. L. K., Sofer, Z., & Pumera, M. (2014). Cytotoxicity of Exfoliated Transition-Metal Dichalcogenides (MoS_2, WS_2, and WSe_2) is Lower Than that of Graphene and Its Analogues. *Chemistry (Weinheim an der Bergstrasse, Germany), 20*(31), 9627–9632. doi:10.1002/chem.201402680 PMID:24976159

Thakur, M. P., Phillips, H. R. P., Brose, U., De Vries, F. T., Lavelle, P., Loreau, M., Mathieu, J., Mulder, C., Van der Putten, W. H., Rillig, M. C., Wardle, D. A., Bach, E. M., Bartz, M. L. C., Bennett, J. M., Briones, M. J. I., Brown, G., Decaëns, T., Eisenhauer, N., Ferlian, O., ... Cameron, E. K. (2020). Towards an integrative understanding of soil biodiversity. *Biological Reviews of the Cambridge Philosophical Society, 95*(2), 350–364. doi:10.1111/brv.12567 PMID:31729831

Thakur, S., Pandey, S., & Arotiba, O. A. (2017). Sol-gel derived xanthan gum/silica nanocomposite-a highly efficient cationic dyes adsorbent in aqueous system. *International Journal of Biological Macromolecules, 103*, 596–604. doi:10.1016/j.ijbiomac.2017.05.087 PMID:28528942

Theron, J., Eugene Cloete, T., & de Kwaadsteniet, M. (2010). Current molecular and emerging nanobiotechnology approaches for the detection of microbial pathogens. *Critical Reviews in Microbiology, 36*(4), 318–339. doi:10.3109/1040841X.2010.489892 PMID:20879848

Thiruvengadam, M., Chung, I. M., Gomathi, T., Ansari, M. A., Gopiesh Khanna, V., Babu, V., & Rajakumar, G. (2019). Synthesis, characterization and pharmacological potential of green synthesized copper nanoparticles. *Bioprocess and Biosystems Engineering, 42*(11), 1769–1777. doi:10.100700449-019-02173-y PMID:31372759

Thompson, R. C., Olsen, Y., Mitchell, R. P., Davis, A., Rowland, S. J., John, A. W. G., McGonigle, D., & Russell, A. E. (2004). Lost at sea, where is all the plastic? *Science, 304*(5672), 838. doi:10.1126cience.1094559 PMID:15131299

Tian, X., Murfin, L. C., Wu, L., Lewis, S. E., & James, T. D. (2021). Fluorescent small organic probes for biosensing. *Chemical Science (Cambridge), 12*(10), 3406–3426. doi:10.1039/D0SC06928K PMID:34163615

Tirumala, M. G., Anchi, P., Raja, S., Rachamalla, M., & Godugu, C. (2021). Novel Methods and Approaches for Safety Evaluation of Nanoparticle Formulations: A Focus Towards In Vitro Models and Adverse Outcome Pathways. *Frontiers in Pharmacology, 12*, 1–12. doi:10.3389/fphar.2021.612659 PMID:34566630

Tiwari, M., Jain, P., Hariharapura, R. C., Narayanan, K., Bhat, U., Udupa, N., & Rao, J. V. (2016). Biosynthesis of copper nanoparticles using copper-resistant Bacillus cereus, a soil isolate. *Process Biochemistry, 51*(10), 1348–1356. doi:10.1016/j.procbio.2016.08.008

Tiwari, R., Tiwari, G., & Lahiri, A. (2021). Localized delivery of drugs through medical textiles for treatment of burns: A perspective approach. *Advanced Pharmaceutical Bulletin, 11*(2), 248. PMID:33880346

Tokuyama, H., Onodera, M., & Ban, T. (2017). Selective temperature-swing adsorption of Pd(II) ions onto apoly(N-isopropylacrylamide-co-allyl mercaptan) gel. *Separation and Purification Technology, 182*, 166–170. doi:10.1016/j.seppur.2017.03.049

Toropova, A. P., & Toropov, A. A. (2022). Nanomaterials: Quasi-SMILES as a flexible basis for regulation and environmental risk assessment. *The Science of the Total Environment, 823*, 153747. doi:10.1016/j.scitotenv.2022.153747 PMID:35149067

Trabelsi, H., Azzouz, I., Sakly, M., & Abdelmelek, H. (2013). Subacute toxicity of cadmium on hepatocytes and nephrocytes in the rat could be considered as a green biosynthesis of nanoparticles. *International Journal of Nanomedicine, 8*, 1121–1128. doi:10.2147/IJN.S39426 PMID:23515701

Trenkel, M. E. (1997). *Controlled-release and stabilized fertilizers in agriculture. Int. Fert.* Ind. Assoc., 234–318.

Tripathi, D. K., Singh, V. P., Prasad, S. M., Chauhan, D. K., & Dubey, N. K. (2015). Silicon nanoparticles (SiNp) alleviate chromium (VI) phytotoxicity in *Pisum sativum* (L.) seedlings. *Plant Physiology and Biochemistry*, *96*, 189–198. doi:10.1016/j.plaphy.2015.07.026 PMID:26298805

Tudu, B. K., Sinhamahapatra, A., & Kumar, A. (2020). Surface modification of cotton fabric using TiO2 nanoparticles for self-cleaning, oil–water separation, antistain, anti-water absorption, and antibacterial properties. *ACS Omega*, *5*(14), 7850–7860. doi:10.1021/acsomega.9b04067 PMID:32309694

Tulinska, J., Krivosikova, Z., Liskova, A., Mikusova, M. L., Masanova, V., Rollerova, E., ... Docekal, B. (2022). Six-week inhalation of lead oxide nanoparticles in mice affects antioxidant defense, immune response, kidneys, intestine and bones. *Environmental Science. Nano*, *9*(2), 751–766. doi:10.1039/D1EN00957E

Tuncel, D., & Demir, H. V. (2010). Conjugated polymer nanoparticles. *Nanoscale*, *2*(4), 484–494. doi:10.1039/b9nr00374f PMID:20644748

Tu, Y., Lv, M., Xiu, P., Huynh, T., Zhang, M., Castelli, M., Liu, Z., Huang, Q., Fan, C., Fang, H., & Zhou, R. (2013). Destructive extraction of Phospholipids from *Escherichia coli* Membranes by Graphene Nanosheets. *Nature Nanotechnology*, *8*(8), 594–601. doi:10.1038/nnano.2013.125 PMID:23832191

Uddin, M. N., Desai, F., & Asmatulu, E. (2020). Engineered nanomaterials in the environment: Bioaccumulation, biomagnification and biotransformation. *Environmental Chemistry Letters*, *18*(4), 1073–1083. doi:10.100710311-019-00947-0

Ude, V. C., Brown, D. M., Viale, L., Kanase, N., Stone, V., & Johnston, H. J. (2017). Impact of copper oxide nanomaterials on differentiated and undifferentiated Caco-2 intestinal epithelial cells; assessment of cytotoxicity, barrier integrity, cytokine production and nanomaterial penetration. *Particle and Fibre Toxicology*, *14*(1), 31. Advance online publication. doi:10.118612989-017-0211-7 PMID:28835236

Ugazio, E., Cavalli, R., & Gasco, M. R. (2002). Incorporation of cyclosporin A in solid lipid nanoparticles (SLN). *International Journal of Pharmaceutics*, *241*(2), 341–344. doi:10.1016/S0378-5173(02)00268-5 PMID:12100861

Umer, A., Naveed, S., Ramzan, N., & Rafique, M. S. (2012). Selection of a suitable method for the synthesis of copper nanoparticles. *Nano*, *7*(05), 1230005. doi:10.1142/S1793292012300058

Upadhyaya, H., Shome, S., Tewari, S., Bhattacharya, M. K., & Panda, S. K. (2015). Effect of Zn nano-particles on growth responses of rice. In B. Singh, A. Kaushik, S. K. Mehta, & S. K. Tripathi (Eds.), *Nanotechnology: Novel Perspectives and Prospects* (pp. 508–512). McGraw Hill Education.

USEPA. (2001). National primary drinking water regulations; arsenic and clarifications to compliance and new source contaminants monitoring" from the environmental protection agency; final rule. *Federal Register*, *66*, 6976–7066.

Uzu, G., Sobanska, S., Sarret, G., Muñoz, M., & Dumat, C. (2010). Foliar lead uptake by lettuce exposed to atmospheric fallouts. *Environmental Science & Technology*, *44*(3), 1036–1042. doi:10.1021/es902190u PMID:20063891

Vafa, Z. N., Sirousmehr, A. R., Ghanbari, A., Khammari, I., & Falahi, N. (2015). Effects of nano zinc and humic acid on quantitative and qualitative characteristics of savory (*Satureja hortensis* L.). *International Journal of Biosciences*, *6*(3), 124–136. doi:10.12692/ijb/6.3.124-136

Valdiglesias, V., Costa, C., Kilic, G., Costa, S., Pasaro, E., Laffon, B., & Teixeira, J. P. (2013). Neuronal cytotoxicity and genotoxicity induced by zinc oxide nanoparticles. *Environment International*, *55*, 92–100. doi:10.1016/j.envint.2013.02.013 PMID:23535050

Valdiglesias, V., Fernandez-Bertolez, N., Kilic, G., Costa, C., Costa, S., Fraga, S., Bessa, M. J., Pasaro, E., Teixeira, J. P., & Laffon, B. (2016). Are iron oxide nanoparticles safe? Current knowledge and future perspectives. *Journal of Trace Elements in Medicine and Biology*, *38*, 53–63. doi:10.1016/j.jtemb.2016.03.017 PMID:27056797

Valsami-Jones, E., & Lynch, I. (2015). How safe are nanomaterials? *Science*, *350*(6259), 388–389. doi:10.1126cience.aad0768 PMID:26494749

Vanamala, K., Bhise, K., Sanchez, H., Kebriaei, R., Luong, D., Sau, S., Abdelhady, H., Rybak, M. J., Andes, D., & Iyer, A. K. (1791). Folate. Functionalized Lipid Nanoparticles for Targeted Therapy of Methicillin-Resistant Staphylococcus aureus. *Pharmaceutics*, *13*(11), 1791. doi:10.3390/pharmaceutics13111791

Vandenberg, L. N., Chauhoud, I., Heindel, J. J., Padmanabhan, V., Paumgartten, F. J., & Schoenfelder, G. (2010). Urinary, circulating and tissue biomonitoring studies indicate widespread exposure to bisphenol A. *Environmental Health Perspectives*, *118*(8), 1055–1070. doi:10.1289/ehp.0901716 PMID:20338858

Vandergheynst, J. S., Scher, H. B., Gou, H. Y., & Schultz, D. L. (2007). Water in oil emulsions that improve the storage and delivery of the biolarvacide Lagenidium giganteum. *Biological Control*, *52*, 207–229.

Vandergheynst, J. S., Scher, H., & Yun, G. H. (2006). Design of formulations for improved biological control agent viability and sequestration during storage. *Industrial Biotechnology (New Rochelle, N.Y.)*, *2*(3), 213–219. doi:10.1089/ind.2006.2.213

Vannini, C., Domingo, G., Onelli, E., Prinsi, B., Marsoni, M., Espen, L., & Bracale, M. (2013). Morphological and proteomic responses of *Eruca sativa* exposed to silver nanoparticles or silver nitrate. *PLoS One*, *8*(7), e68752. doi:10.1371/journal.pone.0068752 PMID:23874747

Varela, J. A., Bexiga, M. G., Åberg, C., Simpson, J. C., & Dawson, K. A. (2012). Quantifyingsize-dependent interactions between fluorescently labeled polystyrene nanoparticles andmammalian cells. *Journal of Nanobiotechnology*, *10*(1), 39. doi:10.1186/1477-3155-10-39 PMID:23006133

Varmazyari, A., Taghizadehghalehjoughi, A., Sevim, C., Baris, O., Eser, G., Yildirim, S., Hacimuftuoglu, A., Buha, A., Wallace, D. R., Tsatsakis, A., Aschner, M., & Mezhuev, Y. (2020). Cadmium sulfide-induced toxicity in the cortex and cerebellum: In vitro and in vivo studies. *Toxicology Reports*, *7*, 637–648. doi:10.1016/j.toxrep.2020.04.011 PMID:32489905

Vasconcelos, C. L. D., Bezerril, P. M., Santos, D. E. S. D., Dantas, T. N. C., Pereira, M. R., & Fonseca, J. L. C. (2006). Effect of molecular weight and ionic strength on the formation of polyelectrolyte complexes based on poly (methacrylic acid) and chitosan. *Biomacromolecules*, *7*(4), 1245–1252. doi:10.1021/bm050963w PMID:16602745

Vasile, C. (2018). Polymeric nanocomposites and nanocoatings for food packaging: A review. *Materials (Basel)*, *11*(10), 1834. doi:10.3390/ma11101834 PMID:30261658

Velmurugan, P., Shim, J., Bang, K. S., & Oh, B. T. (2016). Gold nanoparticles mediated coloring of fabrics and leather for antibacterial activity. *Journal of Photochemistry and Photobiology. B, Biology*, *160*, 102–109. doi:10.1016/j.jphotobiol.2016.03.051 PMID:27104665

Velmurugan, P., Sivakumar, S., Song Young-Chae, S., Seong-Ho, J., Pyoung-In, Y., Jeong-Min, S., & Sung-Chul, H. (2015). Synthesis and characterization comparison of peanut shell extract silver nanoparticles with commercial silver nanoparticles and their antifungal activity. *Journal of Industrial and Engineering Chemistry*, *31*, 51–54. doi:10.1016/j.jiec.2015.06.031

Verma, K. K., Song, X. P., Joshi, A., Tian, D. D., Rajput, V. D., Singh, M., Arora, J., Minkina, T., & Li, Y. R. (2022). Recent trends in nano-fertilizers for sustainable agriculture under climate change for global food security. *Nanomaterials (Basel, Switzerland)*, *12*(1), 173. doi:10.3390/nano12010173 PMID:35010126

Verma, Y., Rana, K., Rani, V., & Rana, S. V. (2021). *Hepatotoxic Manifestations of Arsenic Trioxide Loaded Poly (Lactide-co-Glycolide)*. Nanoparticles in Wistar Rat., doi:10.21203/rs.3.rs-533921/v2

Viet, P. V., Sang, T. T., Ngoc Bich, N. H., & Thi, C. M. (2018). An improved green synthesis method and *Escherichia coli* antibacterial activity of silver nanoparticles. *Journal of Photochemistry and Photobiology. B, Biology*, *182*, 108–114. doi:10.1016/j.jphotobiol.2018.04.002 PMID:29656219

Vigneshwaran, N., Kathe, A. A., Varadarajan, P. V., Nachane, R. P., & Balasubramanya, R. H. (2007). Functional finishing of cotton fabrics using silver nanoparticles. *Journal of Nanoscience and Nanotechnology*, *7*(6), 1893–1897. doi:10.1166/jnn.2007.737 PMID:17654961

Vijayaraghavan, K., & Ashokkumar, T. (2017). Plant-mediated biosynthesis of metallic nanoparticles: A review of literature, factors affecting synthesis, characterization techniques and applications. *Journal of Environmental Chemical Engineering*, *5*(5), 4866–4883. doi:10.1016/j.jece.2017.09.026

Vimercati, L., Cavone, D., Caputi, A., De Maria, L., Tria, M., Prato, E., & Ferri, G. M. (2020). Nanoparticles: An Experimental Study of Zinc Nanoparticles Toxicity on Marine Crustaceans. General Overview on the Health Implications in Humans. *Frontiers in Public Health*, *8*, 192. Advance online publication. doi:10.3389/fpubh.2020.00192 PMID:32509719

Virkutyte, J., & Varma, R. (2010). *Treatment of micropollutants in water and wastewater*. IWA Publishing. doi:10.2166/9781780401447

Viscardi, M., Arena, M., Guadagno, L., Vertuccio, L., & Barra, G. (2018). Multi-functional nanotechnology integration for aeronautical structures performance enhancement. *International Journal of Structural Integrity*.

Viswanathan, S., Radecka, H., & Radecki, J. (2009). Electrochemical biosensor for pesticides based on acetylcholinesterase immobilized on polyaniline deposited on vertically assembled carbon nanotubes wrapped with s-DNA. *Biosensors & Bioelectronics*, *24*(9), 2772–2777. doi:10.1016/j.bios.2009.01.044 PMID:19269805

Vodyanitskii, Y. N. (2013). Contamination of soils with heavy metals and metalloids and its ecological hazard (analytic review). *Eurasian Soil Science*, *46*(7), 793–801. doi:10.1134/S1064229313050153

Volkheimer, G. (1977). Passage of particles through the wall of the gastrointenstinal tract. *Environmental Health Perspectives*, *9*, 215–225. PMID:4470938

Vollath, D. (2008). Nanomaterials an introduction to synthesis, properties and application. *Environmental Engineering and Management Journal*, *7*(6), 865–870.

vomSaal, F. S., Akingbemi, B. T., Belcher, S. M., Birnbaum, L. S., Crain, D. A., Eriksen, M., Farabollini, F., Guillette, L. J. Jr, Hauser, R., Heindel, J. J., Ho, S.-M., Hunt, P. A., Iguchi, T., Jobling, S., Kanno, J., Keri, R. A., Knudsen, K. E., Laufer, H., LeBlanc, G. A., ... Zoeller, R. T. (2007). Chapel Hill bisphenolA expert panel consensus statement: Integration of mechanisms, effects in animals and potential to impact human health at current levels of exposure. *Reproductive Toxicology (Elmsford, N.Y.)*, *24*(2), 131–138. doi:10.1016/j.reprotox.2007.07.005 PMID:17768031

Vrignaud, S., Benoit, J.-P., & Saulnier, P. (2011). Strategies for the nanoencapsulation of hydrophilic molecules in polymer-based nanoparticles. *Biomaterials*, *32*(33), 8593–8604. doi:10.1016/j.biomaterials.2011.07.057 PMID:21831421

Vu, X. H., Duong, T. T. T., Pham, T. T. H., Trinh, D. K., Nguyen, X. H., & Dang, V.-S. (2018). Synthesis and study of silver nanoparticles for antibacterial activity against *Escherichia coli* and *Staphylococcus aureus*. *Advances in Natural Sciences: Nanoscience and Nanotechnology*, *9*(2), 025019. doi:10.1088/2043-6254/aac58f

Waalkes, M. P. (2000). Cadmium carcinogenesis in review. *Journal of Inorganic Biochemistry*, *79*(1-4), 241–244. doi:10.1016/S0162-0134(00)00009-X PMID:10830873

Waldron, D. L., Preske, A., Zawodny, J. M., Krauss, T. D., & Gupta, M. C. (2017). PbSe Quantum Dot Based Luminescent Solar Concentrators. *Nanotechnology*, *28*(9), 095205. Advance online publication. doi:10.1088/1361-6528/aa577f PMID:28060769

Wang, D., Behura, S., Karim, F., Kazi, A., Wangila, G., & Wang, S. (2021). *Nanoscience and Nanotechnology for Food and Agroforestry*. Academic Press.

Wang, B., & Du, Y. (2013). Cadmium and its neurotoxic effects. *Oxidative Medicine and Cellular Longevity*, *898034*, 1–12. Advance online publication. doi:10.1155/2013/898034 PMID:23997854

Wang, E. C., & Wang, A. Z. (2014). Nanoparticles and their applications in cell and molecular biology. *Integrative Biology*, *6*(1), 9–26. doi:10.1039/c3ib40165k PMID:24104563

Wang, F. Y., Wang, L., Shi, Z. Y., Li, Y. J., & Song, Z. M. (2012). Effects of AM inoculation and organic amendment, alone or in combination, on growth, P nutrition, and heavy-metal uptake of tobacco in Pb-Cd-contaminated soil. *Journal of Plant Growth Regulation*, *31*(4), 549–559. doi:10.100700344-012-9265-9

Wang, F., Li, H., Yuan, Z., Sun, Y., Chang, F., Deng, H., Xie, L., & Li, H. (2016). A highly sensitive gas sensor based on CuO nanoparticles synthetized via a sol–gel method. *RSC Advances*, *6*(83), 79343–79349. doi:10.1039/C6RA13876D

Wang, J., Deng, X., Zhang, F., Chen, D., & Ding, W. (2014). ZnO nanoparticle-induced oxidative stress triggers apoptosis by activating JNK signaling pathway in cultured primary astrocytes. *Nanoscale Research Letters*, *9*(1), 117. doi:10.1186/1556-276X-9-117 PMID:24624962

Wang, J., Zhou, G., Chen, C., Yu, H., Wang, T., Ma, Y., Jia, G., Gao, Y., Li, B., & Sun, J. (2007). Acute toxicity and biodistribution of different sized titanium dioxide particles in mice after oral administration. *Toxicology Letters*, *168*(2), 176–185. doi:10.1016/j.toxlet.2006.12.001 PMID:17197136

Wang, K., Wang, X., Januszewski, B., Liu, Y., Li, D., Fu, R., Elimelech, M., & Huang, X. (2022). Tailored design of nanofiltration membranes for water treatment based on synthesis–property–performance relationships. *Chemical Society Reviews*, *51*(2), 672–719. doi:10.1039/D0CS01599G PMID:34932047

Wang, L. J., Wang, Y. H., Li, M., Fan, M. S., Zhang, F. S., Wu, X. M., Yang, W. S., & Li, T. J. (2002). Synthesis of ordered biosilica materials. *Chinese Journal of Chemistry*, 107–110.

Wang, L., Hu, C., & Shao, L. (2017). The antimicrobial activity of nanoparticles: Present situation and prospects for the future. *International Journal of Nanomedicine*, *12*, 1227–1249. doi:10.2147/IJN.S121956 PMID:28243086

Wang, M., Li, S., Chen, Z., Zhu, J., Hao, W., Jia, G., Chen, W., Zheng, Y., Qu, W., & Liu, Y. (2021). Safety assessment of nanoparticles in food: Current status and prospective. *Nano Today*, *39*, 101169. doi:10.1016/j.nantod.2021.101169

Wang, T., Rong, F., Tang, Y., Li, M., Feng, T., Zhou, Q., & Huang, W. (2021). Targeted polymer-based antibiotic delivery system: A promising option for treating bacterial infections via macromolecular approaches. *Progress in Polymer Science*, *116*, 101389. doi:10.1016/j.progpolymsci.2021.101389

Wang, W. N., Tarafdar, J. C., & Biswas, P. (2013). Nanoparticle synthesis and delivery by an aerosol route for watermelon plant foliar uptake. *Journal of Nanoparticle Research*, *15*(1), 1–13. doi:10.100711051-013-1417-8

Wang, X. P., Li, Q. Q., Pei, Z. M., & Wang, S. C. (2018). Effects of zinc oxide nanoparticles on the growth, photosynthetic traits, and antioxidative enzymes in tomato plants. *Biologia Plantarum*, *62*(4), 801–808. doi:10.100710535-018-0813-4

Wang, X., Xia, T., Duch, M. C., Ji, Z., Zhang, H., Li, R., Sun, B., Lin, S., Meng, H., Liao, Y. P., Wang, M., Song, T. B., Yang, Y., Hersam, M. C., & Nel, A. E. (2012). Pluronic F108 coating decreases the lung fibrosis potential of multiwall carbon nanotubes by reducing lysosomal injury. *Nano Letters*, *12*(6), 3050–3061. doi:10.1021/nl300895y PMID:22546002

Wang, Y., Dong, Y., Liu, Y., & Liu, Z. (2018). Arsenic trioxide nanoparticles inhibit acute promyelocytic leukemia cell proliferation and induce apoptosis via PTEN/AKT signalling pathway. *Biomedical Research, 29*(7), 1450–1454. doi:10.4066/biomedicalresearch.29-17-3983

Wang, Y., Xu, C., & Ow, H. (2013). Commercial nanoparticles for stem cell labeling and tracking. *Theranostics, 3*(8), 544–560. doi:10.7150/thno.5634 PMID:23946821

Wang, Z., Xu, L., Zhao, J., Wang, X., White, J. C., & Xing, B. (2016). CuO nanoparticle interaction with *Arabidopsis thaliana*: Toxicity, parent-progeny transfer, and gene expression. *Environmental Science & Technology, 50*(11), 6008–6016. doi:10.1021/acs.est.6b01017 PMID:27226046

Wang, Z., Zhu, W., Qiu, Y., Yi, X., von dem Bussche, A., Kane, A., Gao, H., Koski, K., & Hurt, R. (2016). Biological and Environmental Interactions of Emerging Two-Dimensional Nanomaterials. *Chemical Society Reviews, 45*(6), 1750–1780. doi:10.1039/C5CS00914F PMID:26923057

Wanunu, M., Dadosh, T., Ray, V., Jin, J., McReynolds, L., & Drndic, M. (2010). Rapid electronic detection of probe-specific microRNAs using thin nanopore sensors. *Nature Nanotechnology, 5*(11), 807–814. doi:10.1038/nnano.2010.202 PMID:20972437

Warad, H. C., Ghosh, S. C., Thanachayanont, C., Dutta, J., & Hilborn, J. G. (2004). Highly luminescent manganese doped ZnS quantum dots for biological labelling. *Proc. Int. Conf. on Smart Materials/Intelligent Materials*, 203–206.

Waring, J. F., Gum, R., Morfitt, D., Jolly, R. A., Ciurlionis, R., Heindel, M., Gallenberg, L., Buratto, B., & Ulrich, R. G. (2002). Identifying toxic mechanisms using DNA microarrays: Evidence that an experimental inhibitor of cell adhesion molecule expression signals through the aryl hydrocarbon nuclear receptor. *Toxicology, 181-182*, 537–550. doi:10.1016/S0300-483X(02)00477-8 PMID:12505364

Waser, O., Hess, M., Guntner, A., Novak, P., & Pratsinis, S. E. (2013). Size controlled CuO nanoparticles for Li-ion batteries. *Journal of Power Sources, 241*, 415–422. doi:10.1016/j.jpowsour.2013.04.147

Weir, A., Westerhoff, P., Fabricius, L., Hristovski, K., & von Goetz, N. (2012). Titanium Dioxide Nanoparticles in Food and Personal Care Products. *Environmental Science & Technology, 46*(4), 2242–2250. doi:10.1021/es204168d PMID:22260395

Wei, W., Xu, C., & Wu, H. (2006). Magnetic iron oxide nanoparticles mediated gene therapy for breast cancer-an in vitro study. *Journal of Huazhong University of Science and Technology, 26*(6), 728–730. doi:10.100711596-006-0628-y PMID:17357502

Wen, K.-P., Chen, Y.-C., Chuang, C.-H., Chang, H.-Y., Lee, C.-Y., & Tai, N.-H. (2015). Accumulation and Toxicity of Intravenously-Injected Functionalized Graphene Oxide in Mice. *Journal of Applied Toxicology, 35*(10), 1211–1218. doi:10.1002/jat.3187 PMID:26099253

White, J. C., & Gardea-Torresdey, J. (2018). Achieving food security through the very small. *Nature Nanotechnology, 13*(8), 627–629. doi:10.103841565-018-0223-y PMID:30082813

WHO. (1990). Guidelines for drinking-water quality, 3rd edition: Volume 1 - Recommendations incorporating the first and second addenda. WHO.

Wilson, M. A., Tran, N. H., Milev, A. S., Kannangara, G. S. K., Volk, H., & Lu, G. Q. M. (2008). Nanomaterials in soils. *Geoderma, 146*(1-2), 291–302. doi:10.1016/j.geoderma.2008.06.004

Woggon, U. (1997). Optical properties of semiconductor quantum dots. Springer.

Wongrakpanich, A., Mudunkotuwa, I. A., Geary, S. M., Morris, A. S., Mapuskar, K. A., Spitz, D. R., Grassian, V. H., & Salem, A. K. (2016). Size-dependent cytotoxicity of copper oxide nanoparticles in lung epithelial cells. *Environmental Science. Nano*, *3*(2), 365–374. doi:10.1039/C5EN00271K PMID:27347420

World Health Organization. (2011). *Guidelines for drinking-water quality*. WHO.

Wright, S. L., Thompson, R. C., & Galloway, T. S. (2013). The physical impacts of microplasticson marine organisms: A review. *Environmental Pollution*, *178*, 483–492. doi:10.1016/j.envpol.2013.02.031 PMID:23545014

Wu, L., Jin, T., Chen, H., Shen, Z., & Zhou, Y. (2023). Conductive materials as fantastic toolkits to stimulate direct interspecies electron transfer in anaerobic digestion: New insights into methanogenesis contribution, characterization technology, and downstream treatment. *Journal of Environmental Management*, *326*(PB), 116732. doi:10.1016/j.jenvman.2022.116732

Wudl, F. (2002). Fullerene materials. *Journal of Materials Chemistry*, *12*(7), 1959–1963. doi:10.1039/b201196d

Wu, L., & Liu, M. (2008). Preparation and properties of chitosan-coated NPK compound fertilizer with controlled release and water-retention. *Carbohydrate Polymers*, *72*(2), 240–247. doi:10.1016/j.carbpol.2007.08.020

Wu, N., Wang, X., Das, C. M., Ma, M., Qiao, N., Fan, T., Zhang, H., Xu, G., & Yong, K.-T. (2021). Bioengineering Applications of Black Phosphorus and Their Toxicity Assessment. *Environmental Science. Nano*, *8*(12), 3452–3477. doi:10.1039/D1EN00273B

Wu, S. C., Cao, Z. H., Li, Z. G., Cheung, K. C., & Wong, M. H. (2005). Effects of biofertilizer containing N-fixer, P and K solubilizers and AM fungi on maize growth: A greenhouse trial. *Geoderma*, *125*(1-2), 155–166. doi:10.1016/j.geoderma.2004.07.003

Wu, W., & Ma, B. (2015). Integrated nutrient management (INM) for sustaining crop productivity and reducing environmental impact: A review. *The Science of the Total Environment*, *512-513*, 415–427. doi:10.1016/j.scitotenv.2014.12.101 PMID:25644838

Wu, Y. N., Wu, P. C., Yang, L. X., Ratinac, K. R., Thordarson, P., Jahn, K. A., Chen, D. H., Shieh, D. B., & Braet, F. (2013). The anticancer properties of iron core-gold shell nanoparticles in colorectal cancer cells. *International Journal of Nanomedicine*, *8*, 3321–3331. doi:10.2147/IJN.S47742 PMID:24039416

Wu, Y., Guo, J., Yang, W., Wang, C., & Fu, S. (2006). Preparation and characterization of chitosan-poly (acrylic acid) polymer magnetic microspheres. *Polymer*, *47*(15), 5287–5294. doi:10.1016/j.polymer.2006.05.017

Wu, Z., Xu, E., Chughtai, M. F. J., Jin, Z., & Irudayaraj, J. (2017). Highly sensitive fluorescence sensing of zearalenone using a novel aptasensor based on upconverting nanoparticles. *Food Chemistry*, *230*, 673–680. doi:10.1016/j.foodchem.2017.03.100 PMID:28407965

Xiang, Y., Yang, Z., Zhang, Y., Xu, R., Zheng, Y., Hu, J., Li, X., Jia, M., Xiong, W., & Cao, J. (2019). Influence of nanoscale zero-valent iron and magnetite nanoparticles on anaerobic digestion performance and macrolide, aminoglycoside, β-lactam resistance genes reduction. *Bioresource Technology*, *294*, 122139. doi:10.1016/j.biortech.2019.122139 PMID:31525586

Xiaomei, Z., Rujing, L., Jun, X., Yingying, H., Xinying, Z., & Li, X. (2022). Enhanced methane production by bimetallic metal – organic frameworks (MOFs) as cathode in an anaerobic digestion microbial electrolysis cell. *Chemical Engineering Journal*, *440*, 135799. Advance online publication. doi:10.1016/j.cej.2022.135799

Xiao, Y., Vijver, M. G., Chen, G., & Peijnenburg, W. J. (2015). Toxicity and accumulation of Cu and ZnO nanoparticles in *Daphnia magna*. *Environmental Science & Technology*, *49*(7), 4657–4664. doi:10.1021/acs.est.5b00538 PMID:25785366

Xie, Y., Wang, H., & He, Z. (2021). Recent advances in polyphenols improving vascular endothelial dysfunction induced by endogenous toxicity. *Journal of Applied Toxicology*, *41*(5), 701–712. doi:10.1002/jat.4123 PMID:33251608

Xie, Y., Williams, N. G., Tolic, A., Chrisler, W. B., Teeguarden, J. G., Maddux, B. L., Pounds, J. G., Laskin, A., & Orr, G. (2012). Aerosolized ZnO nanoparticles induce toxicity in alveolar type II epithelial cells at the air-liquid interface. *Toxicological Sciences*, *125*(2), 450–461. doi:10.1093/toxsci/kfr251 PMID:21964423

Xi, G., Peng, Y., Xu, L., Zhang, M., Yu, W., & Qian, Y. (2004). Selected-control synthesis of PbO2 submicrometer-sized hollow spheres and Pb3O4 microtubes. *Inorganic Chemistry Communications*, *7*(5), 607–610. doi:10.1016/j.inoche.2004.03.001

Xiong, J. Q., Kurade, M. B., & Jeon, B. H. (2018). Can microalgae remove pharmaceutical contaminants from water? *Trends in Biotechnology*, *36*(1), 30–44. doi:10.1016/j.tibtech.2017.09.003 PMID:28993012

Xiong, Z., Zhang, X., Zhang, S., Lei, L., Ma, W., Li, D., Wang, W., Zhao, Q., & Xing, B. (2018). Bacterial Toxicity of Exfoliated Black Phosphorous Nanosheets. *Ecotoxicology and Environmental Safety*, *161*, 507–514. doi:10.1016/j.ecoenv.2018.06.008 PMID:29913419

Xu, G. R., Wang, J. N., & Li, C. J. (2013). Strategies for improving the performance of the polyamide thin film composite (PA-TFC) reverse osmosis (RO) membranes: Surface modifications and nanoparticles incorporations. *Desalination*, *328*, 83–100. doi:10.1016/j.desal.2013.08.022

Xu, L., Shoaie, S., Jahanpeyma, F., Zhao, J., Azimzadeh, M., & Al-Jamal, K. T. (2020). Optical, electrochemical and electrical (nano) biosensors for detection of exosomes: A comprehensive overview. *Biosensors & Bioelectronics*, *161*, 112222. doi:10.1016/j.bios.2020.112222 PMID:32365010

Xu, T., & Geng, Z. (2021). Strategies to improve performances of LSPR biosensing: Structure, materials, and interface modification. *Biosensors & Bioelectronics*, *2021*(174), 112850. doi:10.1016/j.bios.2020.112850 PMID:33309521

Yadav, T. P., Yadav, R. M., & Singh, D. P. (2012). Mechanical Milling: A top down approach for the synthesis of nanomaterials and nanocomposites. *Nanoscience and Nanotechnology*, *2*(3), 22–48. doi:10.5923/j.nn.20120203.01

Yang, K., Li, Y., Tan, X., Peng, R., & Liu, Z. (2013). Behavior and Toxicity of Graphene and Its Functionalized Derivatives in Biological Systems. *Small*, *9*(9-10), 1492–1503. doi:10.1002mll.201201417 PMID:22987582

Yang, L., & Watts, D. J. (2005). Particle surface characteristics may play an important role in phytotoxicity of alumina nanoparticles. *Toxicology Letters*, *158*(2), 122–132. doi:10.1016/j.toxlet.2005.03.003 PMID:16039401

Yang, W., Zhou, H., & Cicek, N. (2004). Treatment of organic micropollutants in water and wastewater by UV-based processes: A literature review. *Critical Reviews in Environmental Science and Technology*, *44*(13), 1443–1476. doi:10.1080/10643389.2013.790745

Yang, Y., Wang, J., Xiu, Z., & Alvarez, P. J. (2013). Impacts of silver nanoparticles on cellular and transcriptional activity of nitrogen-cycling bacteria. *Environmental Toxicology and Chemistry*, *32*(7), 1488–1494. doi:10.1002/etc.2230 PMID:23554086

Yang, Z. M., Guo, R. B., & Dong, X. H. (2022). Promoting biomethane production from propionate with Fe2O3@carbon nanotubes composites. *The Science of the Total Environment*, *818*, 151762. doi:10.1016/j.scitotenv.2021.151762 PMID:34800454

Yan, H., Dai, J., Yang, Z., Yang, H., & Cheng, R. (2011). Enhanced and selective adsorption of copper(II) ions on surface carboxymethylated chitosan hydrogel beads. *Chemical Engineering Journal*, *174*(2-3), 586–594. doi:10.1016/j.cej.2011.09.064

Yan, J. X., Guan, H. N., Yu, J., & Chi, D. F. (2013). Acetylcholinesterase biosensor based on assembly of multiwall carbon nanotubes onto liposome bioreactors for detection of organophosphates pesticides. *Pesticide Biochemistry and Physiology*, *105*(3), 197–202. doi:10.1016/j.pestbp.2013.02.003

Yaqoob, S. B., Adnan, R., Rameez Khan, R. M., & Rashid, M. (2020). Gold, silver, and palladium nanoparticles: A chemical tool for biomedical applications. *Frontiers in Chemistry*, *8*, 376. doi:10.3389/fchem.2020.00376 PMID:32582621

Yassen, A., Abdallah, E., Gaballah, M., & Zaghloul, S. (2017). Role of silicon dioxide Nanofertilizer in mitigating salt stress on growth, yield and chemical composition of cucumber (*Cucumis sativus* L.). *International Journal of Agricultural Research*, *12*(3), 130–135. doi:10.3923/ijar.2017.130.135

Yassin, M. A., Elkhooly, T. A., Elsherbiny, S. M., Reicha, F. M., & Shokeir, A. A. (2019). Facile coating of urinary catheter with bio–inspired antibacterial coating. *Heliyon*, *5*(12), e02986. doi:10.1016/j.heliyon.2019.e02986 PMID:31886428

Yigit, M. V., Mazumdar, D., & Lu, Y. (2008). MRI detection of thrombin with aptamer functionalized superparamagnetic iron oxide nanoparticles. *Bioconjugate Chemistry*, *19*(2), 412–417. doi:10.1021/bc7003928 PMID:18173225

Yih, T., & Al-Fandi, M. (2006). Engineered nanoparticles as precise drug delivery systems. *Journal of Cellular Biochemistry*, *97*(6), 1184–1190. doi:10.1002/jcb.20796 PMID:16440317

Yildirimer, L., Thanh, N. T., Loizidou, M., & Seifalian, A. M. (2011). Toxicology and clinical potential of nanoparticles. *Nano Today*, *6*(6), 585–607. doi:10.1016/j.nantod.2011.10.001 PMID:23293661

Yin, I. X., Zhang, J., Zhao, I. S., Mei, M. L., Li, Q., & Chu, C. H. (2020). The antibacterial mechanism of silver nanoparticles and its application in dentistry. *International Journal of Nanomedicine*, *15*, 2555–2562. doi:10.2147/IJN.S246764 PMID:32368040

Yoshimatsu, K., Lesel, B. K., Yonamine, Y., Beierle, J. M., Hoshino, Y., & Shea, K. J. (2012). Temperature-Responsive "Catch and Release" of Proteins by using Multifunctional Polymer-Based Nanoparticles. *Angewandte Chemie*, *124*(10), 2455–2458. doi:10.1002/ange.201107797 PMID:22287126

Younis, A., Khattab, H., & Emam, M. (2020). Impacts of silicon and silicon nanoparticles on leaf ultrastructure and TaPIP1 and TaNIP2 gene expressions in heat stressed wheat seedlings. *Biologia Plantarum*, *64*, 343–352. doi:10.32615/bp.2020.030

Youns, M., Hoheisel, J. D., & Efferth, T. (2011). Therapeutic and diagnostic applications of nanoparticles. *Current Drug Targets*, *12*(3), 357–365. doi:10.2174/138945011794815257 PMID:20955146

You, Y., Tong, X., Wang, W., Sun, J., Yu, P., Ji, H., Niu, X., & Wang, Z. M. (2019). Eco-Friendly Colloidal Quantum Dot-Based Luminescent Solar Concentrators. *Advancement of Science*, *6*(9), 1801967. Advance online publication. doi:10.1002/advs.201801967 PMID:31065522

Yuan, Y. G., Zhang, S., Hwang, J. Y., & Kong, I. K. (2018). Silver nanoparticles potentiates cytotoxicity and apoptotic potential of camptothecin in human cervical cancer cells. *Oxidative Medicine and Cellular Longevity*, *2018*, 6121328. doi:10.1155/2018/6121328 PMID:30647812

Yu, G. X., Wu, W. X., Zhao, Q., Wei, X. Y., & Lu, Q. (2015). Efficient immobilization of acetylcholinesterase onto amino-functionalized carbon nanotubes for the fabrication of highly sensitive organophosphorus pesticides biosensors. *Biosensors & Bioelectronics*, *68*, 288–294. doi:10.1016/j.bios.2015.01.005 PMID:25594160

Yun, Y., Cho, Y. W., & Park, K. (2013). Nanoparticles for oral delivery: Targeted nanoparticles with peptidic ligands for oral protein delivery. *Advanced Drug Delivery Reviews*, *65*(6), 822–832. doi:10.1016/j.addr.2012.10.007 PMID:23123292

Yu, Q., Xiong, X. Q., Zhao, L., Xu, T. T., Bi, H., Fu, R., & Wang, Q. H. (2018). Biodistribution and toxicity assessment of superparamagnetic iron oxide nanoparticles in vitro and in vivo. *Current Medical Science*, *38*(6), 1096–1102. doi:10.100711596-018-1989-8 PMID:30536075

Yu, S. J., Yin, Y. G., & Liu, J. F. (2013). Silver nanoparticles in the environment. *Environmental Science. Processes & Impacts*, *15*(1), 78–92. doi:10.1039/C2EM30595J PMID:24592429

Yu, Y., Zheng, J., Li, J., Lu, L., Yan, J., Zhang, L., & Wang, L. (2021). Applications of Two-Dimensional Materials in Food Packaging. *Trends in Food Science & Technology*, *110*, 443–457. doi:10.1016/j.tifs.2021.02.021

Zahra, Z., Habib, Z., Hyun, S., & Sajid, M. (2022). Nanowaste: Another Future Waste, Its Sources, Release Mechanism, and Removal Strategies in the Environment. *Sustainability*, *14*(4), 2041. doi:10.3390u14042041

Zakaria, B. S., & Dhar, B. R. (2022). Intermittent energization catalyzes direct interspecies electron transfer in electro-anaerobic digestion of sewage sludge. *Chemical Engineering Journal*, *136177*, 136177. Advance online publication. doi:10.1016/j.cej.2022.136177

Zamfir, L. G., Puiu, M., & Bala, C. (2020). Advances in electrochemical impedance spectroscopy detection of endocrine disruptors. *Sensors (Basel)*, *20*(22), 6443. doi:10.339020226443 PMID:33187314

Zbyszewski, M., & Corcoran, P. L. (2011). Distribution and degradation of fresh water plasticparticles along the beaches of Lake Huron, Canada. *Water, Air, and Soil Pollution*, *220*(1-4), 1–8. doi:10.100711270-011-0760-6

Zeng, S., Jin, X., Rajarathnam, D., & Chen, Z. (2019). Enhanced degradation of malachite by iron nanoparticles encapsulated in sodium alginate beads. *Journal of Industrial and Engineering Chemistry*, *77*, 238–242. Advance online publication. doi:10.1016/j.jiec.2019.04.043

Zhang, C., Yan, L., Gu, Z., & Zhao, Y. (2019). Strategies based on metal-based nanoparticles for hypoxic-tumor radiotherapy. *Chemical Science (Cambridge)*, *10*(29), 6932–6943. doi:10.1039/C9SC02107H PMID:31588260

Zhang, D., Ma, X., Gu, Y., Huang, H., & Zhang, G. (2020). Green Synthesis of Metallic Nanoparticles and Their Potential Applications to Treat Cancer. *Frontiers in Chemistry*, *8*, 799. doi:10.3389/fchem.2020.00799 PMID:33195027

Zhang, H., Omer, A. H., Hu, Z., Yang, L., Ji, C., & Ouyang, X. (2019). Fabrication of magnetic bentonite/carboxymethyl chitosan/sodium alginate hydrogel beads for Cu (II) adsorption. *International Journal of Biological Macromolecules*, *135*, 490–500. doi:10.1016/j.ijbiomac.2019.05.185 PMID:31145956

Zhang, L., Chu, M., Ji, C., Tan, J., & Yuan, Q. (2022). Preparation, applications, and challenges of functional DNA nanomaterials. *Nano Research*, 1–18. doi:10.100712274-022-4793-5 PMID:36065175

Zhang, W., Lin, K., Sun, X., Dong, Q., Huang, C., Wang, H., Guo, M., & Cui, X. (2012). Toxicological effect of MPA–CdSe QDs exposure on zebrafish embryo and larvae. *Chemosphere*, *89*(1), 52–59. doi:10.1016/j.chemosphere.2012.04.012 PMID:22595531

Zhang, X., Davidson, E., Mauzerall, D., Searchinger, T. D., Dumas, P., & Shen, Y. (2015). Managing nitrogen for sustainable development. *Nature*, *528*(7580), 51–59. doi:10.1038/nature15743 PMID:26595273

Zhang, X., Zhang, Z., Zhang, S., Li, D., Ma, W., Ma, C., Wu, F., Zhao, Q., Yan, Q., & Xing, B. (2017). Size Effect on the Cytotoxicity of Layered Black Phosphorus and Underlying Mechanisms. *Small*, *13*(32), 1701210. doi:10.1002mll.201701210 PMID:28696584

Zhao, J., Liu, P., Ma, J., Li, D., Yang, H., Chen, W., & Jiang, Y. (2019). Enhancement of radiosensitization by silver nanoparticles functionalized with polyethylene glycol and aptamer As1411 for glioma irradiation therapy. *International Journal of Nanomedicine*, *14*, 9483–9496. doi:10.2147/IJN.S224160 PMID:31819445

Zheng, H., Pan, L., Lv, J., Zhang, Z., Wang, Y., Hu, W., Liu, X., Zhou, P., Wang, Y., & Zhang, Y. (2020). Comparison of immune responses in guinea pigs by intranasal delivery with different nanoparticles-loaded FMDV DNA vaccine. *Microbial Pathogenesis*, *142*, 104061. doi:10.1016/j.micpath.2020.104061 PMID:32061916

Zheng, X. G., Xu, C. N., Tomokiyo, Y., Tanaka, E., Yamada, H., & Soejima, Y. (2000). Observation of charge stripes in cupric oxide. *Physical Review Letters*, *85*(24), 5170–5173. doi:10.1103/PhysRevLett.85.5170 PMID:11102213

Zheng, Z., Zhou, Y., Li, X., Liu, S., & Tang, Z. (2011). Highly-sensitive organo phosphorous pesticide biosensors based on nanostructured films of acetylcholinesterase and CdTe quantum dots. *Biosensors & Bioelectronics*, *26*(6), 3081–3085. doi:10.1016/j.bios.2010.12.021 PMID:21196108

Zhong, X., Wang, Y. X., & Wang, S. C. (1996). Pressure dependence of the volume phase-transition of temperature-sensitive gels. *Chemical Engineering Science*, *51*(12), 3235–3239. doi:10.1016/0009-2509(95)00344-4

Zhou, H., Mayorga-Martinez, C. C., Pané, S., Zhang, L., & Pumera, M. (2021). Magnetically Driven Micro and Nano-robots. *Chemical Reviews*, *121*(8), 4999–5041. doi:10.1021/acs.chemrev.0c01234 PMID:33787235

Zhou, J., Hao, B., Wang, L., Ma, J., & Cheng, W. (2017). Preparation and characterization of nano-TiO_2/chitosan/poly(N-isopropylacrylamide) composite hydrogel and its application for removal of ionic dyes. *Separation and Purification Technology*, *176*, 193–199. doi:10.1016/j.seppur.2016.11.069

Zhou, X., Zhao, L., Luo, J., Tang, H., Xu, M., Wang, Y., Yang, X., Chen, H., Li, Y., Ye, G., Shi, F., Lv, C., & Jing, B. (2019). The Toxic Effects and Mechanisms of Nano-Cu on the Spleen of Rats. *International Journal of Molecular Sciences*, *20*(6), 1469. doi:10.3390/ijms20061469 PMID:30909528

Zhou, Z., Wang, X., Zhang, H., Huang, H., Sun, L., Ma, L., Du, Y., Pei, C., Zhang, Q., Li, H., Ma, L., Gu, L., Liu, Z., Cheng, L., & Tan, C. (2021). Activating Layered Metal Oxide Nanomaterials via Structural Engineering as Biodegradable Nanoagents for Photothermal Cancer Therapy. *Small*, *17*(12), 2007486. doi:10.1002mll.202007486 PMID:33590671

Zhu, G., Yang, K., & Zhang, C. Y. (2015). A single quantum dot-based biosensor for telomerase assay. *Chemical Communications*, *51*(31), 6808–6811. doi:10.1039/C5CC01600B PMID:25788014

Zhu, H. Y., Fu, Y. Q., Jiang, R., Yao, J., Xiao, L., & Zeng, G. M. (2012). Novel magnetic chitosan/poly (vinyl alcohol) hydrogel beads: Preparation, characterization and application for adsorption of dye from aqueous solution. *Bioresource Technology*, *105*, 24–30. doi:10.1016/j.biortech.2011.11.057 PMID:22189074

Zhu, H., Jiang, R., Xiao, L., Chang, Y., Guan, Y., Li, X., & Zeng, G. (2009). Photocatalytic decolorization and degradation of Congo Red on innovative crosslinked chitosan/nano-CdS composite catalyst under visible light irradiation. *Journal of Hazardous Materials*, *169*(1-3), 933–940. doi:10.1016/j.jhazmat.2009.04.037 PMID:19477069

Zhu, M. T., Feng, W. Y., Wang, B., Wang, T. C., Gu, Y. Q., Wang, M., Wang, Y., Ouyang, H., Zhao, Y. L., & Chai, Z. F. (2008). Comparative study of pulmonary responses to nano- and submicron-sized ferric oxide in rats. *Toxicology*, *247*(2-3), 102–111. doi:10.1016/j.tox.2008.02.011 PMID:18394769

Zhu, Y., Xie, Z., Li, J., Liu, Y., Li, C., Liang, W., Huang, W., Kang, J., Cheng, F., Kang, L., Al-Hartomy, O. A., Al-Ghamdi, A., Wageh, S., Xu, J., Li, D., & Zhang, H. (2021). From Phosphorus to Phosphorene: Applications in Disease Theranostics. *Coordination Chemistry Reviews*, *446*, 214110. doi:10.1016/j.ccr.2021.214110

Zhu, Z. J., Wang, H., Yan, B., Zheng, H., Jiang, Y., Miranda, O. R., Rotello, V. M., Xing, B., & Vachet, R. W. (2012). Effect of surface charge on the uptake and distribution of gold nanoparticles in four plant species. *Environmental Science & Technology*, *46*(22), 12391–12398. doi:10.1021/es301977w PMID:23102049

Zrazhevskiy, P., Sena, M., & Gao, X. (2010). Designing multifunctional quantum dots for bioimaging, detection, and drug delivery. *Chemical Society Reviews, 39*(11), 4326–4354. doi:10.1039/b915139g PMID:20697629

Zsiborács, H., Baranyai, N. H., Vincze, A., Zentkó, L., Birknei, Z., Máté, K., & Pintér, G. (2019). Intermittent Renewable Energy Sources: The Role of Energy Storage in the European Power System of 2040. *Electronics (Switzerland), 8*(7), 1–18. doi:10.3390/electronics8070729

Zubris, K. A. V., & Richards, B. K. (2005). Synthetic fibers as indicator of land application ofsludge. *Environmental Pollution, 138*(2), 201–211. doi:10.1016/j.envpol.2005.04.013 PMID:15967553

Zulfiqar, F., Navarro, M., Ashraf, M., Akram, N. A., & Munné-Bosch, S. (2019). Nanofertilizer use for sustainable agriculture: Advantages and limitations. *Plant Science, 289*, 110270. doi:10.1016/j.plantsci.2019.110270 PMID:31623775

Zuverza-Mena, N., Martínez-Fernández, D., Du, W., Hernandez-Viezcas, J. A., Bonilla-Bird, N., López-Moreno, M. L., Komárek, M., Peralta-Videa, J. R., & Gardea-Torresdey, J. L. (2017). Exposure of engineered nanomaterials to plants: Insights into the physiological and biochemical responses-A review. *Plant Physiology and Biochemistry, 110*, 236–264. doi:10.1016/j.plaphy.2016.05.037 PMID:27289187

Zweifel, H. (2001). *Plastics additives handbook* (5th ed.). Carl HanserVerlag.

About the Contributors

Rafiq Lone is presently working as Assistant Professor of Botany at Central University of Kashmir, Jammu and Kashmir- India. He has completed his M. Sc and Ph.D. in Botany from Jiwaji University, Gwalior, Madhya Pradesh, India. Dr. Lone has been awarded National Post-Doctoral Fellowship (NPDF) by SERB, DST, Government of India. Previously he was working as Assistant Professor in Botany at SBBS University, Khiala Jalandhar, Punjab, Lecturer (Academic arrangement) in higher education, Jammu and Kashmir, Teaching Assistant at School of Studies in Botany, Jiwaji University, Gwalior-India, and Project Fellow in a project funded by MPCST-Bhopal-(M.P)-India. Dr. Lone has five years of teaching and ten years of research experience. He has published more than 25 research papers in National and International journals and more than 16 book chapters, and four edited books are also to his credit. Dr. Lone has attended 25 national and international conferences and has presented papers on plant-microbe interaction and plant phenolics. His main research interest focuses on Plant-microbe interaction, Plant Nutrition, Plant Phenolics, and Invasion Biology.

* * *

Richa Arora is working as Microbiologist in the Department of Microbiology, Punjab Agricultural University, Ludhiana, India. She received her M.Sc. in Microbiology and Ph.D. in Life Sciences (Biotechnology). She is having research experience in integrated biorefineries, waste valorization and food fermentations. She has been awarded with Young Scientist Award and International Travel Support to United States. She has filed 1 patent and published over 30 research articles in peer-reviewed journals, book chapters and papers in conference proceedings. Dr. Vijay K. Bharti, received his M.V.Sc. and Ph.D. in Veterinary Sciences. Presently he is working as Scientist 'E' and Group Head and Project Officer at Animal Science Division, DIHAR, Leh. He is having research experience in high altitude animal physiology and nutrition, experimental toxicology, green synthesis of nanomaterial, and water research. He has granted 2 and filed 4 patents, and published over 80 research articles in peer-reviewed journals, and wrote 12 book chapters published by various national and international publishers.

Debraj Biswal is currently employed as Assistant Professor of Zoology in a Government college in West Bengal, India. His teaching career spans for more than 14 years. His research activities for over 15 years mainly include areas in parasitology, helminthology, vector biology and microbiology. He has published a number of articles related to these topics. Additionally, he has keen interests in ecology, environmental biology and eco-toxicology with a few publications in these fields too.

Syed Javid Indrabi is working as Assistant professor in the department of higher education, JKUT. He has pursued his PhD in the field of environmental science and engineering from Indian institute of technology, New Delhi India and qualified National Eligibility Test (NET). His teaching and research interests are in the areas of air pollution, water pollution control and treatment and also has research interests in environmental impact assessment. Dr Andrabi is author of more than thirty scientific articles of international repute. He has received various national and international awards related to publishing his work. Dr Andrabi has been working as editorial board member and reviewer of various international journals.

Mithu Maiti Jana is an associate professor at the department of Physical Sciences in the Sant Baba Bagh Sing University, Jalandhar, India. She was former assistant professor in Lovely Professional University and Dr B R Ambedkar National Institute of Technology Jalandhar, India. She has more than 11 years of teaching experiences. Her research interest includes nano composites, green composites, green films and nanohydrogel. She has published more than 60 papers in national and international repute journal and conferences. She has guided more than 10 MSc Project students at NITJ and SBBSU. Currently 3 PhD students are working under her supervision.

Uzma Kadri is currently working as Technical staff at ICMR-National Institute of Occupational Health (NIOH), Govt. of India, and holds a M.Sc. Degree in Biotechnology from Nirma University, India.

Javid Manzoor is working as Assistant professor in the Department of Environmental Science, GDC, Bhaderwah, J&K, India. He has obtained his Master's degree and doctorate from Jiwaji University, Gwalior, M.P. He has three years of teaching experience. He is an awardee of University Research Scholarship. He is a permanent member of people for animal. His areas of interest include Environmental pollution, Environmental monitoring and assessment. He has published research papers and book chapters in various national and international journals with impact factor. He has been working as editorial board member and reviewer of various national and international journals.

Nibedita Naha, PhD, is Scientist 'E' and Head of the Biochemistry Department under Biological Sciences Division, ICMR-National Institute of Occupational Health (Ministry of Health and Family Welfare, Department of Health Research, Govt. of India). She received PhD in Science from Jadavpur University, India; and Postdoctorate in prostate cancer cell signalling, tumour xenografting, and neurobiochemistry of substance abuse from Gyeongsang National University, South Korea. She also served as a faculty of Govt. Degree Colleges, Medical Colleges, and Universities in India and abroad. She has vast experience on mechanism-based approach of different techniques using pre-clinical animal models, cell lines and human subjects (patient's samples) with respect to occupational and environmental exposure since 2000. Dr. Naha is a recognized PhD Guide of Gujarat University, India; and a member of several PhD Research Committees (RDC) and Board of Studies (BoS) of different universities, as well as the CPCSEA nominee of Animal Ethics Committee (Ministry of Fisheries, Animal Husbandry and Dairying, Govt. of India) for monitoring animal experiments in several research organizations and pharmaceutical companies; and Department of Biotechnology nominee of the Institute Biosafety Committee (Ministry of Science and Technology, Govt. of India). Apart from that, Dr. Naha is also a reviewer and editorial member of several international and national peer-reviewed journals; and invited speakers of different national and international conferences. She is also a trained NABL internal auditor of ISO/

IEC17025:2017 and ISO/IEC15189:2012. She has authored more than 30 research articles, invited reviews and book chapters along with patents on her research work. Dr. Naha is an elected member of the Physiological Society of India (PSI) since 2013, and the advisory board member of some national conferences sponsored by Govt. of India. She has also honoured as a life member of several scientific societies in India and US such as, SRBCE, ISSRF, NESA, ISNA, ISCA, SfN and AAAS etc.; and awardees of many national and international scientific reputes including SfN (USA), Brain Korea 21 (South Korea), PB Sen Gold Medal (Physiological Society of India), Young Scientist Award (Indian Science Congress), National Scholarship (Govt. of India) etc. Dr. Naha considers herself extremely fortunate to contribute in improving the health and wellbeing through her research findings in the respective fields. She may be reached via email at nibedita.72@yahoo.com, or by telephone at +91-79-22688737 (work).

Fernanda C. Romeiro received her degree in Chemistry from Universidade Federal de Uberlandia-Brazil (2012), her Master (2014) and Ph.D. (2018) degrees in Chemistry also from Universidade Federal de Uberlandia. She is currently working as a postdoc researcher at São Paulo State University (UNESP). Her main research interests are on the microwave-assisted hydrothermal synthesis of inorganic materials as well as their electrochemical and photocatalytic applications including CO2 reduction and water splitting.

Nisha Sharma is working as Associate Professor in Chemistry, Department of Physical Sciences at Sant Baba Bhag Singh University, Jalandhar, India since 2017. She has acquired PhD in Organic Chemistry from Himachal Pradesh University Shimla, India in 2010. Currently she is working as a Head of Department of Physical Sciences, SBBSU. She has handled Major Research Project of MOE, Govt of India. She has more than 9 year of teaching & research experience. Her main research interests includes: Biodegradation strategies for plastics, Functionalization of natural and synthetic polymeric systems, Drug delivery systems & nanocomposite formulations, Mathematical Modeling of drug delivery systems, swelling Kinetics and Release Dynamics of Pharmaceutical agents and dyes. Dr Nisha has published more than 23 peer-reviewed research papers and more than 30 research papers in international/national conferences. She has guided 1 PhD , 1 MPhil and 13 MSc Project students. Currently 6 PhD students are working under her supervision. Organized 10 National & International Conference /Seminar/ Short Term Courses/ Winter School and Workshops. She is serving as reviewer for various peer reviewed Journals and is life Member of Asian Polymer Association (Membership no. L-641).

Somi N. Sharma is currently a PhD student in ICMR-National Institute of Occupational Health, Govt. of India. She has a Masters in Biochemistry from Gujarat University, India.

Shweta Singh is working as Associate Professor in Microbiology, Department of Life Sciences & Allied Health Sciences at Sant Baba Bhag Singh University, Jalandhar, India since 2016. She has acquired PhD in Microbiology in the year of 2011 from H.N.B. Garhwal Central University, India. Having more than 9 year of teaching & Research experience. Currently Dr Shweta is working as a Head of Department of Life Sciences & Allied Health Sciences, SBBSU. Her main research interests includes: Clinical Bacteriology, Virology. Handled DBT funded projects "Network Project on Brucellosis" and Short term Project on "Bovine Papilloma Virus." Dr Singh has guided 1 MPhil student, 12 M.Sc Project students and currently 3 PhD students are working under her supervision.

Aline Varella Rodrigues is a professor at Federal University of Technology, Paraná, Brazil (2022). Postdoctoral (2021) in the Analytical, Physical-Chemistry and Inorganic Chemistry Department at the UNESP University of the Institute of Chemistry in Araraquara-Brazil. She did her prior postdoctoral in the Engineering, Physics and Mathematics Department at the same University. Ph.D. in Natural Sciences (Dr. rer. nat.) with emphasis on Inorganic Chemistry (2017) at the Universität Siegen in the Inorganic Chemistry Department in Siegen-Germany. Master in Chemistry (2012) at the UNESP University in the Physical-Chemistry Department. Bachelor in Technological Chemistry (2012) at the same university. She did her traineeships at the Martin-Luther-Universität Halle-Wittenberg in the Engineering Department in Halle an der Saale (2011) and Duisburg-Essen Universität in the Science Department of Nano Engineering in Essen and Duisburg-Germany (2012). Bachelor in Chemistry (2009) at the UNESP University of the Institute of Chemistry in Araraquara-Brazil. She did her traineeships in the Analytical and Biotechnological Departments at the same University.

Caroline Varella Rodrigues has a PhD in Chemistry at the UNESP University of the Institute of Chemistry in Araraquara-Brazil with a CNPq fellowship (2021). Master in Chemistry at the same University with a CAPES fellowship (2016), she worked in the area of waste reuse for energy purposes. Bachelor in Technological Chemistry from the same University (2014). She did her traineeship in Chemistry at the Multinational Chemical Industry "Rhodia- Solvay Group" (2013), working in the Research and Development area in the Solvents sector in Paulínia/SP. She had FAPESP fellowship as Scientific Initiation Scholar (2012) in the Department of Organic Chemistry, developing the project "Chemical and biological study in Piper arboreum aublet (Piperaceae)". She has experience in the field of bioenergy, with emphasis on anaerobic reactors and molecular biology, working mainly on the following research topics: anaerobic co-digestion with the use of agro-industrial residues from the citrus and biodiesel industries (crude glycerol) for the generation of biogas (H2 and CH4) and value-added products (1,3-propanediol and lactic acid); hydrogen production by fermentative bacteria and methane production by methanogenic archaea for the treatment of effluents in anaerobic reactors in sequential phases; applied microbiology for energy generation through the use of pure cultures and anaerobic consortia.

Rafael Vargas-Bernal received a Bachelor's degree in Communications and Electronics Engineering from the University of Guanajuato in 1995, and the degrees of Master of Science and Doctorate in Science with Specialty in Electronics from the National Institute of Astrophysics, Optics and Electronics (INAOE) in 1997 and 2000, respectively. Since January 2002, he has been a professor-researcher at the Higher Technological Institute of Irapuato (ITESI), Mexico and, particularly since 2006, he has worked in the Department of Materials Engineering where he has established himself as a researcher. He has authored 17 articles in journals, 36 chapters books, and about 100 conference articles. He is a member of the National System of Researchers (SNI-Mexico). He regularly serves as a reviewer of scientific articles in RSC Advances and Royal Society Open Science as well as Standards in Semiconductor Equipment and Materials International (SEMI). His research interests include two-dimensional materials, nanomaterials, space materials, composite materials, gas sensors, and biosensors.

Khursheed Ahmad Wani is Assistant Professor in the Department of Environmental Science, Government Degree College Thidim Kreeri, J&K. Dr. Wani served at ITM University Gwalior as Assistant Professor in the Department of Environmental Science from 2011 to 2017. He has completed his M. Sc. and Ph. D. in Environmental Science from Jiwaji University, Gwalior, Madhya Pradesh.

He has also qualified Jammu and Kashmir State Level Eligibility Test (JKSET). His areas of interest include Environmental management, Environmental Impact Assessment and Occupational Health. He has published research papers in various national and international journals and is the author of three books on Environmental Science. He is a member of several national and international organizations and is in the editorial board of various journals. He has organized different programmes on Environmental Awareness and has completed research project sponsored by Madhya Pradesh Council of Science and Technology, Bhopal, India. Dr. Wani has organized various national and international conferences/ seminars/ workshops as well.

Index

A

B

C

D

E

F

G

H

W

X

www.ingramcontent.com/pod-product-compliance
Lightning Source LLC
LaVergne TN
LVHW060412300926
849162LV00006B/1405